POROUS SILICON

From Formation to Application

Biomedical and Sensor Applications, Volume Two

Porous Silicon: From Formation to Application

Porous Silicon: Formation and Properties, Volume One

Porous Silicon: Biomedical and Sensor Applications, Volume Two

*Porous Silicon: Optoelectronics, Microelectronics, and Energy
Technology Applications, Volume Three*

Edited by
Ghenadii Korotcenkov

POROUS SILICON

From Formation to Application

Biomedical and Sensor Applications, Volume Two

CRC Press
Taylor & Francis Group
Boca Raton London New York

CRC Press is an imprint of the
Taylor & Francis Group, an **informa** business

CRC Press
Taylor & Francis Group
6000 Broken Sound Parkway NW, Suite 300
Boca Raton, FL 33487-2742

First issued in paperback 2019

© 2016 by Taylor & Francis Group, LLC
CRC Press is an imprint of Taylor & Francis Group, an Informa business

No claim to original U.S. Government works

ISBN-13: 978-1-4822-6456-2 (hbk)
ISBN-13: 978-0-367-37712-0 (pbk)

Visit the Taylor & Francis Web site at
http://www.taylorandfrancis.com

and the CRC Press Web site at
http://www.crcpress.com

Contents

Preface..xi
Editor...xiii
Contributors..xv

SECTION I—Sensors

Chapter 1 Solid State Gas and Vapor Sensors Based on Porous Silicon..3

Ghenadii Korotcenkov

Chapter 2 Optimization of PSi-Based Sensors Using IHSAB Principles..45

James L. Gole and Caitlin Baker

Chapter 3 Porous Silicon-Based Optical Chemical Sensors..69

Luca De Stefano, Ilaria Rea, Alessandro Caliò, Jane Politi, Monica Terracciano, and Ghenadii Korotcenkov

Chapter 4 PSi-Based Electrochemical Sensors...95

Yang Zhou, Shaoyuan Li, Xiuhua Chen, and Wenhui Ma

Chapter 5 Porous Silicon-Based Biosensors: Some Features of Design, Functional Activity, and Practical Application..............107

Nicolai F. Starodub

Chapter 6 MEMS-Based Pressure Sensors..135

Ghenadii Korotcenkov and Beongki Cho

Chapter 7 Micromachined Mechanical Sensors 147

Ghenadii Korotcenkov and Beongki Cho

Chapter 8 Porous Silicon as a Material for Thermal Insulation in MEMS...167

Pascal J. Newby

Chapter 9 PSi-Based Microwave Detection 185

Jonas Gradauskas and Jolanta Stupakova

SECTION II—Auxiliary Devices

Chapter 10 PSi-Based Ultrasound Emitters (Acoustic Emission)..197

Ghenadii Korotcenkov and Vladimir Brinzari

Chapter 11 Porous Silicon in Micromachining Hotplates Aimed for Sensor Applications............................207

Ghenadii Korotcenkov and Beongki Cho

Chapter 12 PSi-Based Preconcentrators, Filters, and Gas Sources 223
Ghenadii Korotcenkov, Vladimir Brinzari, and Beongki Cho

Chapter 13 Silicon Nanostructures for Laser Desorption/Ionization Mass Spectrometry 239
Sergei Alekseev

Chapter 14 PSi Substrates for Raman, Terahertz, and Atomic Absorption Spectroscopy 265
Ghenadii Korotcenkov and Songhee Han

Chapter 15 PSi-Based Diffusion Membranes 279
Andras Kovacs and Ulrich Mescheder

Chapter 16 Liquid Microfluidic Devices 299
Ghenadii Korotcenkov and Beongki Cho

SECTION III—Biomedical Applications

Chapter 17 Biocompatibility and Bioactivity of Porous Silicon 319
Adel Dalilottojari, Wing Yin Tong, Steven J.P. McInnes, and Nicolas H. Voelcker

Chapter 18 Applications of Porous Silicon Materials in Drug Delivery 337
Haisheng Peng, Guangtian Wang, Naidan Chang, and Qun Wang

Chapter 19 Tissue Engineering 363
Pierre-Yves Collart Dutilleul, Csilla Gergely, Frédérique Cunin, and Frédéric Cuisinier

Chapter 20 Use of Porous Silicon for *In Vivo* Imaging Techniques 381
Igor Komarov and Sergei Alekseev

Index 401

Contents for Porous Silicon: *Formation and Properties, Volume One*

SECTION I—Introduction

Chapter 1   Porous Silicon Characterization and Application: General View
Ghenadii Korotcenkov

SECTION II—Silicon Porosification

Chapter 2   Fundamentals of Silicon Porosification via Electrochemical Etching
Enrique Quiroga-González and Helmut Föll

Chapter 3   Technology of Si Porous Layer Fabrication Using Anodic Etching: General Scientific and Technical Issues
Ghenadii Korotcenkov and Beongki Cho

Chapter 4   Silicon Porosification: Approaches to PSi Parameters Control
Ghenadii Korotcenkov

SECTION III—Properties and Processing

Chapter 5   Methods of Porous Silicon Parameters Control
Mykola Isaiev, Kateryna Voitenko, Dmitriy Andrusenko, and Roman Burbelo

Chapter 6   Structural and Electrophysical Properties of Porous Silicon
Giampiero Amato

Chapter 7   Luminescent Properties of Porous Silicon
Bernard Jacques Gelloz

Chapter 8   Optical Properties of Porous Silicon
Gilles Lérondel

Chapter 9   Thermal Properties of Porous Silicon
Pascal J. Newby

Chapter 10   Alternative Methods of Silicon Porosification and Properties of These PSi Layers
Ghenadii Korotcenkov and Vladimir Brinzari

Chapter 11   The Mechanism of Metal-Assisted Etching of Silicon
Kurt W. Kolasinski

Chapter 12 Porous Silicon Processing
Ghenadii Korotcenkov and Beongki Cho

Chapter 13 Surface Chemistry of Porous Silicon
Yannick Coffinier and Rabah Boukherroub

Chapter 14 Contacts to Porous Silicon and PSi-Based *p-n* Homo- and Heterojunctions
Jayita Kanungo and Sukumar Basu

Index

Contents for Porous Silicon: *Optoelectronics, Microelectronics, and Energy Technology Applications, Volume Three*

SECTION I—Optoelectronics and Photoelectronics

Chapter 1 Electroluminescence Devices (LED)
Bernard Jacques Gelloz

Chapter 2 Photodetectors Based on Porous Silicon
Ghenadii Korotcenkov and Nima Naderi

Chapter 3 PSi-Based Photonic Crystals
Gonzalo Recio-Sánchez

Chapter 4 Passive and Active Optical Components for Optoelectronics Based on Porous Silicon
Raghvendra S. Dubey

SECTION II—Electronics

Chapter 5 Electrical Isolation Applications of PSi in Microelectronics
Gael Gautier, Jérôme Billoué, and Samuel Menard

Chapter 6 Surface Micromachining (Sacrificial Layer) and Its Applications in Electronic Devices
Alexey Ivanov and Ulrich Mescheder

Chapter 7 Porous Silicon as Substrate for Epitaxial Films Growth
Eugene Chubenko, Sergey Redko, Alexey Dolgiy, Hanna Bandarenka, and Vitaly Bondarenko

Chapter 8 Porous Silicon and Cold Cathodes
Ghenadii Korotcenkov and Beongki Cho

Chapter 9 Porous Silicon as Host and Template Material for Fabricating Composites and Hybrid Materials
Eugene Chubenko, Sergey Redko, Alexey Dolgiy, Hanna Bandarenka, Sergey Prischepa, and Vitaly Bondarenko

SECTION III—Energy Technologies

Chapter 10 Porous Si and Si Nanostructures in Photovoltaics
Valeriy A. Skryshevsky and Tetyana Nychyporuk

Chapter 11 PSi-Based Betavoltaics
Ghenadii Korotcenkov and Vladimir Brinzari

Chapter 12 Porous Silicon in Micro-Fuel Cells
Gael Gautier and Ghenadii Korotcenkov

Chapter 13 Hydrogen Generation and Storage in Porous Silicon
Valeriy A. Skryshevsky, Vladimir Lysenko, and Sergii Litvinenko

Chapter 14 PSi-Based Microreactors
Caitlin Baker and James L. Gole

Chapter 15 Li Batteries with PSi-Based Electrodes
Gael Gautier, François Tran-Van, and Thomas Defforge

Chapter 16 PSi-Based Supercapacitors
Diana Golodnitsky, Ela Strauss, and Tania Ripenbein

Chapter 17 Porous Silicon as a Material for Thermoelectric Devices
Androula G. Nassiopoulou

Chapter 18 Porous Silicon Based Explosive Devices
Monuko du Plessis

Index

Preface

In recent decades, porous silicon has been regarded as a means to further increase the functionality of silicon technology. It was found that silicon porosification is a simple and cheap way of nanostructuring and bestowing to silicon properties, which are markedly different from the properties of the bulk material. Because of this, increased interest in porous silicon appeared in various fields, including optoelectronics, microelectronics, photonics, medicine, chemical sensing, and bioengineering. It was established that this nanostructured and biodegradable material has a range of properties, making it ideal for indicated applications. As a result, during the last decade we have been observing an extremely fast evolution of porous Si-based optoelectronics, photonics, microelectronics, sensorics, energy technologies, and biomedical devices and applications. It is predicted that this growth will continue in the near future. However, despite the progress achieved in the field of design of porous silicon–based devices and their applications, it is necessary to note that there is a very limited number of books published in the field of silicon porosification and especially porous silicon (PSi) applications. No doubt such situations should be recognized as unsatisfactory. Thus, it was decided to prepare a set of books devoted to the analysis of the current state of the technology of silicon porosification and the use of these technologies in the development of devices for different applications. While developing the concept of this series, the objective was to collect in one edition information concerning all aspects of the formation and the use of porous silicon. This is of great importance nowadays, due to the speed of technological development and the rate of an appearance of new fields of PSi-based technology applications

This set, Porous Silicon: From Formation to Application, prepared by an international team of expert contributors, well known in the field of porous silicon study and having high qualifications, represents the most recent progress in the field of porous silicon and gives a fascinating report on the state-of-the-art in silicon porosification and the valuable perspective one can expect in the near future.

The set is divided into three books by their content. Chapters in *Porous Silicon: Formation and Properties, Volume One* focus on the fundamentals and practical aspects of silicon porosification by anodization and the properties of porous silicon, including electrical, luminescence, optical, thermal properties, and contact phenomena. Processing of porous silicon, including drying, storage, oxidation, etching, filling, and functionalizing, are also discussed in this book. Alternative methods of silicon porosification using chemical stain and vapor etching, reactive ion etching,

spark processing, and so on are analyzed as well. *Porous Silicon: Biomedical and Sensor Applications, Volume Two* describes applications of porous silicon in bioengineering and various sensors such as gas sensors, biosensors, pressure sensors, optical sensors, microwave detectors, mechanical sensors, etc. The chapters in this book present a comprehensive review of the fabrication, parameters, and applications of these devices. PSi-based auxiliary devices such as hotplates, membranes, matrices for various spectroscopies, and catalysis are discussed as well. Analysis of various biomedical applications of porous silicon including drug delivery, tissue engineering, and *in vivo* imaging can also be found in this book. No doubt, porous silicon is rapidly attracting increasing interest in this field due to its unique properties. For example, the pores of the material and surface chemistry can be manipulated to change the rate of drug release from hours to months.

Finally, *Porous Silicon: Optoelectronics, Microelectronics, and Energy Technology Applications, Volume Three* highlights porous silicon applications in opto- and microelectronics, photonics, and micromachining. Features of fabrication and performances of photonic crystals, fuel cells, elements of integral optoelectronics, solar cells, LED, batteries, cold cathodes, hydrogen generation and storage, PSi-based composites, etc. are analyzed in this volume.

I believe that we have prepared useful books that could be considered a real handbook encyclopedia of porous silicon, where each reader might find the answers to most questions related to the formation, properties, and applications of porous silicon in practically all possible fields. Previously published books do not provide such an opportunity. Recently, several interesting books became available to readers, such as *Porous Silicon in Practice: Preparation, Characterization and Applications* by M.J. Sailor (Wiley-VCH 2011), *Porous Silicon for Biomedical Applications* by Santos H.A. (ed.) (Woodhead Publishing Limited 2014), and *Handbook of Porous Silicon*, by Canham L. (ed.) (Springer 2015). However, in *Porous Silicon in Practice: Preparation, Characterization and Applications*, M.J. Sailor describes mainly features of silicon porosification by electrochemical etching without any analysis of the correlation between parameters of porosification and properties of porous silicon. In the *Handbook of Porous Silicon* most attention was paid to the properties of porous silicon and, as well as in the book of M.J. Sailor, the consideration of PSi applications in devices was brief. At the same time, *Porous Silicon for Biomedical Applications* focuses on only the analysis of biomedical applications of porous silicon. I hope that our books, which cover all of the above-mentioned fields and provide a more detailed analysis of PSi advantages and disadvantages for practically all possible applications, will also be of interest to the reader. Our books contain a great number of various figures and tables with necessary information. These books will be a technical resource and indispensable guide for all those involved in the research, development, and application of porous silicon in various areas of science and technology.

From my point of view, our set will be of interest to scientists and researchers, either working or planning to start activity in the field of materials science focused on multifunctional porous silicon and porous silicon–based semiconductor devices. It also could be useful for those who want to find out more about the unusual properties of porous materials and about possible areas of their application. I am confident that these books will be interesting for practicing engineers or project managers working in industries and national laboratories who intend to design various porous silicon–based devices, but don't know how to do it. They might help select an optimal technology of silicon porosification and device fabrication. With many references to the vast resources of recently published literature on the subject, these books can serve as a significant and insightful source of valuable information and provide scientists and engineers with new insights for better understanding of the process of silicon porosification, for designing new porous silicon–based technology, and for improving performances of various devices fabricated using porous silicon.

I believe that these books can be of interest to university students, post docs, and professors, providing a comprehensive introduction to the field of porous silicon application. The structure of these books may serve as a basis for courses in the field of material science, semiconductor devices, chemical engineering, electronics, bioengineering, and environmental control. Graduate students may also find the books useful in their research and for understanding that porous silicon is a promising multifunctional material.

Finally, I thank all contributing authors who have been involved in the creation of these books. I am also thankful that they agreed to participate in this project and for their efforts in the preparation of these chapters. Without their participation, this project would have not been possible.

I also express my gratitude to Gwangju Institute of Science and Technology, Gwangju, Korea, which invited me and gave me the ability to prepare these books for publication, and especially to Professor Beongki Cho for his fruitful cooperation. Many thanks to the Ministry of Science, ICT, and Future Planning (MSIP) of the Republic of Korea for supporting my research. I am also grateful to my family and my wife, who always support me in all undertakings.

Editor

Ghenadii Korotcenkov earned his PhD in physics and the technology of semiconductor materials and devices from Technical University of Moldova in 1976 and his DrSci degree in the physics of semiconductors and dielectrics from the Academy of Science of Moldova in 1990 (Highest Qualification Committee of the USSR, Moscow). He has more than 40 years of experience as a teacher and scientific researcher. He was a leader of a gas sensor group and manager of various national and international scientific and engineering projects carried out in the Laboratory of Micro- and Optoelectronics, Technical University of Moldova. In particular, during 2000–2007 his scientific team was involved in eight international projects financed by EC (INCO-Copernicus and INTAS Programs), United States (CRDF, CRDF-MRDA Programs), and NATO (LG Program). In 2007–2008, he was an invited scientist at the Korea Institute of Energy Research (Daejeon) in the Brain Pool Program. Since 2008, Dr. Korotcenkov has been a research professor in the Department of Materials Science and Engineering at Gwangju Institute of Science and Technology (GIST) in Korea.

Specialists from the former Soviet Union know Dr. Korotcenkov's research results in the field of study of Schottky barriers, MOS structures, native oxides, and photo-receivers on the base of III-Vs compounds very well. His present scientific interests include material sciences, focusing on metal oxide film deposition and characterization, surface science, porous materials, and gas sensor design. Dr. Korotcenkov is the author or editor of 29 books and special issues, including the 11-volume Chemical Sensors series published by Momentum Press, the 10-volume Chemical Sensors series published by Harbin Institute of Technology Press, China, and the 2-volume *Handbook of Gas Sensor Materials* published by Springer. He has published 17 review papers, 19 book chapters, and more than 200 peer-reviewed articles (h-factor = 33 [Scopus] and h = 38 [Google scholar citation]). A citation average for his papers, included in Scopus, is higher than 25. He is a holder of 18 patents. In most papers, Dr. Korotcenkov is the first author. He has presented more than 200 reports on national and international conferences, and was the co-organizer of several conferences. His research activities were honored by an award of the Supreme Council of Science and Advanced Technology of the Republic of Moldova (2004), a prize of the Presidents of Ukrainian, Belarus and Moldovan Academies of Sciences (2003), a Senior Research Excellence Award of the Technical University of Moldova (2001, 2003, 2005), a fellowship from International Research Exchange Board (1998), and the National Youth Prize of the Republic of Moldova (1980), among others.

Contributors

Sergei Alekseev
Taras Shevchenko National University of Kyiv
Kyiv, Ukraine

Caitlin Baker
Georgia Institute of Technology
Atlanta, Georgia

Vladimir Brinzari
State University of Moldova
Chisinau, Republic of Moldova

Alessandro Caliò
Institute for Microelectronics and Microsystems
Naples, Italy

Naidan Chang
Daqing Campus of Harbin Medical University
Daqing, China

Xiuhua Chen
Yunnan University
Kunming, China

Beongki Cho
Gwangju Institute of Science and Technology
Gwangju, Republic of Korea

Frédéric Cuisinier
BioNano Laboratory
Montpellier University
Montpellier, France

Frédérique Cunin
Institut Charles Gerhardt Montpellier
Ecole Nationale Supérieure de Chimie de Montpellier
Montpellier, France

Adel Dalilottojari
ARC Centre of Excellence in Convergent Bio-Nano
    Science and Technology
Mawson Institute
University of South Australia
Mawson Lakes, South Australia, Australia

Luca De Stefano
Institute for Microelectronics and Microsystems
Unit of Naples-National Research Council
Naples, Italy

Pierre-Yves Collart Dutilleul
BioNano Laboratory
Montpellier University
Montpellier, France

Csilla Gergely
Laboratoire Charles Coulomb
Montpellier University
Montpellier, France

James L. Gole
Georgia Institute of Technology
Atlanta, Georgia

Jonas Gradauskas
Center for Physical Sciences and Technology
and
Vilnius Gediminas Technical University
Vilnius, Lithuania

Songhee Han
Mokpo National Maritime University
Mokpo, Republic of Korea

Igor Komarov
Institute of High Technologies
Taras Shevchenko National University of Kyiv
Kyiv, Ukraine

Ghenadii Korotcenkov
Gwangju Institute of Science and Technology
Gwangju, Rep. of Korea

Andras Kovacs
Furtwangen University
Furtwangen, Germany

Shaoyuan Li
Kunming University of Science and Technology
and
State Key Laboratory of Complex Nonferrous Metal
    Resources
Clean Utilization
Kunming University of Science and Technology
Kunming, China

Wenhui Ma
Kunming University of Science and Technology
Kunming, China

Steven J.P. McInnes
ARC Centre of Excellence in Convergent Bio-Nano
    Science and Technology
Mawson Institute
University of South Australia
Mawson Lakes, South Australia, Australia

Ulrich Mescheder
Furtwangen University
Furtwangen, Germany

Pascal J. Newby
Centre de collaboration MiQro Innovation (C2MI)
Université de Sherbrooke
Bromont, Quebec, Canada

Haisheng Peng
Daqing Campus of Harbin Medical University
Daqing, China

and

Iowa State University
Ames, Iowa

Jane Politi
Institute for Microelectronics and Microsystems
Unit of Naples-National Research Council
Naples, Italy

Ilaria Rea
Institute for Microelectronics and Microsystems
Unit of Naples-National Research Council
Naples, Italy

Nicolai F. Starodub
National University of Life and Environmental
    Sciences of Ukraine
Kiev, Ukraine

Jolanta Stupakova
Vilnius Gediminas Technical University
Vilnius, Lithuania

Monica Terracciano
Institute for Microelectronics and Microsystems
Unit of Naples-National Research Council
Naples, Italy

Wing Yin Tong
ARC Centre of Excellence in Convergent Bio-Nano
    Science and Technology
Mawson Institute
University of South Australia
Mawson Lakes, South Australia, Australia

Nicolas H. Voelcker
ARC Centre of Excellence in Convergent Bio-Nano
    Science and Technology
Mawson Institute
University of South Australia
Mawson Lakes, South Australia, Australia

Guangtian Wang
Daqing Campus of Harbin Medical University
Daqing, China

Qun Wang
Iowa State University
Ames, Iowa

Yang Zhou
Kunming University of Science and Technology
Kunming, China

Sensors

Solid State Gas and Vapor Sensors Based on Porous Silicon

Ghenadii Korotcenkov

CONTENTS

1.1	Introduction	4
1.2	Gas and Vapor Sensors Based on Porous Silicon	4
	1.2.1 Sensors Employing Photoluminescence Quenching	5
	1.2.2 Capacitance Sensors	9
	1.2.3 Conductometric Sensors	14
	1.2.4 Gas and Vapor Sensors Based on Schottky Barriers and Heterostructures	19
	1.2.5 Gas Sensors Based on CPD Measurements	22
	1.2.6 Combined Approach	23
	1.2.7 Field-Ionization Gas Sensors	24
1.3	Disadvantages of PSi-Based Gas Sensors	27
1.4	Improvement of PSi-Based Sensor Parameters through Surface Modification of Porous Semiconductors	32
1.5	Outlook	35
	Acknowledgments	36
	References	36

1.1 INTRODUCTION

In the past decades, porous silicon (PSi) has attracted attention as a promising material for optoelectronic applications (Canham et al. 1996; Hirschman et al. 1996; Canhman 1997; Parkhutik 1999; Bisi et al. 2000; Foll et al. 2006; Kochergin and Foell 2006). Microporous Si became famous because of its unexpected optical property of showing strong luminescence from red-orange to blue, depending on its precise structure. However, exploiting this property for devices proved difficult if not impossible until now (Bisi et al. 2000; Kochergin and Foell 2006). Therefore, at present, the interest extends to other applications of PSi including biomedical, micromachining, and sensor applications (Steiner and Lang 1995; Di Francia et al. 1998; Marsh 2002; Angelescu et al. 2003; Zhu et al. 2005). PSi layers are very attractive from a sensor point of view because of a unique combination of crystalline structure: (1) a large internal surface area of up to 200–500 m^2/cm^2, which ables enhancement of the adsorbate effects, and (2) high activity in surface chemical reactions. Several investigations show that electrical and optical characteristics of PSi may change considerably on adsorption of molecules to their surfaces and by filling the pores (Feng and Tsu 1994; Mares et al. 1995; Canhman 1997; Cullis et al. 1997). This means that surface adsorption and capillary condensation effects in PSi layers can be used for development of effective sensor systems (Anderson et al. 1990; Parkhutik 1999).

Lauerhaas and Sailor (1993) and Di Francia et al. (1998) noted the following advantages of PSi over other porous materials such as ceramics or nano- and polycrystalline films of metal oxides used for sensor design: (1) It is basically a crystalline material and thus, in principle, is perfectly compatible with common microelectronic processes devices. (2) It can be electrochemically fabricated in very simple and cheap equipment. (3) It can be produced in a large variety of morphologies, all exhibiting large values of surface-to-volume ratio. Other advantages of PSi include possibilities of creation of three-dimensional structures and design of multisensors, based on the use of various registration techniques for gas and vapor detection; for example, optical, electrical, luminescent, and so on. Lower power consumption of PSi-based devices in comparison with metal oxide gas sensors due to working at room temperature is another important advantage of those devices.

Thus, because PSi sensors are based on silicon wafers, manufactured using integrated circuit production techniques, and operated at room temperature using relatively low voltages, they can be used for producing compact and low-cost sensor systems on a chip, where both the sensing element and the read-out electronics can be effectively integrated on the same wafer. Some authors contend that the sensors based on PSi are so simple that they could ultimately be mass produced for pennies apiece. This means that those new sensors could be integrated into electronic equipment and used for building sensing arrays.

1.2 GAS AND VAPOR SENSORS BASED ON POROUS SILICON

Numerous research works have shown that many parameters of PSi such as the photoluminescence (PL) intensity, the capacity of the porous layer, the conductance, the reflection coefficient, infrared (IR) absorption, the resonance frequency of a Fabry-Pérot resonator made of the PSi, and so on are sensitive to gas surroundings (Anderson et al. 1990; Ben-Chorin and Kux 1994; Feng and Tsu 1994; Vial and Derrien 1995; Canhman 1997; Blackwood and Akber 2006; Korotcenkov and Cho 2010). For example, repetitive change of the PL band of PSi from green to red and vice versa may be observed while removing the sample from an electrolyte and replacing it (Fellah et al. 1999). Exposing PSi to chemical reagents strongly alters its PL intensity and conductance features (Baratto et al. 2001a,b). This means that the measurement of all parameters of PSi may be the basis for gas sensor design (Mizsei 2007; Ozdemir and Gole 2007; Korotcenkov and Cho 2010). Furthermore, both simple resistor-type structures (Di Francia et al. 2005) and more complex ones, such as transistors (Lazzerini et al. 2013) or heterostructures (Lundstrom et al. 1975), can be used for designing solid state gas and vapor sensors (see Figure 1.1).

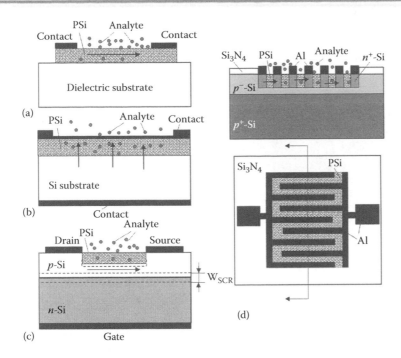

FIGURE 1.1   Variants of PSi-based devices used for gas sensing applications: (a, d) resistor-type structures, (b) Schottky barrier and *p-n* junction-type structure, and (c) transistor-type structure.

1.2.1 SENSORS EMPLOYING PHOTOLUMINESCENCE QUENCHING

Chemical sensors using strong gas surrounding influence on PL of PSi are the most widely studied PSi-based sensors (Ben-Chorin and Kux 1994; Chandler-Henderson et al. 1995; Fisher et al. 1995; Schechter et al. 1995; Harper and Sailor 1996; Kelly et al. 1996; Cullis et al. 1997; Parkhutik 1999; Hedrich et al. 2000; Lukasiak et al. 2000; Salcedo et al. 2004). Research has shown that the appearance of active gas and vapor is usually accompanied by a strong decrease of PL intensity of PSi. However, it is necessary to note that in some cases, the effect may be opposite. For example, Skryshevsky (2000) found that depending on the excitation wavelength and the condition of measurement, the gas adsorption can result in the quenching or increase of PL. Skryshevsky established that the emission efficiency of the as-prepared PSi was shown to decrease at the adsorption of acetone molecules for each excitation wavelength in the 405- to 546-nm range. However, if the 546-nm excitation causes small change of the PL spectrum during exposition, the 405-nm excitation quenches the PL in the ambient air and increases emission efficiency in acetone vapors.

By the results given in Figure 1.2, one can make a judgment about the measurement resources of PSi-based sensors, utilizing PL intensity control for gas vapor detection. It is seen that both the rate of response and reproducibility could be better.

Hedrich et al. (2000) indicated that typical time responses of sensors, whose operating characteristics are presented in Figure 1.2, were characterized by time constants below 10 s. The values are lower for lower alcohol concentrations. However, one can see that the real response time is much bigger. We can also see that for an as-prepared PSi sample, a partial degradation of the PL baseline takes place. For derivatized samples, the degradation is much less pronounced. However, even in this case the nonreproducibility of the baseline is observed. The information related to other gas and vapor sensors designed on the base of PL quenching is presented in Table 1.1.

As seen from the results presented in Table 1.1, the sensitivity of the PL to the change of the surrounding gas is high. For example, the threshold sensitivity for detecting alcohol vapors is only 1–4 ppm. As a rule, the sensor sensitivity rises in the line of alcohols from lower alcohols (methanol) to higher alcohols (hexanol). For NO_2 and SO_2 sensors, the threshold sensitivity is even smaller. This demonstrates that the lower detection limit is competitive with currently available sensors. At the same time, it was established that gases such as CO_2 and N_2O at pressures as

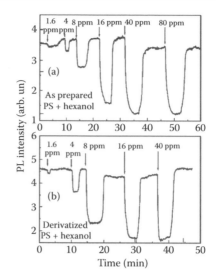

FIGURE 1.2 Successive PL quenching responses of as-prepared (a) and methyl 10-undeacetonate derivatized (b) PSi sensors to various concentrations of *n*-hexanol. Excitation by the 457.9-nm line of an Ar laser, emission wavelength 710 nm. Return to initial conditions is achieved by purging the measurement cell with pure nitrogen. (Reprinted from *Sens. Actuators B*, 84, Hedrich F. et al., 315, Copyright 2000, with permission from Elsevier.)

TABLE 1.1   Typical Parameters of PSi-Based Sensors Using the PL Quenching Effect

Target Gas	Concentration	Parameters	Ref.
Methanol, hexanol (in N_2)	150 ppm	$S_m \sim 0.55–0.6$ $\tau_{res} < 10$ sec	Hedrich et al. 2000
NO (vacuum)	130 ppm	$S_m \sim 0.7; P_I \sim 2$ ppm $S_m \sim 0.9; P_I \sim 70$ ppb	Harper and Sailor 1996
NO_2 (vacuum)	180 ppb	$S_m \sim 0.9; P_I \sim 70$ ppb $\tau_{res} \sim$ few sec, $\tau_{rec} \sim 5$ min	Harper and Sailor 1996
SO_2 (Ar)	3 ppm	$S_m \sim 0.6$ $P_I \sim 400$ ppb	Kelly et al. 1996
Vapor of linear alcohols (MeOH, EtOH, n-BuOH, etc.)	50 µg/ml	$S_m \sim 0.5–0.75$	Rivolo et al. 2004
EtOH, pentanol, Methanol	10 mg/L	$S \sim 0.1$	Chvojka et al. 2004
Butanol	10 mg/L	$S_m \sim 0.15$	Chvojka et al. 2004
Pentanol	10 mg/L	$S_m \sim 0.3$	Chvojka et al. 2004
Hexanol	10 mg/L	$S_m \sim 0.5$	Chvojka et al. 2004
Protein (in Air)		$P_I < 1$ µg/m³	Starodub and Starodub 1999
Oxygen	10–100%	$S_m \sim 0.7$	Green and Kathirgamanathan 2000

Note: S_m—sensor response ($S_m = \Delta I/I_o$); P_I—threshold of sensitivity.

low as those demonstrated for the nitrogen oxides do not have an effect on the PL of PSi. With 760 Torr of CO_2 or N_2O, no change in the PL intensity was perceptible.

The gas CO displays a minimal amount of PL quenching, ~5%, at 760 Torr as well (Harper and Sailor 1996). This means that those gases would not interfere with NO or NO_2 detection. However, it is necessary to note that as a rule the strong dependence of the PL intensity from the partial pressure of gas (P_{gas}), for which the linear dependence in the form $\Delta I/I_o = f(P_{gas})$ is typical, is observed in a narrow enough range of partial pressure of target gas. Further growth of partial pressure leads to a saturation of the dependence $\Delta I/I_o = f(P_{gas})$ (see Figure 1.3), which excludes the possibility of the sensors under discussion being used for gas concentration estimation in this range. However, one can admit that in alarm systems that respond only to the appearance of dangerous gases, sensors with such working characteristics could be used without limitations.

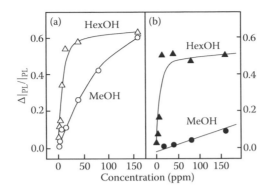

FIGURE 1.3   (a) Typical curves of relative PL quenching as a function of methanol (circles) and hexanol (triangles) concentration for as-prepared PSi (open symbols) and (b) for metal 10-undeacetonate derivatized PSi (filled symbols). (Reprinted from *Sens. Actuators B*, 84, Hedrich F. et al., 315, Copyright 2000, with permission from Elsevier.)

Among other disadvantages of the PSi-based sensors under consideration, strongly restricting their practical applicability for real devices, it is necessary to mark out insufficient stability and low selectivity of the PL quenching response. For example, Harper and Sailor (1996) showed that after 11 quenching cycles (i.e., exposure to gas and then evacuation) with pressures of NO below $7.6 \cdot 10^{-3}$ Torr, the intensity of PL from the PSi sample (in vacuum) is reduced by ~10%.

Exposure of the porous Si samples to higher partial pressures of NO results in a correspondingly lower degree of reversibility. In an attempt to form a more passive surface oxide and thus obtain more completely reversible NO quenching isotherms, freshly etched PSi was oxidized in air at 110°C for 10 min. It was established that the sensors became more stable. However, these samples were less sensitive to a given amount of NO than nonoxidized samples. For instance, to achieve the same degree of PL quenching on thermally oxidized PSi as was observed on freshly etched PSi, five times the amount of NO was required. Exposure of these mildly oxidized samples to NO still resulted in incomplete reversibility; that is, incomplete recovery of the initial (vacuum) intensity of the PSi samples after each successive exposure/evacuation cycle. Thus, even mild air oxidation does not appear to improve the stability of PSi toward oxidation by NO (Harper and Sailor 1996).

Since the appearance of the first reports in this field, the mechanism responsible for the PL quenching in PSi has been the subject of numerous investigations and discussions as well as controversy because this effect has strong dependence on the nature of the gas and vapor interacting with the PSi (Vrkoslav et al. 2005). For example, it was established that the effect of PL is partially reversible in the case of $C_2(CN)_4$ and completely reversible (by vacuum drying at room temperature) for the adsorption of ethyl alcohol. It was found that the dipole momentum of the solvent plays an important role in the effect of PL quenching. Chvojka et al. (2004) showed that the strength of PL quenching is directly determined by the dielectric constant and concentration of analyte in the liquid phase, whereas in the gas phase it primarily depends on the effective concentration of analyte inside the PSi matrix. It was concluded that the thermodynamic equilibrium concentration of the analyte inside the PSi matrix is controlled by a capillary condensation effect. In experiments carried out by Chvojka et al. (2004), a very good correlation between gas phase concentration and room temperature saturated vapor pressure of the studied analytes and PSi PL quenching response was obtained. It was shown that as a rule the maximum of PL intensity is blue shifted. The blue shifting or the decrease of the emission in the long wavelength region was observed after the treatment in the boiling CCl_4 (Hory et al. 1995), organoamine molecules (Coffer et al. 1993), C_2H_5OH (Dittrich et al. 1995a), methanol (Ben-Chorin and Kux 1994), and many other gases and vapors. However, there is experimental evidence of anomalous behavior of PL quantum efficiency at the adsorption of some molecules (Li et al. 1994; Baratto et al. 2000; Skryshevsky 2000).

Salcedo et al. (2004) as well as Sailor and Wu (2009) believed that electron transfer (charge and energy transfer) also participates in the PL quenching effect (see Figure 1.4). For example, according to research carried out by Harper and Sailor (1996), an adsorption mechanism is operative

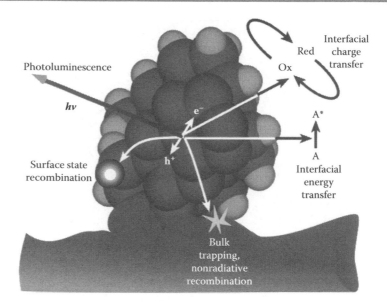

FIGURE 1.4   Schematic diagram depicting the nonradiative de-excitation pathways available to a nanocrystallite in porous Si. (Sailor M.J. and Wu E.C.: *Adv. Funct. Mater.* 2009. 19. 3195. Copyright Wiley-VCH Verlag GmbH & Co. KGaA. Reproduced with permission.)

during PSi interaction with NO and NO_2 at low partial pressure. In this range of partial pressure, the PL quenching effect is reversible and follows Stern–Volmer behavior (Turro 1991). Thus, plots of I_o/I (measured at a given wavelength) versus quencher concentration yield straight lines, whose slope is related to the efficiency of the quenching process via Equation 1.1:

(1.1)
$$\frac{I_o}{I} = 1 + \frac{k_q[Q]}{k_r + k_{nr}},$$

where I_o is the PL intensity in the presence of pure solvent, I is the PL intensity in the presence of a concentration of quencher Q, k_r is the radiative rate constant, k_{nr} is the intrinsic nonradiative rate constant (not dependent on quencher), and k_q is the rate constant for quenching by the molecular species. The constant k_q can also represent an equilibrium-binding constant, or a product of binding constant and kinetic rate constant instead of a purely kinetic constant. Consistent with the dynamic Stern–Volmer quenching model, the time-resolved PL decays usually show a distinct decrease in excited state lifetime with added quencher. Adsorbed molecules through the capture of electrons or holes can reduce radiative recombination channels. Adsorbates are also able to form adducts with surface-active centers, thus substituting Si-OH bonds.

Adsorption of oxygen produces an irreversible quenching effect due to chemical bonding of oxygen with surface silicon atoms. The oxidation reaction presumably introduces efficient non-radiative surface recombination centers that account for the lower PL intensity and reduced sensitivity of oxidized PSi. The dissociation with subsequent oxidation reaction with the PSi surface can be used for explanation of the partial irreversible quenching effect observed at high NO_X pressure or long exposure times (Harper and Sailor 1996). Some authors believe that surface traps introduced by surface oxidation reactions can account for the irreversible quenching of PSi PL. However, we note that the mechanism responsible for reversible PL quenching is not known and further work in order to address this question is required. In this context, the results given by Green and Kathirgamanathan (2000) are of interest. It was found that at low intensity of exciting radiation (~0.3 mW/cm²) the effect of oxygen on the PL of PSi is reversible. However, it was established that this process is very slow, with characteristic response and recovery times exceeding 1000 min. This means that the surface reactions, accompanied by the change in concentration of adsorbed oxygen, could be a reason for a temporal drift of the PSi-based sensor characteristics. The quenching effect from interaction with oxygen has both fast and slow components, and their contribution in transient characteristics depends on the wavelength at which the PL was

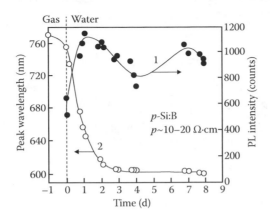

FIGURE 1.5   The change of PL intensity of PSi and the wavelength position of the PL peak with time of aging. PSi were formed using 1:1 HF (48%) to ethanol (96%) electrolyte ($J = 60$ mA/cm^2, $t = 10$ or 50 min). (With kind permission from Springer Science+Business Media: *J. Nanopart. Res.*, 12, 2907, 2010, Balaguer M. and Matveeva E.)

observed (Green and Kathirgamanathan 2000). Such behavior of PL spectra and their sensitivity to the intensity of exciting radiation testifies to the complicated character of oxygen interaction with the silicon surface, which should be taken into account when analyzing gas-sensing effects observed in the oxygen-containing atmosphere.

Molecules that have no accessible energy or redox states can also quench PL (Sailor and Wu 2009). For example, adsorption of a wide range of inert molecules such as benzene, hexane, or dichloromethane results in a loss of PL intensity that is recovered on removal of the chemical. These molecules quench luminescence without effecting a net chemical transformation on the PSi surface. The PL quenching mechanism followed by these molecular adsorbates is still not well understood. A number of mechanisms have been invoked for explanation of this effect: in particular, it was assumed that (i) the presence of a dielectric medium outside the silicon nanocrystallites increases carrier recombination rates (Chazalviel et al. 1994), (ii) the adsorption of molecules at the surface induces strain, which could cause native defects to become more efficient nonradiative traps (Bellet and Dolino 1994), and (iii) the adsorption of molecules could enhance nonradiative vibronic coupling to surface vibrational modes (Rehm et al. 1995). Further researches are still needed.

Regarding the appearance of PL-based gas and vapor sensors on the sensor market, we have to say that the aging effect (see Figure 1.5) and irreversibility of PL quenching observed with most chemicals, especially oxidants, is a severe limitation for such application (Sailor and Wu 2009). In addition, (1) they are not able to detect the target analyte in the varied mix of background compounds present at much larger concentrations in the environmental matrix; (2) they are subject to temporal drift due to fouling or chemical degradation; (3) their response is highly dependent on fluctuations in temperature or relative humidity; and (4) they do not supply predictable responses when challenged with interferents that were not present in the calibration or training set. Sailor and Wu (2009) believe that a single-use "canary in a coalmine" sensor used to detect the release of corrosive gases in an industrial environment is a more promising potential application. However, the inability to test or calibrate such systems and the drift in signal that occurs due to air oxidation of PSi are very undesirable factors. Medical applications are more relevant for single-use sensors.

1.2.2 CAPACITANCE SENSORS

As shown in numerous research works, capacitance sensors on the base of PSi may be successfully used for detection of humidity, gases, and various vapors. Examples of such sensors can be found in the literature (Anderson et al. 1990; Kim et al. 2000a,b; Archer et al. 2005; Ozdemir and Gole 2007; Harraz et al. 2014) and in Figure 1.6a. For example, Kim et al. (2000a) established that PSi-based alcohol sensors had considerable sensitivity even at low alcohol concentrations (see Figure 1.6b). The capacitance characteristics exhibited a slope of 2.5% against the alcohol

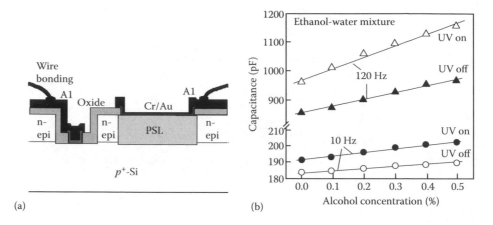

(a)   (b)

FIGURE 1.6   (a) The cross-section and (b) gas sensing characteristics of the fabricated sensor based on p+-Si(100) wafers with 0.02 Ω·cm. PSi with porosity 35% and thickness 4 μm were formed in 25% HF solution mixed with ethanol at a $J = 13$ mA/cm² for 120 s. Cr/Au film had 30 nm thickness. (With kind permission from Springer Science+Business Media: *J. Solid State Electrochem.* 4, 2000, 363, Kim S.-J. et al.)

concentration increment of 0.1% under ambient conditions. In addition, it was observed that CO_2 and N_2 gas concentrations had little effect on the capacitance responses. Hasan et al. (2010) also believe that PSi is a promising platform for capacitive detection of organic solvent vapors at the ppm level. The same conclusion was made by Harraz et al. (2014), who successfully fabricated and tested a capacitive chemical sensor based on meso-PSi layers (30 nm pore size and 4.5 μm pore length). They have found that the PSi layers were easily infiltrated with and cleared of various organic solvents without damaging the porous matrix. Changes in capacitance of the device could be readily and repeatedly observed. They also have shown that the sensing response toward polar solvents at room temperature was highly sensitive, and the process was reversible and reliable. Further, the sensing of nonpolar solvents was sensitive but the process was not reversible. The overall change in capacitance on exposure to organic solvents can be understood as the impact of charge redistribution on the pore walls altering the electric field inside the pores and leading finally to a gain or a reduction of measured capacitance. The polarizability plays a role in sensing response because it defines the molecule orientation with respect to the electric field and surface, which in turn affects the net surface charge.

One should note that in addition to the traditional configuration of PSi-based capacitance sensors, used in the above-mentioned articles, innovative approaches to the application of PSi also could be used for the development of capacitive chemical sensors. For example, Kavalenka et al. (2012) proposed to use ultrathin PSi membranes metallized with gold as flexible conductive electrodes in polymer-based capacitive vapor sensor (see Figure 1.7). They believe that such use of a porous electrode simplifies the conventional parallel-plate design of typical sensors; the very

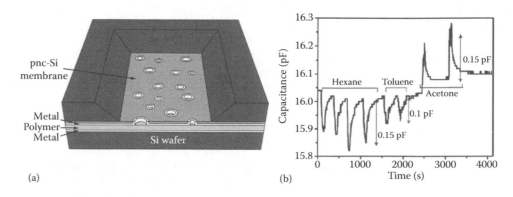

(a)   (b)

FIGURE 1.7   (a) Schematic of the PSi membrane based capacitive sensor. (b) Capacitive response of the sensor on exposure to hexane, toluene, and acetone vapors (800–1000 ppm). (Reprinted from *Sens. Actuators B* 162, Kavalenka M.N. et al., 22, Copyright 2012, with permission from Elsevier.)

thin porous membrane allows fast analyte vapor permeation to the underlying polymer material that serves as receptor. The swelling caused by vapor permeation was measured by optical profilometry. PSi-based membranes had a thickness of 15 nm, pore sizes ranging from 5 to 50 nm, and porosities from 1 to 2%. Membranes were fabricated using standard silicon semiconductor processing techniques. Kavalenka et al. (2012) have shown that such PSi membranes are stable, and no measurable plastic deformation was observed.

However, we note that most research related to PSi-based capacitance sensors focuses on the design of humidity sensors. In general, for commercial devices, the sensing materials utilized to detect humidity are either metal oxides (Al_2O_3, TiO_2, etc.) or polymer films. A good humidity sensor has to fulfill several requirements, such as high sensitivity over a wide range of humidity, short response time, small hysteresis, and especially good long-term stability. These are difficult to meet concurrently (Björkqvist et al. 2004a,b). When the ease of manufacture and low cost are added on the list, the situation becomes even more complicated.

One proposed and, to some extent, studied alternative is to use (oxidized) PSi as a miniaturized humidity sensor material. PSi-based humidity sensors operating on the capacitive principle have been demonstrated by a few authors (Kim et al. 2000b; Connolly et al. 2002, 2004a,b; Salonen et al. 2002; Fürjes et al. 2003; Xu et al. 2005; Björkqvist et al. 2004a, 2006). One advantage of PSi as a sensor material is the possibility to integrate it with already existing silicon technology. In some applications, even a coarse control of operation environments with a simple and cheap humidity sensor is desirable (e.g., inside a mobile phone).

Humidity sensors can be fabricated in both one-sided and two-sided configurations. Electrodes with a comb structure cover only a small part of the porous layer surface. An example of the measuring abilities of a humidity sensor designed based on PSi is presented in Figure 1.8.

Although the principle of humidity sensing of porous semiconductors is not very clear, the relative electrical permittivity of these layers would undergo a change on exposure to a humid atmosphere, due to the adsorption of water molecules in its micropores (Kim et al. 2000b). Experimental study has shown that the humidity sensing of a PSi capacitor structure is based on the alteration of its dielectric permittivity resulting from the infiltrated water molecules. The typical pore size in Si layers used for sensors is in the range of 2–4 to 12–15 nm. Therefore, water transport in PSi is described by Knudsen diffusion. In addition to adsorption-chemisorption processes, capillary condensation phenomena are involved when considering the fine structure of the porous matrix. According to the basic theory of adsorption on porous ceramic sensors (Adamson and Gast 1997), when the vapor molecules are first physicosorbed onto the porous surface, capillary condensation will occur if the micropores are narrow enough. The critical size of pores for a capillary effect is characterized by Kelvin radius, which is

(1.2)
$$r_K = \frac{2\gamma M}{\rho RT \ln(P_S/P)},$$

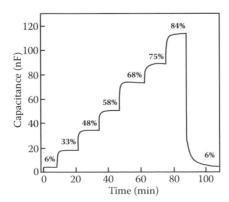

FIGURE 1.8 Dynamic response of the capacitance of a TC-PSi humidity sensor. Electrical parameters were measured using 85 Hz frequency. (Reprinted from *Sens. Actuators A* 112, Björkqvist M. et al., 244, Copyright 2004, with permission from Elsevier.)

where γ is the surface tension of vapor in the liquid phase, M is the molecular mass of vapor, ρ is the density of vapor in liquid phase, P is the vapor pressure, and P_S is the saturation vapor pressure. Therefore, the performance of a humidity sensor is determined by its nano- and microscopic dimensions, including pore size, thickness of the porous layer, size distribution of the surface structural unit, regularity of the surface morphology, and electrode distance (Kim et al. 2000b; Di Francia et al. 2002). This fact indicates that the regularity and controllability of porous semiconductor structures are of great importance in sensor application. For example, due to water condensation in nanosize pores, it is difficult to prepare rapid humidity sensors based on nanoporous material (Björkqvist et al. 2004a).

The characteristics of PSi-based humidity sensors are strongly affected by the hydrophilic/hydrophobic properties of the walls (Yarkin 2003; Björkqvist et al. 2004b). It has been assumed that only the hydrophilic regions (water adsorbing sites) on a sample adsorb polar water molecules through hydrogen bonding, whereas the hydrophobic regions, which contain very weak dispersion forces, do not adsorb water (Raghavan et al. 2001). It should be noted that the functional groups of hydrophilic surfaces, such as NH_2, OH, and COOH, are highly hydrogen bonded and tend to associate weakly with other chemical species except strong electron donors such as water. For SiO_2 surface, two-thirds of the SiOH groups are hydrogen bonded, and water preferentially reacts at these sites (Iler 1979). Once water is adsorbed, the site then has free active hydrogen for adsorption through hydrogen bonding. This water-induced increase of the hydrogen bonding sites explains the interaction increase with increasing relative humidity (RH) for silicon substrate.

It is well known that a hydrogen-covered silicon surface is hydrophobic, whereas the surface covered with imperfect oxide is hydrophilic (Okorn-Schmidt 1999). Based on the Kelvin equation, the Kelvin radius becomes smaller when the surface becomes more hydrophobic. Modification of the contact angle (hygroscopicity) of the PSi surface is possible using various treatment temperatures (Buckley et al. 2003). For example, it was established that the oxidation strongly promotes capillary condensation of water vapors in pores. Modification of surface properties may be used for influence on operating characteristics of humidity sensors since the condensation of water on the hydrophobic surface occurs at higher values of relative air humidity. However, as Björkqvist et al. (2006) established, in that case the configuration of sensors must be optimized because the hydrophilic electrodes also have effects on water condensation in porous medium. This influence could lead to anomalous behavior of the sensor.

Analysis of fabricated devices has shown that porous semiconductor-based humidity sensors in capacitance configuration efficiently operate only in the range of relative humidity higher than 10–20% (O'Halloran et al. 1999; Connolly et al. 2002, 2004a,b; Salonen et al. 2002; Fürjes et al. 2003; Björkqvist et al. 2004a,b, 2006; Xu et al. 2005). The same situation takes place during gas and vapor detection. As indicated earlier, the capacity effects are attributed to the variation of the dielectric constant due to the filling of the pores by the gaseous species. For this reason, PSi-based capacitance sensors are sensitive only to saturated vapors. For comparison, the PL quenching and conductance variations of a PSi in the presence of gaseous species are appreciable even at lower gas and vapor concentrations because a charge transfer mechanism is involved (Salcedo et al. 2004).

Slow response, due to the limited mass transportation and capillary condensation, is another disadvantage of capacitance-type sensors (Connolly et al. 2004b). According to Connolly et al. (2002), even the fastest SiC-based humidity sensors have response time of about 120 s when the RH changed from 10 to 90%. As shown by Fujikura et al. (2000), an excessive thickness of the PSi layer would make the absorption and desorption processes much more difficult and lead to a long response time. Therefore, the optimization of porous layer thickness is needed for achievement of optimal parameters of humidity sensors.

Poor linearity of operating characteristics can be added to the disadvantages of porous semiconductor-based RH sensors. Research carried by O'Halloran et al. (1999) showed that this parameter of humidity sensors can be improved by changing the current density used to form the porous layer. It has been shown that it is possible to form a device that is linear over the humidity range from 10 to 90% RH. However, PSi layers formed at such low current density values show a long response time and poor sensitivity in comparison to layers formed using higher values of current density.

Large hysteresis is also a common problem in humidity-sensing based on adsorption (Björkqvist et al. 2004a,b) (see Figure 1.9a). It is related to the size and geometry of the pores and how the

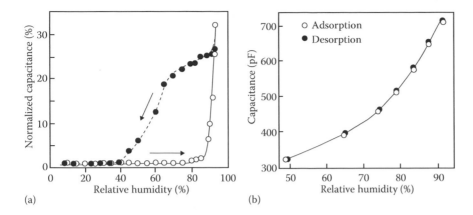

FIGURE 1.9 (a) Hysteresis effect in as-prepared PSi-based humidity sensors without any heating. (Reprinted from *Sens. Actuators B* 95, Fürjes P. et al., 140, Copyright 2003, with permission from Elsevier.) (b) Hysteresis loop for PSi-based humidity sensors with larger pores. Humidity conditions are changed every 4 min. (Data extracted from Björkqvist M. et al., *IEEE Sensors J.* 6(3), 542, 2006.)

presence of moisture changes this geometry. Widening of the pore size in PSi structures might solve the hysteresis problem (Björkqvist et al. 2004a,b). For example, a TCP Si humidity sensor with larger pores designed in Björkqvist et al. (2006) showed only slight hysteresis above 75% RH (see Figure 1.9b). In addition, larger pores should reduce the response time. Unfortunately, that reduces the sensitivity, too. Because of that, the pore size in the range 2–10 nm is optimal for effective humidity sensing in the low RH range, whereas material with larger (20–100 nm) pores is more preferable for the design of humidity sensors for high RH values (Seiyama et al. 1983; Connolly et al. 2002).

As follows from the above, humidity sensors on the basis of PSi have disadvantages that can restrain their commercialization (Connolly et al. 2004b). However, research carried out during recent years has shown that operating parameters of PSi-based humidity sensors can be considerably improved. For example, Xu et al. (2005) showed that for PSi-based humidity sensors prepared by hydrothermal etching, when the RH changed from 11 to 95%, the capacitance showed an increment over 1500% at a frequency of 100 Hz. Only about 15 and 5 s were needed for the capacitance to reach 90% of its final/initial values during an RH-increasing process and an RH-decreasing process, respectively.

For hydrothermal etching, samples usually are placed in a closed stainless steel tank, which makes it possible to carry out treatments for a long time at high temperature without loss of etchant. In this case, the nonuniformity in sample etching due to the insulating effect of hydrogen bubbles that form on the wafer surface because of electrochemical etching can be avoided in the hydrothermal process due to the supercritical condition. Xu et al. (2005) believed that the fast response to humidity of Si-NPPA sensors might be due to the regular morphology and suitable thickness of the sensing layer.

Carbonization of the PSi surface is another promising approach to optimization of humidity sensors. The long-term stability studies of differently stabilized PSi samples have shown that the thermal carbonization (TC) of PSi is an even more efficient stabilizing method than thermal oxidation (Björkqvist et al. 2006). Thermal carbonization is a stabilizing method that exploits the dissociation of acetylene at high temperatures (Salonen et al. 2002). A thermally carbonized PSi surface has been found stable in humid air and even in harsh environments. This treatment also changes the originally hydrophobic PSi surface to hydrophilic, thus improving its humidity sensing properties. Moreover, the sensitivity of TC-PSi is presumably better due to its larger specific surface area. A recent report on TC-PSi humidity sensor showed good sensitivity and repeatability of the sensor but also inappropriate hysteresis (Björkqvist et al. 2004a,b).

Another method contributing to the improvement of PSi-based humidity sensors' parameters was proposed by Allongue et al. (1997) and Björkqvist et al. (2006). It was found that both sensitivity and recovery time extremes might be improved through annealing of PSi structures at 200°C. It was established that the continuous heating reduced the hysteresis drastically but also decreased the sensitivity. With periodical refreshing, the sensitivity was nearly the same as without heating, but the hysteresis was negligible. This means that the application of chemical sensor refreshing before every measuring step is highly recommended. Investigations have shown that the application of internal heating filaments allowed a more rapid reaching of the desired temperature values around the active part of the chip than for heaters placed outside of the PSi structure (Fürjes et al. 2003). The humidity sensors realized in the frame of the proposed approach usually consist of three parts: (1) a humidity-sensitive capacitor with a PSi dielectric, (2) two integrated thermoresistors, and (3) a refresh resistor (Rittersma and Benecke 1999). The capacitor is formed between a meshed top electrode and a low-Ohmic backside of the sensor.

It is necessary to note that in each mentioned case, reduced hysteresis is mostly based on a smaller amount of condensed water (Björkqvist et al. 2006). In a capacitive-type sensor, this also means lower sensitivity. However, the sensitivity of the PSi humidity sensor even after the proposed modifications is still adequate for accurate humidity measurements.

1.2.3 CONDUCTOMETRIC SENSORS

Several realizations of chemical sensors based on electrical conductivity effects in PSi have been reported (Schechter et al. 1995a; Baratto et al. 2001a,b; Han et al. 2001; Kim et al. 2001; Green and Kathirgamanathan 2002; Seals et al. 2002; Barillaro et al. 2003, 2005; Oton et al. 2003; Pancheri et al. 2003, 2004; Iraji zad et al. 2004; Di Francia et al. 2005; Lewis et al. 2005; Luongo et al. 2005; Rahimi et al. 2006; Salgado et al. 2006; Li et al. 2013a). Typical configurations of conductometric type gas sensors designed based on porous materials are shown in Figures 1.1 and 1.10.

Measurement of both direct current (DC) and alternating current (AC) electrical conductivity of PSi in such structures has shown that the properties of PSi change dramatically as a result of the environmental impact. In particular, a strong effect was observed during a change of air humidity. For decreasing this affect, Schechter et al. (1995a) proposed the use of PSi with a hydrophobic surface. Experiments carried out by Schechter et al. (1995a) showed that samples, prepared in such a way that restored the original hydrophobic nature, could have no response to water vapor, but a large effect was induced by methanol. It was found that the humidity response could be easily changed by simple pretreatment or aging of the samples. Water vapor affected the conductivity of hydrophobic PSi only at a high humidity level and prolonged exposure.

Absorption of other chemical species at the surface of PSi also produced an essential impact onto their electrical properties. For example, Baratto et al. (2001a) showed that PSi-based conductometric sensors might be very sensitive to NO_2. These room temperature (RT) sensors had threshold sensitivity to NO_2 on the level of ~20 ppb. Moreover, these sensors had negligible response to ethanol (1000 ppm), ozone (200 ppb), CH_4 (15,000 ppm), CO (1000 ppm), ethylene (200 ppm), benzene (20 ppm), and methanol (1000 ppm). Response and recovery times equaled 20–30 min.

It was found that a sub-ppm sensor for benzene using PSi membrane may be fabricated as well (Schechter et al. 1995a). PSi membranes showed good sensing characteristics in a sensor of oxygen dissolved in an electrolyte. A PSi membrane was prepared by double-stage anodization of an Si wafer in ethanol/HF solution and mounted in an industrial sensor of oxygen. Seals et al. (2002)

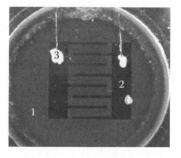

FIGURE 1.10 Typical view of PSi-based sensor with interdigital (2,3) contacts. (Reprinted from *Sens. Actuators B* 111–112, Di Francia G. et al., 135, Copyright 2005, with permission from Elsevier.)

introduced a PSi gas sensor for detection of HCl, NH_3, and NO at the 10-ppm level. A few similar PSi-Si structures have been introduced as resistor type ethanol (Green and Kathirgamanathan 2000; Han et al. 2001; Iraji zad et al. 2004), organic vapor (Barillaro et al. 2003, 2005; Salgado et al. 2006), and H_2 (Luongo et al. 2005; Rahimi et al. 2006) sensors. Recently, NO_2 sensors based on PSi have been developed that can detect concentrations of NO_2 in air near and even below the attention level (Baratto et al. 2001b; Pancheri et al. 2003, 2004). For example, Pancheri et al. (2003) have proposed a PSi sensor that can detect NO_2 at the 50 ppb level at any relative humidity between 0 and 70%, thus demonstrating that the necessary sensitivity for NO_2 pollution monitoring is achievable. It was shown that NO_2 sensors based on porous silicon membranes (PSM) were capable of detecting nitrogen dioxide at room temperature with negligible interference from organic vapors and other environmental pollutants such as benzene, ozone, and CO (Oton et al. 2003). Conductometric PSi sensors consisting of a sensitive surface layer that is conducive to the rapid and reversible transduction of sub-ppm levels of the analyte gas have been developed as well (Lewis et al. 2005). Several new fabrication and testing methods allowed them to detect a number of analytes including CO (<5 ppm), NO_x (<1 ppm), SO_2 (<1 ppm), and NH_3 (500 ppb). Several operating characteristics of NH_3 sensors designed by Li et al. (2013b) are shown in Figure 1.11. Some important parameters of PSi-based conductometric gas and vapor sensors are collected in Table 1.2.

PSi structures were also tested as humidity sensors (Di Francia et al. 2005; Jalkanen et al. 2012). Experiments carried out by Jalkanen et al. (2012) have shown that PSi-based sensors provided good sensitivity and reproducible results throughout the entire relative humidity range

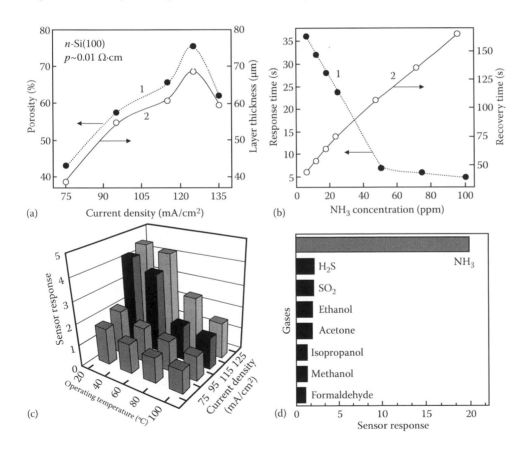

FIGURE 1.11   (a) The average porosity and layer thickness of PSi samples as a function of etching current density. PSi layers were formed using electrolyte containing 60 mL 40 wt% aqueous HF and 300 mL deionized water; (b) response and recovery times of the PSi sensor as a function of NH_3 concentrations; (c) the histogram showing the relationship between the sensor response of PSi toward 50 ppm NH_3 with etching current density as well as operating temperature; and (d) the selectivity histogram of PSi sensor to 100 ppm various gases at RT. (Reprinted from *Electrochim. Acta* 108, Li M. et al., 167, Copyright 2013, with permission from Elsevier.)

TABLE 1.2   Parameters of Conductometric Gas Sensors Based on Porous Materials

Structure	Detected Gas	T_{oper}	Parameters	Refs.
PSi	Ethanol (4000 ppm)	RT	$S_m \sim 19$	Han et al. 2001
PSi (fresh)	NO_2 (50 ppb)	RT	$S_m \sim 10$	Pancheri et al. 2004
PSi (aged)	NO_2 (100 ppb)	RT	$S_m \sim 1.5$	
Psi	NO_2 (200 ppb)	RT	$S_m \sim 110$	Baratto et al. 2001b
PSi:(Au;Sn)	NO_x	RT	$P_g < 1$ ppm	Lewis et al. 2005
	CO		$P_g < 5$ ppm	
	NH_3		$P_g < 500$ ppb	
	SO_2		P_g 1 ppm	
PSi (fresh)	NH_3 (6–100 ppm)	RT	$S_m \sim 4.3$ (50 ppm)	Li et al. 2013b
			$P_g < 1$ ppm	
PSi	Humidity (35–70 %RH)	RT	$S_m \sim 2$–100	Di Francia et al. 2005
PSi (fresh)	Ethanol (4%)	RT	$S_m \sim 32$	Iraji zad et al. 2004
	Methanol (4%)		$S_m \sim 40$	
	Acetone (4%)		$S_m \sim 40$	
PSi-Si (APSFET)	Butanol (1000 ppm)	RT	$S_m \sim 3$	Barillaro et al. 2003, 2005
	Isopropanol (1000 ppm)		$S_m \sim 1.3$	
PSi-Si (APSFET)	Acetic acid (1000 ppm)	RT	$S_m \sim 3$	Barillaro et al. 2003
PSi:Pd	H_2 (1%)	RT	$S_m \sim 2$	Luongo et al. 2005
PSi	O_2 (1–100 Torr)	RT	$S_m \sim 1.32$	Green and Kathirgamanathan 2002

Note: APSFET—adsorption porous silicon-based field effect transistor; P_g—threshold of sensitivity; RT—room temperature; S_m—maximum sensitivity $|R_{air}/R_{gas}|$; T_{oper}—operation temperature.

(see Figure 1.12a). In addition, negligibly small hysteresis (Figure 1.12b) accompanied by relatively fast response and recovery times was also demonstrated. For achievement of such results, the surface of PSi layer was stabilized using thermal hydrocarbonization (THC). In order to turn the surface hydrophilic, the THC PSi films were functionalized with undecylenic acid by immersing the films in an undecylenic acid solution for 12 h at 110°C. As it was indicated in the previous section, conventional humidity sensors based on micro- and mesoporous materials usually display larger hysteresis, which is due to the intrinsic properties of the material causing capillary condensation of the adsorbate gas to occur in the nanometer scale pores. Jalkanen et al. (2012) believe that the low hysteresis, observed in Figure 1.12b, is a result of applied surface functionalization, which minimized the influence of capillary condensation and increased the role of surface effects.

It is necessary to note that monitoring the conductance variation of porous semiconductors is the simplest and cheapest way to realize a gas or vapor sensor with PSi. However, some

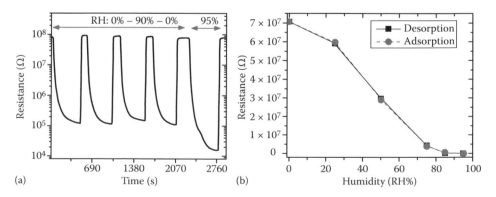

FIGURE 1.12   (a) Sensor response to RH modulation between 0% and 90%. Final humidity pulse to 95 RH% demonstrates that the sensitivity is well below 5 RH%, even for higher humidity values. (b) Sensor response to a step-wise relative humidity sweep ranging from 95 to 0 RH%, and the corresponding hysteresis curve determined from the measurement. (Reprinted with permission from Jalkanen T. et al., *Appl. Phys. Lett.* 101, 263110, 2012. Copyright 2012, American Institute of Physics.)

difficulties need to be overcome. For example, a reliable contact on PSi using a microwelder cannot be obtained due to the fragility of the material.

The optimization of the sensor performance also needs a better control of the thickness and porosity of the PSi membrane and its wetting ability. Salgado et al. (2006) concluded that a very thin PSi layer is not enough to detect some appreciable resistance changes and for thick PSi layers the diffusion time of the molecules plays an important role to have fast responses. Results presented in Figure 1.13 show how important such optimization can be. It was found that maximum sensitivity corresponds to PSi with maximum porosity (Iraji zad et al. 2004).

The same result, concerning the strong sensitivity dependence on porous microstructure as well as on the thickness of the porous layer, was obtained by Pancheri et al. (2003, 2004). They observed a large microstructural transformation in PSi even for an HF concentration change between 13 and 15%. Pancheri et al. (2003, 2004) assumed that the observed different sensitivity most likely originated in the degree of interconnection between the conducting microchannels. In the structures with lower degree of interconnection, high resistance paths, which are highly sensitive to gas, dominate the overall resistance, as in a series arrangement. On the other hand, in structures with higher interconnections, low resistance and insensitive paths frequently allow local bypasses across the sensitive paths (parallel arrangement). Thus, in the latter case, the sensitivity is dramatically inhibited.

Another opportunity of manufacturing the conductometric sensors based on PSi was realized by Barillaro et al. (2000, 2003, 2005), who suggested to control of the I–V characteristics of (n^+)-polysilicon/(n^+)-silicon heterojunctions fabricated on a p-Si substrate. Recently, similar sensors were fabricated and characterized by Borini et al. (2007) and Archer et al. (2005) as well. A picture of a sensor design based on the above-mentioned approach is shown in Figure 1.14.

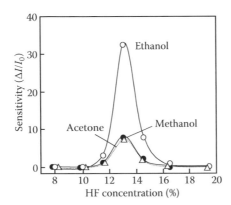

FIGURE 1.13 Sensitivity ($\Delta I/I_0$) of PSi-based conductometric sensors versus HF concentration for ethanol, methanol, and acetone. (Reprinted from *Sens. Actuators B* 100, Iraji zad A. et al., 341, Copyright 2004, with permission from Elsevier.)

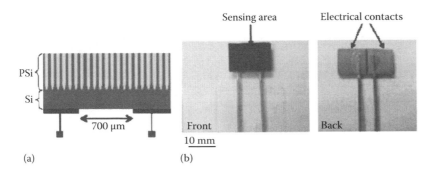

FIGURE 1.14 (a) Schematic cross-sectional view of a PSi sensor. The electrical contacts are placed on the back part of the layer (c-Si) by aluminum evaporation or colloidal silver paint 700 μm apart. (b) Pictures of the front and back sides of the device. (Reprinted from *Sens. Actuators B* 106, Archer M. et al., 347, Copyright 2005, with permission from Elsevier.)

This solution solves the compatibility problems, improving the electrical properties of the Ohmic contacts as well.

It was found that current-voltage characteristics are sensitive to the presence of different organic vapors in the surrounding atmosphere (Schechter et al. 1995a). Based on the conducted research, it was concluded that in the structures mentioned previously (see Figure 1.11), the change of the sensed quantity (current) is caused by a change of free carrier concentration in the silicon channel directly below the porous sensing layer, with advantages in terms of measurement and sensitivity. According to Archer et al. (2005), due to interaction of gas molecules with the PSi surface, the porous layer becomes a charged layer that can modulate the field in the Si channel (see Figure 1.14). This modulation can be carried out by two mechanisms: (1) change in the space charge region by charge redistribution and (2) change in the width of the conductive channel.

There are several approaches to explain the behavior of the current in PSi with the adsorbed gas (Schechter et al. 1995a; Green and Kathirgamanathan 2002). The most popular is associated with changes of the dielectric constant due to the condensation of the gas and vapor in the pores. Moreover, the formation of a condensed film can introduce a parallel ionic conductivity. Nevertheless, Ben-Chorin and Kux (1994) have ruled out this hypothesis. Indeed, they observed that the variation of conductivity versus the methanol pressure started below the condensation conditions.

Luongo et al. (2005) reported on a Pd-doped PSi as a resistor-type sensor for H_2 at room temperature. The authors stated that the basis of this sensor operation is the bulk change of the Pd particles dispersed on a PSi layer, reducing the layer impedance due to closer contact. Stievenard and Deresmes (1995) proposed a model where adsorbed gas or vapor molecules at the PSi surface change the width of a depleted region inside the silicon nanocrystals, thus modifying the overall conductance of the PSi layer. Interestingly, though an increase in current is observed for ethanol, exposure to acetic acid decreases both the conductance and the saturation current.

However, the adsorption mechanism and its influence on the carrier concentration in PSi are still not clear (Schechter et al. 1995a; Barillaro et al. 2003). For example, Stievenard and Deresmes (1995) removed the hydrogen atoms from the surface of PSi by placing it in boiling CCl_4 and observed that the sensitivity toward different gases was annihilated. They concluded that the existence of hydrogen on the surface plays an important role. However, Glass et al. (1995) removed H atoms from the surface of PSi inside a Fourier transform infrared (FTIR) spectroscopy chamber and observed that the surface adsorbed methanol, but when it was covered by hydrogen, there was no visible adsorbed methanol at the surface of the PSi. They suggested that the dangling bonds are responsible for gas surface reactions. Therefore, based on current experimental results, we can conclude only that the surface of PSi plays a crucial role in its electrical behavior. This means that there are no full explanations of the operation and theoretical basis of PSi-based gas sensors (Mizsei 2007).

From our point of view, the absence of a realistic model of PSi-based gas sensors is connected with the lack of understanding of the conductivity mechanism in porous materials. At present, too many assumptions are being used for this phenomenon explanation (Yamana et al. 1990; Ben-Chorin and Kux 1994; Ben-Chorin et al. 1994; Fejfar et al. 1995; Lubianiker and Balberg 1998; Lee et al. 2000; Shi et al. 2000; Remaki et al. 2001; Garrone et al. 2003; Sukach et al. 2003), which is probably a consequence of the presence of many factors influencing the current percolation in porous structures. For example, how the structure of porous materials may be important for their conductivity can be estimated based on dependencies shown in Figure 1.15.

Figure 1.15 depicts the temperature dependence of the dark conductivity of freestanding PSi films with different porosities. One can see that, as the porosity increases, the conductivity decreases and the activation energy E_a of the Arrhenius relation $\sigma \sim \exp(-2E_a/kT)$ increases, indicating that apparently the porosity has an influence on the charge transport mechanism. The above-mentioned regularity is similar to dependencies observed for metal oxides during grain size decrease. For explanation of the observed effect, a model was proposed that assumed the increased role of surface processes in the conductivity of polycrystalline metal oxides. However, Lee et al. (2000) believed that this effect is related to the increase of activation energy of deep levels in PSi.

Based on impedance measurements, Fonthal et al. (2006) concluded that the measured impedance of PSi is the series combination of a resistance-capacitance (RC) network related to the PSi

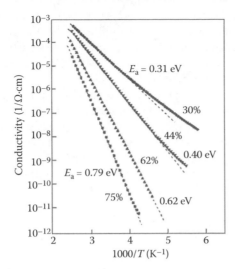

FIGURE 1.15 Temperature dependence of the dark conductivity of PSi with different porosities. (Reprinted from *Sol. St. Commun.* 113, Lee W.H. et al., 519, Copyright 2000, with permission from Elsevier.)

layer (geometric capacitance), an RC network related to the *p*-Si wafer (depletion capacitance), and a resistance-inductance (RL) network related to the wiring. Apparently, it is necessary to take into account a large number of trapping centers in the volume and at the surface of the PSi, which participate in the trapping and transport of charge carriers. Remaki et al. (2001) believed that the electrical conduction in mesoporous silicon layers of 100-μm thickness and 50% porosity at high frequencies (above 10 kHz) is mainly controlled by hopping transport on localized states in the chaotic porous structure. According to Yamana et al. (1990), those defects have high concentration and various ionization energies. Thus, based on this analysis, we can state that we do not have full understanding of the operating mechanism of the sensors discussed (Schechter et al. 1995a). The full mechanism of response is only a topic of continuing study. Therefore, for a detailed understanding, many structural and electric properties of PSi sensors require intensive investigations. At present, we can assert only that the observed change in conductivity on exposure to different organic solvents and other molecules such as oxygen suggests the existence of at least two response mechanisms. The first one relies on a change in the charge distribution within the crystallites due to the alignment of polar molecules on the surface (Ben-Chorin and Kux 1994; Schechter et al. 1995b). The second one involves charge transfer reactions mediated by surface traps during adsorption (Fejfar et al. 1995) and oxidation of molecules on the surface of PSi (Schechter et al. 1995b).

1.2.4 GAS AND VAPOR SENSORS BASED ON SCHOTTKY BARRIERS AND HETEROSTRUCTURES

Schottky barrier-type gas and vapor sensors (see Figure 1.1) are typical sensors fabricated based on standard semiconductors (Lundstrom et al. 1975; Fogelberg et al. 1995; Lin et al. 2004b; Mazet et al. 2004; Yam and Hassan 2007). Such types of sensors operated at high temperatures are fast devices that make it possible to control in situ the appearance of H_2 in the surrounding atmosphere. However, for PSi, such types of sensors are not as widely propagated compared to standard semiconductors. We found only a few papers in this field (Fastykovsky and Mogilnitsky 1999; Hedborg et al. 1999; Kwok et al. 1999; Litovchenko et al. 2004; Gabouze et al. 2006; Arakelyan et al. 2007). Schottky barriers were formed by deposition of Pt (Hedborg et al. 1999), Pd (Litovchenko et al. 2004), Au (Pancheri et al. 2003), or Al (Kwok et al. 1999). The thickness of deposited metal films was varied in the range of 20–500 nm. Values for the barrier height reported heretofore ranged from 1.2 to 0.3–0.4 eV.

As a rule, the maximum sensor response was obtained for reverse current. However, in some cases, Schottky diode sensors can operate both in forward and in reverse voltage. It was established that besides H_2 (Hedborg et al. 1999; Litovchenko et al. 2004; Arakelyan et al. 2007), the

Schottky-type sensors under discussion can detect the appearance of NO, NO_2 (Pancheri et al. 2003), and vapors of acetone, propanol, methanol (Arakelyan et al. 2007), CO, and even CO_2 (Gabouze et al. 2006) in the surrounding atmosphere. For comparison, sensors based on a nonporous semiconductor are sensitive to H_2 only. As was established (Fastykovsky and Mogilnitsky 1999; Litovchenko et al. 2004), PSi-based devices are sensitive to H_2O and O_2 as well (Yamana et al. 1990; Arakelyan et al. 2007). For example, a three-orders-of-magnitude increase of conductivity in an Au/PSi/c-Si diode was observed by Yamana et al. (1990) when the relative humidity of the ambient atmosphere increased from 10 to 100%. We believe that such sensitivity of PSi-based sensors is conditioned by the porosity of the metal layer and the participation of the uncovered PSi surface in the gas sensing effects. It is necessary to note that the best variants of H_2 sensors designed based on Schottky contacts are fast enough (see Figure 1.16). As a rule, the current increases linearly with an increase of the gas or vapor concentration. This last feature is important for the design of the measurement devices.

As was established by Salehi and Kalantari (2007), using porous instead of nonporous semiconductors opens new measurement possibilities for Schottky barrier–type sensors. Comparison of sensing characteristics of different Schottky Au/GaAs-fabricated structures based on porous and nonporous n-type GaAs wafers has shown that the indicated structures have different sensitivities and selectivities toward various gases. For example, the Au/porous-GaAs Schottky contact was highly sensitive to polar CO and NO gases with high selectivity, whereas the Au/GaAs contacts responded negligibly to the gases. The negligible response of the Schottky-based sensors based on nonporous semiconductors toward CO is clear because there are no pores on the Au layer. Therefore, the gas molecules cannot be adsorbed on the surface and hence can induce no changes at the interface. Experiments confirmed this statement (Hedborg et al. 1999; Salehi and Kalantari 2007). For example, Salehi and Kalantari (2007) established that the sensor response to CO and NO strongly depended on the Au film thickness. A sensor with an Au layer of 20 nm thickness exhibited larger responses to both NO and CO gases than that of a sensor with an Au layer of 50 nm. It was assumed that the lower response to the polar gases of sensors with thicker Au films was because the thicker Au layer covered the pores and produced resistance to gas penetration inside the porous matrix. So, in order to improve the sensitivity and selectivity to gases that do not produce detectable hydrogen atoms on a thick metal gate, a thin layer of catalytic metal must be used as the electrode in Schottky diodes or MOS structures.

It is important that using porous instead of nonporous semiconductors usually be accompanied by an increase of sensor response. For example, Yam and Hassan (2007) showed that the sensor signal, that is, the ratio of the current after and before introduction of 2% H_2 in N_2 for sensors fabricated based on bulk and porous GaN, is approximately 11 and 1050 times, respectively. This means that the sensor signal for porous GaN is about two orders of magnitude larger than the sensors based on nonporous GaN. According to Hedborg et al. (1999), the gas sensitivity in such sensors is strongly dependent on the microstructure of the film in addition to a dependence on the metal layer thickness. The optimum gas sensitivity was achieved for a metal film that was

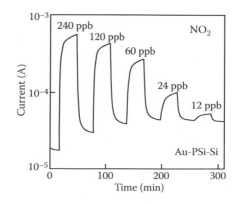

FIGURE 1.16   Dynamic response of a sensor based on an Au-PSi-Si structure to different concentrations of NO_2 in dry air. (Reprinted from *Sens. Actuators B* 89, Pancheri L. et al., 237, Copyright 2003, with permission from Elsevier.)

thick but had as high a density of cracks. This means that the metal film must be thick enough to give sufficient coverage on the surface and thin enough so that it does not cover the pores. According to Hedborg et al. (1999), suitable metal film thicknesses are in the range of 10–50 nm.

For explanation of the observed effects, a well-known model designed for Pd-semiconductor gas sensors is commonly used (Lundstrom et al. 1975, 1989; Johansson et al. 1998). According to this model, the sensor response is determined by the change of potential barrier height caused by a change of gas environment. Hydrogen or water molecules rapidly dissociate into hydrogen atoms and hydroxyl species by the catalytic property of the Pd metal and then the hydrogen atoms penetrate through the Pd metal with a high diffusion coefficient. These hydrogen atoms are trapped at the interface between Pd metal and semiconductor and, due to forming a dipole layer, give rise to an electrical polarization, ΔV (Johansson et al. 1998):

(1.3)
$$\Delta V = \frac{\mu N_i \theta_i}{\varepsilon}$$

where μ is the effective dipole moment, N_i is the number of sites per area at the interface, θ_i is the coverage of hydrogen atoms at the interface, and ε is the dielectric constant. It is found that the higher the hydrogen concentration, the larger the ΔV; that is, the larger change of the energy barrier height at Pd-semiconductor interface. As a result, the current through the Schottky barrier is changed as well. The current through the Schottky barrier is described by the equation:

(1.4)
$$J \sim \exp\left(-\frac{e\varphi_b}{kT}\right)\left[\exp\left(\frac{eU}{kT}\right)-1\right]$$

where φ_b is the height of potential barrier at the metal-semiconductor interface and U is applied voltage (Rhoderick 1978). The mechanism of hydrogen dissociation and diffusion in Pd can be found in Kroes et al. (2002) and Hong and Rahman (2007). However, it is necessary to note that the above-mentioned model is too simplified. Due to the porous structure of both metal film and semiconductor, inside the pores we have a semiconductor surface uncovered by metal, which also can participate in gas-sensing phenomena. This means that direct gas–PSi interactions must be taken into account for a correct explanation of the observed effects. Moreover, we believe that the modulation of the PSi conductivity along the pores, due to gas interaction with PSi surface inside the pores, plays a dominant role in achievement of the extremely high sensitivity of PSi-based sensors. The rectifying properties of Schottky contacts based on porous materials are worse than are those based on single-crystal semiconductors. Therefore, the interaction with oxygen and water also has a strong influence on the characteristics of Schottky diodes based on porous materials (Fastykovsky and Mogilnitsky 1999; Strikha et al. 2001; Litovchenko et al. 2004; Salehi and Kalantari 2007).

Another approach to the manufacturing of Schottky barrier–type sensors based on PSi was realized by Belhousse et al. (2004, 2005) and by Gabouze et al. (2006). The authors of these papers have suggested the control of I–V characteristics of heterostructures CH$_x$/PSi-p-Si. The porous samples were coated with hydrocarbon groups deposited in a methane/argon plasma. It was experimentally demonstrated that the CH$_x$/PSi-p-Si devices have a long lifetime and can be used for detecting a low concentration of ethylene, ethane, propane, CO$_2$, and H$_2$ gases. Moreover, Gabouze et al. (2006) showed that those sensors could work easily in air without any influence of the oxygen environment that explains the high stability and nonoxidation of the sensor surface. Gabouze et al. (2007) also showed that at the presence of Pd on the top of this structure, the sensitivity of the sensor is improved by more than two orders of magnitude compared with the uncoated CH$_x$/PSi/Si device.

A variant of PSi gas sensors designed based on heterostructures was fabricated as well (Arakelyan et al. 2007; Subramanian et al. 2007). The gas sensors had Au-TiO$_{2-x}$(Pt)-PSi and Pd:SnO$_2$/PSi/p-Si structures. Thin TiO$_{2-x}$ films were deposited by electron beam evaporation (Arakelyan et al. 2007). Experiments showed that after injection of hydrogen into the measuring chamber, the resistance of samples changed and became stable. After the removal of the hydrogen, the resistance of samples reverted to values before hydrogen injection and became stable again. Results of measurements over a period of 6 months indicated that a noticeable change in the measured characteristics was absent, which implied that the evaporated metal oxide layers on top of the PSi layer effectively protected it from the ambient. This implies that it is possible to obtain samples that are sensitive to hydrogen at room temperature and that can be characterized by durability. For an explanation of the observed gas-sensing effects, Arakelyan et al. (2007) used a model similar to the model discussed earlier in this section. They believed that the hydrogen molecules, which pass through the Pt layer, split into hydrogen ions, pass through the TiO_{2-x} layer, and are absorbed onto the PSi surface, where they capture free electrons that exist in the PSi skeleton. Therefore, the total current that flows through the device—that is, flows through the PSi layer—decreases due to the decrease in its electronic part.

Sensors based on Pd:SnO_2/PSi/p-Si structures were tested with NO_2 and liquefied petroleum gas (LPG) (Subramanian et al. 2007). The response and recovery characteristics of the sensor devices at different operating temperatures showed a short response time for LPG. As follows from the studies, maximum sensitivity and optimum operating temperature of the devices toward LPG and NO_2 gas sensing have been estimated as 69% at 180°C and 52% at 220°C, respectively. The developed sensor devices show short response times of 25 and 57 s for sensing LPG and NO_2 gases, respectively.

Therefore, the presented results show that metal oxide-PSi (MeOx-PSi) heterostuctures can be used for gas sensor design. However, it is necessary to note that in such sensors, in all probability, the porous material plays only a passive role. The metal oxide covers the pores and therefore porous material does not participate in the gas-sensing effects. In the above-mentioned structures, the metal oxide probably controls the observed gas sensing characteristics.

Due to rectifying behavior of the structure of Si-PSi, gas and vapor sensors based on current control through the interface Si-PSi can be ascribed to heterostructure-type sensors as well. The gas-sensing characteristics of such sensors were discussed in several papers (Galeazzo et al. 2003; Tucci et al. 2004). Galeazzo et al. (2003) established that PSi-Si vapor sensors had reversible and reproducible response to polar molecules (acetone and ethanol). Tucci et al. (2004) found that the maximum response for sensors based on amorphous/porous silicon heterojunction was conserved for interaction with molecules with higher dipolar moments.

1.2.5 GAS SENSORS BASED ON CPD MEASUREMENTS

As one can see from experimental results, gas sensors based on semiconductors could be designed based on the influence of the surrounding gas on the contact potential difference (CPD).

However, this approach was used in just a few works (Souteyrand et al. 1995; Polishchuk et al. 1998). The effect of hydrogen on the CPD of PSi structures was tested using a vibrating capacitor of the Kelvin-Zissman method. The vibrating capacitor method has been used for a long time to investigate semiconductor gas sensor surfaces (Mizsei 2005). This technique does not require the use of the whole MOSFET or GasFET devices and, due to sensitivity to the change of the work function of the gas sensor surface, has an advantage for studying various gas–semiconductor surface interactions (Nicolas et al. 1997; Mizsei 2007). All measurements were performed at room temperature, and zero calibration was carried out in 20% O_2 + 80% N_2 (Polishchuk et al. 1998). These experiments showed that sensors based on CPD control are characterized by a considerable sensitivity in the range of 200–4000 ppm H_2 concentrations. The dynamic CPD response of the above-mentioned sensors is shown in Figure 1.17. It is seen that the response time is in the minutes range and the recovery time takes tens of minutes when operated at room temperature.

However, one should admit that in spite of acceptable sensors parameters, the prospects of their application in real devices are low because of the peculiarities of their construction. Besides, it is necessary to know that if metal completely covers the porous surface, the continuous metal layer completely shields the influence of electromagnetic field on the potentials in the layers below (Mizsei 2007). In that case, the results of CPD measurements characterize only the metal surface, and there is no evidence of the importance of the PSi layer in observed gas-sensing effects. Apparently, the vibrating capacitor method has value just for the study of the interaction between semiconductor surfaces and the surrounding gas (Mizsei et al. 2004).

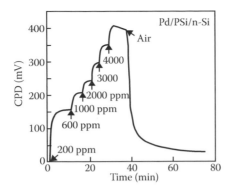

FIGURE 1.17 Dynamic CPD response of Pd/PSi/n-Si structure exposed to successively injected H_2 concentrations in 20% O_2 + 80% N_2. (Reprinted from *Anal. Chim. Acta* 375, Polishchuk V. et al., 205, Copyright 1998, with permission from Elsevier.)

1.2.6 COMBINED APPROACH

As shown before, the electrical conductivity, PL, and effective refractive index of PSi are sensitive to the presence of different gases. Because they depend on different physical properties of the gas, the simultaneous monitoring of all three parameters gives independent information about the gas. For example, a strong correlation between the PL intensity and the static dipole moment of gas molecules has been demonstrated. The refractive index of the gas, and thus of the PSi, depends, on the other hand, on the dynamic response of the gas molecules at optical frequencies. Therefore, having access to several gas properties, multiparametric PSi sensors are expected to have better gas selectivity than single-parameter sensors. This approach was successfully used for NO_2 detection in humid air. Oton et al. (2003) showed that simultaneous measurement of electrical conductivity and the index of PSi refraction gives the possibility for independent control of NO_2 and air humidity. As established by Oton et al. (2003), the effective index of refraction of the porous layer depends on the water content inside the porous layer but not on the NO_2 concentration (see Figure 1.18).

This approach has received further development in the research of Baratto et al. (2002), who suggested comparing the sensor responses related to the gas influence on the PL intensity,

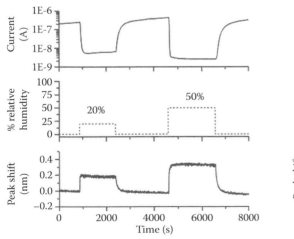
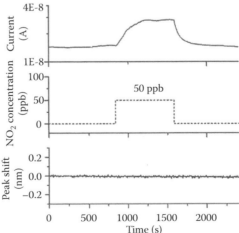

FIGURE 1.18 Simultaneous measurement of optical and electrical responses to humidity and NO_2. The electrical current is measured on the 30-μm deep monolayer employing the three-terminal configuration, and the reflectance peak shift is measured on the microcavity. Left plot: response to two different humidity values in the absence of NO_2. Right plot: response to 50 ppb of NO_2 in the presence of 20% humid air. (Oton C.J. et al.: *Phys. Stat. Sol.* (a). 2003. 197. 523. Copyright Wiley-VCH Verlag GmbH & Co. KGaA. Reproduced with permission.)

TABLE 1.3   The Influence of Gases on Three Parameters of a Porous Silicon Microcavity

	Sensor Parameter		
Analyte	DC Current	PL Intensity	Peak Shift
NO_2	Increase	Decrease	None
Humidity	Decrease	Vary	Red shift
Ethanol	Decrease	Vary	Red shift

Source: Reprinted from Baratto C. et al., *Sensors* 2, 121, 2002. Published by MDPI as open access.

electrical conduction, and effective refractive index for resolving the problem of the selectivity of gas sensors based on PSi. It was shown that in a PSi microcavity, the PL intensity, the electrical conduction, and the resonance peak of the cavity are affected by the presence of NO_2, ethanol, and water in different ways (see Table 1.3).

One can see that NO_2 increases DC current and PL intensity, but no peak shift is produced. Humidity and ethanol cause a peak shift and decrease in DC current. The peak shift is higher with ethanol than with water due to the higher value of refractive index, whereas the conductance decrease is higher with humidity than with ethanol. In contrast with the case of nitrogen dioxide, the peak intensity variation is small for both humidity and ethanol. Thus, the use of porous silicon in multiparametric sensors can help to discriminate the presence of NO_2, which is a dangerous pollutant, from interfering gases like humidity and ethanol. The decrease in DC current can be due to the decrease in NO_2 concentration or to the increase in relative humidity. If a concomitant peak shift and intensity variation is observed, it happens due to the presence of humidity. If no peak shift is observed, the variation must be ascribed to NO_2 concentration variations. The same discrimination can be obtained with ethanol. An appropriate pattern recognition step could improve preliminary results of gas species identification and quantification.

1.2.7 FIELD-IONIZATION GAS SENSORS

As shown by studies, field-ionization gas sensors can also be designed based on PSi. As it is known, there are two types of gas sensors: chemical and physical. All sensors discussed previously concern chemical-type devices. However, the application of the chemical-type gas sensor is limited by several disadvantages, such as the potential difficulties in detecting gases with low adsorption energies, poor diffusion kinetics, or poor charge transfer with PSi. It is also challenging to use this technique to distinguish between gases or gas mixtures; gases in different concentrations could produce the same net change in conductance as produced by a single pure gas. Chemical type sensors are also very sensitive to changes in environmental conditions (moisture and temperature), and chemisorption could cause irreversible changes in properties of gas-sensing materials (Modi et al. 2003).

Field-ionization gas sensors do not have indicated disadvantages because these devices work by fingerprinting the ionization characteristics of distinct gases. Field ionization (FI) is a quantum tunneling phenomenon whereby an atom or a molecule is ionized in the presence of a very high electric field, typically of order 10^{10} V/m (Gomer 1961). This type of sensor can detect gases regardless of their adsorption energies. However, conventional gas ionization sensors are limited by their huge, bulky architecture, high power consumption, and risky high-voltage operation. Research carried out during the last decade has shown that indicated disadvantages of gas ionization sensors can be overcome by the use of 1D structures (Modi et al. 2003; Hui et al. 2006; Wang et al. 2011). A schematic diagram illustrating the construction of modern gas ionization sensors is shown in Figure 1.19a.

It was established that the sharp tips of nanotubes generate very high electric fields at relatively low voltages, lowering breakdown voltages several-fold in comparison to traditional electrodes, and thereby enabling compact, battery-powered, and safe operation of such sensors (Modi et al.

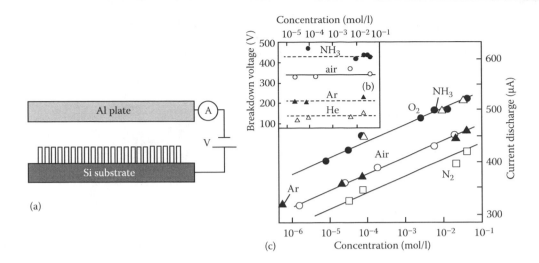

FIGURE 1.19 (a) Schematic diagram of gas-ionized sensor with 1-D electrodes. Controlled DC voltage is applied between the anode (vertically aligned nanotube film) and the cathode (Al sheet), which are separated by an insulator. The distance between the two electrodes could be adjusted in a range of 20–200 μm by the thickness of the plastic film or glass. (b, c) Effect of gas concentration on (b) electrical breakdown and (c) current discharge in CNTs-based sensors. It is seen that breakdown voltages vary only slightly with gas concentration and the discharge current varies logarithmically with concentration. (Reprinted by permission from Macmillan Publishers Ltd. *Nature*, Modi A. et al., 424, 171, copyright 2003.)

2003; Riley et al. 2003). For example, physical gas sensors based on carbon nanotubes demonstrated low breakdown voltage and showed good sensitivity and selectivity. In many researches, carbon MWNTs were grown by chemical vapor deposition (CVD) on an Si substrate and had 25–30 nm in diameter and length from 30 mm (Modi et al. 2003) to 2 μm (Hui et al. 2006). Several curves illustrating operating characteristics of field-ionization gas sensors are shown in Figure 1.19b,c. Experiments carried out by Nikfarjam et al. (2010) have shown that PSi also can be used as a substrate for CNTs deposition. Their research has shown that gas ionization sensors based on CNTs/PSi had high sensitivity and selectivity for inert gases. For example, the sensitivity of 28,000 was obtained by inserting 516 ppm He.

It was established that these sensors could detect many gases, such as Air, He, Ar, and gas mixtures without gas separation (Modi et al. 2003). In addition, it was found that the simple, low-cost sensors described here could be deployed for a variety of applications, such as environmental monitoring, sensing in chemical processing plants, and gas detection for counterterrorism. Unfortunately, the sensitivity of field-ionization gas sensors is low in comparison with conductometric gas sensors. For example, below 1% concentration, the breakdown of Ar ceases and the breakdown voltage rises sharply to the value for pure air. Higher sensitivity up to 25 ppm was achieved only with gas chromatograph using.

However, the same research established that the CNT-based ionization gas sensors show poor stability because carbon nanotubes could easily be oxidized and degraded in the oxygen-contained atmosphere (Wang et al. 2005; Liao et al. 2008). In addition, we have to say that though strong decreasing the gaseous breakdown voltages up to a range of 100–450 V (Modi et al. 2003; Hui et al. 2006; Wang et al. 2011), such voltages are still hazardous to employ. Moreover, CNT-based field-ionization sensors operate in corona discharge mode. Corona discharges are difficult to control and they generate excessive heat that may destroy sharp and slender CNTs. For resolving this problem, Sadeghian and Kahrizi (2007, 2008) proposed to use a sparse array of vertically aligned gold nanorods as substitutes for CNTs. One should note that porous silicon could be used as a template for synthesis of such vertically oriented gold nanowires and nanorods. Sadeghian and co-workers (2008, 2011) have shown that silicon nanowires also can be used for gold nanorod assisted gas ionization. Using the indicated approach, they have obtained threshold field-ionization voltages of 0.2–9 V (depending on the measured gas) by incorporating whisker-covered gold nanorods on the silicon-based nanowires (see Figure 1.20). With such an approach, sensors with

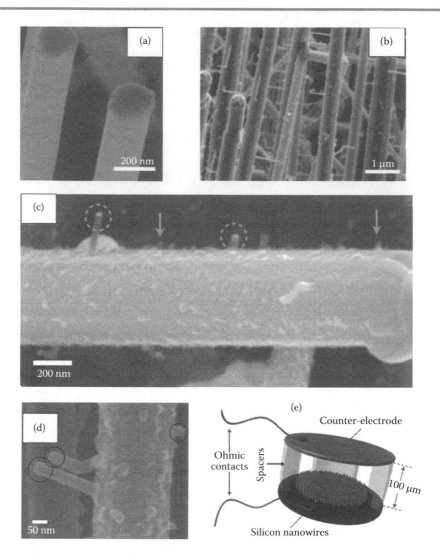

FIGURE 1.20 Nanowires used to measure anomalous semiconductor-assisted gas ionization: (a) a close-up SEM image of smooth silicon nanowires after annealing in HCl; (b,c) SEM micrographs of a forest of, and a single, whiskered silicon nanowire that showed low-voltage FI; (d) SEM images of whiskered nanowires after removal of gold catalyst for the tips; and (e) a three-dimensional schematic illustration of the device used to measure gas ionization on both types of nanowire. Note that the nanowires were planted at the anode. $d_{gap} = 100$ μm is the spacing between anode and cathode. (Reprinted by permission from Macmillan Publishers Ltd., *Nature Mater.*, Sadeghian R.B. and Islam M.S., 10, 135, copyright 2011.)

high sensitivity, high selectivity, long durability, and (ultra)-low-power operation ($P < 10$ μW with ionization currents ≤ 1 μA) are made possible.

Liao et al. (2008) and Wang et al. (2011) have shown that metal oxide nanowires as well as silicon nanowires and macroporous silicon, which are stable at room temperatures, can also be used for stable field-ionization gas sensors instead of CNTs. In particular, it was established that nanotips with curvature radii of $r_c \leq 10$ nm (see Figure 1.21), which is sufficient for a large field enhancement effect, can be easily fabricated using macroporous silicon structures subjected to the procedure of sharpening (Gesemann et al. 2011; Wehrspohn et al. 2013). For example, Gesemann et al. (2011) achieved this effect after just two oxidation steps at low temperature. To functionalize the silicon tips for increased ionization or electron emission rates, the tips can be covered with the different materials according to the specific application.

Spindt (1968, 1992) and Ghodsian et al. (1998) have shown that micromachining technology also gives possibility to design low-cost field-ionization gas sensors with low working voltage.

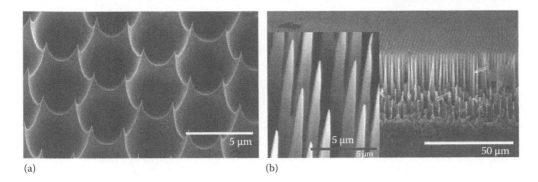

(a)

(b)

FIGURE 1.21   Scanning electron micrograph: (a) additional tip sharpening results after different steps of thermal oxidation + HF dips. (b) Needle-shaped tips with high aspect ratio and covered with platinum for surface ionization (insert). (Reprinted from *C.R. Chimie* 16, Wehrspohn R.B. et al., 51, Copyright 2013, with permission from Elsevier.)

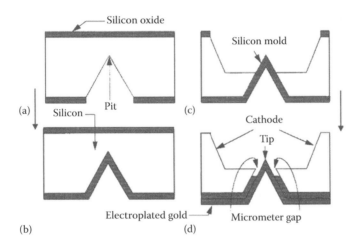

FIGURE 1.22   Fabrication sequence for one single tip proposed by (a) Spindt (1968, 1992) and (b) Ghodsian et al. (1998).

A technological process designed by Spindt (1968) and Ghodsian et al. (1998) (see Figure 1.22) allowed achieving a high-resolution gap between the tip and a second electrode with a mask that has millimeter resolution and this gap can simply be controlled accurately with the oxide thickness. Due to the small distance between electrodes, the gas sensor had a possibility both to operate with only 5 V applied voltage and to detect the sample gas down to 14 ppm. Since the technology is based on bulk micromachining, sensors designed can easily be integrated with the micromachined gas chromatography system. Ghodsian et al. (1998) believe that this will in turn improve the yield and manufacturing cost of micromachined gas chromatography system for remote environmental monitory.

1.3 DISADVANTAGES OF PSi-BASED GAS SENSORS

Research has shown that gas sensors based on porous semiconductors have extremely high sensitivity, allowing, for example, the detection of various gases and vapors on the ppm and even ppb levels in the presence of normal atmosphere. However, the same researches (Canhman 1997; Torchinskaya et al. 1997; Parkhutik 1999; Holec et al. 2002; Iraji zad et al. 2004; Pancheri et al. 2004; Di Francia et al. 2005; Islam and Saha 2007) have shown that gas sensors based on porous materials have one important disadvantage. As a chemically active material, PSi can be oxidized easily. For example, it is well known that at room temperature and ambient environment about

7 nm of Si native oxide can be formed with time. However, oxidation, as reported by Björkqvist et al. (2003), is a very slow process at room temperature and thus leads to continuous changes in the structure of PSi and its physical parameters. Therefore, parameters of PSi are not stable in time, especially at initial stages of exploitation after manufacturing (see Figure 1.23).

Moreover, aging depends heavily on the environmental storage, operating conditions, and how well the sensor materials are isolated from the environment (Islam and Saha 2007). For example, degradation of PSi luminescence in an oxygen atmosphere is strongly enhanced by light illumination (Tischler et al. 1992) and annealing (Torchinskaya et al. 2001). Other ambient gases such as N_2, Ar_2, and H_2 did not produce an essential change of PSi properties (Xu et al. 1992). It is also interesting to note that an as-anodized PSi sample is hydrophobic in nature but the oxidized PSi sample is hydrophilic in nature, thereby leading to significant drift in sensor output with time (Dittrich et al. 1995b; Björkqvist et al. 2004a,b; Archer et al. 2005). In particular, Iraji zad et al. (2004) showed that annealing in air or boiling in CCl_4 strongly decreases the sensing properties of PSi-based conductometric gas sensors. XPS measurements have shown that the surface of PSi is partially oxidized after the indicated treatments.

Pancheri et al. (2004) observed the same effect for PSi-based NO_2 sensors. High sensitivity to NO_2 was achieved only for fresh samples. Extremely poor stability was also observed for PSi-based ethanol sensors (Han et al. 2001). The aging of PSi in different oxidizing environments affects the stability of the PL property as well (Dittrich et al. 1995a,b; Björkqvist et al. 2003). For example, Kim et al. (2001) found that the exposure of freshly formed PSi films to humid atmosphere (30 Torr) at room temperature results in a gradual increase of the PL yield during the first two days of storage. Simultaneously, the maximum spectral yield shifts toward red. The initial luminescence intensity may be restored by evacuating the PSi sample at a temperature of 150°C. Treatment of PSi in humid atmosphere in excess of two days results in irreversible decrease of PL intensity. It was established that the aging time might require more than 5–10 days (Setzu et al. 1998). For example, in the research of Holec et al. (2002), because of temporal instability, PSi samples were stored for about 1 month before sensing characterization. It was established that when a PSi is stored in oxidizing environments like water and H_2O_2, the formation of Si-O-Si bonds replacing Si-H_x bonds takes place (Hossain et al. 2000). The same effect was observed during PSi surface carbonization as well (Mahmoudi et al. 2007). It was observed that the rate of oxidation of the PSi layer depends on the concentration of OH– groups on the surface and holes in the valence band of the PSi.

The PL parameters of porous layers turned out to be sensitive to thermo-vacuum treatments (Lisachenko and Aprelev 2001). It was shown that vacuum annealing at temperatures below 600°C led to a decrease of the PL intensity (Balagurov et al. 1996) in the region of 720–880 nm. For example, Shin et al. (2003) observed that after PSi annealing in vacuum at about 550°C, the PL centered at about 720 nm was completely quenched.

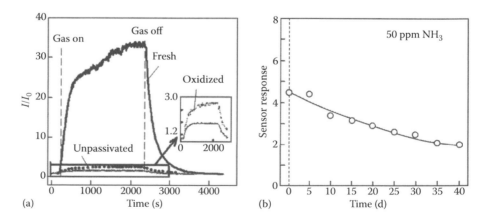

FIGURE 1.23   (a) Response of fresh, oxidized, and unpassive (boiling in CCl_4) PSi samples in the presence of 4% ethanol. (Reprinted from *Sens. Actuators B* 100, Iraji zad A. et al., 341, Copyright 2004, with permission from Elsevier.) (b) The long-term stability of the PSi sensor response to 50 ppm NH_3 at RT. (Reprinted from *Electrochim. Acta* 108, Li M. et al., 167, Copyright 2013, with permission from Elsevier.)

The observed decrease in the PL intensity has been explained by the processes such as the hydrogen desorption from monohydride and dihydride species, a decrease of the PSi band gap, and an increase of the silicon dangling bonds (Robinson et al. 1992; Kovalev et al. 1994; Ludwig 1996). IR spectroscopy data also testify to the exit of hydrogen atoms from the PSi surface. Dimitrov et al. (1997) established that the hydrogen desorption follows the second-order chemical kinetics. It was also observed that the creation of nonradiative dangling bonds increases the cross-conductivity of the PSi-Si structures.

The appearance in PSi of a large number of defects, which are the consequence of broken silicon bonds on the surface of nanostructures, is registered by the method of electron paramagnetic resonance (Meyer et al. 1993). The slow formation of native oxide on all the pore walls stimulates strong optical effects as well (Kochergin and Foell 2006). The loss of transparency of the mesoporous silicon in the mid- and far-IR range was observed during the aging of freshly prepared PSi. The above-mentioned properties indicate that PSi-based sensors have great limitations for application in real devices. For practical field applications, the long-term drift of a sensor sometimes needs more attention than the sensitivity (Björkqvist et al. 2003). Nevertheless, the cheap mass production and the possibility of integration of the sensor element with driving and read-out electronics on the same chip pushes the development toward silicon-based processes.

The strong dependence of sensor parameters on air humidity is another important disadvantage of PSi-based devices. For example, Pancheri et al. (2003) established for PSi-based NO_2 sensors that a switch from 40% RH to dry air, without introducing NO_2, leads to a conductance increase of around one order of magnitude, which is about half of the change obtained adding 50 ppb of NO_2 without changing the RH (kept at 40%) (Pancheri et al. 2003). In particular, Figure 1.24 shows sensor response and recovery by applying 50 ppb of NO_2 at different RH levels.

One can see that an increase in RH leads to a decrease in conductivity and, therefore, in sensor current. This effect was observed for all kinds of analyzed sensors. For example, it was established by Massera et al. (2004) that the increase of air humidity in the range of 0–60% RH decreases the sensitivity of PSi-based conductometric NO_2 sensors more than four times. The observed behavior could be qualitatively explained by the donor-like character of water molecules adsorbed at PSi surface defects. The relative change of conductivity, $\Delta G/G$, in the presence of NO_2 (50 ppb) decreases with increasing relative humidity, passing from 11.7 for 0% RH, to 2.2 for 40% RH, to 1.2 for 70% RH.

For comparison, the change of relative humidity in the range of 0–60% was accompanied by the change of PSi resistance more than 20 times (Baratto et al. 2001b). Clearly, such behavior is undesirable because the conductometric measurement alone cannot distinguish between

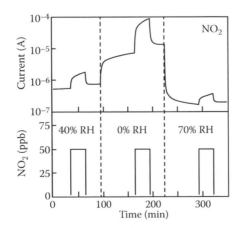

FIGURE 1.24 Dynamic response of a sensor based on Au-PSi-Si structures in the presence of different relative humidity levels and NO_2 concentrations. The graph above represents the sensor response and the graph below represents NO_2 concentration as a function of time. The dashed vertical lines divide regions with different relative humidity values. (Reprinted from *Sens. Actuators B* 89, Pancheri L. et al., 237, Copyright 2003, with permission from Elsevier.)

variations caused by changes in humidity or gas concentration. The decrease in sensitivity could be due to a partial compensation effect between the acceptor-like NO_2 molecules and donor-like water molecules. Such behavior of PSi-based sensors requires the elaboration of special methods of stabilizing the parameters of gas sensors based on porous materials, which considerably narrows the area of their application. The use of preaged treatment by oxidation was one of such elaborations suggested for aging effect prevention. It was found that oxidized PSi may be more stable and may contain fewer interface states affecting the Fermi-level pinning. For example, it was established that PSi could be converted to porous SiO_2 within 60 min at 950°C under a water vapor atmosphere. Connolly et al. (2002) proposed using the burn-in process, which involves heating the device to ~55°C, and repeated cycling of the RH between 5 and 95% for sensor parameter stabilization.

It was found that this treatment improved humidity sensor stability as well as reduced hysteresis effects dramatically. Pancheri et al. (2004) showed that in aged PSi samples, the changes in resistivity of the porous layer, associated with exposure to NO_2, have improved the reversibility compared to fresh samples. The stabilization of the current is also faster in aged than in fresh samples (see Figure 1.25). It is seen that the baseline shift (1) is absent in the aged sample and significant in the fresh sample and the signal stabilization (2) is good in the aged, thin sample and poor in the fresh, thick sample.

Both results suggest that sample aging is a potentially good strategy for use of PSi in NO_2 conductometric sensors because reversibility and signal stabilization are still major unsolved limitations for such applications of freshly etched PSi-based sensors. For example, for stabilization of NO_2 sensor parameters, Massera et al. (2004) proposed using prolonged exposure to high concentrations of NO_2. It was established that after such treatments, including a 10-h exposure to 2 ppm of NO_2 at 20% RH in synthetic air, followed by 2 weeks of stabilization in ambient air, the devices were stable in the sub-ppm range and their electrical characteristics were greatly improved. The sensors had faster dynamics and reduced hysteresis. For the same purposes, Ben-Chorin and Kux (1994) proposed storing porous samples in a highly oxidizing medium like H_2O_2 for 48 h. It was established that initial oxidation in an H_2O_2 solution could stabilize the PSi layer to a greater extent. Holec et al. (2002) showed that methyl 10-undeacetonate derivatization of the PSi surface increases the PL time stability of PSi-based sensors as well. Boiling in HNO_3 can also be used for PSi properties stabilization (Kochergin and Foell 2006). However, it is necessary to note that the problem of PSi-based sensor parameter instability is still unresolved. None of the above-mentioned techniques succeeded in complete prevention of the aging effect. Even after aging, sensors have temporal drift (Islam and Saha 2007). Islam and Saha (2007) believed that the problem of temporal drift could be partially resolved through compensation using soft computing techniques. However, we need more fundamental understanding of the nature of the drift effect, which requires additional research. According to Pancheri et al. (2004), further efforts should be aimed at the development of an oxidation treatment, which could reproduce the effects of aging on PSi conductivity. At the same time, Lewis et al. (2005) have asserted that their

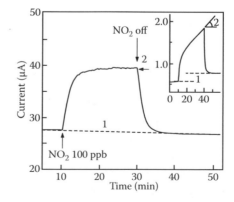

FIGURE 1.25   Effect of exposure to NO_2 (100 ppb) on thin (3 μm) aged PSi samples (main plot) and on thick fresh samples (inset). The relative humidity is 30% in both cases. T = 22°C. (Reprinted from *Sens. Actuators B* 97, Pancheri L. et al., 45, Copyright 2004, with permission from Elsevier.)

devices continued working for months after their initial fabrication without requiring cleaning or recalibration.

The low selectivity of PSi-based sensors is another important disadvantage of such kinds of sensors. Due to the sensing mechanism, this kind of device is not able to identify the components of a complex mixture. However, it is necessary to note that this disadvantage is peculiar to all adsorption type sensors, including metal oxide-based gas sensors (Korotcenkov 2005, 2007). In order to enhance the sensor selectivity through specific interactions, some researchers have proposed to chemically or physically modify the PSi surface. These results will be discussed in Section 1.4. Unfortunately, any appreciable success in this direction was not achieved.

The fabrication of stable contacts with low resistance is a great problem for PSi as well (Lewis et al. 2005). It was established that the physical methods of metal deposition do not minimize contact resistance due to a small contact area, whereas during chemical methods of metal deposition a strong degradation of gas-sensing characteristics could occur. Some technical approaches for resolving these problems were proposed (Gole et al. 2000; Tucci et al. 2004; Lewis et al. 2005). In particular, in research carried out by Lewis et al. (2005), the optimization of technology of contact forming allowed considerable improvement of the operating characteristics of PSi-based sensors, including their stability. Using a designed metallization process, Gole and his collaborators (2000) dramatically reduced the resistance of the electrodes, allowing their PSi-based sensors to operate at between 1 and 10 mV. This is very good result because sensors based on PSi that have been built before usually required an operating voltage of as much as 3–5 V due to high resistance in the electrodes connected to the PSi.

Relatively slow response and recovery due to operation at low temperatures are also disadvantages (see Figure 1.21). The other important disadvantage of the devices based on PSi is strong nonreproducibility of the properties of PSi. The varieties of the morphology of PSi are too high, and the dependence of its parameters on the layer structure is too strong (Foll et al. 2002). One can understand how relevant the problem of the irreproducibility of parameters of porous films is from the fact that the samples prepared by different research groups are hardly comparable even if the preparation conditions are apparently the same (Parkhutik 1999). In many published papers, there is no information relating to a detailed structural characterization of the studied porous semiconductors. With such an approach, it is hard to expect sufficient progress in the understanding of the nature of phenomena that provide the gas sensitivity of porous semiconductors. Therefore, the elaboration of new technologies promoting the essential reproducibility of the porous films morphology, and, as a result, the parameters of sensors designed on their basis, is the task of the greatest importance.

From our point of view, the unification of the porous semiconductor parameters, which need to be controlled for correct description of the studied objects, is extremely necessary. At present, several approaches are being employed for the characterization of porous materials, which might use such parameters as average pore and silicon branch widths, porosity, pore and branch orientation, and layer thickness. Unfortunately, because of the complexity of the porous material structure, the indicated parameters are not able to characterize correctly the porous layers used in experiments individually. For example, the pore diameter does not contain any information about the pore morphology.

The term *pore morphology* is used for such properties as shape (smooth, branched, faceted, etc.), orientation, interconnection of pores, and so on. The morphology is the least quantifiable aspect of PSi. We understand that it is difficult to systematically characterize the morphology of PSi, which has extremely rich details with respect to variations in pore size, shape, and spatial distribution. For example, microporous and mesoporous silicon typically exhibit a sponge-like structure with densely and randomly branched pores. The tendency to branching increases with a decrease of the pore diameter.

In contrast, macroporous silicon can have discrete pores with smooth walls, with either short branches or dendritic branches. Undoubtedly, all those peculiarities of porous films morphology have an effect on the gas-sensing characteristics of sensors and therefore they should be taken into account when designing and testing gas sensors. The better the structural characterization, the deeper our understanding of the nature of processes controlling gas-sensing characteristics of PSi-based devices will be. The improvement of characterization can be attained only at the expense of enlarging the number of methods used for this purpose. At present, the techniques used to assess these properties include only various microscopy techniques (pore diameter, microstructure, and layer thickness), gravimetric analysis (Brumhead et al. 1993) (porosity and layer thickness), and gas adsorption isotherms (Herino et al. 1987) (pore diameter).

1.4 IMPROVEMENT OF PSi-BASED SENSOR PARAMETERS THROUGH SURFACE MODIFICATION OF POROUS SEMICONDUCTORS

For attainment of the essential parameters of chemical sensors, it is necessary to use porous layers with optimal thickness and porosity, which in most cases are established experimentally. For example, Connolly et al. (2002) established that if the maximum sensitivity of PSi and polysilicon based devices to humidity was achieved using 30% HF, the best result for humidity sensitivity of porous silicon carbide (PSiC) was obtained using 73% HF.

The layer microstructure in terms of pore shape, pore size, and pore distribution is very important for the capillary condensation mechanisms. However, it should not be forgotten that the surface chemistry of the inner walls of the pores would influence the adsorption of gases and vapors as well as the capillary condensation. Therefore, for the design of sensors based on porous materials, the opportunity to control those processes should not be excluded. As demonstrated earlier for metal oxide-based sensors, such an approach gives the possibility to optimize considerably the parameters of chemical sensors. The results of numerous experiments presented in Table 1.4 have shown that such an approach for the design of gas sensors based on PSi is effective as well. For example, Sharma et al. (2007) demonstrated that the texturization of a silicon surface before porosification is a simple and effective method for the formation of highly porous, highly luminescent, thick films of PSi with reduced stress, improved stability, and superior mechanical properties. It was also established that the surface chemistry could be changed by thermal oxidation, silanization, carbonization, or functionalization by the covalent binding of functional groups (O'Halloran et al. 1997; Buriak et al. 1998; Stewart and Buriak 1998; Salonen et al. 2000;

TABLE 1.4 Surface Treatments Applied for Optimization PSi-Based Sensors

Surface Treatment	Type of Sensor	Effect	Ref.
Methyl 10-undeacetonate (derivatization)	PL	Stability improvement	Hedrich et al. 2000
Co phalocyanine (immersion)	PL	Stability improvement	Vrkoslav et al. 2005
Electrochemical deposition of ZnO	PL	Increase sensitivity and selectivity to NO_2	Yan et al. 2014a,b
Burn-in process (heating at 55°C and repeated cycling of RH between 3 and 95%)	Capacitance RH sensor	Increase sensitivity to RH Decrease response time	Connolly et al. 2002
Thermal oxidation	Capacitance Vapor sensor	Stability improvement	Aggarwal et al. 2014
Carbonization	Capacitance RH sensor	Improvement of stability Decrease of hysteresis	Björkqvist et al. 2004
Au (electroless)	Conduct.	Increase sensitivity to NH_3	Lewis et al. 2005
Sn (electroless)	Conduct.	Increase sensitivity to NO, CO	Lewis et al. 2005
Pd (el. beam evaporation, annealing at 900°C, $t = 60$ min.)	Conduct.	Increase H_2 sensitivity Improvement response time	Luongo et al. 2005
Cu/Pd (magnetron sputtering)	Conduct.	H_2 sensitivity	Litovchenko et al. 2004
Au (8–40 nm, DC sputtering)	Conduct.	Increase sensitivity to NO_2	Baratto et al. 1999
Surface texturing	Conduct.	Increase sensitivity to EtOH	Brumhead et al. 1993
	PL	Increase sensitivity to EtOH	Sharma et al. 2007
Cu_xS (chemical deposition)	Conduct.	Sensitivity to NH_3	Setkus et al. 2001
ZnO; WO_3 nanoparticles (dip-coating)	Conduct.	Increase sensitivity to NO_2	Yan et al. 2014a,b
Pd (evaporation, d ~ 20 nm)	CPD	Increase sensitivity to H_2	Polishchuk et al. 1998
CH_x group (thermal process), $CH_x/PSi/Si$		O_2 sensors	Mahmoudi et al. 2007
CH_x group (RF sputtering), $CH_x/PSi/Si$	Conduct. Capacitance	CO_2, H_2, and C_3H_8 sensors	Belhousse et al. 2005; Gabouze et al. 2006

Björkqvist et al. 2004a,b; Mahmoudi et al. 2007). The long-term stability studies of differently stabilized PSi samples have shown that thermal carbonization of PSi is even a more efficient stabilizing method than that thermal oxidation (Björkqvist et al. 2003). A thermally carbonized porous silicon (TC-PSi) surface is at least as stable in humid atmosphere as a thermally oxidized PSi surface. TC-PSi is also stable even in chemically harsh environments. Moreover, the sensitivity of TC-PSi is presumably better due to its larger specific surface area.

The deposition of polymer layers on the porous structure had similar effects and changed the sensitivity as well. Vrkoslav et al. (2005) showed that impregnation of PSi with cobalt phthalocyanine (CoIIPc) is an effective way to improve the stability of the PL quenching response. For CoIIPc-modified PSi films, the shortening of the PL quenching time and prolongation of the PL recovery time for a homological set of linear alcohols also was observed. According to results obtained by Vrkoslav et al. (2005), the protection of the PSi surface with CoIIPc results in a substantial increase of resistance against slow ambient temperature oxidation. Regarding the accelerating of PL quenching in modified PSi, Vrkoslav et al. (2005) believed that the reduction of pore size and the increase of PSi surface polarity due to CoIIPc impregnation were the main reasons for these effects. Hedrich et al. (2000) established that methyl 10-undeacetonate derivatization of the PSi surface also increased the PL time stability of PSi-based sensors. However, the better stability of sensor parameters is paid for by a decrease of sensor response.

For example, PSi surface stabilization by CoIIPc was accompanied by a decrease of sensitivity by a factor of 1.5–2.2. Interesting results were reported by Rocchia et al. (2003). Experiments carried out by the authors of this article have shown that due to a surface modification by 3-amino-1-propanol, one can fabricate PSi-based conductometric devices that are sensitive to CO_2. Unfortunately, the chemical nature of the surface species both before and after the binding of CO_2 is not clear at this time. Besides, the detection limit shown in this work is still far from the market requirements, but the reversibility and low cost of this system represent a starting point for the future development of PSi-based gas sensors.

Research has shown that good results may be also obtained by the partial oxidation of PSi (Fürjes et al. 2003). It was found that because of partial oxidation, the sensitivity of both surface potential and electroconductivity of the PSi to the change of H_2S concentration considerably increased. Fürjes et al. (2003) supposed that the positive influence of PSi oxidation was connected with a decrease of density of local electron states at the interface of PSi-SiO_2. Unfortunately, thermally oxidized PSi is not completely stable, especially under high humidity (Björkqvist et al. 2003). It was shown that the improvement of sensitivity and selectivity can be achieved by deposition of palladium (Polishchuk et al. 1998) and copper (Arwin et al. 2003) on PSi layers. Arwin et al. (2003) established that the sensitivity to methanol was increased by two times. Similar experiments with ethanol and propanol did not show an increase in sensitivity.

For PSi-based Pd-catalyzed H_2 sensors, it has been established that (1) the distribution of Pd over the porous skeleton plays a significant role dictating the dynamics of the sensor response (Rahimi et al. 2005) and (2) the response time and stability of the sensor are influenced by the complex Pd-H interactions (McLellan 1997), which in turn critically depend on the concentration of binding sites (pores).

Tsamis et al. (2002) established that Pd-doped PSi films showed enhanced catalytic activity to hydrogen oxidation into water, which implies an increase of the reaction rate. It was also found that the catalytic activity of Pd-doped porous silicon at 160°C is significantly higher than that of a planar surface covered with Pd. Rahimi et al. (2005) observed that sensors made with porous silicon and palladium nanoparticles demonstrated a significant decrease in resistivity with respect to time when exposed to hydrogen.

The Pd nanoparticles also decreased the adsorption and desorption times, which increases the sensitivity, and response and regeneration times of the sensor. The optimizing effect from PSi coating by Pd clusters was observed for optical H_2 sensors as well (Lin et al. 2004a). H_2 gas at a concentration as low as 0.17 vol.% can be detected in a few seconds by monitoring either the optical thickness change or the change in reflected light intensity obtained from the interferometric reflectance spectrum. Baratto et al. (1999) found that gold deposited by sputtering catalyzed the PSi response toward NO_2. The response toward interfering gases such as CO, CH_4, ethanol, and methanol was negligible. The gold penetrated into the pores instead of forming a continuous layer. Lewis et al. (2005) established also that the surface modification by Au promotes an increase of sensor response to NH_3 and SO_2, whereas surface modification by tin is more effective for improvement of the sensitivity to NO and CO (see Figure 1.26). For surface modification of PSi, the electroless process was used. Modified sensors had the possibility to detect a number of analytes with very low concentrations including CO (<5 ppm), NO_x (<1 ppm), SO_2 (<1 ppm), and NH_3 (500 ppb). Andsager et al. (1994) have shown that surface modification by metals such as Cu, Ag, and Au can also be used for stabilization of PL properties of PSi.

It was established that performance of PSi sensors based on PL quenching can also be strongly improved after surface modification by HfO_2 (Gan et al. 2014) and LaF_3 (Mou et al. 2011) layers. At optimal thickness of the covering, enhanced PL and a slightly decreased PL lifetime were achieved. However, with the increasing thickness of LaF_3 layer PL intensity of PSi was decreasing along with a small blue-shift. According to Gan et al. (2014), the compact coating of an HfO_2 layer effectively passivates the surface states and induces the Purcell effect, which suppresses the nonradiative recombination rates and increases the radiative recombination rates. Furthermore, the coating of HfO_2 modifies the surface properties and thus improves the sensing stability of the PSi-based devices (see Figure 1.27). Mou et al. (2011) established that LaF_3 coatings do not have such pronounced stabilizing properties. It was observed that all the coated PSi samples showed degradation in PL intensity with time, but annealing at 400°C could recover and stabilize the degraded PL. Mou et al. (2011) believe that LaF_3 layer prevents oxidation of PSi.

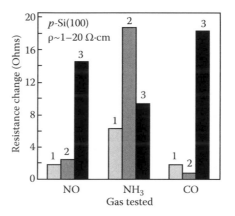

FIGURE 1.26   Comparison of responses of PSi-based sensors that are (1) untreated, (2) treated with electroless gold, or (3) treated with electroless tin, and tested with 30 repeat pulses of 20 ppm NO_x, NH_3, or CO. Their average impedance change is given. PSi layers ($d \sim 20$ μm) were formed using electrolyte containing 1 M H_2O, 1 M HF, and 0.1 M TBAP in acetonitrile. ($J = 4$–8 mA/cm², $t = 5$–30 min). (Reprinted from *Sens. Actuators B* 110, Lewis S.E. et al., 54, Copyright 2005, with permission from Elsevier.)

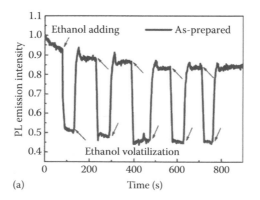

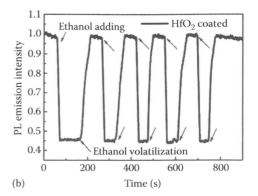

FIGURE 1.27   PL sensor responses with ethanol: (a) as-prepared PSi NW arrays, (b) HfO_2-coated porous Si NW arrays. The arrows indicate the addition and volatilization of ethanol. (From Gan L. et al., *Phys. Chem. Chem. Phys.* 16, 890, 2014. Reproduced by permission of The Royal Society of Chemistry.)

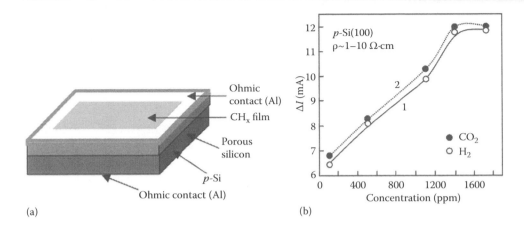

FIGURE 1.28 (a) Schematic diagram and (b) operating characteristics of the CH_x/PSi-Si sensor. The size of the realized sensor is $(4 \times 4)\cdot10^{-3}$ m^2. Applied bias was equal to 5 V. The growing of the porous layers was provided by the anodization in the solution of an HF (49% wt): C_2H_5OH (1:1), at current density of 100–500 A/m^2. Porosity from 40% to 85%. The PSi thickness was varied from 1 to 10 μm. (Reprinted from *Thin Solid Films* 482, Belhousse S. et al., 253, Copyright 2005, with permission from Elsevier. Reprinted from *Vacuum* 80, Gabouze N. et al., 986, Copyright 2006, with permission from Elsevier.)

Gabouze and co-workers (Belhousse et al. 2005; Gabouze et al. 2006; Zouadi et al. 2015) have found that the hydrocarbon groups (CH_x) deposition on PSi surface provides the fabrication of CH_x/PSi/Si structures for C_3H_8, CO_2 and H_2 sensing (see Figure 1.28). Hydrocarbon groups have been deposited by plasma of methane (Gabouze et al. 2006) and by laser ablation method (Zouadi et al. 2015). The stability of the sensors was demonstrated by performing electrical characterizations after 1 year. Belhousse et al. (2005) believe that the CH_x layer blocks the oxidation of the PSi layer and thus promotes the stability of gas-sensor parameters.

1.5 OUTLOOK

A resume of the recent research and development activities in the field of PSi demonstrates that the material has strong potential for practical applications in the field of gas and vapor sensors design. However, PSi is promising first in the micromachining technology of gas sensors (Korotcenkov and Cho 2010), and applications as a sensing material would depend on the improvement of stability and reproducibility of the gas-sensing characteristics. It is necessary to recognize that for realization of the potential advantages of either PSi or other porous semiconductors, as practically important materials, further research is necessary. This will allow better understanding of important physical and chemical factors, determining their growth and properties: first the mechanism of the PSi growth, and second the mechanism of the electrical conduction in PSi layers. The last factor is important for the design of a realistic model of gas-sensing effects in porous materials because the porous semiconductor cannot be inserted easily into the classification of porous materials used for polycrystalline materials. A PSi itself is somewhat similar to the nanocrystals but usually is formed over a heavily doped silicon surface. Thus, a macroscopic potential barrier (Schottky barrier), which is present at an intergrain interface in a polycrystalline material, exists at the PSi crystalline structures, whereas the potential distribution in the PSi layer is not well known, in spite of many attempts to reveal it by I–V characteristics, vibrating capacitor, and surface photovoltage measurements (Mizsei 2007). Moreover, due to the large density of the traps (Skryshevsky et al. 2006), there is a possibility of Fermi-level pinning at the surface of the PSi layer.

We have very limited conception about structural parameters of porous semiconductors that exert a dominating influence on the operating characteristics of gas sensors. It is clear that the set of structural parameters would be specific for each type of sensor. For example, one can suppose that for capacitance-type humidity sensors, the diameter, the length of pores, and their surface density are the main structural parameters controlling their working characteristics. For conductometric gas sensors, parameters that are more important would be the thickness of walls and the porosity of the gas-sensing matrix. An experimentally sound determination of those parameters is a hard task. However, the results of that research could become the basis for further development of the sensor technologies using porous semiconductors. Therefore, a detailed classification of structural peculiarities of porous semiconductors in correlation with the working characteristics of the devices, elaborated based on PSi, would be an important step toward both a deeper understanding of the nature of gas-sensing effects on porous semiconductors and successful commercialization of gas sensors fabricated on their base.

The influence of the concentration of free charge carries in the initial material (bulk Si) on operating characteristics of PSi-based sensors has not been analyzed well. By analogy with gas sensors based on the metal oxides one could expect their strong correlation, especially with the condition that the thickness of walls becomes commensurable with the Debye length. One can understand that the level of doping of a semiconductor has a strong influence on the structure of the layers formed because of porosification. However, the establishment of a correlation between electrophysical properties of silicon and parameters of PSi-based sensors could be very useful.

It is necessary to note that the above-mentioned situation in the field of gas sensors based on PSi reflects the approach used during the design and study of those devices. As a rule, for sensor fabrication, the porous layers formed at fixed parameters without any variations are used. Only in a small number of works can one find attempts to establish some correlations between the conditions of porosification and sensor parameters. Therefore, the results presented in a majority of articles can be considered just as ascertaining the presence of this or that gas sensing effect. Using this approach, it is almost impossible to create a database of experimental results suitable for any generalization taking into account the specific characters of porous materials.

The problem of irreversible surface deviation that leads to the instability of optical and electrical properties of porous material during sensor aging with ambient gases also remains unsolved. Several attempts have been made to stabilize a PSi surface against oxidation through the substitution or surface layer protection of highly reactive Si-H bonds and it has been shown that this problem can be partially resolved. Likewise, binding of the compounds with structurally recognizing properties was found to enhance the selectivity of PSi sensor response. However, a suitable all-purpose surface modification procedure for improving stability of a PSi surface and its selectivity and sensitivity toward structurally different analytes has not been found. Therefore, the detailed study of aging mechanisms of freshly formed material is urgently needed.

Adaptation of new measuring approaches using specific porous materials would lead to a considerable widening of the area of application of PSi-based gas sensors. For example, the possibilities of PSi-based gas sensors would be significantly wider if one could use them for detection of gas adsorption FTIR spectroscopy, analysis of the multiple luminescence peak structures, and effects of optical birefringence of PSi.

ACKNOWLEDGMENTS

This work was supported by the Ministry of Science, ICT and Future Planning (MSIP) of the Republic of Korea.

REFERENCES

Adamson, A.W. and Gast, A.P. (1997). *Physical Chemistry of Surface*. Wiley, New York.

Aggarwal, G., Mishra, P., Joshi, B., Harsh, and Islam S.S. (2014). Porous silicon surface stability: A comparative study of thermal oxidation techniques. *J. Porous Mater.* 21, 23–29.

Allongue, P., Henry, de Villeneuve, C., Bernard, M.C., Peo, J.E., Boutry-Forveill,e A., and Levy-Clement, C. (1997). Relationship between porous silicon formation and hydrogen incorporation. *Thin Solid Films* 297, 1–4.

Anderson, R.C., Muller, R.S., and Tobias, C.W. (1990). Investigation of porous silicon for vapour sensing. *Sens. Actuators A* 21–23, 835–839.

Andsager, D., Hilliard, J., and Nayfeh, M.H. (1994). Behavior of porous silicon emission spectra during quenching by immersion in metal ion solutions. *Appl. Phys. Lett.* 64, 1141–1143.

Angelescu, A., Kleps, I., Mihaela, M. et al. (2003). Porous silicon matrix for applications in biology. *Rev. Adv. Mater. Sci.* 5, 440–449.

Arakelyan, V.M., Galstyan, V.E., Martirosyan, Kh.S., Shahnazaryan, G.E., Aroutiounian, V.M., and Soukiassian, P.G. (2007). Hydrogen sensitive gas sensor based on porous silicon/TiO_{2-x} structure. *Physica E* 38, 219–221.

Archer, M., Christophersen, M., and Fauchet, P.M. (2005). Electrical porous silicon chemical sensor for detection of organic solvents. *Sens. Actuators B* 106, 347–357.

Arwin, H., Wang, G., and Jansson, R. (2003). Gas sensing based on ellipsometric measurement on porous silicon. *Phys. Stat. Sol.* (a) 197, 518–522.

Balaguer, M. and Matveeva, E. (2010). Quenching of porous silicon photoluminescence by molecular oxygen and dependence of this phenomenon on storing media and method of preparation of pSi photosensitizer. *J. Nanopart. Res.* 12, 2907–2917.

Balagurov, L.A., Yarkin, D.G., Petrova, E.A., Orlov, A.F., and Karyagin, S.N. (1996). Effects of vacuum annealing on the optical properties of porous silicon. *Appl. Phys. Lett.* 69, 2852.

Baratto, C., Sberveglieri, G., Comini, E. et al. (1999). Gold catalysed porous silicon sensor for nitrogen dioxide. In: *CD Proceeding of the 13th European Conference on Solid-State Transducers, EUROSENSORS XIII*, September 12–15, 1999, The Hague, the Netherlands. Abstract 4P10, p. 105.

Baratto, C., Comini, E., Faglia, G. et al. (2000). Gas detection with a porous silicon based sensor. *Sens. Actuators B* 65, 257–259.

Baratto, C., Faglia, G., Comini, E. et al. (2001a). A novel porous silicon sensor for detection of sub-ppm NO_2 concentrations. *Sens. Actuators B* 77, 62–66.

Baratto, C., Faglia, G., Sberveglieri, G., Boarino, L., Rossi, A.M., and Amato, G. (2001b). Front-side micromachined porous silicon nitrogen dioxide gas sensor. *Thin Solid Films* 391, 261–264.

Baratto, C., Faglia, G., Sberveglieri, G. et al. (2002). Multiparametric porous silicon sensors. *Sensors* 2, 121–126.

Barillaro, G., Diligenti, A., Nannini, A., and Pieri, F. (2000). Organic vapours detection using an integrated porous silicon device. In: *Proceeding of the 14th European Conference on Solid-State Transducers, EUROSENSORS XIV*, August 27–30, 2000, Copenhagen, Denmark. Abstract M2P39, p. 197–199.

Barillaro, G., Nannini, A., and Pieri, F. (2003). APSFET: A new, porous silicon-based gas sensing device. *Sens. Actuators B* 93, 263–270.

Barillaro, G., Diligenti, A., Marola, G., and Strambini, L.M. (2005). A silicon crystalline resistor with an adsorbing porous layer as gas sensor. *Sens. Actuators B* 105, 278–282.

Belhousse, S., Cheraga, H., and Gabouze, N. (2004). Characterization of a new sensing device based on hydrocarbon groups (CH_x) coated porous silicon. *Sens. Actuators B* 100, 250–255.

Belhousse, S., Gabouze, N., Cheraga, H., and Henda, K. (2005). CH_x/PS/Si as structure for propane sensing. *Thin Solid Films* 482, 253–257.

Bellet, D. and Dolino, G. (1994). X-ray observation of porous-silicon wetting. *Phys. Rev. B* 50, 17162–17165.

Ben-Chorin, M. and Kux, A. (1994). Adsorbate effects on photoluminescence and electrical conductivity of porous silicon. *Appl. Phys. Lett.* 67, 481–483.

Ben-Chorin, M., Moller F., and Koch F. (1994). Nonlinear electrical transport in porous silicon. *Phys. Rev. B* 49, 2981–2984.

Bisi, O., Ossicini, S., and Pavesi, L. (2000). Porous silicon: A quantum sponge structure for silicon based optoelectronics. *Surf. Sci. Rep.* 38, 1–126.

Björkqvist, M., Salonen, J., Paski, J., and Laine, E. (2003). Comparison of stabilizing treatments on porous silicon for sensor applications. *Phys. Stat. Sol. (a)* 197, 374–377.

Björkqvist, M., Salonen, J., Paski J., and Laine, E. (2004a). Characterization of thermally carbonized porous silicon humidity sensor. *Sens. Actuators A* 112, 244–247.

Björkqvist, M., Salonen, J., and Laine, E. (2004b). Humidity behavior of thermally carbonized porous silicon. *Appl. Surf. Sci.* 222, 269–274.

Björkqvist, M., Paski, J., Salonen, J., and Lehto, V.-P. (2006). Studies on hysteresis reduction in thermally carbonized porous silicon humidity sensor. *IEEE Sensors J.* 6 (3), 542–547.

Blackwood, D.J. and Akber, M.F.B.M. (2006). In-situ electrochemical functionalization of porous silicon. *J. Electrochem. Soc.* 153, G976–G980.

Borini, S., Boarino, L., and Amato, G. (2007). Coulomb blockade sensors based on nanostructured mesoporous silicon. *Physica E* 38, 197–199.

Brumhead, D., Canham, L.T., Seekings, D.M., and Tufton, P.J. (1993). Gravimetric analysis of pore nucleation and propagation in anodised silicon. *Electrochim. Acta* 38, 191–197.

Buckley, D.N., O'Dwyer, C., Harvey, E. et al. (2003). Anodic behavior of InP: Film growth, porous layer formation and current oscillations. In: *Proceeding of 203 Meeting of Electrochemical Society*, April 27–May 2, 2003, Paris, France. Abstract 776. The Electrochemical Society, Inc. Pennington, New Jersey.

Buriak, J.M. and Allen, M.J. (1998). Lewis acid mediated functionalization of porous silicon with substituted alkenes and alkynes. *J. Am. Chem. Soc.* 120, 1339–1340.

Canham, L.T., Cox, T.I., Loni, A., and Simons, A.J. (1996). Progress towards silicon optoelectronics using porous silicon technology. *Appl. Surf. Sci.* 102, 436–441.

Canhman, L. (ed.) (1997). *Properties of Porous Silicon.* INSPEC, London, pp. 44–86.

Chandler-Henderson, R.R., Sweryda-Krawiec, B., and Coffer, J.L. (1995). Steric considerations in the amine-induced quenching of luminescent porous silicon. *J. Phys. Chem.* 99, 8851–8855.

Chazalviel, J.N., Ozanam, F., and Dubin, V.M. (1994). On the correct convergence of complex Langevin simulations for polynomial actions. *J. Phys. I* 4, 1325–1330.

Chvojka, T., Vrkoslav, V., Jelinek, I., Jindrich, J., Lorenc, M., and Dian, J. (2004). Mechanisms of photoluminescence sensor response of poroussilicon for organic species in gas and liquid phases. *Sens. Actuators B* 100, 246–249.

Coffer, J.L., Lilley, S.C., Martin, R.A., and Files-Sesler, L.A. (1993). Surface reactivity of luminescent porous silicon. *J. Appl. Phys.* 74, 2094.

Connolly, E.J., O'Halloran, G.M., Pham, H.T.M., Sarro, P.M., and French, P.J. (2002). Comparison of porous silicon, porous polysilicon and porous silicon carbide as materials for humidity sensing applications. *Sens. Actuators A* 99, 25–30.

Connolly, E.J., Pham, H.T.M., Groeneweg, J., Sarro, P.M., and French, P.J. (2004a). Relative humidity sensors using porous SiC membranes and Al electrodes. *Sens. Actuators B* 100, 216–220.

Connolly, E.J., Timmer, B., Pham, H.T.M. et al. (2004b). Porous SiC as an ammonia sensor. In: *Proceedings of IEEE Sensors* 1(M2P-P.21), 178–181.

Cullis, A.G., Canham, L.T., and Calcott, P.D.G. (1997). The structural and luminescence properties of porous silicon. *J. Appl. Phys.* 82, 909–965.

Di Francia, G., De Filippo, F., La Ferrara, V. et al. (1998). Porous silicon layers for the detection at RT of low concentrations of vapoura from organic compounds. In: White N.M. (ed.) *Proceeding of European Conference Eurosensors XII*, Sept. 13–16, 1998, Southampton, UK, IOP, Bristol, 1, 544–547.

Di Francia, G., Noce, M.D., Ferrara, V.L., Lancellotti, L., Morvillo, P., and Quercia, L. (2002). Nanostructured porous silicon for gas sensor application. *Mater. Sci. Technol.* 18, 767–771.

Di Francia, G., Castaldo, A., Massera, E., Nasti, I., Quercia, L., and Rea, I. (2005). A very sensitive porous silicon based humidity sensor. *Sens. Actuators B* 111–112, 135–139.

Dimitrov, D.B., Papadimitriou, D., and Beshkov, G. (1997). Photoluminescence spectra of high temperature vacuum annealed porous silicon. In: *Proceedings of 21 International Conference on Microelectronics*, September 14–17, 1997, Nis, Yugoslavia, 1, 91.

Dittrich, T., Konstantinova, E.A., and Timoshenko, V.Y. (1995a). Influence of molecule adsorption on porous silicon photoluminescence. *Thin Solid Films* 255, 238–240.

Dittrich, T., Flietner, H., Timoshenko, V.Y., and Kashkarov, P.K. (1995b). Influence of the oxidation process on the luminescence of HF-treated porous silicon. *Thin Solid Films* 255, 149–151.

Fastykovsky, P.P. and Mogilnitsky, A.A. (1999). Effect of air humidity on the metal-oxide-semiconductor tunnel structures' capacitance. *Sens. Actuators B* 57, 51–55.

Fejfar, A., Pelant, I., Sípek, E. et al. (1995). Transport study of self-supporting porous silicon. *Appl. Phys. Lett.* 66, 1098–1100.

Fellah, S., Wehrspohn, R.B., Gabouze, N., Ozanam, F., and Chazalviel, J.N. (1999). Photoluminescence quenching of porous silicon in organic solvents: Evidence for dielectric effects. *J. Lumin.* 80, 109–113.

Feng, Z.C. and Tsu, R. (eds.) (1994). *Porous Silicon.* Word Science, Singapore.

Fisher, D.L., Harper, J., and Sailor, M.J. (1995). Energy transfer quenching of porous Si photoluminescence by aromatic molecules. *J. Am. Chem. Soc.* 117, 7846–7847.

Fogelberg, J., Eriksson, M., Dannetun, H., and Petersson, L.G. (1995). Kinetic modeling of hydrogen adsorption/absorption in thin films on hydrogen-sensitive field-effect devices: Observation of large hydrogen-induced dipoles at the Pd-SiO$_2$ interface. *J. Appl. Phys.* 78, 988–996.

Foll, H., Christophersen, M., Carstensen, J., and Hasse, G. (2002). Formation and application of porous silicon. *Mater. Sci. Eng.* R 39, 93–142.

Foll, H., Carstensen, J., and Frey, S. (2006). Porous and nanoporous semiconductors and emerging applications. *J. Nanomater.* 2006, 91635 (1–10).

Fonthal, F., Trifonov, T., Rodriguez, A., Marsal, L.F., and Pallares, J. (2006). AC impedance analysis of Au/porous silicon contacts. *Microelectronic Eng.* 83, 2381–2385.

Fujikura, H., Liu, A., Hamamatsu, A., Sato, T., and Hasegawa, H. (2000). Electrochemical formation of uniform and straight nano-pore arrays on (001) InP surfaces and their photoluminescence characterizations. *Jpn. J. Appl. Phys.* 39, 4616–4620.

Fürjes, P., Kovács, A., Dücso, Cs., Ádám, M., Müller, B., and Mescheder, U. (2003). Porous silicon-based humidity sensor with interdigital electrodes and internal heaters. *Sens. Actuators B* 95, 140–144.

Gabouze, N., Belhousse S., Cheraga H. et al. (2006). CO$_2$ and H$_2$ detection with a CH$_x$/porous silicon-based sensor. *Vacuum* 80, 986–989.

Gabouze, N., Cheraga, H., Belhoisse, S. et al. (2007). Influence of Pd layer on the sensitivity of CH$_x$/PSi/Si as structure for H$_2$ sensing. *Phys. Stat. Sol. (a)* 204(5), 1412–1416.

Galeazzo, E., Peres, H.E.M., Santos, G., Peixoto, N., and Ramirez-Fernandez, F.J. (2003). Gas sensitive porous silicon devices: Responses to organic vapors. *Sens. Actuators B* 93, 384–390.

Gan, L., He H., Sun, L., and Ye, Z. (2014). Improved photoluminescence and sensing stability of porous silicon nanowires by surface passivation. *Phys.Chem.Chem.Phys.* 16, 890–894.

Garrone, E., Borini, S., Rivolo, P., Boarino, L., Geobaldo, F., and Amato, G. (2003). Porous silicon in NO_2: A chemisorption mechanism for enhanced electrical conductivity. *Phys. Stat. Sol. (a)* 197, 103–106.

Gesemann, B., Wehrspohn, R., Hackner, A., and Muller, G. (2011). Large-scale fabrication of ordered silicon nanotip arrays used for gas ionization in ion mobility spectrometers. IEEE *Transactions Nanotechnol.* 10(1), 50–52.

Ghodsian, B., Parameswaran, M., and Syrzycki, M. (1998). Gas detector with low-cost micromachined field ionization tips. *IEEE Electron Dev. Lett.* 19(7), 241–243.

Glass, J.A. Jr., Wovchko, E.A., and Yates, J.T. Jr. (1995). Reaction of methanol with porous silicon. *Surf. Sci.* 338, 125–137.

Gole, J.L., Seals, L.T., and Lillehei, P.T. (2000). Patterned metallization of porous silicon from electroless solution for direct electrical contact. *J. Electrochem. Soc.* 147, 3785–3789.

Gomer, R. (1961). *Field Emission and Field Ionization.* Harvard University Press, Cambridge, MA.

Green, S. and Kathirgamanathan, P. (2000). The quenching of porous silicon photoluminescence by gaseous oxygen. *Thin Solid Films* 374, 98–102.

Green, S. and Kathirgamanathan, P. (2002). Effect of oxygen on the surface conductance of porous silicon: Towards room temperature sensor applications. *Mater. Lett.* 52, 106–113.

Han, P.G., Wong, H., and Poon, M.C. (2001). Sensitivity and stability of porous polycrystalline silicon gas sensor. *Colloids Surf. A: Physicochem. Eng. Asp.* 179, 171–175.

Harper, J. and Sailor, M.J. (1996). Detection of nitric oxide and nitrogen dioxide with photoluminescent poprous silicon. *Annal. Chem.* 68, 3713–3717.

Harraz, F.A., Ismail, A.A., Bouzid, H., Al-Sayari, S.A., Al-Hajry, A., and Al-Assiri, M.S. (2014). A capacitive chemical sensor based on porous silicon for detectionof polar and non-polar organic solvents. *Appl. Surf. Sci.* 307, 704–711.

Hasan, P.M.Z., Islam, S.S., Islam, T., Azam, A., and Harsh, (2010). Capacitive detection of organic vapours at low ppm level by porous silicon: Role of molecular structure in sensing mechanism. Sensor Rev. 30(4), 336–340.

Hedborg, E., Björklund, R., Eriksson, M., Martensson, P., and Lundström, I. (1999). Metal-oxide-semiconductor field-effect gas sensors based on nanostructured SiO_2. In: *CD Proceeding of the 13th European Conference on Solid-State Transducers, EUROSENSORS XIII*, September 12–15, 1999, The Hague, the Netherlands. Abstract 8A5, pp. 301–304.

Hedrich, F., Billat, S., and Lang, W. (2000). Structuring of membrane sensors using sacrificial porous silicon. *Sens. Actuators B* 84, 315–323.

Herino, R., Bomchil, G., Barla, K., and Bertrand, C. (1987). Porosity and pore size distribution of porous silicon layers. *J. Electrochem. Soc.* 134, 1994–2000.

Hirschman, K.D., Tsybeskov, L., Duttagupta, S.P., and Fauchet, P.M. (1996). Silicon-based visible light-emitting devices integrated into microelectronic circuits. *Nature* 384, 338–341.

Holec, H., Chvojka, T., Jelinek, I. et al. (2002). Determination of sensoric parameters of porous silicon in sensing of organic vapours. *Mater. Sci. Eng.* C 19, 251–254.

Hong, S. and Rahman, T.S. (2007). Adsorption and diffusion of hydrogen on Pd(211) and Pd(111): Results from first-principles electronic structure calculations. *Phys. Rev. B* 75(15), 155405.

Hory, M.A., Herino, R., Ligeon, M. et al. (1995). Fourier transform IR monitoring of porous silicon passivation during post-treatments such as anodic oxidation and contact with organic solvents. *Thin Solid Films* 255, 200–203.

Hossain, S., Chakraborty, S., Dutta, S.K., Das, J., and Saha, H. (2000). Stability in hotoluminescence of porous silicon. *J. Lumin.* 91, 195–202.

Hui, G., Wu, L., Pan, M.. Chen, Y., Li, T., and Zhang, X. (2006). A novel gas-ionization sensor based on aligned multi-walled carbon nanotubes. *Meas. Sci. Technol.* 17, 2799–2805.

Iler, R.K. (1979). *The Chemistry of Silica.* John Wiley, New York, p. 622.

Iraji zad, A., Rahimi, F., Chavoshi, M., and Ahadian, M.M. (2004). Characterization of porous poly-silicon as a gas sensor. *Sens. Actuators B* 100, 341–346.

Islam, T. and Saha, H. (2007). Study of long-term drift of a porous silicon humidity sensor and its compensation using ANN technique. *Sens. Actuators A* 133, 472–479.

Jalkanen, T., Makila, E., Maattanen, A. et al. (2012). Porous silicon micro- and nanoparticles for printed humidity sensors. *Appl. Phys. Lett.* 101, 263110.

Johansson, M., Lundstrom, I., and Ekedahl, L.-G. (1998). Bridging the pressure gap for palladium metal-insulator-semiconductor hydrogen sensors in oxygen containing environments. *J. Appl. Phys.* 84(1), 44–51.

Kavalenka, M.N., Striemer, C.C., DesOrmeaux, J.-P.S., McGrath, J.L., and Fauchet, P.M. (2012). Chemical capacitive sensing using ultrathin flexible nanoporous electrodes. *Sens. Actuators B* 162, 22–26.

Kelly, M.T., Chun, J.K.M., and Bocarsly, A.B. (1996). A silicon sensor for SO_2. *Nature* 382, 214–215.

Kim, S.-J., Jeon, B.-H., Choi, K.-S., and Min, N.-K. (2000a). Capacitive porous silicon sensors for measurement of low alcohol gas concentration at room temperature. *J. Solid State Electrochem.* 4, 363–366.

Kim, S.J., Park, J.Y., Lee, S.H., and Yi, S.H. (2000b). Humidity sensors using porous silicon layer with mesa structure. *J. Phys. D: Appl. Phys.* 33, 1781–1784.

Kim, S.J., Lee, S.H., and Lee, C.J. (2001). Organic vapour sensing by current response of porous silicon layer. *J. Phys. D: Appl. Phys.* 34, 3505–3509.

Kochergin, V.R. and Foell, H. (2006). Novel optical elements made from porous Si. *Mater. Sci. Eng. R* 52, 93–140.

Korotcenkov, G. (2005). Gas response control through structural and chemical modification of metal oxides: State of the art and approaches. *Sens. Actuators B* 107, 209–232.

Korotcenkov, G. (2007). Metal oxides for solid state gas sensors. What determines our choice? *Mater. Sci. Eng. B* 139, 1–23.

Korotcenkov, G. and Cho, B.K. (2010). Porous semiconductors: Advanced material for gas sensor applications. *Crit. Rev. Sol. St. Mater. Sci.* 35(1), 1–37.

Kovalev, D.I., Yaroshetzkii, I.D., Muschik, T., Petrova-Koch, V., and Koch, F. (1994). Fast and slow visible luminescence bands of oxidized porous Si. *Appl. Phys. Lett.* 64, 214–216.

Kroes, G.-J., Gross, A., Baerends, E.-J., Schwffler, M., and Mccormack, D.A. (2002). Quantum theory of dissociative chemisorption on metal surfaces. *Acc. Chem. Res.* 35, 193–200.

Kwok, W.M., Bow, Y.C., Chan, W.Y., Poon, M.C., Wan, P.G., and Wong, H. (1999). Study of porous silicon gas sensors. In: *Proceedings of IEEE Electron Devices Meeting*, June 26, 1999, Hong Kong, IEEE, pp. 80–83.

Lauerhaas, J.M. and Sailor, M.J. (1993). Chemical modification of the photoluminescence quenching of porous silicon. *Science* 261, 1567–1568.

Lazzerini, G.M., Strambini, L.M., and Barillaro, G. (2013). Self-tuning porous silicon chemitransistor gas sensors. In: *Proceedings of IEEE Sensors Conference*, SENSORS, 2013, November 3–6, Baltimore, MD, doi: 10.1109/ICSENS.2013.6688369.

Lee, W.H., Lee, C., Kwon, Y.H., Hong, C.Y., and Cho, H.Y. (2000). Deep level defects in porous silicon. *Sol. St. Commun.* 113, 519–522.

Lewis, S.E., DeBoer, J.R., Gole, J.L., and Hesketh, P.J. (2005). Sensitive, selective, and analytical improvements to a porous silicon gas sensor. *Sens. Actuators B* 110, 54–65.

Li, K., D. Diaz, D.C., Campbell, J.C., and Tsai, C. (1994). Porous silicon: Surface chemistry. In: Fengs Z. C. and Tsu R. (eds.) *Porous Silicon*. World Scientific, Singapore, pp. 261–274.

Li, M., Hu, M., Liu, Q., Ma, S., and Sun, P. (2013a). Microstructure characterization and NO_2-sensing properties of porous silicon with intermediate pore size. *Appl. Surf. Sci.* 268, 188–194.

Li, M., Hu, M., Zeng, P., Ma, S., Yan, W., and Qin, Y. (2013b). Effect of etching current density on microstructure and NH_3-sensingproperties of porous silicon with intermediate-sized pores. *Electrocim. Acta* 108, 167–174.

Liao, L., Lu, H.B., Shuai, M., Li, J.C., Liu, Y.L., Liu, C., Shen, Z.X., and Yu, T. (2008). A novel gas sensor based on field ionization from ZnO nanowires: Moderate working voltage and high stability. *Nanotechnology* 19, 175501.

Lin, H., Gao, T., Fantini, J., and Sailor, M.J. (2004a). A porous silicon-palladium composite film for optical interferometric sensing of hydrogen. *Langmuir* 20, 51043-6 5108.

Lin, K.W., Chen, H.I., Chuang, H.M. et al. (2004b). Characteristics of Pd/InGaP Schottky diode hydrogen sensors. IEEE Sens. J. 4, 72–79.

Lisachenko, A.A. and Aprelev, A.M. (2001). The effect of adsorption complexes on the electron spectrum and luminescence of porous silicon. *Techn. Phys. Lett.* 27(2), 134–137.

Litovchenko, V.G., Gorbanyuk, T.I., Solntsev, V.S., and Evtukh, A.A. (2004). Mechanism of hydrogen, oxygen and humidity sensing by Cu/Pd-porous silicon–silicon structures. *Appl. Surf. Sci.* 234, 262–267.

Lubianiker, Y. and Balberg, I. (1998). A comparative study of the Meyer–Neldel rule in porous silicon and hydrogenated amorphous silicon. *J. Non-Cryst. Solids* 227–230, 180–184.

Ludwig, M.H. (1996). Optical properties of silicon-based materials: A comparison of porous and spark-processed silicon. *Cr. Rev. Sol. St. Mat. Sci.* 21, 265–351.

Lukasiak, Z., Wyrzykowski, Z., Sylwisty, J., and Bala, W. (2000). Influence of gas adsorption on photoluminescence properties of porous silicon layers. *Electron Technol.* 33, 207–209.

Lundstrom, L., Shivaraman, M.S., Svensson, C.M., and Lundkvist, L. (1975). A hydrogen–sensitive MOS field–effect transistor. *Appl. Phys. Lett.* 26, 55–57.

Lundstrom, L., Armgarth, M., and Petersson, L.G. (1989). Physics with catalytic metal gate chemical sensors. *CRC Crit. Rev. Sol. State Mater. Sci.* 15, 201–278.

Luongo, K., Sine A., and Bhansali, S. (2005). Development of a highly sensitive porous Si-based hydrogen sensor using Pd nano-structures. *Sens. Actuators B*, 111–112, 125–129.

Mahmoudi, B., Gabouze, N., Haddadi, M. et al. (2007). The effect of annealing on the sensing properties of porous silicon gas sensor: Use of screen-printed contacts. *Sens. Actuators B* 123, 680–684.

Mares, J.J., Kristofik, J., and Hulicius, E. (1995). Influence of humidity on transport in porous silicon. *Thin Solid Films* 255, 272–275.

Marsh, G. (2002) Porous silicon. A useful imperfection. *MaterialsToday* 5(1), 36–41.

Massera, E., Nasti, I., Quercia, L., Rea, I., and Di Francia, G. (2004). Improvement of stability and recovery time in porous-silicon-based NO_2 sensor. *Sens. Actuators B* 102, 195–197.

Mazet, L., Varenne, C., Pauly, A., Brunet, J., and Germain, J.P. (2004). H_2, CO and high vacuum regeneration of ozone poisoned pseudo-Schottky Pd–InP based gas sensor. *Sens. Actuators B* 103, 190–199.

McLellan, R.B. (1997). The kinetic and thermodynamic effects of vacancy interstitial interactions in Pd–H solutions. *Acta Mater.* 45, 1995–2000.

Meyer, B.K., Hofmann, D.M., Stadler, W. et al. (1993). Defects in porous silicon investigated by optically detected and by electron paramagnetic resonance techniques. *Appl. Phys. Lett.* 63(15), 2120–2122.

Mizsei, J. (2005). Vibrating capacitor method in the development of semiconductor gas sensors. *Thin Solid Films* 490, 17–21.

Mizsei, J. (2007) Gas sensor applications of porous Si layers. *Thin Solid Films* 515, 8310–8315.

Modi, A., Koratkar, N., Lass, E., Wei, B., and Ajayan, P.M. (2003). Miniaturized gas ionization sensors using carbon nanotubes. *Nature* 424, 171–174.

Mou, S.S., Islam, Md. J., and Md. Ismail, A.B. (2011). Photoluminescence properties of LaF_3-coated porous silicon. *Mater. Sci. Appl.* 2, 649–653.

Nicolas, D., Souteyrand, E., and Martin, J.R. (1997). Gas sensor characterization through both contact potential difference and photopotential measurements. *Sens. Actuators B* 44, 507–511.

Nikfarjam, A., Iraji zad, A., Razi, F., and Mortazavi, S.Z. (2010). Fabrication of gas ionization sensor using carbon nanotube arrays grown on porous silicon substrate. *Sens. Actuators A* 162, 24–28.

O'Halloran, G.M., Kuhl, M., Trimp, P.J., and French, P.J. (1997). The effect of additives on the adsorption properties of porous silicon. *Sens. Actuators A* 61, 415–420.

O'Halloran, G.M., van der Vlist, W., Sarro, P.M., and French, P.J. (1999). Influence of the formation parameters on the humidity sensing characteristics of a capacitive humidity sensor based on porous silicon. In: *CD Proceeding of the 13th European Conference on Solid-State Transducers, EUROSENSORS XIII*, September 12–15, 1999, The Hague, the Netherlands. Abstract 4P13, p. 117–120.

Okorn-Schmidt, H.F. (1999). Characterization of silicon surface preparation processes for advanced gate dielectrics. *IBM J. Res. Dev.* 43, 351–365.

Oton, C.J., Pancheri, L., Gaburro, Z. et al. (2003). Multiparametric porous silicon gas sensors with improved quality and sensitivity. *Phys. Stat. Sol. (a)* 197, 523–527.

Ozdemir, S. and Gole, J.L. (2007). The potential of porous silicon gas sensors. *Curr. Opin. Solid St. Mater. Sci.* 11, 92–100.

Pancheri, L., Oton, C.J., Gaburro, Z., Soncini G., and Pavesi L. (2003). Very sensitive porous silicon NO_2 sensor. *Sens. Actuators B* 89, 237–239.

Pancheri, L., Oton, C.J., Gaburro, Z., Soncini, G., and Pavesi, L. (2004). Improved reversibility in aged porous silicon NO_2 sensors. *Sens. Actuators B* 97, 45–48.

Parkhutik, V. (1999). Porous silicon-mechanisms of growth and applications. *Sol. St. El.* 43, 1121–1141.

Polishchuk, V., Souteyrand, E., Martin, J.R., Strikha, V.I., and Skryshevsky, V.A. (1998). A study of hydrogen detection with palladium modified porous silicon. *Anal. Chim. Acta* 375, 205–210.

Raghavan, D., Gu, X., Nguyen, T., and Vanlandingham, M. (2001). Characterization of chemical heterogeneity in polymer systems using hydrolysis and tapping-mode atomic force microscopy. *J. Polymer Sci. B: Polymer Phys.* 39, 1460–1470.

Rahimi, F. and Iraji zad, A. (2006). Glucose oxidase, lactate oxidase, and galactose oxidase enzyme electrode based on polypyrrole with polyanion/PEG/enzyme conjugate dopant. *Sens. Actuators B* 115, 164–169.

Rahimi, F., Irajizad, A., and Razi, F. (2005). Characterization of porous polysilicon impregnated with Pd as a hydrogen sensor. *J. Phys. D: Appl. Phys.* 38, 36–40.

Rehm, J.M., Mclendon, G.L., Tsybeskov, L., and Fauchet, P.M. (1995). How methanol affects the surface of blue and red emitting porous silicon. *Appl. Phys. Lett.* 66, 3669.

Remaki, B., Perichon, S., Lysenko, V., and Barbier, D. (2001). Electrical transport in porous silicon from improved complex impedance analysis. In: *Microcrystalline and Nanocrystalline Semiconductors*, Fauchet P.M., Buriak J.M., Canham L.T., Koshida N., and White B.E. Jr. (eds.) MRS Proc. Vol. 638, MRS, Warrendale, PA, pp. F321–E326.

Rhoderick, E.H. (1978). *Metal-Semiconductor Contacts*. Clarendon, Oxford.

Riley, D.J., Mann, M., MacLaren, D.A., Dastoor, P.C., Allison, W., Teo, K.B.K., Amaratunga, G.A.J., and Milne, W. (2003). Helium detection via field ionization from carbon nanotubes. *Nano Lett.* 3, 1455–1458.

Rittersma, Z.M. and Benecke, W. (1999). A novel capacitive porous silicon humidity sensor with integrated thermo- and refresh resistors. In: *CD Proceeding of the 13th European Conference on Solid-State Transducers, EUROSENSORS XIII*, September 12–15, 1999, The Hague, the Netherlands. Abstract 11A5, pp. 371–374.

Rivolo, P., Pirasteh, P., Chaillou, A. et al. (2004). Oxidised porous silicon impregnated with Congo Red for chemical sensing applications. *Sens. Actuators B* 100, 99–102.

Robinson, M.B., Dillon, A.C., Haynes, D.R., and George, S.M. (1992). Effect of thermal annealing and surface coverage on porous silicon photoluminescence. *Appl. Phys. Lett.* 61, 1414–1416.

Rocchia, M., Garrone, E., Geobaldo, F., Boarino, L., and Sailor, M.J. (2003). Sensing CO_2 in a chemically modified porous silicon film. *Phys. Stat. Sol. (a)* 197, 365–369.

Sadeghian, R.B. and Kahrizi, M. (2007). A novel miniature gas ionization sensor based on freestanding gold nanowires. *Sens. Actuators A* 137, 248–255.

Sadeghian, R.B. and Kahrizi, M. (2008). A novel miniature gas ionization sensor based on freestanding gold nanowires. *IEEE Sensors J.* 8(2), 161–169.

Sadeghian, R.B. and Islam, M.S. (2011). Ultralow-voltage field-ionization discharge on whiskered silicon nanowires for gas-sensing applications. *Nature Mater.* 10, 135–140.

Sailor, M.J. and Wu, E.C. (2009). Photoluminescence-based sensing with porous silicon films, microparticles, and nanoparticles. *Adv. Funct. Mater.* 19, 3195–3208.

Salcedo, W.J., Ramirez Fernandez, F.J., and Rubim, J.C. (2004). Photoluminescence quenching effect on porous silicon films for gas sensors application. *Spectrochim. Acta A* 60, 1065–1070.

Salehi, A. and Kalantari, D.J. (2007). Characteristics of highly sensitive Au/porous-GaAs Schottky junctions as selective CO and NO gas sensors. *Sens. Actuators B* 122, 69–74.

Salgado, G.G., Becerril, T.D., Santiesteban, H.J., and Andrés, E.R. (2006). Porous silicon organic vapor sensor. *Opt. Mater.* 29, 51–55.

Salonen, J., Lehto, V.P., Björkqvist, M., Laine, E., and Niinistö, L. (2000). Studies of thermally-carbonized porous silicon surfaces. *Phys. Stat. Sol. (a)* 182, 123–126.

Salonen, J., Laine, E., and Niinistö, L. (2002). Thermal carbonization of porous silicon surface by acetylene. *J. Appl. Phys.* 91 (1), 456–461.

Schechter, I., Ben-Chorin, M., and Kux, A. (1995). Gas sensing properties of porous silicon. *Anal Chem.* 67, 3727–3732.

Seals L., Gole J.L., Tse L.A., and Hesketh P.J. (2002). Rapid, reversible, sensitive porous silicon gas sensor. *J. Appl. Phys.* 91, 2519–2523.

Seiyama, T., Yamazoe, N., and Arai, H. (1983). Ceramic humidity sensors. *Sens. Actuators* 4, 85–96.

Setkus, A., Galdikas, A., Mironas, A. et al. (2001). The room temperature ammonia sensor based on improved Cu_xS-micro-porous-Si structure. *Sens. Actuators B* 78, 208–215.

Setzu, S., Letant, S., Solsona, P., Romenstain, R., and Vial, J.C. (1998). Improvement of the luminescence in p-type as-prepared or dye impregnated porous silicon microcavities. *J. Lumin.* 80, 129–132.

Sharma, S.N., Bhagavannarayana G., Kumar U., Debnath R., and Mohan S.C. (2007). Role of surface texturization on the gas-sensing properties of nanostructured porous silicon films. *Physica E* 36, 65–72.

Shi, F.G., Mikrajuddin, and Okuyama, K. (2000). Electrical conduction in porous silicon: Temperature dependence. *Microelectron. J.* 31, 187–191.

Shin, H.J., Lee, M.K., Hwang, C.C. et al. (2003). Photoluminescence degradation in porous silicon upon annealing at high temperature in vacuum. *J. Korean Phys. Soc.* 42 (6), 808–813.

Skryshevsky, V.A. (2000). Photoluminescence of inhomogeneous porous silicon at gas adsorption. *Appl. Surf. Sci.* 157, 145–150.

Skryshevsky, V.A., Zinchuk, V.M., Benilov, A.I., Milovanov, Yu.S., and Tretyak, O.V. (2006). Overcharging of porous silicon localized states at gas adsorption. *Semicond. Sci. Technol.* 21, 1605–1608.

Souteyrand, E., Nicolas, D., and Martin, J.R. (1995). Influence of surface modification on sensor gas sensor behaviour. *Sens. Actuators B* 26–27, 174–178.

Spindt, C.A. (1968). A thin-film field-emission cathode. *J. Appl. Phys.* 39(7), 3504–3505.

Spindt, C.A. (1992). Microfabricated field-emission. *Surf. Sci.* 266, 145–154.

Starodub, V.M. and Starodub, N.F. (1999). Optical immune sensors for the monitoring protein substances in the air. In: *CD Proceedings of the 13th European Conference on Solid-State Transducers, EUROSENSORS XIII*, The Hague, the Netherlands, p. 181.

Steiner, P. and Lang, W. (1995). Micromachining applications of porous silicon. *Thin Solid Films* 255, 52–58.

Stewart M.P. and Buriak J.M. (1998). Photopatterned hydrosilylation on porous silicon. *Angew Chem. Int. Ed. Eng.* 37, 3257–3259.

Stievenard, D. and Deresmes, D. (1995). Are electrical properties of an aluminium-porous silicon junction governed by dangling bonds? *Appl. Phys. Lett.* 67, 1570–1572.

Strikha, V., Skryshevsky V., Polishchuk V., Souteyrand E., and Martin J.R. (2001). A study of moisture effects on Ti/porous silicon/silicon Schottky barrier. *J. Porous Mater.* 7, 111–114.

Subramanian, N.S., Sabaapathy, R.V., Vickraman, P., Kumar, G.V., Sriram, R., and Santhi, B. (2007). Investigations on $Pd:SnO_2$/porous silicon structures for sensing LPG and NO_2 gas. *Ionics* 13, 323–328.

Sukach, G.A., Oleksenko, P.F., Smertenko, P.S., Evstigneev, A.M., and Bogoslovskaya, A.B. (2003). Study of charge flow mechanisms in metal-porous silicon structures by photoluminescent and electrophysical techniques. In: *Optics and Photonics: Optical Diagnostics of Materials and Devices for Opto-, Micro-, and Quantum Electronics*, Svechnikov S.V. and Valakh M.Y. (eds.), *Proc. SPIE*, 5024, 72–79.

Tischler, M.A., Collins, R.Y., Stathis, J.H., and Tsang, J.C. (1992). Luminescence degradation in porous silicon. *Appl. Phys. Lett.* 60, 639–641.

Torchinskaya, T.V., Korsunskaya, N.E., Khomenkova, L.Y. et al. (1997). Complex studies of porous silicon aging phenomena. In: *Proceeding of International Semiconductor Conference*, October 6–10, 1997, Sinaia, Romania, IEEE, p. 173–176.

Torchinskaya, T.V., Korsunskaya, N.E., Khomenkova, L.Y., Dhumaev, B.R., and Prokes, S.M. (2001). The role of oxidation on porous silicon photoluminescence and its excitation. *Thin Solid Films* 381, 88–93.

Tsamis, C., Tsoura, L., Nassiopoulou A.G. et al. (2002). Hydrogen catalytic oxidation reaction on Pd-doped porous silicon. *IEEE Sensors J.* 2 (2), 89–95.

Tucci, M., La Ferrara, V., Della Noce, M., Massera, E., and Quercia, L. (2004). Bias enhanced sensitivity in amorphous/porous silicon heterojunction gas sensors. *Non-Cryst. Solids* 338–340, 776–779.

Turro, N.J. (1991). *Modern Molecular Photochemistry*. University Science Books, Mill Valley, CA.

Vial, J.C. and Derrien, J. (eds.) (1995). *Porous Silicon: Science and Technology*, Les Editions de Physique, Les Ulis. Springer, Berlin.

Vrkoslav, V., Jelınek, I., Matocha, M., Kral, V., and Dian, J. (2005). Photoluminescence from porous silicon impregnated with cobalt phthalocyanine. *Mater. Sci. Eng. C* 25, 645–649.

Wang, M.S., Peng, L.M., Wang, J.Y., and Chen, Q. (2005). Electron field emission characteristics and field evaporation of a single carbon nanotube. *J. Phys. Chem. B* 109,110.

Wang, H., Zou, C., Tian, C., Zhou, L., Wang, Z., and Fu, D. (2011). A novel gas ionization sensor using Pd nanoparticle-capped ZnO. *Nanoscale Res. Lett.* 6, 534.

Wehrspohn, R.B., Schweizer, S.L., Gesemann, B. et al. (2013). Macroporous silicon and its application in sensing. *C.R. Chimie* 16, 51–58.

Xu, Z.Y., Gal, M., and Gross, M. (1992). Photoluminescence studies on porous silicon. *Appl. Phys. Lett.* 60, 1375–1377.

Xu, Y.Y., Li, X.J., He, J.T., Hai, X.H., and Wang, Y. (2005). Capacitive humidity sensing properties of hydro-thermally-etched silicon nano-porous pillar array. *Sens. Actuators B* 105, 219–222.

Yam, F.K. and Hassan, Z. (2007). Schottky diode based on porous GaN for hydrogen gas sensing application. *Appl. Surf. Sci.* 253, 9525–9528.

Yamana, M., Kashiwazaki, N., Kinoshita, A., Nakano, T., Yamamoto, M., and Walton, C.W. (1990). Porous silicon oxide layer formation by the electrochemical treatment of a porous silicon layer. *J. Electrochem. Soc.* 137, 2925–2927.

Yan, D., Hu, M., Li, S., Liang, J., Wu, Y., and Ma, S. (2014a). Electrochemical deposition of ZnO nanostructures onto porous silicon and their enhanced gas sensing to NO_2 at room temperature. *Electrochim. Acta* 115, 297–305.

Yan, W., Hu, M., Zeng, P., Ma, S., and Li, M. (2014b). Room temperature NO_2-sensing properties of WO_3 nanoparticles/porous silicon. *Appl. Surf. Sci.* 292, 551– 555.

Yarkin, D.G. (2003). Impedance of humidity sensitive metal/porous silicon/n-Si structures. *Sens. Actuators A* 107, 1–6.

Zhu, Z., Zhang, J., and Zhu, J. (2005). An overview of Si-based biosensors. *Sensor Lett.* 3, 71–88.

Zouadi, N., Messaci, S., Sam, S., Bradai, D., and Gabouze, N. (2015). CO_2 detection with CN_x thin films deposited on porous silicon. *Mater. Sci. Semicond. Proces.* 29, 367–371.

Optimization of PSi-Based Sensors Using IHSAB Principles

James L. Gole and Caitlin Baker

2

CONTENTS

2.1 The Framework of Selective Extrinsic PSi Semiconductors and the IHSAB Principle   46

2.2 The IHSAB Principle as the Basis for Nanostructure-Directed Physisorption (Electrotransduction): The Use of Nanostructured Metal Oxide Island Sites   49

2.3 *In Situ* Conversion of Metal Oxide Sites via Direct Nitration and Sulfur Group Functionalization   55

2.4 Light-Enhanced Conductometric Sensing   59

2.5 Application and Extension of the IHSAB Concept to Alternate Extrinsic Semiconductors   60

2.6 Extension of the IHSAB Concept to Nanowire Configurations   60

2.7 Extension of the Nitration Concept to the Development of Basic Catalyst Sites and Frustrated Lewis Acid/Base Pairs   62

    2.7.1 Amine Doped Oxides   62

    2.7.2 Frustrated Lewis Acid/Base Pairs   63

2.8 Outlook   63

References   65

This chapter is dedicated to Yukio Ogata, not only a great scientist but also a wonderful and caring friend.

2.1 THE FRAMEWORK OF SELECTIVE EXTRINSIC PSi SEMICONDUCTORS AND THE IHSAB PRINCIPLE

There is a wide variety of gas detection devices that include infrared spectroscopic, mass sensitive, and solid-state sensors. Several solid-state devices detect analytes by transducing an electrical property associated with a detection interface and from this measurement evaluating an interacting gas. In this chapter, we focus on highly sensitive conductometric gas sensors that can be made to consist of a sensitive interface layer decorated by nanostructure island sites, which (1) do not interact electronically, (2) can be easily constructed without a requirement of complex self-assembly, (3) are much easier to implement and reproduce than film depositions, and (4) require far fewer starting materials.

Two key areas that need improved understanding are our ability to sense chemicals in a wide range of environments and our need to improve catalytic processes, especially those that use less energy and produce less waste. A combination of uniquely defined active interfaces and the ability to confine processes at the nanoscale, coupled with the ability to manipulate nanostructured materials and their interactions at select interfaces, offers a special opportunity to develop economically viable and energy efficient and sensitive modes of detection for chemical species. Recent advances in the ability to control charge transport at porous semiconductor micro/nanoporous interfaces, driven by nanostructure-focused Brönsted and Lewis acid-base chemistry can play a critical role in the development of highly responsive and sensitive (ppb) conductometric sensor arrays and novel catalysts (Gole et al. 2004; Chen et al. 2005; Kumar et al. 2005; Lewis et al. 2005; Gole and Lewis 2007; Ogden et al. 2008; Gole and Ozdemir 2010).

The most common type of solid-state sensor is the metal oxide sensor. These sensors (Barsan et al. 1999; Comini 2006) use the metal oxides of tungsten (WO_3), nickel (NiO), copper (Cu_xO), aluminum (Al_2O_3), titanium (TiO_2), tin (SnO_x), and zirconium (ZrO_2). Of the methods used to prepare these oxides, those that are most commonly cited in the literature require the use of thin films ($< 1~\mu m$). However, in some cases thick films are used, doped with noble metals or various nanoshapes designed to effect grain boundaries (Yoon and Choi 1997; Moulzolf et al. 2001; LeGore et al. 2002; Jiménez et al. 2003; Ponce et al. 2003; Rothschild and Komem 2004; Comini 2006; Rani et al. 2007). These metal oxide sensors must be heated to elevated temperatures that range from 100 to 600°C (Morrison 1987; Watson 1994; Comini 2006; Yamazoe and Shimanoe 2007) and, in many cases, for the effective monitoring of a given analyte, must be precisely controlled.

The required temperature control of typical metal oxide sensors and the necessity to generate the sensing interface from film technology are two key problems that can be overcome with the technology that we outline in this monograph. We describe a conductometric gas sensor that can be made to consist of an easily designed sensitive hybrid nano/microporous porous silicon (PSi) interface that can be transformed through the introduction of select nanostructures. Although the current interface is designed from PSi, the distinct approach that we employ can be extended to any extrinsic semiconductor onto which the hybrid nanopore coated microporous structure can be generated. This hybrid structure facilitates rapid Fickian (Kottke et al. 2009; Gole et al. 2010a) diffusion into a microporous framework whose nanoporous wall covering serves to provide a phase match for the selectively deposited nanostructures. In contrast to both thin and thick films, which may respond in minutes (Schechter et al. 1995) due to slow diffusion, the current structure, with which analyte interaction is diffusion dominated, allows a response in less than 2 s (Gole et al. 2010).

The creation of novel, highly active, and selective micro/nanostructured porous *extrinsic* semiconductor interfaces and their ability to be transformed with select nanostructure interactions enable new paradigms for sensing based on energy transfer and transduction, while also providing a road map to form select sites for heterogeneous catalysis. Nanopore coated, microporous arrays not only enable enhanced diffusion to active sites (Kottke et al. 2009), but also the nanopores provide a "phase matching" region with which modifying active nanostructured island sites can be made to interact in a controlled manner. Because the deposited nanostructures undergo minimal sintering, we promote a controllable and variable interface sensitivity. We have recently developed a new concept (Gole et al. 2010a, 2012), the inverse hard and soft acid and base (IHSAB) concept, which expands on the tenants of the previously developed HSAB (Pearson 1963, 1988, 1990, 1997, 2005) concept and allows the design of novel sensors and catalysts. The IHSAB principle incorporates the coupling of analyte/interface acid-base chemistry, an approach to the balance and separation of surface physisorption (electron transduction) and chemisorption, and the ability of active nanostructures to utilize these differences. Here, the concept of electron transduction is defined as the transfer of electrons to or from an interface without the formation of a chemical bond. At the heart of the concept, based on experimental observations, is the effective transfer of electrons to acidic or from basic molecules (analytes) at a nanostructure modified semiconductor interface. The nanostructures focus the coupling of an interaction with the majority charge carrier concentration of an extrinsic p- or n-type semiconductor.

The nanostructure-decorated configuration depicted schematically in Figure 2.1 has several properties that are superior to traditional metal oxide thin film designs. First, the nanostructures are readily deposited to the microporous interface from a variety of solution-based sources (e.g., electroless metal, sol-gel generated nanostructures, soluble metal chlorides), which require no heating. Second, once deposited, the nanostructures can be readily functionalized, *in situ*, predictably changing their interaction with a given analyte. This is exemplified as we convert the metal oxides to oxynitrides (Gole et al. 2012; Laminack and Gole 2013b) or as we functionalize the nanostructured metal oxides with S-H$_z$(CH$_x$)$_y$(z = 0,1) groups (Gole and Laminack 2013; Laminack et al. 2015). The important consideration is that the nanostructured sites are easily modified. The outlined attributes suggest enhanced energy efficiency both in the transduction mechanism and in the simplicity of a design involving minimal manufacturing time.

We provide an example of this ready deposition in Figure 2.2 (Gole 2015). Note that the deposition *does not require time consuming and costly self-assembly* or the application of costly and time-intensive lithographic assembly. The nanostructured island sites once deposited to the interface are sustained on that interface in the size range 10–30 nm. There is little or no evidence for the sintering of the nanostructures. The only requirement is that the concentration of the nanostructures be maintained at a level to avoid cross talk between these structures, which will degrade the conductometric response of the interface. In other words, there is an optimum deposition concentration for each nanostructure deposition. However, the deposition process requires no additional control and does not demand that the nanostructured islands be placed precisely at the same points in the micropores. This simplicity of design follows a process that is much more energy efficient than is thin or thick film design. With this form of interface preparation, we provide ready reproducibility, a factor that is not easily obtained in film preparation.

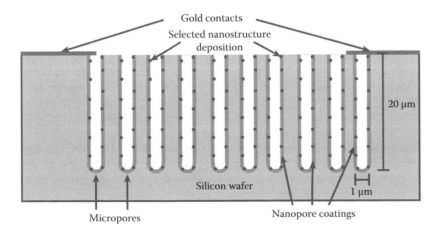

FIGURE 2.1   Schematic diagram of hybrid etched nanopore-covered microporous PSi array deposited with selected metal oxide nanostructures. Au contacts are deposited onto the PSi over a SiC insulation layer on the c-Si surface. (Gole J.L. and Ozdemir S.: *ChemPhysChem*. 2010. **11**. 2573–2581. Copyright Wiley-VCH Verlag GmbH & Co. KGaA, Reproduced with permission.)

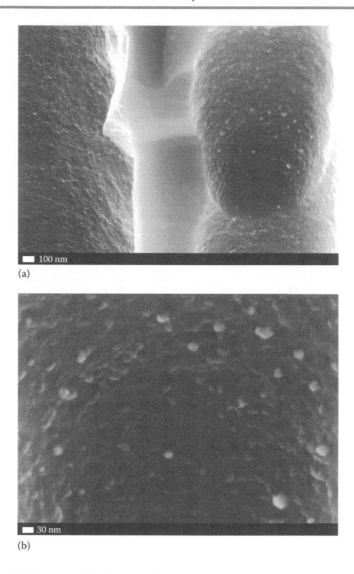

(a)

(b)

FIGURE 2.2 (a) SEM image of thiol-treated TiO_2 decorated PSi at 87.36 KX magnification. (b) SEM image of thiol-treated TiO_2 decorated PSi at 274.17 KX magnification. The lighter images in the micrographs correspond to sulfur-based moiety functionalized titanium oxide. (Reprinted from *Talanta*, 132, Gole J.L., 87–95, Copyright 2015, with permission from Elsevier.)

We have examined the IV response associated with the PSi sensor substrates, testing for linearity and bias. Based on a linear IV curve (Gole 2015) the continuous power consumption is between 25 and 60 µW/s over the range of 1 to 1.5 V. The IV curves in concert with the analysis of Ahlers et al. (2003) (Figure 2.3) demonstrate that the nanostructure directed sensor systems require considerably less power for operation than do thick or thin film sensors. Mueller et al. (2003) following Ahlers et al. (2003) have recently discussed a MEMS toolkit for metal oxide-based sensing systems. Using silicon micromachining technologies, they have developed metal oxide gas sensor elements with very small heat consumption. The power of a single micromachined sensor element is found to be an order of magnitude less (Figure 2.3) than that of thick film devices. In their Table 1, these authors quote heating powers and necessary overhead heating requirements, which sum to 1–1.5 W for a thick-film sensor (column 1; Figure 2.3) and 0.4–0.6 W (0.8 W including sensor) for a micromachined sensor array. The required power for the nanostructure directed PSi systems we have considered is virtually in the noise level of Figure 2.3. We emphasize that the described system competes quite favorably with those micromachined sensor systems whose construction may be more costly and time consuming.

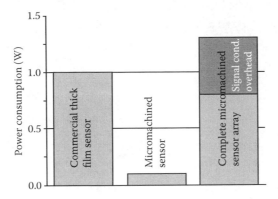

FIGURE 2.3   A comparison of power consumptions. A full-fledged micro-machined sensor array with heater driver requires as much power as a single commercial thick-film device. Additional power for signal conditioning electronics for a micro-machined sensor array is also indicated. (From Ahlers S., Kreisl P., and Müller G. (2003). *Proceedings of 203rd ECS Meeting*, April 28, Paris, France, paper 2873. With permission.)

2.2 THE IHSAB PRINCIPLE AS THE BASIS FOR NANOSTRUCTURE-DIRECTED PHYSISORPTION (ELECTROTRANSDUCTION): THE USE OF NANOSTRUCTURED METAL OXIDE ISLAND SITES

The concept of chemical hardness/softness first developed by Pearson (Gole and Ozdemir 2010) is based on the nature of metal ion complexation in aqueous solution and is a generalization of the Lewis acid/base concept. The tenants of this concept are that hard acids combine with hard bases to form strong ionic bonds, whereas soft acids combine with soft bases to form covalent bonds. In contrast, the IHSAB (inverse hard/soft acid/base) concept, designed for reversible sensor response, seeks to avoid the bonding interaction. This dictates that hard acids combine with soft bases and vice versa. However, the IHSAB and HSAB concepts are complementary and correlate with the same basic parameters associated with electronic structure theory. The HSAB concept, as it is correlated with chemical reactivity theory (CRT), was given a deep foundation in terms of density function theory by Parr and co-workers (Parr et al. 1978; Yang et al. 1984; Yang and Parr 1985; Parr and Yang 1989; Geerlings et al. 2003), following an initial correlation with molecular properties established by Pearson and Parr (Pearson 1963, 1988, 1990, 1997, 2005; Parr et al. 1978). More recently, conflicts underlying the correlation of the DFT and CRT theories have been largely resolved by Cohen and Wasserman (2007) to obtain a further refinement of the concepts of electronegativity and hardness.

The properties of acids and bases can be described as hard and soft based on the correlation of several atomic/molecular properties, which include the ionization potential, I, the electron affinity, A, and the chemical potential, μ. These are associated with the HOMO-LUMO gap concept from molecular orbital (MO) theory as the Kohn-Sham orbitals replace MOs (Parr et al. 1978; Yang et al. 1984; Yang and Parr 1985; Geerlings et al. 2003).

Examples in terms of the groups of hard, borderline, and soft acids and bases are given in Table 2.1 (Pearson 1988). For a soft acid, the acceptor atom is of low positive charge, large size, and has easily polarizable outer electrons. The acceptor atom of a hard acid is of small size and not easily polarized. In a soft base, in precise contrast to a hard base, the donor atom is of low electronegativity, easily oxidized, and highly polarizable with low-lying empty molecular orbitals. The HSAB principle was initially based on empirical observations. However, as it groups acids and bases, the HSAB basis has been developed in terms of DFT. The basis follows the principle that soft–soft combinations correspond mainly to covalent bonding and hard–hard combinations correspond mainly to ionic bonding. Further, the HSAB principle states that hard acids prefer to coordinate to hard bases whereas soft acids prefer to coordinate to soft bases (Gole et al. 2011).

TABLE 2.1 Hard and Soft Acids and Bases

	Hard	Borderline	Soft
Acids	H^+, Li^+, Na^+, K^+	Fe^{2+}, Co^{2+}, Ni^{2+}	Cu^+, Au^+, Ag^+, Tl^+, Hg^+
	Be^{2+}, Mg^{2+}, Ca^{2+}	Cu^{2+}, Zn^{2+}, Pb^{2+}	Pd^{2+}, Cd^{2+}, Pt^{2+}, Hg^{2+}
	Cr^{2+}, Cr^{3+}, Al^{3+}	BBr_3, Sn^{+2}, NO_2	BH_3, NO
	SO_3, BF_3, Sn^{+4}, Ti^{+4}		
Bases	F^-, OH^-, H_2O, NH_3	NO_2, SO_3^{2-}, Br^-	H^-, R^-, CN^-, CO, I
	CO_3^{2-}, NO_3^-, O^{2-}	N_3^-, N_2, H_2S, SO_2	SCN^-, R_3P, C_6H_5
	SO_4^{2-}, PO_4^{3-}, ClO_4^-	C_6H_5N, SCN	R_2S, NO

Source: Gole J.L. et al.: *ChemPhysChem*. 2012. **13**. 549. Copyright Wiley-VCH Verlag GmbH & Co. KGaA. Reproduced with permission.

Within the HSAB-DFT framework, the electronic chemical potential (Pearson 1963, 1988, 1990, 1997, 2005; Parr et al. 1978; Yang et al. 1984; Yang and Parr 1985; Parr and Yang 1989; Geerlings et al. 2003; Zhan et al. 2003) given by Equation 2.1,

$$(2.1) \qquad \mu = \frac{\partial E(N)}{\partial N}\upsilon_e = \frac{\partial E}{\partial \rho}\upsilon_e$$

is a global quantity. Here $E(N)$ is the ground state energy of a system of N electrons in the electrostatic potential energy, υ_e, due to its nuclei (fixed) and E is a functional (Equation 2.1) of the electron density, ρ. The three-point finite difference approximation for $\partial E(N)/\partial N$ gives $\mu \approx -(I + A)/2$ with, I, the ionization potential and, A, the electronegativity so that μ is the negative of the Mulliken electronegativity χ_M (Equation 2.2):

$$(2.2) \qquad \chi_M \approx -\mu \approx \frac{(I+A)}{2}$$

The absolute hardness of a species, η (Pearson 1963, 1988, 1990, 1997, 2005; Parr et al. 1978; Yang et al. 1984; Yang and Parr 1985; Parr and Yang 1989; Geerlings et al. 2003; Zhan et al. 2003), is defined by Equation 2.3:

$$(2.3) \qquad \eta = \left(\frac{\partial^2 E(N)}{\partial N^2}\right)\upsilon_e = \left(\frac{\partial \mu}{\partial N}\right)\upsilon_e \approx (I-A)$$

and the absolute softness (Yang and Parr 1985) is the inverse of the hardness (Equation 2.4):

$$(2.4) \qquad S = \eta^{-1} = \left(\frac{\partial N}{\partial \mu}\right)\upsilon_e$$

The approximation in Equation 2.3 arises from the use of the finite difference formula. Unlike the chemical potential, the hardness is not constrained to be constant everywhere throughout a system, having local values for which η is simply a global average. Parr and co-workers (Parr et al. 1978; Yang et al. 1984; Yang and Parr 1985; Parr and Yang 1989; Geerlings et al. 2003) have defined a local hardness that corresponds to the change in chemical potential with electron density in different parts of a molecule, complex, or simply a system. Cohen and Wasserman (2007), in their formulation of CRT, define a generalization to include a hardness matrix that incorporates both the self-hardness of individual species and the mutual hardness for pairs of species combining in a system. They also provide a description of local softness as they demonstrate how the reactivity of a species depends on its chemical context. Within this context, as interacting constituents separate, the hardness matrix becomes diagonal in the self-hardness. It is possible to establish a more general description of electronegativity (Equation 2.2) equalization. Cohen and Wasserman's (2007) generalization realizes that reactivity indices are chemical-content dependent, not unique properties of an isolated system as an electron can move on and off of a species, interacting with its chemical environment. The electron need be associated with that species only part of the time as, for example, when a Lewis acid/Lewis base interaction takes place. Of equal importance are the correlations, which define the connection between CRT and DFT theories as they can be used to provide a description of those (MOs) involved in the process of electron transfer from an acid to a base.

Within the context outlined, if two systems B and C are brought together, electrons will flow from the system of lower χ (Equation 2.2) to that of higher χ to equilibrate the chemical potentials. If we consider that, in solid–solid interactions, the equilibration of the Fermi levels is the analog of the chemical potential, it is not difficult to envision the extrapolation of these concepts to the interaction of a molecule with an interface.

Within the context of interacting molecular systems B and C, as a first approximation to an acid–base interaction, the fractional number of electrons transferred can be defined by (Gole et al. 2011)

$$(2.5) \qquad \Delta N = \frac{(\chi_C - \chi_B)}{2(\eta_C + \eta_B)}$$

where the difference in electronegativity drives the electron transfer and the sum of the hardness parameters acts as a resistance. This expression, while approximate, is useful because it expresses the nature of the initial interaction between B and C using properties of the isolated systems and providing the backdrop for the first order categorizations given in Table 2.1. Whereas the absolute chemical potential and hardness are molecular parameters, the flow of electrons is from a specific occupied molecular orbital of B to a specific empty orbital in C so that the overlap between the exchanging orbitals will be critical in determining energy change and the nature of chemical interaction.

The correlation of hardness and softness with molecular orbital theory follows readily from the Frontier orbital concept of chemical reactivity theory (Fukui et al. 1952). Within the context of Koopman's theorem, the frontier orbital energies can be correlated with the expressions for chemical potential (Equations 2.1 and 2.2), hardness (Equation 2.3), and softness (Equation 2.4) as

$$(2.6) \qquad -\epsilon_{HOMO} = I, -\epsilon_{LUMO} = A$$

where now the concept of hardness reduces to the statement: hard molecules have a large HOMO-LUMO gap and soft molecules have a small HOMO-LUMO gap (Pearson 1988). This concept must be carefully applied. In formal DFT, the location to which the electron transits corresponding to the electron affinity is not the LUMO but actually the first excited state. In practice, the application of the concept in this manner usually applies, but it is not formally rigorous (Dixon 2010). A further issue in MO theory is that for an infinite basis set, if the EA is negative, the HOMO-LUMO gap is equal to the IP. This means that the outlined concept must be applied carefully to systems that do not have a positive EA and bind an electron (Dixon 2010). The criteria that hard acids prefer to coordinate to hard bases and soft acids to soft bases is, in one sense, a HOMO-LUMO matching criteria. Politzer (1987) has shown that the softness of atoms correlates with their polarizability whereas Huheey (1978) has shown that softness is the ability to accept charge. Within this framework, we promote a HOMO-LUMO mismatch to induce physisorption/electron transduction.

We have outlined but it is possible to link chemical selectivity and the balance of physisorption and chemisorption and their associated impact on electron dynamics at doped semiconductor interfaces with the mechanism of sensor response and catalyzed chemical transformations. Nanopore-coated microchannels, which combine optimized analyte diffusion with maximum interface interaction, provide a unique phase match for the subsequent fractional deposition of select nanostructure islands that decorate the microchannel. These nanostructures can be carefully chosen to guide a controlled balance of physisorbtive versus chemisorbtive interactions at the decorated doped semiconductor interface. Further, the study of conductometric sensor response can be used to guide the choice of heterogeneous catalytic sites. The selection of the nanostructures and the variable and controllable interaction they direct dictates the nature of electron dynamics at the interface and the coupling to the majority charge carriers of the semiconductor. The synthetic approach is unique in that the nanostructures are deposited fractionally to the nanopore-coated semiconductor micropores. This fractional deposition does not require any time-consuming self-assembly within the pores, and is far simpler to implement than traditional thin film or alternative coating techniques. The deposition of nanostructured (e.g., metal oxide) islands, whose response is dictated in large part by their acidic character, is simply monitored to avoid electronically based cross talk between these structures. Within this constraint, precise reproduction of the nanostructure island deposition is not required. This technology facilitates consistently reproducible sensor responses, in contrast to the diversity of responses obtained with similar film coating techniques. These fractionally deposited systems not only display a more sensitive, reversible response than traditional metal oxide systems that require tightly controlled auxiliary heating, but also they operate at room temperature or across a wide temperature range applicable, for example, to flue gas environments. They are far simpler to implement than multiply coated metal oxide systems or electrochemical (especially solid state) devices. The nanostructured metal oxide islands can be readily transformed from acidic character to character that is more basic. This has been demonstrated by direct *in situ* nitridation or sulfidization to form the oxynitrides or sulfur compound functionalized oxysulfides (Laminack and Gole 2012, 2013b). A broad range of nanostructured interfaces can be obtained through the creation of an array of nanostructured island responses, and this technology provides the potential to readily transform surface sites from acidic to basic character for applications in heterogeneous catalysis or to more readily produce the analogs of frustrated Lewis acid/base pairs (FLPs).

The fractional deposition of nanostructured material centers can be used to create inexpensive, microfabricated interfaces, which can serve as selective interfacial platforms that can be developed for applications of focused electron transduction using materials selection tables built on the IHSAB model. We have carried out initial evaluations of the variable sensor response matrices, which can be developed from a diversity of materials at the nanoscale (Gole et al. 2010b, 2012; Laminack et al. 2012). The reversible, sensitive, and selective interaction of nanostructures with Lewis basic analytes depends on their Lewis acid strength. Materials including MgO > nanotitania > SnO_2 > nanoalumina > NiO > Cu_xO > Au_xO (where the progression is from strong to weak acidity) have already been demonstrated for the detection of gases including NH_3, PH_3, CO, NO_x, H_2S, and SO_2 in an array-based format at the sub-ppm level. In current applications, the deposition of metal oxide nanostructures (Figure 2.1) introduces new selective sites, which provide an enhanced and variable conductometric response relative to an untreated interface. This response is in direct relation to the degree of basicity or acidity of the analyte (Gole et al. 2010a). The nanopore-covered microporous structure of the interface has been created specifically to facilitate efficient gaseous diffusion (Fickian) (Gole and Lewis 2007) to the highly active nanostructure modified nanoporous coating. The surface-attached nanoparticles possess unique size-dependent and electronic structure properties that form a basis for changing the sensitivity for exposure to specific gases. In current applications, this exposure alters the conductivity of the PSi (now measured by microprobe circuitry) attached to the gold contacts shown in Figure 2.1.

When operated in the electron transduction mode, the transfer of electrons to an *n*-type PSi interface, as would occur with a basic analyte, increases the majority charge carriers, which are electrons, decreases the conductometric resistance, and increases conductance. The removal of electrons, as would occur with an acidic analyte, decreases the majority charge carriers and the conductance and increases resistance. The opposite behavior will be observed for a *p*-type semiconductor interface. For a selection of *p*-type interfaces, Table 2.2 displays the change in response relative to an untreated PSi interface for sensors decorated with deposited metal oxide nanoparticles and interacting with multiple gases. This data represents the change in response for 1 ppm of the different gases before and after depositing the metal oxide nanoparticles. Equation 2.7 was used to generate the numbers in Table 2.2, where R_0 is the baseline of the sensor before exposure to the gas and ΔR is the resistance change in the senor due to the exposure.

$$(2.7) \qquad \Delta = \frac{\Delta R(\text{deposited})/R_0(\text{deposited})}{\Delta R(\text{untreated})/R_0(\text{untreated})}$$

TABLE 2.2   Increase in the Signal for *p*-Type Silicon for Various Analyte Gases after Decorating with Different Metal Oxide Nanoparticles Relative to an Undecorated PSi Surface

	Tin (SnO$_2$)	Nickel (NiO)	Copper (Cu$_x$O)	Gold (Au$_x$O)
PH$_3$	2	2.5	4	5
NO	7–10	3.5	1	1.5
NH$_3$	1.5	(1.5–2)	(2–2.5)	≈3
SO$_2$	4	2	1+	2

Source: Gole J.L. et al.: *ChemPhysChem.* 2012. **13**. 549. Copyright Wiley-VCH Verlag GmbH & Co. KGaA. Reproduced with permission.

For a selection of *n*-type interfaces, Table 2.2 again represents the change in response for 1 ppm of the different gases before and after depositing the metal oxide nanoparticles. Equation 2.7 was used to generate the numbers in Table 2.3.

Table 2.4 reveals other important aspects of the interaction of the cataloged analytes with a metal oxide decorated extrinsic semiconductor surface. As Figure 2.4 demonstrates, amphoteric NO represents a weak acid, which extracts electrons from an *n*-type PSi interface leading to

TABLE 2.3   Relative Increase or Decrease in Resistance (Decrease or Increase in Conductance) of TiO$_2$, SnO$_x$, NiO, Cu$_x$O, and Gold Clustered Oxide, Au$_x$O, Treated *n*-Type PSi Interfaces

	TiO$_2$	SnO$_2$	NiO	Cu$_x$O	Au$_x$O
NO	−12*	−2*	4	1.2	1.5–2
NO$_2$	0.75	0.5**	(0.9–1)	1	1.5–2**
NH$_3^*$	(3.5–4)	2.5	1.5	2	3
PH$_3$	2–2.5				

Source: Gole J.L. et al.: *ChemPhysChem.* 2012. **13**. 549. Copyright Wiley-VCH Verlag GmbH & Co. KGaA. Reproduced with permission.

Note: The table constitutes a response matrix for the gases NO, NO$_2$, and NH$_3$.

* indicates decrease in resistance with analyte exposure; ** indicates initial response. See text for discussion.

TABLE 2.4   Summary of Relative Oxynitride Treated versus Oxide Responses for the Interactions of NO and NH$_3$ with PSi, TiO$_2$ Treated PSi, SnO$_x$ Treated PSi, NiO Treated PSi, and CuO$_x$ Treated PSi

	NO	NH$_3$
PSi nitrated 15 s	Weaker	Stronger
PSi nitration 1 h	–	Very strong
TiO$_2$	Weaker	Weaker
SnO$_x$	Weaker	Weaker
NiO	Weaker	Stronger
CuO$_x$	Stronger	Stronger

Source: Laminack W.I. and Gole J.L.: *ChemPhysChem.* 2014. **15**. 2473–2484. Copyright Wiley-VCH Verlag GmbH & Co. KGaA. Reproduced with permission.

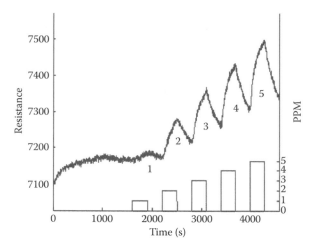

FIGURE 2.4   Response of *n*-type PSi micro/nanostructured interface to 1–5 ppm NO. NO was pulsed onto the PSi interface with a 300-s half-cycle followed by a 300-s half-cycle UHP cleaning. (Gole J.L., Goude E.C., and Laminack W.: *ChemPhysChem*. 2012. **13**. 549–561. Copyright Wiley-VCH Verlag GmbH & Co. KGaA. Reproduced with permission.)

an increase in resistance. However, if the surface is treated with the strong acids $TiO_2 > SnO_2$, the process of electron extraction is completely reversed (Gole et al. 2012), as demonstrated for TiO_2 in Figure 2.5. Electrons are removed from NO dramatically increasing conductance. Similar changes can be observed, as a function of TiO_2 interface deposition, with NO_2. Here, at first NO_2, as a moderately strong acid, extracts electrons from a lightly TiO_2 deposited PSi interface. With increased TiO_2 concentration, dynamic effects come into play as the direction of electron transfer fluctuates as a function of analyte concentration. At even higher TiO_2 concentration, the TiO_2 decorated surface dominates NO_2 and the electron transduction process reverses (Laminack et al. 2012).

By evaluating the reversible interaction of a given analyte with the nanostructure deposited metal oxides as outlined above, it is possible to construct the acid/base interaction diagram depicted in Figure 2.5. The IHSAB principle evaluates the interaction of surface dominating metal oxide sites. The observed responses should be consistent with any previous study in which nano-structured metal oxide clusters are formed on a substrate surface. Baratto et al. (2000) studied gold catalyzed PSi for NO_x sensing. These authors attempted to form thin gold films finding that the gold did not form a continuous layer but rather clusters penetrating into the pores of *n*-type porous silicon. For both NO and NO_2, these authors observed a dynamic electrical response for 1, 2, 4, and 5 ppm NO_x at 20% relative humidity and room temperature, which demonstrated a decrease of conductance (~increase in resistance). This observation is completely consistent with expectations based on the IHSAB model.

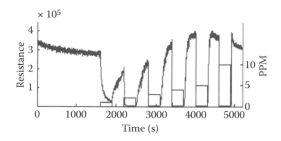

FIGURE 2.5   Comparison of responses for 1, 2, 3, 4, 5, and 10 ppm NO for an *n*-type PSi inter-face treated with TiO_2. (Gole J.L., Goude E.C., and Laminack W.: *ChemPhysChem*. 2012. **13**. 549–561. Copyright Wiley-VCH Verlag GmbH & Co. KGaA. Reproduced with permission.)

2.3 *IN SITU* CONVERSION OF METAL OXIDE SITES VIA DIRECT NITRATION AND SULFUR GROUP FUNCTIONALIZATION

There are further important attributes to this approach to interface modification and the development of an initial materials positioning diagram (Figure 2.6) dictated by the IHSAB concept. Not only is it feasible to greatly expand the metal oxide database, but also extensive evidence has been obtained for the ready *in situ* transformation of the metal oxides to their corresponding oxynitrides and functionalized oxysulfides (Laminack and Gole 2013b). This apparent ability for *in situ* transformation at the nanoscale not only enhances the array of distinct responses, which can be developed and extended to form materials sensitivity matrices for a given analyte, but also provides a route to the conversion of acidic to basic sites. Thus, the generation of the material selection tables, with extrapolation, suggests an efficient means to form basic catalytic sites as well as a potentially simpler way of forming frustrated Lewis acid/base pairs. The degree of nitridation can be used to introduce progressively increasing site basicity at the nanoscale (Gole and Laminack 2012). The *in situ* formation of the oxynitrides shifts the transformed oxides toward the soft acid side of Figure 2.6, which adds a notable flexibility to the materials selectivity table.

Figure 2.7 demonstrates that while the strong (hard) acid TiO_2 increases the sensitivity of an untreated *n*-type PSi interface, the oxynitride, $TiO_{2-x}N_x$, once formed through *in situ* treatment of the TiO_2 deposited surface has gained considerable basic character as evidenced by a considerably decreased conductometric response to NH_3. The observed results are consistent with the observed effect of *in situ* nitridation as it modifies the response of the sensor interface within the IHSAB format. The data in Figure 2.7 compares the response of an untreated *n*-type PSi interface, measuring 2 mm by 5 mm, upon exposure to 2–10 and 20 ppm of NH_3, that for the interface treated with a deposition of "acidic" TiO_2 nanostructures, and this same interface where the deposited nanostructures are converted *in situ* from TiO_2 to the more basic $TiO_{2-x}N_x$. TiO_2, as a strong acid, enhances the capture of electrons, transferring these electrons to increase conductance (decrease resistance) relative to the undecorated PSi interface. The more basic oxynitride does not facilitate electron transduction as efficiently and the sensor response corresponds to a conductance decrease relative to the untreated interface. The *in situ* nitridation of TiO_2 shifts the

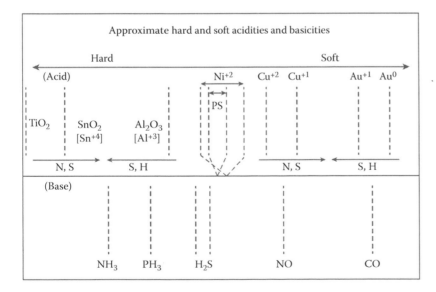

FIGURE 2.6   Hard and soft acidities and basicities estimated based on resistance change relative to a *p*- and *n*-type PSi interface. Acidic metal oxides decorating a semiconductor interface can be modified through *in situ* treatment with nitrogen or sulfur compounds, decreasing their Lewis acidity. In addition, the thiols contribute a hydrogen that promotes an increase in Lewis acidity. The analytes remain as positioned. A horizontal line separates the metal oxides used to modify the semiconductor interface (above) and the analytes (below) in the figure. (From Laminack W.I. and Gole, J.L. (2013) *Nanomaterials*, **3**, 469–485, Creative Commons License. With permission.)

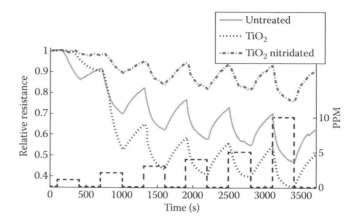

FIGURE 2.7   Response corresponding to decreasing resistance as NH_3 contributes electrons to an untreated PSi, TiO_2, and $TiO_{2-x}N_x$ treated PS interfaces. The $TiO_{2-x}N_x$ treated interface is basic relative to the PS and TiO_2 treated PS acidic sites. (From Laminack W.I. and Gole, J.L. 2013. *Nanomaterials*, **3**, 469–485, Creative Commons License. With permission.)

nature of this metal oxide nanostructure toward the soft acid side of Figure 2.6, closer to ammonia. The IHSAB principle dictates that the response of the $TiO_{2-x}N_x$ interface should decrease relative to TiO_2 as it indeed does. However, the nitridation process does not simply increase the basic character of the nanostructure surfaces but modifies the molecular electronic structure and interaction consistent with the IHSAB principle. This means that the sensitivity of the weaker metal oxides is enhanced by nitridation. The IHSAB principle dictates that the orbital matchup with NH_3 is enhanced and therefore the response of the $TiO_{2-x}N_x$ interface should decrease relative to TiO_2 as it does (Gole and Laminack 2013; Laminack and Gole 2013b, 2014).

Similar decreases in the observed sensor response are observed as nitridated SnO_2 interacts with NH_3 and NO, where the molecular orbital makeup is now more closely aligned for the nitridated interface. Initial results for the nitridation of NiO also lead to a decrease in response for NO; however, the reversible response for interaction with NH_3 increases (Figure 2.8). Nitridation shifts the NiO nanostructures away from ammonia and toward NO. This corresponds to a greater orbital mismatch with NH_3 versus an increase in the HOMO (donor)-LUMO (accepter) orbital matchup and interaction for NO.

Figure 2.9 presents comparable data as 1–10 ppm of ammonia interacts with a copper oxide treated *n*-type PSi interface converting this interface *in situ* to a copper oxynitride interface. Again, the nitridation of Cu_xO forms sites that are more basic and shifts the response of the modified nanostructures further to the soft acid side of Figure 2.6. It is tempting to hypothesize that the formation of the oxynitride should simply increase the basicity of the nanostructure surface and thus should decrease the response to NH_3. However, this does not occur. The nitridated copper oxide is shifted further to the soft acid side of ammonia in Figure 2.6, dictating a greater mismatch of molecular orbitals. The IHSAB principle suggests, counter to intuition, that the response of the *in situ* treated nitridated copper oxide interface should increase relative to that of Cu_xO, precisely as is observed. In Figure 2.7, NO is positioned directly under the copper oxides. Nitridation shifts the copper oxides to the soft acid side and away from NO, leading to an increase in molecular orbital mismatch and the reversible response of the oxynitride to NO.

There are different ways to control the size of the interaction of those molecules that are to be sensed with variably doped metal oxide-sensing interfaces. If the orbital orientation at the surface is not correctly configured, there can be little binding with the lone pairs of the incoming molecules. In addition, a combination of molecular and surface steric effects could also block the interaction at the surface by preventing orbital overlap. If the donor orbital energy (highest occupied molecular orbital, HOMO) is not well matched with the acceptor (lowest unoccupied molecular orbital, LUMO), then the interaction will be weak. As the HOMO (donor) LUMO (acceptor) energy gap decreases, there can be more charge transfer between the molecule and the sensor interface, leading to a stronger Lewis acid–base interaction. For Lewis acid–base bonds, the donor retains the electron pair, a prototypical example being BH_3NH_3, with a B−N bond dissociation energy (BDE)

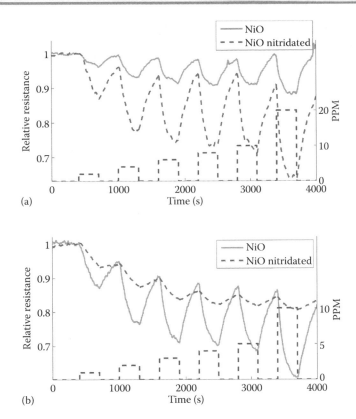

FIGURE 2.8   (a) Relative resistance response to NH_3 of NiO and nitridated NiO treated *n*-type PS and (b) relative resistance response to NO of NiO and nitridated NiO treated *n*-type PSi. (Laminack W.I. and Gole J.L.: *ChemPhysChem*. 2014. **15**. 2473–2484. Copyright Wiley-VCH Verlag GmbH & Co. KGaA. Reproduced with permission.)

of 26 kcal/mol (Dixon and Gutowski 2005). At the other extreme is the interaction of an anion and a cation forming an ionic bond with a much larger BDE. If the electrons are fully shared, leading to the formation of a covalent bond, this can also lead to a large BDE. The IHSAB principle is, in large part, based on controlling the size of the Lewis acid–base bond dissociation energy.

Within the framework of integral nanostructured island sites, the behavior of the interfaces that are generated appears to be well represented by the newly developing IHSAB model. Here, we have begun to expand the versatility inherent to the metal oxides and the range of sensor response

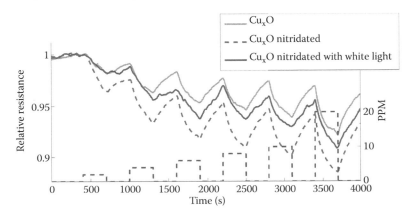

FIGURE 2.9   Response corresponding to decreasing resistance as NH_3 contributes electrons to a Cu_xO treated PSi (gray) and nitridated Cu_xO nanostructure treated PSi (dashed). The nitridated Cu_xO treated interface is basic relative to the PSi and Cu_xO treated PSi acidic sites. This interface becomes more acidic on exposure to white light (black). (Laminack W.I. and Gole J.L.: *ChemPhysChem*. 2014. **15**. 2473–2484. Copyright Wiley-VCH Verlag GmbH & Co. KGaA. Reproduced with permission.)

through *in situ* amination, converting to the more basic oxynitrides, or through *in situ* interaction with the basic sulfides or acidic thiols (Laminack and Gole 2012; Laminack et al. 2015) to produce the more basic oxysulfides or their corresponding hydrogen-functionalized acidic counterparts. It is significant that these results can be obtained with a simple *in situ* treatment at the nanoscale.

The relative response matrices of the nitridated metal oxide deposited PSi interfaces change significantly relative to those for the nanostructured metal oxides. Table 2.4 summarizes the effects of nitridation on TiO_2, SnO_x, NiO, and Cu_xO decorated PSi interfaces. Here, we summarize whether nitridation produces a weaker or stronger response than that of the untreated metal oxide. The interaction of a given analyte with the four acids we have studied is summarized by a group of responses, which are distinctly different from those of an untreated PS interface. The relative responses of the decorated metal oxide interfaces and their corresponding nitridated counterparts are summarized in Table 2.4. The resistance change per base resistance for the nanostructure deposited versus untreated PS interfaces is again given by Equation 2.8. We perform relative measurements with the identical previously characterized untreated sensor when evaluating the changes resulting as nanoparticles are deposited to the PSi interface and as the decorated interfaces are nitridated.

The data in Tables 2.5 and 2.6 demonstrate that the response matrices for the metal oxides and their nitridated counterparts are distinct and that the nitridation process creates additional distinct response matrix elements that can be used to test for a given analyte. These response matrix elements are determined by the acid/base strength of the analyte relative to that of the nanostructure deposited interface.

TABLE 2.5 Relative Responses for Nitridated Metal Oxides in Comparison to Nanostructured Metal Oxides for Conductometric Responses to NO and NH$_3$

	NO	NH$_3$
TiO_2	0.025	0.11
SnO_x	0.28	0.14
NiO	0.5	3
CuO_x	2	1.5

Source: Laminack W.I. and Gole J.L.: *ChemPhysChem*. 2014. **15**. 2473–2484. Copyright Wiley-VCH Verlag GmbH & Co. KGaA. Reproduced with permission.

TABLE 2.6 Relative Responses of Nitridated Metal Oxides versus Untreated PSi for Conductometric Responses to NO and NH$_3$

	NO	NH$_3$
PSi nitrated 15 sec	0.5	2
PSi hour nitration	–	2.2
TiO_2	−0.25*	0.25
SnOx	−0.5*	0.33
NiO	2	4.5
CuO_x	2.4	4

Source: Laminack W.I. and Gole J.L.: *ChemPhysChem*. 2014. **15**. 2473–2484. Copyright Wiley-VCH Verlag GmbH & Co. KGaA. Reproduced with permission.

* The negative numbers correspond to the reversal of the signal for the untreated PSi sensor.

The dependence is dictated by the manner in which the nitridation process modifies the molecular orbital makeup of the metal oxides as it adjusts the HOMO (donor)–LUMO (acceptor) and enhances or diminishes the molecular orbital matchup of the analytes and decorating nanostructures.

The IHSAB principle can be used as an important distinguishing principle of sensor response and the transformation from electron transduction to chemisorption. The simplicity of the *in situ* nitridation process (Gole and Lewis 2007; Gole and Ozdemir 2010; Gole and Laminack 2013; Laminack and Gole 2013b, 2014) provides an important means of enhancing interface modification and selection. The concept of *in situ* nitridation has been extended to functionalization to form the substituted oxysulfides. This suggests the possibility of a readily applied novel and general approach to the formation of sulfidated interfaces, which may find application in the functionalization of biomolecules.

We suggest that it is appropriate to expand the selective deposition and *in situ* modification of nanostructured materials to create inexpensive microfabricated sensor platforms and develop materials selection tables built on the IHSAB model. It is desirable to expand the range of interactions forming materials sensitivity matrices for a given analyte. This will also enhance the capacity to sense a wider range of analytes and their mixtures.

2.4 LIGHT-ENHANCED CONDUCTOMETRIC SENSING

Laminack and Gole (2013a) took advantage of the visible-light absorbing properties of nitridated TiO_2 nanoparticles supported on a PSi interface to create a solar-pumped conductometric gas sensor. TiO_2 and $TiO_{2-x}N_x$ are fractionally deposited to a nanopore-coated microporous silicon sensor, fabricated by electrochemical etching of *n*-type silicon wafers (Gole et al. 2012) and tested in the presence of a range of low ppm concentrations of NH_3. The resistance of the conductometric sensor decreases on contact with the analyte gas because, as a base, NH_3 donates electrons to the *n*-type PSi, increasing the majority charge carriers. Prior to deposition of the titanium dioxide nanoparticles, both white light and UV light were shown to have no effect on the resistance response of the PSi sensor to NH_3 (Laminack and Gole 2013a). As described earlier in this chapter, deposition of the hard acid TiO_2 shifted the PSi sensor interface to the hard acid side of the IHSAB scale and increased the response to NH_3. *In situ* conversion of the metal oxide nanoparticles to $TiO_{2-x}N_x$ caused the interface to obtain a more basic character, reducing the capture of the NH_3 donated electrons and thus reducing sensor response. Upon irradiation with UV light, the response of both the TiO_2 and $TiO_{2-x}N_x$ decorated PS sensors improved by over 100%. As expected, only the $TiO_{2-x}N_x$ modified PSi sensor was improved in the presence of visible light (Laminack and Gole 2013a). Laminack and Gole (2013a) explain that the UV and white light excitation causes the $TiO_{2-x}N_x$ to become more acidic, enhancing the electron withdrawing power of the PSi interface with respect to NH_3 and thus photocatalytically enhancing the sensor response.

The approach that we describe is not only applicable for the optimization of all sensor systems based on PSi but also to any extrinsic semiconductor in which one can generate the configuration outlined in Figure 2.1. We require that one is able to clearly produce a substrate on which electronically separated nanostructures are deposited. Several metal oxide sensor configurations in the literature can benefit from or may have already applied a near variant of the approach that we describe here. We refer to the nanostructured sensor configurations that may well have clusters on the surface interface (Comini 2006) and show effects, which are closely aligned with the IHSAB concept. However, in order to make this claim, the authors must have provided sufficient evidence that the clusters on the surface, at present, are clearly separated electronically on a distinct substrate. In the examples that are given next, especially the nanowires, the authors clearly state that the surfaces are deposited with separated island sites. Baratto et al. (2000) clearly indicate that their gold clusters are well separated and on a PSi substrate. The mutual electronic isolation of the surface clusters (nanostructured island sites) on PSi is important as is the question of a heterogeneous versus homogeneous environment. If the substrate and the surface clusters are identical, the question of establishing a definite IHSAB framework becomes more difficult. One must have well-separated nanostructured metal oxide interaction in an electron transduction mode with the substrate. This is distinct from a homogeneous jumble of overlapping nanowires or clustered nanostructures that make up a surface, or any coating technology. Thus, we suggest that several previous sensor studies may have observed responses consistent with the IHSAB concept, but we choose to interpret only a few clear examples.

2.5 APPLICATION AND EXTENSION OF THE IHSAB CONCEPT TO ALTERNATE EXTRINSIC SEMICONDUCTORS

We note that there is nothing unique about the silicon substrate for the IHSAB concept. There are alternate extrinsic semiconductors (e.g., GaP, InP, CdTe) onto which a porous nanostructure coated microstructure can be generated (Lévy-Clément 2007). In theory, the PSi structures, devices, and methods described in the following discussion can be replaced with these surrogate porous semiconductors.

2.6 EXTENSION OF THE IHSAB CONCEPT TO NANOWIRE CONFIGURATIONS

The IHSAB concept can also predict the response of extrinsic semiconductor sensors in a nanowire configuration if these nanowires are decorated with electrically noninteracting nanoparticles. In one case, Li et al. (2010) placed Pd on SnO_2 nanowires, seen in Figure 2.10, to enhance the sensing to H_2S. This is in line with the IHSAB concept because Pd is a soft acid and H_2S is a moderate base. The orbital mismatch will enhance the electron transduction interaction on the surface. Thus, the nanowire becomes more sensitive. This response would be less if the doping material were Ni or other moderately hard acids with which the orbital mismatch would be smaller. Since the SnO_2 is considered an n-type system, the interaction with the base H_2S will increase the major carriers (electrons) and thus the sensor resistance will decrease. This predicted response matches the group's results (Li et al. 2010).

Kim et al. (2010) have also studied nanowires for sensing applications. They used In_2O_3 wires to detect the presence of O_2. When they added small amounts of Pt forming islands instead of coating the entire wire with Pt as seen in Figure 2.11, the response increased significantly. Since O_2 is a moderate base, the soft acid platinum has an orbital mismatch with O_2 increasing electron transduction and the O_2 signal. In addition, since the In_2O_3 is a p-type semiconductor, the increased electrons decrease the number of holes causing a resistance increase when sensing O_2 (Kim et al. 2010).

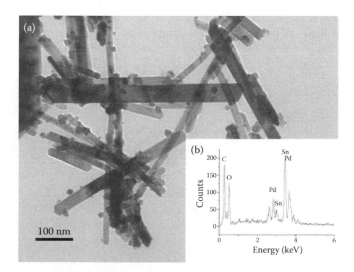

FIGURE 2.10   (a) Transmission electron microscopy (TEM) image and (b) energy-dispersive X-ray spectroscopy (EDS) (inset) of Pd nanoparticles deposited to SnO_2 nanowires. (Reproduced from Talanta, Li H., Xu J., Zhu Y., Chen X., and Xiang Q., **82**, 458–463, Copyright 2010, with permission from Elsevier.)

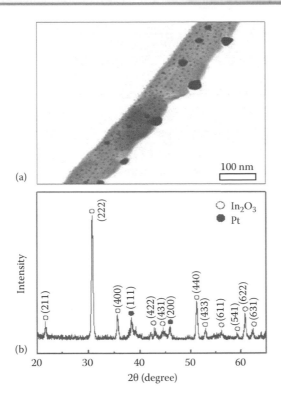

FIGURE 2.11 (a) Low-magnification transmission electron microscopy (TEM) image of an 800°C-annealed, core-shell nanowire. (b) Corresponding X-ray diffraction (XRD) spectrum. (From Kim S.S. et al., *Nanotechnology* **21**(41), 415502, 2010. Copyright IOP Publishing. With permission.)

Finally, we note that Kolmakov et al. (2005) studied the response of a Pd-functionalized SnO_2 nanostructure to sequential oxygen and hydrogen pulses at 473 and 543 K (Figure 2.12). There was a strong response to hydrogen at both temperatures where the resistance increased. This resistance increase was enhanced by the Pd. No response was observed for oxygen at 473 K, but a drop in resistance was monitored at 543 K. These results can be readily explained within

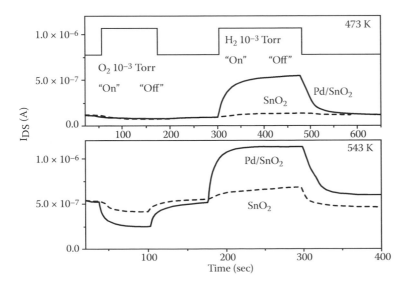

FIGURE 2.12 (a) Schematic view of the formation of electron-depleted regions beneath and in the immediate vicinity of two Pd nanoparticles. (b) Response of a pristine (dashed line) and Pd-functionalized (solid line) nanostructure to sequential oxygen and hydrogen pulses at 473 K (top pane) and 543 K (bottom). (Reproduced with permission from Kolmakov A., Klenov D.O., LilachY., Stemmer S., and Moskovits M. (2005). *Nano Lett.* **5**(4), 667–673. Copyright 2005 American Chemical Society.)

the IHSAB model. The hard acid hydrogen interacts with n-type SnO_2, removing electrons and leading to an increase in resistance. At the higher temperature, O_2^-, a moderate to hard base, is created on the SnO_2 surface. This base provides additional electrons to increase conductance, decreasing the resistance. Again, all interactions are enhanced by the presence of the soft acid Pd nanostructured islands. Since the H_2 has a much larger mismatch in its energy orbital with Pd than with O_2, the relative increase in the signal is seen to be greater as well (Kolmakov et al. 2005).

2.7 EXTENSION OF THE NITRATION CONCEPT TO THE DEVELOPMENT OF BASIC CATALYST SITES AND FRUSTRATED LEWIS ACID/BASE PAIRS

As the nitridation process creates the ability to modify acidic oxides to considerably more basic oxynitrides, the process can be carried out in stages. It is possible to examine several of the metal oxide and oxynitride samples for the qualitative aspects of their surface chemistry, using the methanol decomposition reaction as a probe. The reaction as it has been effectively used to characterize the surface of bifunctional, mixed-metal oxides (Badlani and Wachs 2001; Lee and Wachs 2008). They demonstrated the utility to evaluate redox, acid, and base sites on surface opened reaction manifolds, as described by Equations 2.8 through 2.10, leading to different products.

$$(2.8) \qquad\qquad 2CH_3OH \rightarrow CH_3OCH_3 + H_2O: \text{acid sites}$$

$$(2.9) \qquad\qquad CH_3OH \rightarrow HCHO + H_2: \text{redox sites}$$

$$(2.10) \qquad\qquad CH_3OH \rightarrow CO + 2H_2: \text{base sites (also } CO_2 \text{ production)}$$

Using these probe reactions, it is possible to demonstrate that the nitridation process offers the opportunity to convert the acidic metal oxide acid sites to considerably more basic metal oxynitride surface sites. This can have implications for the development of intermediate acidity, proximal acid-base, and basic sites for use in heterogeneous catalysis. Preliminary work has suggested the existence of acid and base sites present in titania samples (Baker et al. 2015).

Solid acid catalysts have been extensively studied and applied in numerous reactions due to the demand and great progress of petroleum and petrochemical industry in the past 40 years. However, fewer efforts have been made to study solid base catalysts. From a statistical survey made by Tanabe and Hölderich (1999) for industrial processes reported before 1999, the classification of the types of catalysts into solid acid, solid base, and solid acid–base bifunctional catalysts were 103, 10, and 14, respectively. Obviously, the total number of available solid base-related catalysts, including solid base and acid–base bifunctional catalysts, is much less than that of solid acid catalysts. However, many reactions (isomerizations, alkylations, condensations, additions, and cyclizations) are carried out industrially using liquid bases as homogeneous catalysts. The replacement of liquid bases by solid base catalysts, which are nonstoichiometric, noncorrosive, and reusable represents a significant material savings. This makes them attractive for environmental and economic reasons (Hattori 1995). Some potential applications can be summarized as follows.

2.7.1 AMINE DOPED OXIDES

A low-temperature amine treatment can also lead to changes in the surface properties (site chemistry and surface area) of faujasite-Y, alternate zeolites, titania, and alternate metal oxide (e.g., nitridated copper oxides) samples (Baker et al. 2015) with potential catalytic applications as these oxides are converted to their more basic oxynitrides. Through treatment with triethylamine, which can be extended to additional alkyl amines (e.g., n-ethyl amine, n-diethyl amine) and aryl amines (pyridine), it is possible to change the properties of faujasite-Y. One may ask whether these changes can be related to the strength of the bases in terms of both the Brönsted and Lewis scales. One may also ask whether triethylamine will cause changes in the properties of other aluminosilicates, such as ZSM-5, mordenite, zeolite-beta, and so on, and aluminophosphates ($AlPO_4$-5) and silicoaluminophosphates (SAPO$_4$-5). This list of crystalline materials explores (1) the systematic change in pore size/structure (faujasite-Y, ZSM-5, mordenite, zeolite-beta) and (2) the change in T-atoms for crystalline materials having the same pore size (ZSM-5, AlPO$_4$-5, SAPO$_4$-5). Within one family member of crystalline materials (Y-faujasite), it is possible to examine the effect of silica/alumina ratio on the resulting number and strengths of acid and base sites.

2.7.2 FRUSTRATED LEWIS ACID/BASE PAIRS

An extrapolation of the nitridation process suggests that FLPs (Douglas 2008) might be created from simple Lewis base molecular analytes as they interact with high surface area heterogeneous catalysts having Lewis acid sites on their surface. The potential cooperation between Lewis acid and Lewis base functionalities can be explored in traditional acid catalysts, as they are selectively decorated with Lewis bases. The opportunity addressed here is the economically feasible synthesis of high surface area interfaces dominated by having strong surface acid and base sites located near each other, thus enabling the use of heterogeneous catalysts for organic synthesis reactions catalyzed by both acid and base sites, that is, the development of FLPs in a heterogeneous system. The technological advantages of solid catalysts (robustness for operation at high temperatures, lack of corrosion, and ease of separation of products) might then be combined with the advantages of soluble catalysts (e.g., selectivity). Issues with catalyst degradation might also be eliminated. The decarboxylative/dehydrative coupling of organic acids to form ketones and substituted benzenes is a class of reactions that would benefit from a high surface area catalyst having both acid and base sites.

2.8 OUTLOOK

Within the IHSAB/HSAB framework, we have exemplified the modification of the porous silicon interface with nanostructures, which represent a selection of several metal oxides that can be produced and used to modify the PSi extrinsic semiconductor framework. As we have noted, the processes are not limited to PSi but can be extended to any interface to which electron transfer is facilitated. Any metal oxide whose synthesis is well established can be selected to produce reversibly interacting nanostructures at the PSi active interfaces. Many, if not the vast majority, of these metal oxides can be deposited using solution-based techniques, eliminating the necessity of complex lithography or self-assembly processing on the PSi. As we have exemplified, selective nanostructured metal oxides can be deposited to the PSi interface using electroless gold, tin, copper, and nickel solutions to form Au$_x$O ($x \gg 1$), SnO$_2$, Cu$_x$O ($x = 1, 2$), and NiO, and by directly depositing nanoalumina, nanotitania, and nanozirconia particles (Figure 2.1). XPS and SEM measurements on several nanostructure deposited sensors demonstrate the Sn-, Ni-, Cu-, and Au-based nanostructured oxide deposits (Ozdemir and Gole 2010), which interact with the PSi substrate but are themselves electronically isolated. The electroless techniques can be extended to the Pt and Pd oxides. In addition, the nitridating process is readily applied at the nanoscale (Laminack and Gole 2012, 2014). In addition, colloidal solutions, in concert with the microporous structure of the PSi, are uniquely suited to incorporate and uniformly disperse metals, metal ions, and select compounds as seed materials to produce a porous oxynitride product, which can be readily deposited to an effective modification of both the PSi and metal oxide sites. Further, if we deposit the oxides of the transition metals including Co^{+2}, Fe^{+2}, and Ni^{+2} within the PSi pore structure, it is possible to enhance sensing using very small magnetic fields (Baker et al. 2014). In fact, these transition metal oxides can be used to produce facile room temperature phase transformations (Gole et al. 2008) when combined with a TiO$_2$ matrix deposited to PSi. In order to augment the nanostructured detection matrices, it is readily possible to extend the PSi interface modification to at least the oxides of Mg^{+2}, Ca^{+2}, B^{+3}, Zn^{+2}, and Ge^{+3}. As well, it is possible to make use of phase transformation and charge transfer (Gole et al. 2008) techniques to produce Fe^{+3} and Ni^{+3} decorated PSi.

The control of the interaction of targeted analytes with a specific material such as decorated PSi and the degree to which this dictates the tailoring of interfaces offers a uniquely defined approach to enable the selection of device materials, and a generalized approach to advanced silicon-based sensor platforms and catalytic microreactors. A unique mechanism that combines the basic tenants of acid/base chemistry and semiconductor physics is relatively simple to follow and implement, potentially leading to cost-effective, deployable silicon-based devices sensitive to the ppb level. This clearly defined prescription for design, as it is carefully quantified and extrapolated to a broad range of applicable interfaces, can provide not only complementarity but also a significant superiority to alternate technologies. In contrast to traditional metal oxide systems, a dual interface is created as the nanostructured islands on PSi and this extrinsic semiconductor act separately but are coupled. The design and its corresponding "open channel" variant add considerable flexibility not possible in a singly or multiply "coated" metal oxide interface. Typical metal oxide sensors (Figure 2.13) have marginal sensitivity, high power requirements, and require operation at *precisely controlled* elevated temperatures. The latter requirement is a drawback for several reasons. First, a power-consuming heating element must be provided with the sensor housing to precisely control the temperature of the sensor element. In large part, this is intimately tied to the correct identification of the gas of interest. Distinguishing one gas from another thus requires that the heating element and sensor be well separated (channel) from the remaining electronics. This means that this configuration can be greatly affected by an impinging combustion or flue gas, rendering difficult the correct identification of gaseous species in an analyte flow. In contrast, the PSi configuration depicted in Figure 2.1 and again in Figure 2.13 operates at room temperature, consumes less power, is far simpler, and does not require the complexity of a system separated sensor/heater configuration. In a heat sunk environment, the PSi sensor is capable of operation in a high temperature gas flow.

Finally, we note that the sensitivity obtained for the conductometric responses exemplified in Figures 2.4, 2.5, and 2.7 through 2.9 are not only greater than that for a typical metal oxide deposition but also compare very favorably with other modes of detection. X-ray photoelectron (XPS) spectra taken for the nitridated (Figures 2.7 through 2.9) and sulfur functionalized PS interfaces are sensitive to 0.1% (Laminack and Gole 2013b; Gole 2015). The typical time frame for the depositions used to obtain the conductometric data in Figures 2.7 through 2.9 is 15–30 s. To obtain the XPS data, we use deposition times that are at least 5 min. This corresponds approximately to an order of magnitude increase in concentration to generate signals in the 0.1% range. This indicates the significant sensitivity of the conductometric sensor responses. The magnitudes of the observed reversible interactions are well described by the developing IHSAB model.

Despite significant effort, the basic science and technology linking sensor design for detection and the transformation of gas phase analytes has not yet been fully developed. Current technology for chemical sensing is characterized by stand-alone sensors or sensor arrays that have specific chemical selectivity and known interferences. The alternative approach is to develop miniature analytical instruments, such as mass-spectrometers, gas chromatography systems, or infrared

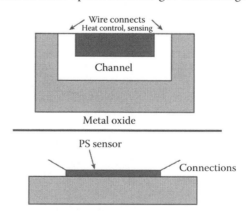

FIGURE 2.13   Comparison of a metal oxide (usually SnO_2 or WO_3) elevated temperature (150–500°C) heat-controlled sensors separated from their electronics by a channel with a heat sink with a PSi sensor operating at room temperature and capable of operation to temperatures in excess of 200°C. (From Gole, J.L. and Laminack W.I. (2013). *Beilstein Journal of Nanotechnology*, **3**, 469–485, Creative Commons License. With permission.)

spectrophotometers. However, these instruments, although portable, are not yet miniaturized to the extent necessary for distributed systems, have high power consumption, require careful maintenance and recalibration, are not always readily used in the field or in harsh operating environments, and they are expensive. Sensor/microreactors (Campbell et al. 2008) developed on PSi within the IHSAB/HSAB framework offer an alternative based on arrays of selective and orthogonal sensors that have the distinct advantage of low power consumption, room temperature operation, and ease of use in a broad range of environments. The proposed PSi-based sensor frameworks hold the additional advantage of operation across a broad temperature range and include reliable and rapid false-positive identification with incorporated fast Fourier transform (FFT) techniques (Lewis et al. 2007; Gole and Laminack 2012). As an example, an improved PSi-based detection system that can simultaneously test for NO, NO_2, and NH_3 in a device that can be maintained in the room of an asthmatic and can be used to detect the rapid increase of NO and the subsequent formation of NO_2 during an asthma attack, allowing the patient to take medication early on. This can save a \$6000 overnight hospital stay.

REFERENCES

Ahlers, S., Kreisl, P., and Müller, G. (2003). A low-power metal-oxide based gas sensing system. In: *Proceedings of 203rd ECS Meeting*, April 28, Paris, France, paper 2873.

Badlani, M. and Wachs, I.E. (2001). Methanol: A "smart" chemical probe molecule. *Catal. Lett.* **75**, 137–149. doi: 10.1023/A:1016715520904.

Baker, C., Laminack, W., Tune, T., and Gole, J.L. (2014). Magnetically induced enhancement of reversibly responding conductometric sensors. *J. Appl. Phys.* **115**(16), 164312. doi: 10.1063/1.4874183.

Baker, C., Gole, J.L., Graham, S., Hu, J., Kenvin, J.C., D'Amico, A., and White, M.G. (2015). Chemical activity of nitrogen-doping of titania and zeolite Y dosed with triethylamine. *Langmuir* (submitted).

Baratto, C., Sberveglieri, G., Comini, E. et al. (2000). Gold-catalysed porous silicon for NO_x sensing. *Sens. Actuators B* **68**, 74–80. doi: 10.1016/S0925-4005(00)00464-0.

Barsan, N., Schweizer-Berberich, M., and Göpel, W. (1999). Fundamental and practical aspects in the design of nanoscaled SnO_2 gas sensors: A status report. *Fresenius' J. Anal. Chem.* **365**, 287–304. doi: 10.1007/s002160051490.

Campbell, J., Corno, J.A., Larsen, N., and Gole, J.L. (2008). Development of porous-silicon-based active microfilters. *J. Electrochem. Soc.* **155**, D128–D132. doi: 10.1149/1.2811868.

Chen, X., Lou, Y., Samia, A.C.S., Burda, C., and Gole, J.L. (2005). Formations of oxynitride as the photocatalytic enhancing site in nitrogen-doped titania nanocatalysts comparisons to a commercial nanopowder. *Adv. Funct. Mater.* **15**, 41–49. doi: 10.1002/adfm.200400184.

Cohen, M.H. and Wasserman, A. (2007). On the foundations of chemical reactivity theory. *J. Phys. Chem. A* **111**, 2229–2242. doi: 10.1021/jp066449h.

Comini, E. (2006). Metal oxide nano-crystals for gas sensing. *Anal. Chim. Acta* **568**, 28–40. doi: 10.1016/j.aca.2005.10.069.

Dixon, D.A. (2010). Private discussions.

Dixon, D.A. and Gutowski, M. (2005). Thermodynamic properties of molecular borane amines and the $[BH_4^-][NH_4^+]$ salt for chemical hydrogen storage systems from *ab Initio* electronic structure theory. *J. Phys. Chem. A* **109**(23), 5129–5135. doi: 10.1021/jp0445627.

Douglas, S.W. (2008). "Frustrated lewis pairs": A concept for new reactivity and catalysis. *Org. Biomol. Chem.* **6**, 1535–1539. doi: 10.1039/b802575b.

Fukui, K., Yonezawa, T., and Shingu, H. (1952). A molecular orbital theory of reactivity in aromatic hydrocarbons. *J. Chem. Phys.* **20**, 722. doi:10.1063/1.1700523.

Geerlings, P., De Proft, F., and Langenaeker, W. (2003). Conceptual density functional theory. *Chem. Rev.* **103**, 179; doi: 10.1021/cr990029p.

Gole, J.L. (2015). Increasing energy efficiency and sensitivity with simple sensor platforms. *Talanta* **132**, 87–95. doi: 10.1016/j.talanta.2014.08.038.

Gole, J.L. and Lewis, S.E. (2007). Porous silicon—Sensors and future applications. In: Kumar S. (ed.) *Nanosilicon*. Elsevier, London, pp. 147–192.

Gole, J.L. and Ozdemir, S. (2010). Nanostructure directed physisorption vs. chemisorption at semiconductor interfaces: The inverse of the hard-soft acid-base (HSAB) concept. *ChemPhysChem* **11**, 2573–2581. doi: 10.1002/cphc.201000245.

Gole, J.L. and Laminack, W.I. (2012). General approach to design and modeling of nanostructure modified semiconductor and nanowire interfaces for sensor and microreactor applications. In: Korotcenkov G. (ed.) *Chemical Sensors: Simulation and Modeling, Vol. 3- Solid State Sensors*. Momentum Press, New York, 87–136.

Gole, J.L. and Laminack, W.I. (2013). Nanostructure-directed chemical sensing: The IHSAB principle and the dynamics of acid/base-interface interaction. *Beilstein J. Nanotechnology* **4**, 20–31. doi: 10.1002/cphc.201100712.

Gole, J.L., Stout, J., Burda, C., Lou, Y., and Chen, X. (2004). Highly efficient formation of visible light tunable TiO$_{2-x}$N$_x$ photocatalysts and their transformation at the nanoscale. *J. Phys. Chem. B* **108**, 1230–1240. doi: 10.1021/jp030843n.

Gole, J.L., Prokes, S.M., and Glembocki, O.J. (2008). Efficient room-temperature conversion of anatase to rutile TiO$_2$ induced by high-spin ion doping. *J. Phys. Chem. C* **112**(6), 1782–1788. doi: 10.1021/jp075557g.

Gole, J.L., Ozdemir, S., and Osburn, T.S. (2010a). Novel concept for the formation of sensitive, selective, rapidly responding conductometric sensors. *ECS Trans.* **33**, 239–244. doi: 10.1149/1.3484127.

Gole, J.L., Ozdemir, S., Prokes, S.M., and Dixon, D.A. (2010b). Active nanostructures at interfaces for photocatalytic reactors and low-power consumption sensors. In: *Symposium O–Multifunctional Nanoparticle Systems-Coupled Behavior and Applications*, Bao, Y., Dattelbaum, A.M., Tracy, J.B. and Yin, Y. (eds.) *MRS Proc.* 1257, O09–04. doi: 10.1557/PROC-1257-O09-04.

Gole, J.L., Goude, E.C., and Laminack, W. (2012). Nanostructure driven analyte-interface electron transduction: A general approach to sensor and microreactor design. *ChemPhysChem* **13**, 549–561.

Hattori, H. (1995). Heterogeneous basic catalysis. *Chem. Rev.* **95**, 537–558. doi: 10.1021/cr00035a005.

Huheey, J.E. (1978). *Inorganic Chemistry*. 2nd ed. Harper and Row, New York.

Jiménez, I., Arbiol, J., Dezanneau, G., Cornet, A., and Morante, J.R. (2003). Crystalline structure, defects and gas sensor response to NO$_2$ and H$_2$S of tungsten trioxide nanopowders. *Sens. Actuators B* **93**, 475–485. doi: 10.1016/S0925-4005(03)00198-9.

Kim, S.S., Park, J.Y., Choi, S. et al. (2010). Significant enhancement of the sensing characteristics of In$_2$O$_3$ nanowires by functionalization with Pt nanoparticles. *Nanotechnology* **21**(41), 415502. doi: 10.1088/0957-4484/21/41/415502.

Kolmakov, A., Klenov, D.O., Lilach, Y., Stemmer, S., and Moskovits, M. (2005). Enhanced gas sensing by individual SnO$_2$ nanowires and nanobelts functionalized with Pd catalyst particles. *Nano Lett.* **5**(4), 667–673. doi: 10.1021/nl050082v.

Kottke, P.A., Federov, A.G., and Gole, J.L. (2009). Multiscale mass transport in porous silicon gas sensors. In: Schlesinger M. (ed.) *Modern Aspects of Electrochemistry*, Vol. 43. Springer, New York, 139–168.

Kumar, S., Fedorov, A.G., and Gole, J.L. (2005). Photodegredation of ethylene using visible-light responsive surfaces prepared from titania nanoparticle slurries. *Appl. Catal. B: Env.* **57**, 93–107. doi: 10.1016/j.apcatb.2004.10.012.

Laminack, W.I. and Gole, J.L. (2013a). Light enhanced electron transduction and amplified sensing at a nanostructure modified semiconductor interface. *Adv. Funct. Mater.* **23**(47), 5916–5924. doi: 10.1002/adfm.201301250.

Laminack, W.I. and Gole, J.L. (2013b). Nanostructure directed chemical sensing: The IHSAB principle and the effect of nitrogen, and sulfur functionalization on metal oxide decorated interface response. *Nanomaterials* **3**, 469–485. doi:10.3390/nano3030469.

Laminack, W.I. and Gole, J.L. (2014). Direct in-situ nitridation of nanostructured metal oxide deposited semiconductor interfaces: Tuning the response of reversibly interacting sensor sites. *ChemPhysChem* **15**, 2473–2484. doi:10.1002/cphc.201402108.

Laminack, W.I., Pouse, N., and Gole, J. L. (2012). The dynamic interaction of NO$_2$ with a nanostructure modified porous silicon matrix: The competition for donor level electrons. *ECS J. Solid State Sci. Tech.* **1**, Q25–Q34. doi: 10.1149/2.002202jss.

Laminack, W.I., Baker, C., and Gole, J.L. (2015). Sulphur-H$_z$(CH$_x$)$_y$(z = 0,1) functionalized metal oxide nanostructure decorated interfaces: Evidence of Lewis base and Brönsted acid sites—Influence on chemical sensing. *J. Mater. Chem. Phys.* **160**, 20–31. doi: 10.1016/j.matchemphys.2015.03.070.

Lee, E.L. and Wachs, I.E. (2008). Surface chemistry and reactivity of well-defined multilayered supported M$_1$O$_x$/M$_2$O$_x$/SiO$_2$ catalysts. *J. Catal.* **258**, 103–110. doi: 10.1016/j.jcat.2008.06.002.

LeGore, L.J., Lad, R.J., Moulzolf, S.C., Vetelino, J.F., Frederick, B.G., and Kenik, E.A. (2002). Defects and morphology of tungsten trioxide thin films. *Thin Solid Films* **406**, 79–86. doi: 10.1016/S0040-6090(02)00047-0.

Lévy-Clément, C. (2007). Macroporous microstructures including silicon. In: *Encyclopedia of Electrochemistry*. Wiley-VCH Verlag, Weinheim. doi: 10.1002/9783527610426.bard060302.

Lewis, S.E., DeBoer, J.R., Gole, J.L., and Hesketh, P.J. (2005). Sensitive, selective, and analytical improvements to a porous silicon gas sensor. *Sens. Actuators B* **110**, 54–65. doi: 10.1016/j.snb.2005.01.014.

Lewis S.E., DeBoer J.R., and Gole J.L. (2007). A pulsed system frequency analysis for device characterization and experimental design: Application to porous silicon sensors and extension. *Sens. Actuators B* **122**(1), 20–29. doi: 10.1016/j.snb.2006.04.113.

Li, H., Xu, J., Zhu, Y., Chen, X., and Xiang, Q. (2010). Enhanced gas sensing by assembling Pd nanoparticles onto the surface of SnO$_2$ nanowires. *Talanta* **82**, 458–463. doi: 10.1016/j.talanta.2010.04.053.

Morrison, S.R. (1987). Selectivity in semiconductor gas sensors. *Sens. Actuators* **12**, 425–440. doi: 10.1016/0250-6874(87)80061-6.

Moulzolf, S.C., Ding, S., and Lad, R.J. (2001). Stoichiometry and microstructure effects on tungsten oxide chemiresistive films. *Sens. Actuators B* **77**, 375–382. doi: 10.1016/S0925-4005(01)00757-2.

Mueller, G., Friedberger, A., Kreisl, P., Ahlers, S., Schulz, O., and Becker, T. (2003). A MEMS toolkit for metal-oxide-based gas sensing systems. *Thin Solid Films* **436**, 34–45. doi: 10.1016/S0040-6090(03)00523-6.

Ogden, A., Corno, J.A., Hong, J.I., Fedorov, A., and Gole, J.L. (2008). Maintaining particle size in the transformation of anatase to rutile titania nanostructures. *J. Phys. Chem. Solids* **69**, 2898-2906. doi: 10.1016/j.jpcs.2008.07.016.

Ozdemir, S. and Gole, J.L. (2010). A phosphine detection matrix using nanostructure modified porous silicon gas sensors. *Sens. Actuators B* **151**(1), 274–280. doi: 10.1016/j.snb.2010.08.016.

Parr, R.G. and Yang, W. (1989). *Density Functional Theory of Atoms and Molecules*. Oxford University Press, New York.

Parr, R.G., Donnelly, R.A., Levy, M., and Palke, W.E. (1978). Electronegativity: The density functional viewpoint. *J. Chem. Phys.* **68**, 3801. doi: 10.1063/1.436185.

Pearson, R.G. (1963) Hard and soft acids and bases. *J. Am. Chem. Soc.* **85**, 3533–3539. doi: 10.1021/ja00905a001.

Pearson, R.G. (1988). Absolute electronegativity and hardness: Application to inorganic chemistry. *Inorg. Chem.* **27**, 734–740. doi: 10.1021/ic00277a030.

Pearson, R.G. (1990). Hard and soft acids and bases—The evolution of a chemical concept. *Coord. Chem. Rev.* **100**, 403–425. doi: 10.1016/0010-8545(90)85016-L.

Pearson, R.G. (1997). *Chemical Hardness.* John Wiley VCH, Weinheim.

Pearson, R.G. (2005). Chemical hardness and density functional theory. *J. Chem. Sci.* **117**(5), 369–377. doi: 10.1007/BF02708340.

Politzer, P. (1987). A relationship between the charge capacity and the hardness of neutral atoms and groups. *J. Chem. Phys.* **86**, 1072. doi:10.1063/1.452296.

Ponce, M., Aldao, C., and Castro, M. (2003). Influence of particle size on the conductance of SnO_2 thick film. *J. Eur. Ceram. Soc.* **23**(12), 2105–2111. doi: 10.1016/S0955-2219(03)00037-2.

Rani, S., Roy, S.C., and Bhatnagar, M. (2007). Effect of Fe doping on the gas sensing properties of nanocrystalline SnO_2 thin films. *Sens. Actuators B* **122**, 204–210. doi: 10.1016/j.snb.2006.05.032.

Rothschild, A. and Komem, Y. (2004). The effect of grain size on the sensitivity of nanocrystalline metal-oxide gas sensors. *J. Appl. Phys.* **95**, 6374–6380. doi: 10.1063/1.1728314.

Schechter, I., Ben-Chorin, M., and Kux, A. (1995). Gas sensing properties of porous silicon, *Anal. Chem.* **67**, 3727. doi: 10.1021/ac00116a018.

Tanabe, K. and Hölderich, W.F. (1999). Industrial application of solid acid–base catalysts. *Appl. Catal. A* **181**, 399–434. doi: 10.1016/S0926-860X(98)00397-4.

Watson, J. (1994). The stannic oxide gas sensor. *Sensor Rev.* **14**, 20–23. doi: 10.1108/EUM0000000004248.

Yamazoe, N. and Shimanoe, K. (2007). Overview of gas sensor technology. In: Aswal D.K. and Gupta S.K. (eds.) *Science and Technology of Chemiresistor Gas Sensors.* Nova Science Publishers Inc., New York, pp. 1–31.

Yang, W. and Parr, R.G. (1985). Hardness, softness, and the Fukui function in the electronic theory of metals and catalysis. *Proc. Natl. Acad. Sci. USA* **82**, 6723–6726.

Yang, W., Parr, R.G., and Pucci, R. (1984). Electron density, Kohn– Sham frontier orbitals, and Fukui functions. J. Chem. Phys. **81**, 2862–2863. doi: 10.1063/1.447964.

Yoon, D.H. and Choi, G.M. (1997). Microstructure and CO gas sensing properties of porous ZnO produced by starch addition. *Sens. Actuators B* **45**, 251–257. doi: 10.1016/S0925-4005(97)00316-X.

Zhan, C.G., Nichols, J.A., and Dixon, D.A. (2003). Ionization potential, electron affinity, electronegativity, hardness, and electron excitation energy: Molecular properties from density functional theory orbital energies. *J. Phys. Chem. A* **107**, 4184–4195. doi: 10.1021/jp0225774.

Porous Silicon-Based Optical Chemical Sensors

3

Luca De Stefano, Ilaria Rea, Alessandro Caliò, Jane Politi,
Monica Terracciano, and Ghenadii Korotcenkov

CONTENTS

3.1	Introduction	70
3.2	Different Approaches in Porous Silicon-Based Optical Chemical Sensors Designing	70
3.3	Porous Silicon Optical Transducers for Reflectivity-Based Gas and Vapor Sensors	74
	3.3.1 Fabry-Pérot Interferometer	75
	3.3.2 Bragg Mirror	75
	3.3.3 Optical Microcavity	76
	3.3.4 Photonic Quasi-Crystal	78
	3.3.5 Rugate Filter	78
3.4	Gas and Vapor Monitoring by Optical Reflectivity: A Case Study	79
3.5	Examples of Optical PSi-Based Chemical Sensors Realization	83
3.6	Integrated Porous Silicon Microsystems for Optical Sensing	85
3.7	Approaches to Optimization of Porous Silicon-Based Optical Sensors	87
	3.7.1 Sensitivity	87
	3.7.2 Selectivity	88
	3.7.3 Stability	90
3.8	Conclusions and Future Trends	91
References		91

3.1 INTRODUCTION

Gas and vapor sensing is an important task in the industrial field, both for production and safety, but this item is relevant also in social ground: from soil extraction to terrorism surveillance, there is always a volatile substance to be monitored and quantified. There are many transducing methods to detect gaseous substances and the most common is the so-called electrical nose: an array of gas sensitive semiconductors, such as ZnO, WO_3, or other composite oxides, generates a current/voltage change on exposure to pure or mixed substances, and complex software analyzes the sensor response and quantifies the atmosphere composition. Even if it is a very diffuse and powerful device, the electronic nose and, in general, all the electrical gas sensors, suffers from several limitations. First, the oxide semiconductors are sensitive only at high temperature, so these devices are power consuming; moreover, the most severe limit is that electric power and high temperature cannot be used at all for monitoring explosive or flammable substances. From this point of view, optical sensors easily overcome these difficulties: light is contactless and can be generated with very low consumption (it is the case of light emitting diodes [LED]). In addition, optical sensors do not require electric contacts that may cause explosions or fire during sample preparation. Moreover, optical sensing can be remote, assuring total safety for instruments and operators. Therefore, in the presence of a harsh environment, optical sensing is by far the best technique for *in situ* monitoring of hazardous compounds. Response times can also be very fast.

Gas and vapor sensing optical systems gained great advantage by the advent of nanoporous materials, which are used in many academic research and industrial applications just for their particular morphology: the presence of nanometric pores can be exploited for molecule separation, isolation and purification, ion exchange, enhancing chemical reaction by catalysis, and for sensing purposes. The ability to adsorb and interact with chemical and biological species strongly depends on their large interior surfaces and on the sized pore space. Many nanostructured porous materials can be easily found in nature, just by mining proper soils, such as zeolites, carbon, and diatomite, which are low-cost and do not require industrial production equipment. On the other hand, design and realization of materials for specific applications needs some effort of synthesis and fabrication. In this view, polymers with nanocavities and fabricated zeolites have been produced and largely used in gas detection, filtration, and sequestration. Porous silicon (PSi) is one of the most promising fabricated materials because of its huge specific area (which can be several hundred meters square for a centimeter square), the surface chemistry (from hydrophobic to hydrophilic, depending on the treatment applied), and the possibility of integration in small, compact microelectronic systems. PSi is fabricated by a simple, but nontrivial, computer-controlled, electrochemical dissolution process of doped silicon in aqueous solution of hydrofluoridic acid. The most fascinating skill of PSi is the tunability of pores' dimension and disposition: by changing production parameters, such as etch time, current density, acid concentration, and wafer doping level, pore size can be tuned from nanometric to micrometric size. Detailed analysis of the relationship between parameters of Si porosification and parameters of PSI can be found in *Porous Silicon: Formation and Properties*.

3.2 DIFFERENT APPROACHES IN POROUS SILICON-BASED OPTICAL CHEMICAL SENSORS DESIGNING

At present, there are three main approaches to the design of PSi-based chemical sensors based on optical principles: luminescence, infrared absorption, and refractivity. Each one has its pros and cons, which will briefly discussed in the following text.

Reflectivity-based chemical sensors form a large group of devices, detecting an analyte on the basis of the refractive index changes of PSi due to the partial substitution of the air in the pores by the chemicals to be detected. Of course, the strongest effect was observed in the detection of liquids because when the liquid replaces air, the average dielectric constant, and thus the optical properties, changes dramatically. The sensing mechanism in this case depends on the refractive index value of the gas liquid phase, and also on how it can fill the pores: in smaller pores, the hydraulic resistance (which depends on liquid-PSi walls interaction) is too high, so that the liquid cannot penetrate into them.

In the case of the gas phase, the changes in dielectric constant are not so significant. However, in PSi with a certain pore size, there are conditions for capillary condensation phenomenon, which significantly increases the sensitivity of the optical properties of PSi to the influence of a volatile substance. The condensation conditions depend not only on the average pore size, distribution, and shape, but also on the strength of the interaction between the fluid and the pore walls (De Stefano et al. 2004c). Once the pore's shape and the surface chemistry are fixed, a one-to-one correspondence exists between the condensation conditions and the pore diameters given by the Kelvin equation:

$$(3.1) \qquad k_B T \rho_l \ln\left(\frac{p_{sat}}{p}\right) = 2\gamma_{\mathrm{lg}} \cos\frac{\theta}{R}$$

where ρ_l is the density of the liquid phase, γ_{lg} is the liquid-gas surface tension at temperature T, R is the radius of the pores, p/p_{sat} is the relative vapor pressure into the pore, and θ is the contact angle. From Equation 3.1 it is easy to see that the relative pressure increases with the average radius of pores R.

The interaction of PSi sensors with the chemical species induces a variation of effective refractive index of PSi layers, thus a shift of the multilayer reflectivity spectrum (Ouyang and Fauchet 2005). The average refractive index of the PSi layer, n_p, can be determined in the near infrared range by using the Bruggemann effective medium approximation for a heterogeneous mixture of components (nanocrystalline silicon and pore contents):

$$(3.2) \qquad (1-p)\frac{n_{Si}^2 - n_p^2}{n_{Si}^2 + 2n_p^2} + (1-p-\Lambda)\frac{n_{air}^2 - n_p^2}{n_{air}^2 + 2n_p^2} + \Lambda\frac{n_{ch}^2 - n_p^2}{n_{ch}^2 - 2n_p^2} = 0$$

where Λ is the layer liquid fraction (LLF), that is, the volume filled by the chemical species with refractive index n_{ch}, p is the porosity of layer, and n_{Si}, and n_{air} are the refractive indices of silicon and air. From Equation 3.2 the relative variation of the refractive index, $\Delta n_p/n_p$ as function of p, Λ, and n_{ch} can be numerically determined. In Figures 3.1a and b, the behavior of $\Delta n_p/n_p$ as a function of Λ and n_{ch} for a PSi layer with $p = 0.5$ is reported: the relative change of the average

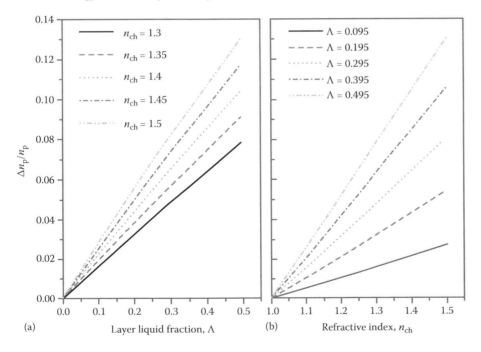

FIGURE 3.1   Relative variation of PSi refractive index layer with $P = 0.5$ as a function of layer liquid fraction for different refractive indices n_{ch} (a); as a function of n_{ch} for different layer liquid fractions (b). (Reprinted with permission from Moretti L. et al., *Appl. Phys. Lett.* 90, 191112. Copyright 2007, American Institute Physics.)

refractive index of the layer has a linear dependence on the filling factor and the refractive index of the chemical species:

(3.3)
$$\frac{\Delta n_p}{n_p} = c_p(n_{ch} - 1)\Lambda$$

The constant c_p depends on the layer porosity.

Based on this kind of physics, many optical detection schemes have been proposed in the past, and new ones are published every year. One big category, which is deeply discussed in the next paragraph, is based on resonant optical structures, in the sense that there is one or more characteristic wavelengths of light reflected/transmitted by the PSi photonic device that can be used as a marker to qualitatively or quantitatively monitor the gas presence.

Other effective methods are based on coupling the properties of PSi and some classical optical read-outs like ellipsometry, photoluminescence (PL), birefringence, polarization interferometry, and infrared absorption. In these photonic transducers, the mechanisms of sensitivity in optical and luminescence gas sensors are different. Optical gas sensors can be classified as refractometers, that is, they are based on change of refractive index due to pore filling; luminescence gas sensors work mainly on the charge transfer due to gas interaction with the surface, accompanied by the change of space charge region and the change of carrier recombination rates. The success of PSi is due to the discovery of a very bright emission of visible light under illumination of ultraviolet radiation (Canham 1990). Following this enthusiastic discovery, some groups used PSi PL as a signal to detect gas in the atmosphere: both intensity and maximum wavelength changes, on exposure to volatiles, can be used to sense them (Setzu et al. 1998; Snow et al. 1999; Mulloni and Pavesi 2000; Naderi et al. 2012). The drawback is the need of a pumping source and also fluctuation of the PL emission, which limits the resolution of the technique. This type of PSi-based optical gas sensor was discussed in Chapter 1 of this book and, therefore, luminescence-type sensors will not be discussed in this chapter.

Other groups relied on gas and vapor sensing on a very sensitive optical technique, which is normally used in thin film characterization—the ellipsometry. The basic principle of the ellipsometry is that the polarization change of light reflected by a sample is measured and analyzed (Zangooie et al. 1997; Arwin et al. 2003). The measured parameters are the two ellipsometric (Stokes) parameters ψ and Δ, which characterize the change of amplitude (ψ) and the phase (Δ) of the light beam (usually laser beam) on reflection. With precise measurements of ψ and Δ, it is possible to record small changes in pore filling of a PSi layer on gas exposure. In particular, Zangooie et al. (1997) have shown that ellipsometry can be used for detection of water, ethanol, and acetone vapors. The illustration of such measurements for acetone vapor control is shown in Figure 3.2. From the results presented in Figure 3.2, one can see that gas sensors designed on the basis of the above-mentioned method had good enough operating characteristics. The technique is very sensitive, fast, and accurate. For example, the detection limit threshold for acetone vapor for these sensors was smaller than 12 ppm (Zangooie et al. 1997). However, it is necessary to admit that an ellipsometer is, in general, a relatively complicated instrument. In addition, the technique requires a bench instrument that cannot be integrated in a small sensor system. This means that the method mentioned above could be widely adopted when a simplified low-cost instrument without moving parts is designed. The latest research in this field indicates that this task can be resolved successfully (Arwin et al. 2003).

The standard PSi fabrication process proceeds from top surface toward the bottom, so that pores are somehow vertical: this method introduces a preferential direction, that is, an optical axis, and the main result is that a PSi layer is optically anisotropic. The birefringency affects the polarization of light crossing the PSi membrane so that any change in the refractive index of the layer can be measured by polarization interferometry (Liu et al. 2002; Alvarez et al. 2012). This means that polarization interferometry also may be the basis for elaboration of the optical PSi-based gas sensors. A schematic diagram for characterization of optical anisotropy in PSi samples is shown in Figure 3.3. As any other interferometric technique, this optical scheme has great sensitivity: the detection limits reported are of the order of 10^{-6} in refractive index unit, which corresponds to ppm or sub-ppm gas concentration. Nevertheless, it cannot be omitted that the

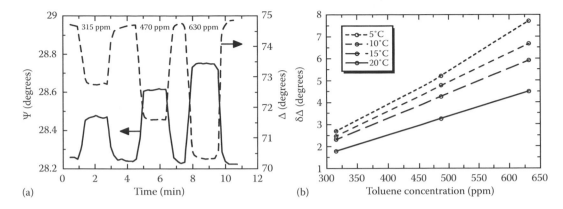

(a)   Time (min)   (b)   Toluene concentration (ppm)

FIGURE 3.2   (a) Variations in ψ and Δ caused by exposing a PSFP sample to three different toluene vapor concentrations: 315, 470, and 630 ppm at 20°C. (b) Ellipsometric response in terms of the absolute change in Δ caused by exposure of toluene vapor at three different concentrations (315, 470, and 630 ppm) at different temperatures. (Reprinted with permission from Zangooie S. et al., *J. Appl. Phys.* 86(2), 850. Copyright 1999, American Institute of Physics.)

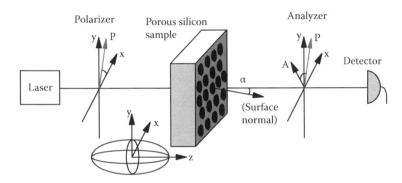

FIGURE 3.3   Schematic diagram for characterization of optical anisotropy in PSi samples. (Reprinted from *Sens. Actuators B* 87, Liu R. et al., 58, Copyright 2002, with permission from Elsevier.)

optical setup in this case is very complex; for example, it always require the use of a lock-in amplifier, and for this reason it cannot be thought of for field use or portable devices.

Interaction between a gaseous species and PSi has also been monitored by infrared absorption: this sensing scheme, much simpler than interferometry, exploits the correlation existing between gas chemisorbed on PSi surface and the population of electric carrier generated. This is the case of nitrogen dioxide that chemically binds the PSi surface releasing an electron: since the infrared absorption coefficient is proportional to the free carrier concentration, a dramatic increase of absorption is observed (Geobaldo et al. 2001). A nontrivial advantage of this method is a good selectivity against many common interference substances, such as water, oxygen, carbon oxide, and carbon dioxide. Due to both high surface area and the effect of preconcentration, PSi-based spectroscopic gas sensors can have considerably shorter interaction path and simpler construction in comparison with conventional gas sensors (Lambrecht et al. 2007; Zhao et al. 2011). Figure 3.4a shows the conventional gas sensor using multireflection cells to increase sensitivity of gas sensing, but it is unfit for practical application because of the strict fabrication process and it is easily influenced by outer circumstances. Figure 3.4b shows the gas sensor by employing the PSi to replace the traditional gas cell.

In the next section, main attention will be focused on refractive-type optical sensors: this class of transducer requires the simplest and cheapest readout of all and is the most used in laboratories around the world. Some of them can be read by the naked eye because they change color on exposure to gases; others just need an optical fiber and an optical spectrum analyzer.

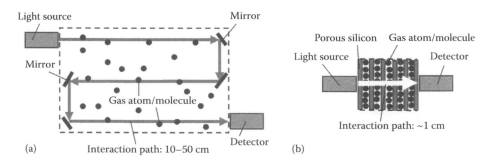

(a)   Interaction path: 10–50 cm   Detector   (b)   Interaction path: ~1 cm

FIGURE 3.4   Configuration of (a) conventional and (b) PSi-based spectroscopic gas sensors. (Idea from Lambrecht et al. 2007.)

3.3 POROUS SILICON OPTICAL TRANSDUCERS FOR REFLECTIVITY-BASED GAS AND VAPOR SENSORS

Experiment and simulations have shown that the most effective reflectivity-based optical sensors can be developed based on resonant photonic structures such as Fabry-Pérot interferometers (Dancil et al. 1999), Bragg reflectors (Snow et al. 1999), optical microcavities (Mulloni and Pavesi 2000; De Stefano et al. 2003), Thue-Morse sequences (Moretti et al. 2007), rugate files (Ruminski et al. 2010), and optical waveguides (Arrand et al. 1998; Kim and Murphy 2013), which during the last decades have been intensively studied by several research groups in particular for their photonic properties as interference filters. As a rule, the above-mentioned resonant photonic structures are multilayer structures with varying refractive index. As previously shown in *Porous Silicon: Formation and Properties*, electrochemical etching allows forming silicon layers with different porosity, which is an important advantage of this technology for producing such structures. Since the dissolution is self-stopping, it is possible to produce multilayer structures in a single run by using data reported in Figure 3.5. Figure 3.5a shows the dependence of the real part of the complex refractive index of the PSi on porosity given by the Bruggeman model, and Figure 3.5b shows the dependences of the porosity on the current density for PSi fabricated using highly doped p^+-silicon, <100> oriented, 0.01 $\Omega \cdot$cm resistivity, 400 µm thick. Thus, the refractive index profile of a PSi multilayered structure can be realized by choosing the proper current density

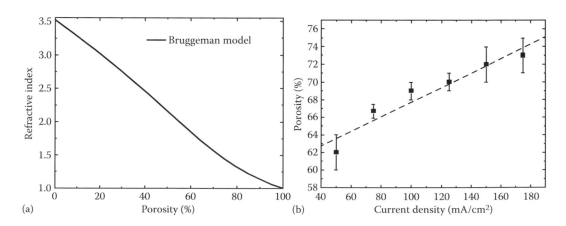

(a)   Porosity (%)   (b)   Current density (mA/cm²)

FIGURE 3.5   (a) Dependence of real part of the PSi refractive index (real part) on the porosity given by the Bruggeman effective medium approximation. (b) Dependence of the porosity on the current density for p^+ <100> silicon using a solution of 15% hydrofluoric acid in ethanol. The electrochemical etching of crystalline silicon was performed in the dark and at room temperature. The values of porosity and etch rate have been estimated by variable angle spectroscopic ellipsometry (UVISEL, Horiba, Jobin–Yvon). (Reprinted with permission from Moretti L. et al., *Appl. Phys. Lett.* 90, 191112. Copyright 2007, American Institute of Physics.)

profile during the electrochemical etching of crystalline silicon in minutes (Lehman 2002). At that, one should note that the geometric characteristic of PSi structures perfectly combines with very good optical properties: the surface and all the interfaces between layers are so smooth that the photonic spectra of PSi devices are always of very good optical quality, that is, optical losses due to pores scattering are negligible, at least for pore dimensions under 200 nm.

3.3.1 FABRY-PÉROT INTERFEROMETER

A single layer of PSi optically acts as a Fabry-Pérot interferometer. A Fabry-Pérot porous silicon film is perhaps the simplest of the structures mentioned above, with reflected spectra exhibiting thin film optical interference. These thin films are often referred to as "straight" etches, in contrast to the corrugated pores of rugate structures. The uniform porosity of the relatively straight pore channels results in a uniform refractive index of the layer with depth. Light striking the porous sensor reflects off the air PSi and PSi-substrate interfaces, with the path difference of the two reflected rays resulting in thin film interference (see Figure 3.6a). In Figure 3.6b, the reflectivity spectrum of a PSi layer under white light illumination is reported. The maxima in the reflectivity spectrum appear at wavelengths λ_m, which satisfy:

(3.4) $$m = 2nd/\lambda_m$$

where m is an integer, d is the film thickness, and n is the average refractive index of the layer (Lin et al. 1997; Anderson et al. 2003).

Assuming that the refractive index is independent of the wavelength over the considered range, the maxima are equally spaced in the wavenumber $(1/\lambda_m)$. When m maxima are plotted as a function of the wavenumber, each point lies on a straight line whose slope is two times the optical path of the interferometer, as it is shown in Figure 3.7.

On exposure to gases, the change in average refractive index of the layer reflects in a change of the optical path, that is, half the slope of the straight line in Figure 3.7: ordering maxima as function of wavenumber allows the measurement of a single gas concentration.

3.3.2 BRAGG MIRROR

The Bragg mirror is a periodic structure made of alternating layers of high (n_H) and low (n_L) refractive index, whose thicknesses satisfy the relation $2(n_H \cdot d_H + n_L \cdot d_L) = m\lambda_B$, where m is the

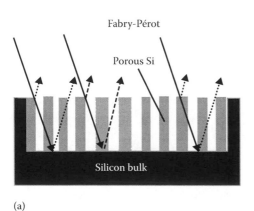

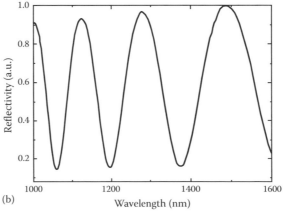

(a)   (b)

FIGURE 3.6   (a) Schematic illustration of a Fabry-Pérot interferometer. Incident rays are shown at an angle for illustration: spectra are usually taken at normal incidence. (b) Reflectivity spectrum of a PSi layer realized by the electrochemical etching of p^+ crystalline silicon in a solution of 15% hydrofluoric acid applying a current density of 115 mA/cm^2 for 11 s ($P = 69\%$; $n = 1.593$ at 1.2 µm; $d = 2.9$ µm).

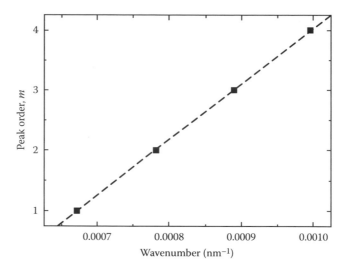

FIGURE 3.7   *m*-Order peaks are plotted as a function of the wave number. The optical path of the interferometer has been estimated to be (4620 ± 40) nm.

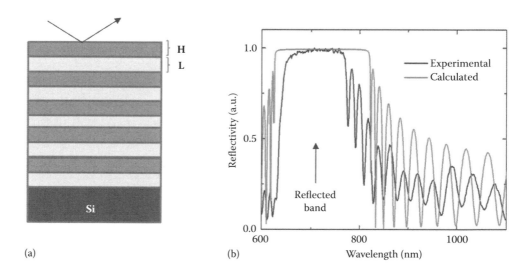

(a)   (b)

FIGURE 3.8   (a) Schematic of a Bragg mirror. (b) Experimental normal incidence reflectivity spectrum from a Bragg mirror (black line) compared with the calculated one (gray line).

order of the Bragg condition (Figure 3.8a). The layer stack is usually denoted as $[HL]^N$, where *N* is the number of periods. The periodicity gives to the structure a photonic band gap (PBG) behavior characterized by the property to forbid the propagation of the light at fixed wavelengths. The reflectivity spectrum of a Bragg mirror is characterized by the presence of a stop band centered on the Bragg wavelength λ_B (Figure 3.8b). For a given number of periods, the height and width of the reflectivity stop band increases by increasing the index ratio H/L. A low index contrast can be compensated by a higher number of periods.

The Bragg wavelength, that is, the middle of the stop band, is an optical resonance because the transmitted optical field is concentrated everywhere but the stop band: on exposure to gas, the optical spectrum in Figure 3.8b shifts toward higher wavelengths and the stop band also translates, so that the position of the Bragg wavelength is a measure of the gas concentration.

3.3.3 OPTICAL MICROCAVITY

An optical microcavity is a λ/2 layer sandwiched between two distributed Bragg mirrors (Figure 3.9a). The reflectivity spectrum of a microcavity is characterized by a transmittance peak in the

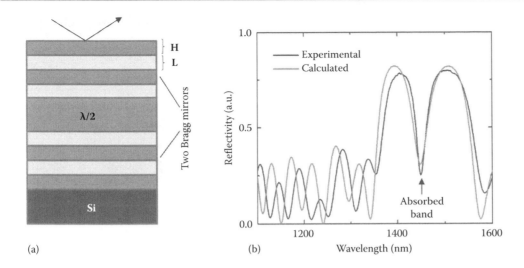

(a)   (b)

FIGURE 3.9 (a) Schematic of an optical microcavity. (b) Experimental normal incidence reflectivity spectrum of a microcavity (black line) compared with the calculated one (gray line).

photonic stop band (Figure 3.9b). The Q factor of the microcavity is defined as $Q = \lambda/\Delta\lambda$, where λ is the wavelength of the resonance peak and $\Delta\lambda$ is the full width half maximum (FWHM) of the resonance. This parameter is used to evaluate how the light is confined in the PBG structure.

The calculated reflectivity spectra of the structures reported in Figure 3.10 have been reproduced by a transfer matrix method (Muriel and Carballar 1997), also taking into account the wavelength dispersion of silicon. The gas sensing is just the same in the case of a Bragg reflector, but the existence of a very sharp resonance, that is, the dip inside the stop band, makes the measure more accurate and of higher resolution because it is more easy to follow the shift of the dip

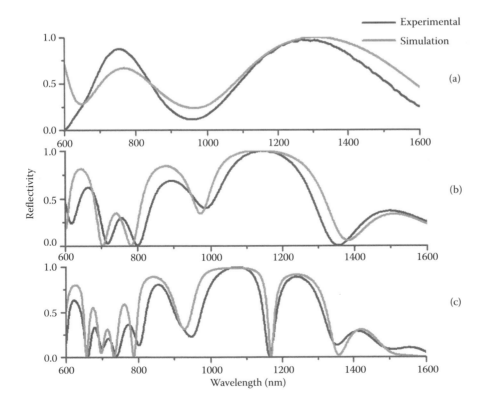

FIGURE 3.10 Experimental (black line) and calculate (gray line) reflectivity for S3 T-M structure (a), S4 T-M structure (b), and S5 T-M structure (c). The measurements have been taken at normal incidence.

(there is less uncertainty in defining the resonance wavelength because less wavelengths can be associated to it) on exposure to gas.

3.3.4 PHOTONIC QUASI-CRYSTAL

A quasi-crystal (QC) does not have a geometrical periodicity but is still deterministically generated. Even if these structures do not have a translational symmetry, they show several interesting physical properties such as the photonic band gaps, some resonance frequencies, and some highly localized states (Soukoulis and Economou 1982). Thue-Morse (T-M) (Liu 1997) sequence is one of the most common examples of one-dimensional QC. The T-M one-dimensional structure is constituted by the sequence of two layers A and B with refractive index n_A (n_B) and thickness d_A (d_B). Applying the substitution rules A→AB and B→BA, all subsequent orders can be deduced, as follows: $S_0 = A$, $S_1 = AB$, $S_2 = ABBA$, $S_3 = ABBABAAB$, $S_4 = ABBABAABBAABABBA$, and so on. The layer number of S_N is $2N$, where N is the T-M order.

In Figure 3.10, the experimental (black line) and calculated (gray line) reflectivity spectra are shown in the case of S_3 (x.3.5-a), S_4 (x.3.5-b), and S_5 (x.3.5-c) T-M structures. In these experiments, the high porosity layers were characterized by a porosity $P_A = 81\%$, with an average refractive index $n_A \cong 1.3$ and a thickness $d_A \cong 135$ nm, and the low porosity layers were characterized by a porosity $P_B = 56\%$, with an effective refractive index $n_B \cong 1.96$ and a thickness $d_B \cong 90$ nm. The thickness d_i of each layer was designed to satisfy the Bragg condition $n_i d_i = \lambda_0/4$ where n_i is the average refractive index and $\lambda_0 = 700$ nm.

As it is seen, the spectrum of the S_3 T-M multilayer is characterized by two band gaps separated by a large transmission peak at 1000 nm. On increasing the order of T-M sequence, the PBG splits and very narrow transmission peaks appear (FWHM about 6 nm). The main difference between a Bragg mirror or a microcavity and a T-M sequence is the presence of the so-called fractal optical resonances in the T-M optical spectrum, which are not in correspondence with Bragg resonances and could be very useful to quantify gas monitoring.

The agreement between the measured and calculated spectra demonstrates that the fabrication process of the devices using electrochemical etching provides good control. The not perfect matching can be ascribed to nonuniformities of thickness and porosities of layers along the etching direction.

3.3.5 RUGATE FILTER

Another very used optical PSi multilayer is the rugate filter (Figure 3.11a). This structure exploits the unique property of PSi of a continuous modulation of its porosity, and hence of its refractive index, whereas all other standard dielectric materials, such as silicon nitrides or titanium oxide, only allow a discrete alternation of different layers. The stepwise anodic current used for the fabrication of Bragg reflectors can be substituted by a continuous sinusoidal current with maximum and minimum values equal to the two Bragg currents. The result is an optical spectrum with a resonant peak located at the Bragg wavelength, small sidelobe bands, and suppression of higher harmonics (Lorenzo et al. 2005). Typical spectrum is shown in Figure 3.11b. As a rule, PSi photonic crystals are fabricated from highly doped p-type silicon. Due to these important optical properties, rugate filters are currently used not only as gas sensors but also for biochemical monitoring, after a proper functionalization (Sciacca et al. 2011; Li et al. 2012).

If rugate structures exhibit a sinusoidal variation of their refractive index with depth (Lorenzo et al. 2005), an optical reflection band is centered at a wavelength determined by the period and amplitude of the refractive index profile. In the simplest case, a sinusoidal variation of their refractive index can be achieved by a sinusoidal variation in the current density of the etch with time, which will be accompanied by a sinusoidal change in the porosity of the PSi layer with depth. The sensing principal of rugate structures is similar to the sensing principal of the previously mentioned optical structures based on PSi. Infiltration of chemical vapors into the porous matrix displaces air from the porous voids, increasing the refractive index of the film, and resulting in

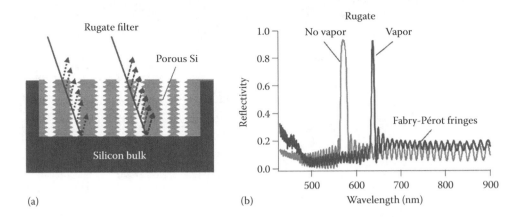

FIGURE 3.11   (a) Schematic diagram of a rugate filter. A rugate layer consists of cyclically varying regions of high and low porosity with depth. Incident rays are shown at an angle for illustration: spectra are usually taken at normal incidence. (b) Typical reflectivity spectra of rugate PSi optical layers before and after exposure to a high concentration of vapor. (Idea from King 2010.)

a red shift of the optical stop band wavelength that enables sensing by monitoring the reflectivity peak wavelength over time (Sailor and Link 2005). Rugates offer advantages over other PSi optical structures that include the ability to monitor sensing with the naked eye, their retention of the sensing characteristics of Fabry-Pérot interference fringes (Imenes and McKenzie 2006; Pacholski and Sailor 2007), and the ability to construct photonic crystals with tailored optical stop bands that allow their use as optical filters for the indirect sensing of target compounds.

More complex PSi-based resonant transducers have been published that are worth mentioning: integrated interferometer and two- or three-dimensional photonic crystals (Chow et al. 2004; Kim and Murphy 2013; Pacholski 2013 and references therein). Without entering in a detailed discussion of these structures (which all show good sensing skills), it should be emphasized that all require a strong effort in design and realization because a standard fabrication flow must be set: the simple one-step electrochemical process that produces the porous layer is only one among all the other required for integration, such as spinning, baking, developing, removing, and so on. In other words, the exceptional simplicity of PSi optical devices fabrication is lost, and on the other hand, performances obtained are not so extraordinary.

3.4 GAS AND VAPOR MONITORING BY OPTICAL REFLECTIVITY: A CASE STUDY

The sensitivity is a key issue of a sensor, so that several experimental works investigating the sensitivity of the different PSi structures have been reported in literature (Neimark and Ravikovitch 2001). Ouyang et al. (2005, 2006) focused their research on the sensitivity of PSi microcavities as a function of the material properties, such as pore size, porosity, and number of layers. In our previous work, we proposed (Moretti et al. 2007) a simple model to study the behavior of PSi multi-layered structures on exposure to different compounds and to determine their response curve. The two multilayered structures analyzed are a one-dimensional periodic multilayer, the Bragg Mirror (BM), and an aperiodic multilayer, the Thue-Morse Sequence (TMS) (see Figure 3.12). Both the PSi structures are composed of 64 layers, 32 with high (H) refractive index (low-porosity), and 32 with low (L) refractive index (high-porosity). The layers thicknesses are $d_H = \lambda_0/4n_H$ and $d_L = \lambda_0/4n_L$, respectively. The different spatial order of the layers is the only difference between the two structures.

It is well known that the refractive index change, due to the interaction of the PSi multilayers with external agents, preserves the shape of the reflectivity spectrum, so that it is still possible to individuate the resonant characteristics, that is, the transmittance peaks of the TMS or the high reflectivity stop band of the BM. The shape of reflectivity spectrum depends on the phase

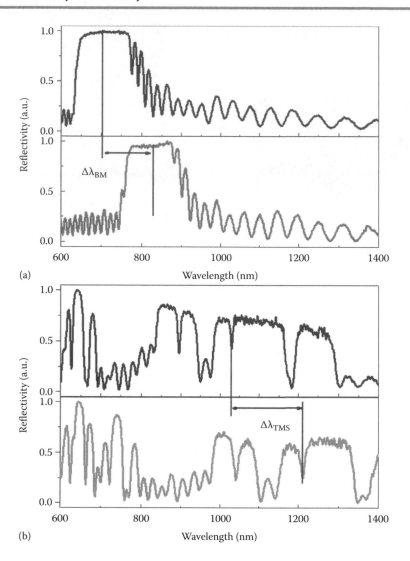

FIGURE 3.12 Experimental reflectivity spectra of Bragg multilayer (a, black line) and Thue-Morse multilayer (b, black line) composed of 64 layers. In the gray line are reported the reflectivity spectra after exposure to methanol. $\Delta\lambda_{BM}$ and $\Delta\lambda_{TMS}$ are the monitor wavelengths for Bragg and Thue-Morse multilayer, respectively. (Reprinted with permission from Moretti L. et al., *Appl. Phys. Lett.* 90, 191112. Copyright 2007, American Chemical Society.)

modulation of each layer $\phi_i = 2\pi n_i d_i/\lambda$; for a couple of layers, the phase modulation is $\phi = \phi_L + \phi_H$. The reflectivity can be factorized as a product of two contributions:

$$(3.5) \qquad R = A\left(\frac{n_L}{n_H}\right) \cdot \Re(\phi)$$

where the function A takes into account the value of the reflectivity due to the refractive index contrast, and $\Re$ is a shape factor due to the different optical paths of the light into the layers. If the reflectivity is simply shifted on a wavelength range without changing the shape during the measurements process, it is possible to write:

$$(3.6) \qquad \Re[\phi(\lambda_r, n_i, d_i)] \cong \Re[\phi(\lambda_r + \Delta\lambda_r n_i + \Delta n_i d_i)]$$

where λ_r is the characteristic wavelength used as a reference to measure the spectral shift, and Δn_i is the variation of the layer refractive index due to the interaction of the devices with the chemical species.

The equality $\phi(\lambda_r, n_i, d_i) = \phi(\lambda_r + \Delta\lambda_r, n_i + \Delta n_i, d_i)$ can be deduced by Equation 3.9. By evaluating the variation of ϕ, an expression for $\Delta\lambda_r$ as function of layer refractive index variations, Δn_L, and Δn_H, can be deduced:

$$(3.7) \qquad \Delta\lambda_r = 2\frac{\lambda_r}{\lambda_0}(d_L\Delta n_L + d_H\Delta n_H) = \frac{\lambda_r}{2}\left(\frac{\Delta n_L}{n_L} + \frac{\Delta n_H}{n_H}\right)$$

This formula is a powerful tool in the design of all resonant optical sensors, based on the average refractive index change.

Combining Equation 3.7 with Equation 3.4, it is possible to completely characterize the optical response of whatever PSi multilayer

$$(3.8) \qquad \frac{\Delta\lambda_r}{\lambda_r} = \frac{(n_{ch}-1)}{2}(c_L\Lambda_L + c_H\Lambda_H)$$

It is clear that the sensitivity of PSi multilayer depends strictly on the filling capability of the layers. The BM and the TMS have the same response as a function of the layer liquid fraction. In Figure 3.13 are shown the reflectivity spectra of BM (a) and TMS (b) when unperturbed, and on exposure to a methanol ($n_{ch} = 1.328$) saturated atmosphere.

The BM reflectivity shows a classic photonic band gap centered at the Bragg wavelength 2 ($n_L d_L + n_H d_H$) = 712 nm. This wavelength is a natural candidate as monitor wavelength λ_r^{BM} because it is simply recognizable after the interaction process. On the other side, the TMS spectrum shows a more complex photonic band gap structure due to the aperiodic sequence of the layers: three photonic band gaps can be observed in wavelength intervals centered at 640 nm, 890 nm, and 1120 nm, and three resonant transmittance peaks at 894 nm, 1030 nm, and 1184 nm. In this case, it is possible to choose λ_r among the one of the resonant transmittance peaks. In particular, the spectral shift of the resonance transmittance peak at 1030 nm ($\lambda_r^{TMS} = 1030\,$nm) is monitored. In Figure 3.13, using Equation 3.8, the relative wavelength shift, $\Delta\lambda_r/\lambda_r$ as a function of Λ on exposure to methanol obtained in the case of TMS and BM is reported. The filling can be assumed to proceed uniformly into the entire multilayer stack until the low porosity layers are completely filled ($\Lambda_L = \Lambda_H = \Lambda$ for $\Lambda < p_L$), then the filling process proceeds only in high porosity

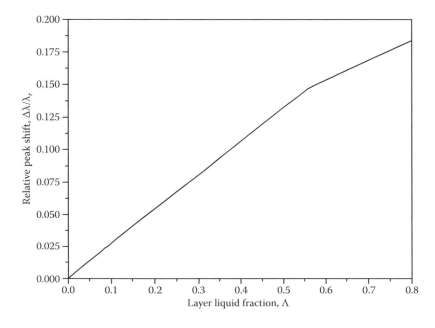

FIGURE 3.13   Calculated relative wavelength shift $\Delta\lambda_r/\lambda_r$ for BM and TMS as a function of layer liquid fraction Λ after exposure of methanol ($n_{ch} = 1.328$). The two curves coincide. (Reprinted with permission from Moretti L. et al., *Appl. Phys. Lett.* 90, 191112. Copyright 2007, American Chemical Society.)

layers (Snow et al. 1999; Allcock and Snow 2001): the filling curves show a slope change when $\Lambda = p_L$; the values of c_L and c_H are 0.927 and 0.677, respectively. In Table 3.1, the experimental wavelength shifts and the layer liquid fractions extrapolated from the curve of Figure 3.13 for several compounds are reported. First, it is worth noting that the pore filling is a characteristic parameter of the multilayer sequence, and is almost invariant with respect to the compound investigated, in agreement with Gurvitsch's rule, which states that the volume of liquid adsorbed should be the same for all adsorptives on a given porous solid.

In Figure 3.14, it is reported that the response curve of both structures, the sensitivities, normalized to the reference wavelengths, are respectively, $S^{TMS} = 0.51(0.05)$ RIU^{-1} (refractive index units) and $S^{BM} = 0.41(0.05)$ RIU^{-1}; the TMS higher sensitivity can be ascribed to higher filling capability. This effect can be explained by considering the number of L-H interfaces in the different multilayers. The different spatial order of layers between BM and TMS can be observed:

$$\text{TMS: LHHLHLLHHLLHLHHL...}$$

$$\text{BM: LHLHLHLHLHLHLHLH...}$$

It is possible to conclude that the periodic arrangement of the BM induces a greater number of porosity gradients due to the presence of a greater number of L-H interfaces ($n = 63$), than in

TABLE 3.1   Chemical Organic Substances and Its Refractive Index Used in Sensing Experiment

Solvent	n_{ch}	$\Delta\lambda_r^{TMS}$ (nm)	Λ^{TMS}	$\Delta\lambda_r^{BM}$ (nm)	Λ^{BM}
Methanol	1.328	180	0.745	108	0.599
Pentane	1.358	199	0.761	118	0.599
Isopropanol	1.377	209	0.761	125	0.599
Isobutanol	1.396	215	0.745	127	0.467

Note: $\Delta\lambda_r^{TMS}$ ($\Delta\lambda_r^{BM}$) and Λ^{TMS} (Λ^{BM}) are the wavelength shift and the layer liquid fraction for Thue-Morse (Bragg) structures. The reference wavelength λ_r is 1030 nm for TMS and 712 nm for BM.

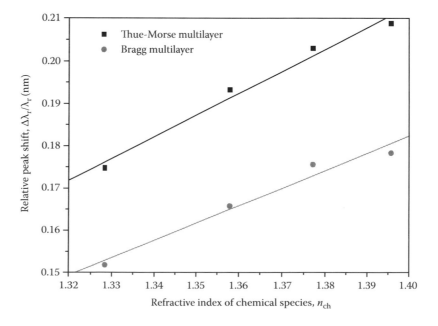

FIGURE 3.14   Experimental response curves of TMS and BM. (Reprinted with permission from Moretti L. et al., *Appl. Phys. Lett.* 90, 191112. Copyright 2007, American Institute of Physics.)

TMS ($n = 42$). These interfaces produce an inhomogeneity in the pore network, which obstacles the propagation of the liquid phase and thus reduces the filling capability of structures. Therefore, after the condensation process or the capillary penetration, the liquid phase finds a more homogeneous pore network in TMS with respect to BM.

It should be noted that despite the fact that optical sensing with the previously mentioned photonic structures is achieved by monitoring the shift of reflectance spectrum, caused by a change in refractive index, the signal processing is specific to each type of structure. For example, in Fabry-Pérot structures, the change in refractive index results in a change in the optical thickness. The optical thickness (OT, in nm) is related to the refractive index (n) and thickness of the porous film (L, in nm) by the following equation:

(3.9)
$$OT = nL$$

A fast Fourier transform (FFT) of the interference spectrum of the PSi film results in an FFT peak whose position is two times the optical thickness (Gao et al. 2002a). A Gaussian fit of the FFT peak is performed to determine the peak position. Therefore, for a PSi Fabry-Pérot sensor, the sensor response at any given time is measured by doing an FFT of the reflectance spectrum and a Gaussian fit of the resulting FFT peak to calculate the OT. For rugate filters, however, an FFT is not required because the resulting photonic crystal acts as a physical Fourier transform. The wavelength position in nanometers of the photonic stop band peak (λ) is related to the total refractive index of the porous matrix (n) and the thickness in nanometers of one period in the multilayered rugate structure (d) by the following equation:

(3.10)
$$\lambda = 2nd$$

As with the FFT peak, a Guassian fit of the photonic stop band is performed to determine the peak position. Therefore, one advantage of PSi rugate films is that the sensor response can be measured without the need to perform an FFT.

3.5 EXAMPLES OF OPTICAL PSi-BASED CHEMICAL SENSORS REALIZATION

For realization of PSi-based optical sensors, various approaches can be used. Several examples of possible configuration of PSi-based optical sensors are shown in Figure 3.15.

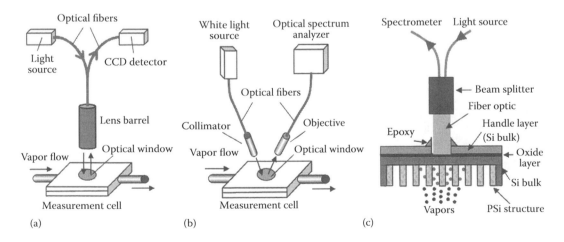

FIGURE 3.15 Sensor configurations of PSi-based optical sensors: (a) a light impinges on the PSi layer through an optical fiber and the output was collected by an CCD detector; (b) a white light impinges, through an optical fiber and a collimator, on the PSi layer and the output was collected by an objective and coupled into a multimode fiber; and (c) a light impinges on the PSi layer, through an fiber optic cable directly integrated into the bulk Si, and the output was collected by an spectrometer. (Idea from Karacali et al. 2013.)

In particular, De Stefano et al. (2006) for optical detection used a white light source (Figure 3.15b). A white light impinges, through an optical fiber and a collimator, on the PSi layer and the output was collected by an objective and coupled into a multimode fiber. The signal was sent to an optical spectrum analyzer (Ando, Mod. AQ-6315B) and measured with a 0.1–0.2 nm resolution. Sailor's group in many designs used an LED as the spectral source and a photodiode or phototransistor as detector (King 2010). Another approach was used by Karacali et al. (2013). To improve reproducibility of the monitoring of optical property changes, Karacali et al. (2013) have developed a fabrication process to integrate a fiber optic cable directly into the bulk Si (see Figure 3.15c). The new design features more robust reflectivity measurements by eliminating measurement error from variation in fiber optic cable alignment angle, shortening experiment set-up time, and allowing for flexible positioning of the sensor. The first fabrication step involves hole milling by reactive ion etching (RIE) of the insulator backside of a silicon on insulator (SOI) wafer. A thin oxide layer between the silicon and insulator is used to reliably stop the RIE process. PSi rugate sensor was fabricated using electrochemical anodic etching with sinusoidally modulated current. The fiber optic cable is then integrated into the PSi sensor by fixing with epoxy at an optimized position.

Of course, PSi layers can be adhered directly to the end of silica optical fibers (see Figure 3.16a). King (2010) used such an approach. In this case, PSi layers were detached from the bulk Si. Freestanding PSi was attached to optical fiber with a partially cured mixture of epoxy resin and the amine curing agent, and allowed to cure for two days before use. The ratio of resin to curing agent was determined using the amine value (curing agents) and epoxide equivalent weights of each component from the respective certificates of analysis. These optical fiber sensors have a small footprint of <1 mm diameter and are impervious to electrical interference because the spectral detector may be placed at a distance from the sensor head. King (2010) has shown that such sensors are suitable for remote sensing applications.

Experiment has shown that PSi-based optical sensors can be applied for detection of gases, humidity, and vapor-phase chemicals, mainly volatile organic compounds (VOCs) (methanol, ethanol, acetone, acetonitrile, iso-propanol, heptane, ethylene dichloride, toluene, etc.), with high sensitivity. In particular, PSi-based sensors have been used to optically sense alcohol vapor concentrations as low as 250 ppb, VOCS and alkanes at ppm levels (Zangooie et al. 1999; Gao et al. 2000, 2002a,b; Dorvee and Sailor 2005; Ruminski et al. 2010) (Figure 3.16b), and organophosphate

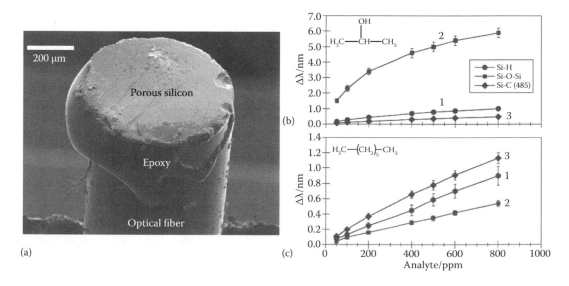

FIGURE 3.16 (a) Scanning electron microscope (SEM) image of an optical fiber capped with a PSi vapor sensor (rugate filter). The PSi layer was prepared as a freestanding film and attached to the glass fiber with epoxy. (b,c) Dose-response curves for fiber-mounted optical PSi sensors exposed to the analytes isopropanol (b) and heptane (c). Surface chemistries on the PSi samples were: (1) hydrogen-terminated (Si-H, circles), (2) thermally oxidized (Si-O-Si, squares), and (3) thermally reacted with acetylene at 485°C (Si-C, diamonds). (Ruminski A.M. et al.: *Adv. Funct. Mater.* 20(17) 2874, 2010. Copyright 2010: Wiley InterScience. With permission.)

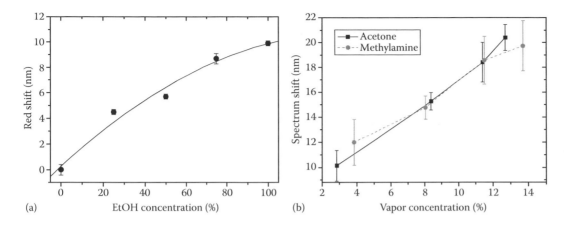

(a) EtOH concentration (%)   (b) Vapor concentration (%)

FIGURE 3.17   (a) The calibration curve for $\lambda/2$ Fabry–Pérot optical microcavity in case of an ethanol/ deionized water binary mixture. (From De Stefano L. et al., *J. Phys.: Condens. Matter.* 19, 395008. Copyright 2007, IOP. With permission.) (b) The spectral redshifts for PSi-based Bragg mirror measured under acetone and methylamine vapor exposure. (From Jalkanen T. et al., *Opt. Expr.* 17(7), 5446. Copyright 2009, The Optical Society. With permission.)

chemical agents at concentrations of tens of ppm (Dorvee and Sailor 2005; Jang et al. 2008) and even hundreds of ppb (Jang et al. 2008). However, we should recognize that the most sensors provide detection of vapor at a significantly higher concentration, usually in a concentration range of more than 1% (Snow et al. 1999; Jalkanen et al. 2009) (see Figure 3.17).

3.6 INTEGRATED POROUS SILICON MICROSYSTEMS FOR OPTICAL SENSING

Measurement cells used in PSi-based optical sensors can have different configurations. In particular, De Stefano et al. (2006) designed for these purposes integrated PSi microsystems, which can be used as an element of a Lab-on-Chip (LOC). LOC is a device integrating laboratory functions on a single chip of a few square centimeters in size. In recent years, there was a fast and intensive interest in LOC for sensing applications due to several factors such as very limited sample consumption and short analysis time. Figure 3.18 shows the cross-section of the basic element of an integrated PSi-based optical sensor, obtained integrating a PSi optical transducer and a glass slide, which ensures sealing and the interconnections for fluid inlet and outlet.

The reaction microchamber has been realized by a two-step electrochemical etching of the silicon. The first step is the electropolishing of the material obtained by means of a high current density electrochemical etch, which creates a microwell. The second step is a consecutive electrochemical etching used to fabricate a PSi layer on the bottom of the microchamber. After the mechanical drilling of the flow channels, the glass slide has been cleaned and activated for the AB process following standard RCA and H_2O_2 procedures. Due to the highly reactive PSi nature, the standard cleaning procedures had to be changed with a soft cleaning procedure based on trichloroethylene, acetone

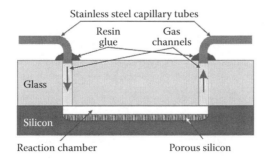

FIGURE 3.18   Schematic of the sensor device for LOC-application. (Reprinted from *Sens. Actuators B* 114, De Stefano L. et al., 625, Copyright 2006, with permission from Elsevier.)

and ethanol, and deionised water. Silicon-etched wafer and glass-top prefabricated components have been anodically bonded together with mutual alignment at a temperature of 200°C, voltage of 2.5 kV, and a process time of 2 min. The reflectivity spectra in the VIS-IR wavelength region have been recorded with an experimental set-up similar to the set-up shown in Figure 3.15b: a white source illuminates, through an optical fiber and a collimator, the PSiat nearly normal incidence. The reflected light is collected by an objective and coupled into a multimode fiber.

Time-resolved measurements conducted using the laser beam from an IR source have shown that due to chamber miniaturization, the response time (τ_{resp} = 2 s), the time interval between the 10% and 90% of the maximum signal, is significantly shorter. Results of these experiments are presented in Figure 3.19. In this figure, the result of time-resolved measurement is compared to the data acquired on the same PSi layer before the integration process. It is known that the value of response time depends not only on the physical phenomena involved (i.e., equilibrium between adsorption and desorption in the PSi layer), but also on the geometry of the test chamber and on the measurement procedure, that is, static or continuous flow mode. In static condition, the response time is mainly determined by the diffusion of the gas into the chamber volume: in fact, when vapor is in contact with the PSi surface, the capillary condensation takes place instantaneously. For the same reason, the recovery time is longer (τ_{rec} = 8 s): as soon as nitrogen is introduced into the μ-chamber, the conditions for capillary condensation are not still valid so that the liquid phase disappears, depending on the atmosphere rate exchange. As it is shown in the graphic of Figure 3.19, the sensor response is completely reversible.

One should note that due to miniaturization, this integrated cell could be used for flow injection analysis (FIA). FIA is a versatile technique to perform quantitative chemical analysis. FIA was invented at the Department of Chemistry at Technical University of Denmark (DTU) in 1975 by Ruzicka and Hansen (1975, 1998). FIA is based on injecting by a valve of small and well-defined volume of sample into a continuously flowing carrier stream to which appropriate auxiliary reagent streams can be added. The sample disperses and reacts with the components of the carrier in a reactor, forming a species that is sensed by a detector and recorded. A schematic of the FIA is reported in Figure 3.20.

Thus, in contrast to conventional continuous flow procedures, FIA does not rely on complete mixing of sample and reagent. Combined with the inherent exact timing of all events, it is not necessary to wait until all chemical reactions are in equilibrium. These feats, which allow transient signals to be used as the readout, do not only permit the procedures to be accomplished in a very short time, but also have opened new ways to perform an array of chemical analytical assays, which are very difficult and in many cases directly impossible to implement in a traditional way. Thus, in FIA, it is possible to base the assay on the measurement of metastable compounds, which exhibit particularly interesting analytical characteristics. The concept of FIA depends on a combination of three factors: reproducible sample injection volumes, controllable sample

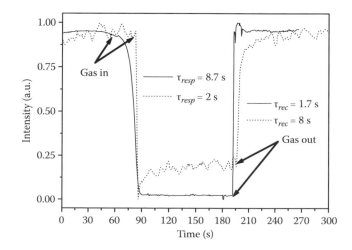

FIGURE 3.19 Time-resolved measurements of PSi layer during the monitoring of acetone in 0.4 l test chamber (solid line) and in the integrated 10 μl micro-chamber (dashed line).

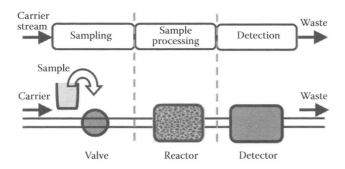

FIGURE 3.20   Schematic description of the flow injection analysis.

dispersion, and reproducible timing of the injected sample through the flow system. The system is ready for instant operation as soon as the sample is introduced. FIA offers several advantages in terms of considerable decrease in sample (normally using 10 to 50 µL) and reagent consumption, high sample throughput (50 to 300 samples per hour), reduced residence times (reading time is about 1 to 40 s), shorter FIA reaction times (1 to 60 s), easy switching from one analysis to another (manifolds are easily assembled or exchanged), reproducibility, reliability, low carryover, high degree of flexibility, and ease of automation (Valcorcel and Luque de Castro 1987). Perhaps the most compelling advantage of the FIA technique is the great reproducibility in the results obtained by this technique that can be set up without excessive difficulties and at very low cost of investment and maintenance. These advantages have led to an extraordinary development of FIA, unprecedented in comparison to any other technique. De Stefano et al. (2006) believe that the evolution of the integrated system previously described using the FIA method largely improves the performances of the PSi-based sensor.

3.7 APPROACHES TO OPTIMIZATION OF POROUS SILICON-BASED OPTICAL SENSORS

3.7.1 SENSITIVITY

As it follows from previous discussions, the degree of spectral shift, and thus sensitivity, of a PSi sensor is determined by both the amount of analyte adsorbed and the refractive index of the analyte (Salem et al. 2006). For a liquid analyte, there is full infiltration of the porous matrix and the PSi sensor acts as a refractometer (see Figure 3.21).

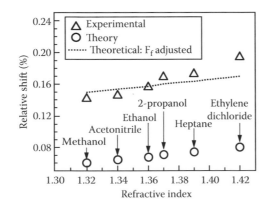

FIGURE 3.21   Plot of the theoretical and observed relative shifts in resonant wavelength of anodically oxidized PSi rugate filter (PSRF) versus refractive index of the liquid immersed into the pores. (Reprinted with permission from Salem M.S. et al., *J. Appl. Phys.* 100(8), 083520. Copyright 2006, American Institute of Physics.)

However, for a gas phase analyte, the spectral shift of PSi is determined mainly by surface adsorption effects, which govern the amount of analyte adsorbed, rather than the refractive index of the material. For example, Ruminski et al. (2010) established that despite the fact that isopropanol and heptane have similar refractive indexes, the response toward 500 ppm of isopropanol was ~9 times as large as the response toward 500 ppm of heptane (see Figure 3.16b). Only in the case of capillary condensation of vapors into PSi matrix, the refractive index of the material becomes a factor that determines the sensor response. Microcapillary condensation is a nanoscale phenomenon that provides an additional means of increasing sensitivity by spontaneously concentrating volatile molecules (Casanova et al. 2007). Experiment has shown that the extent of both adsorption and capillary condensation of vapors into PSi optical sensors is influenced by the surface affinity of analytes to the porous matrix. For example, freshly etched PSi possesses a hydride-terminated surface that is generally unstable and does not provide good binding sites for organic vapors (Ruminski et al. 2010). Thus, freshly etched PSi is generally modified by diverse chemical treatments to provide more stable surfaces and to impart sensitivity or specificity for the intended analyte. For example, thermal oxidation of Si produces a more hydrophilic surface with much stronger binding sites for isopropanol than for heptane (Ruminski et al. 2010). This means that the sensitivity toward specific analytes can be achieved by modification of the sensor surface's chemistry through simple or extensive surface functionalization (Gao et al. 2000, 2002a,b; Buriak 2002).

Decreasing the pore size of the PSi sensor is also a key factor in improving its sensitivity toward vapor analytes. Although analyte vapors will adsorb onto the surface of meso- and macroporous silicon, nanoscale pores possess an additional capability to concentrate a vapor, known as microcapillary condensation (Israelachvili 1992). The smaller the pore radius, the lower the partial pressure at which condensation can occur at a given temperature (Adamson 1990). Capillary condensation is responsible for the visibly observable, large (~100 nm) spectral shifts of the rugate stop band on vapor exposure. As it was shown in *Porous Silicon: Formation and Properties*, pore size can be controlled by changing both the current density during silicon porosification and the dopant level of the crystalline silicon wafer. Chemical stain etching (Kolasinski 2005), which degrades PSi without the application of an electrochemical bias, has also been used to create microporous layers with small pore diameters.

Experiment has shown that temperature control also gives possibility to influence the sensitivity of PSi-based sensors. In particular, Zangooie et al. (1999) demonstrated that vapor capture is promoted at lower temperatures, resulting in an increase of the ellipsometric response (see Figure 3.2b). Furthermore, decreasing the temperature resulted in the shifts in the ellipsometric parameters occurring at lower partial pressures.

The roughness between adjacent layers is another important parameter for the optical response of the multilayer-type PSi-based devices. According to conclusions made by Anderson et al. (2003), the roughness of interfaces between layers is one of the reasons for the observed difference between results of experiment and simulations. Huanca et al. (2008) found that by using a double electrochemical cell for device fabrication, the roughness at interfaces could be minimized. Another possible route to a reduction in the interface roughness might be to use low-temperature anodization as discussed by Setzu et al. (1998). These results can be explained by the changes taking place in the kinetics of electrochemical reactions at the solid/semiconductor interface.

3.7.2 SELECTIVITY

In principle, PSi-based optical sensors are not selective sensors. This means that it is not possible to quantify a complex gas mixture in one measurement, as it happened in gas chromatography. Experiment has shown that the improvement of selectivity can be achieved only via using a number of special approaches. They are as follows:

Lock and key approach. In this approach to specificity, PSi can be conjugated to a specific capture probe for a target species (Rickert et al. 1996; Albert et al. 2000). Specificity comes with a tradeoff because a separate sensor must be constructed for each new target analyte, and capture probes that tightly bind their targets can have poor reversibility (Rickert et al. 1996). This approach is mainly used in PSi-based bio-sensors to specifically bind a target analyte such as a protein.

Surface reactivity control via surface fuctionalizing. The surface of PSi has been extensively tailored due to the large library of reactions available for modifying silicon-hydrogen and silicon-silicon bonds (Song and Sailor 1999; Salem et al. 2006; Ruminski et al. 2010). Methods include hydrosilylation (Buriak 1999), radical coupling (Lewis 1998), Grignard reactions (Chazalviel 1987), electrochemical reduction of alkyl halides (Gurtner et al. 1999), and thermal carbonization with acetylene (Salonen et al. 2000, 2004). Because of these treatments, the PSi surface can directly react with some chemical species. For example, amine-terminated PSI provides effective detection of CO_2 (Rocchia et al. 2003). The carbonization of PSi surface increases the sensitivity to hydrophobic VOCs (heptane) in comparison with hydrophilic VOCs (isopropanol) (Ruminski et al. 2010).

Selectivity PSi interaction with analyte can be improved also by loading with catalysts that react with specific compounds. For instance, Rocchia et al. (2003) modified ozone-oxidized PSi with 3-amino-1-propanol, with the resulting amine groups on the sensor surface reactive toward carbon dioxide to form carbamate species. Hydrogen has been sensed with PSi by inclusion of a palladium hydrogen trap in the porous layer (Lin et al. 2004).

Using specific chemical reactions. Another approach for obtaining selectivity is to use a surface termination that reacts with certain gases in an irreversible fashion. For example, monitoring the corrosion of as-anodized and oxidized PSi surfaces caused by HF or Cl_2 (Letant and Sailor 2000; Ruminski et al. 2011) gases has been shown to be a feasible method for achieving highly selective single-use sensors. PSi reaction with HF or Cl_2 degrades the oxidized porous layer leading to a change in porosity, and therefore the average refractive index and spectral properties of the sensor. Jang et al. (2008) used another approach. They incorporated copper(II) sulfate in a PSi Bragg mirror, and using an LED as a light source, achieved extremely high sensitivity to triethyl phosphate. The detection limit of the sensor for triethyl phosphate was lowered from 1.4 ppm to 150 ppb. However, sensing schemes utilizing irreversible reactions are restricted to specific gases, and are thus difficult to generalize.

Physical control of surface reactions. As it was shown before, the porous photonic crystal surface chemistry has a profound effect on analyte vapor sorption, but sorption can also be modulated physically by heating or cooling the porous layer. Methods of thermal cycling to discriminate organic vapors have been employed with metal oxide and other sensor types (Heilig et al. 1997; Nakata et al. 1998; Lee and Reedy 1999), but have not been widely investigate with PSi. There are only two papers related to this topic (King et al. 2011; Kelly et al. 2011). The necessity of the spectral curves measurement limits the ability of measurements in dynamic mode. However, King (2010) has shown that a thermal pulse can be successfully applied to PSi optical sensors to refresh the sensor response after exposure to low volatility vapor analytes. As it is known, PSi sensors have long recovery times for low volatility organic vapors, and the gradual accumulation of vapor molecules in the porous matrix leads to reduced sensitivity and dynamic range of the sensors (Salem et al. 2006; Ruminski et al. 2008). It was established that thermal pulse with heating already up to 160°C rapidly refreshes the oxidized silica sensors to their initial baseline (King 2010).

Size exclusion. The ability to construct layers of PSi with different pore geometries has led to molecular selectivity by size exclusion. However, this approach was used mostly in biosensors. For instance, Orosco et al. (2009) constructed in enzyme sensors a double layer of mesoporous silicon for this purpose. Collins et al. (2002) have shown that in addition to stacks of PSi layers of different pore sizes, lateral pore diameter gradients of PSi can also be used for the size separation of molecules.

Using sensors arrays. This approach is the basis for the development of various electronic noses. Usually this approach uses the arrays, which vary the material or coating of each sensor element but use a single transduction methodology. In the case of PSi-based sensors, the change in pore size can replace the use of different materials. Every element in this sensor array has a specific response to analyte. Therefore, processing of the sensor array responses using specialized computer software can produce multidimensional response profiles that can be thought of as a digital "smellprint" or "fingerprint" that represents the chemical complexity of the gas mixture (Oton et al. 2003; Hutter and Ruschin 2010).

Multiparameters sensing. This approach uses multiple, differing transduction methods. In particular, multiparameters control can combine responses from sensors that use different operating principles, like reflectance sensing with PSi, PL quenching, Fourier transform infrared spectroscopy, Raman spectroscopy, fluorescence detection, ion mobility spectroscopy, mass spectroscopy, surface Plasmon resonance, or other methods (Letant et al. 2000; Mulloni et al. 2000; Baratto et al. 2002; Park et al. 2010). In addition to selectivity and sensitivity, this provides a way to rule out false positives, which increases the detection capability of the sensor. For instance, simultaneously monitoring thin film reflectance and PL enabled Letant et al. (2000) to discriminate among a large number of vapors, with the results being comparable to the performance of a commercial electronic nose. The group of Lorenzo Pavesi has used PSi optical microcavities and by monitoring electrical current and PL, they showed that it is possible to differentiate between NO_2 and humidity (Baratto et al. 2002).

3.7.3 STABILITY

Instability of PSi-based optical sensors' parameters has two main reasons. First, due to the high surface reactivity of PSi, there is an aging of PSi, which is accompanied by changes in its optical properties. The reasons for this aging and the approaches used for the stabilization of PSi properties were considered earlier in Chapter 12 in *Porous Silicon: Formation and Properties*. The experiment showed that the oxidation of PSi is the most effective method for resolving this problem (Salem et al. 2006). Ruminski et al. (2010) established that the sensors with ozone-oxidized PSi layer are the most stable. However, it is necessary to take into account that silicon oxide has a refractive index lower than crystalline silicon, so that the average refractive index of oxidized PSi films strongly changes. This means that the reflectance spectrum will be changed during PSi oxidation (Figure 3.22). The design of optical devices must take care of this.

The second reason for the possible instability of the PSi-based optical sensor parameters is the fluctuations in probe light intensity, absorbing or scattering species in the optical path, and changes in temperature, pressure, or humidity, which can all affect the sensor response of PSi optical sensors and lead to false positives (Ruminski et al. 2008; King 2010). It should be noted that this is a common problem for all optical sensors. For its solution, for example, fiber optic sensor systems typically employ temporal, frequency, spatial, or wavelength separated referencing (Grattan and Meggitt 2000). The last two methods are most commonly used with PSi optical sensors (King 2010). In spatially separated referencing, a reference channel is physically distinct. Numerous configurations are possible, but typically involve splitting light from the illumination source to a separate reference detector, bypassing the sensor, or using a duplicate interrogation source, sensor, and detector in a physically separate reference channel, with the sensing element in the reference channel sealed and unable to interact with analyte vapors but experiencing temperature or other changes to be mitigated by referencing (Grattan and Meggitt 2000; Wolfbeis 2006). In wavelength-separated referencing, the signal and reference channels often share the same optical path, but correspond to different spectral features at separated wavelengths. This simplifies the referencing approach, and ensures that interferents or probe light fluctuations are experienced by both the signal and reference channels.

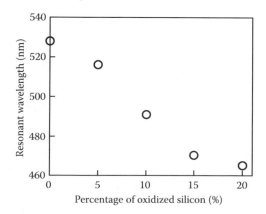

FIGURE 3.22   Simulated resonant wavelength of oxidized PSi rugate filter (PSRF) at different volume fractions of oxidized silicon. (Reprinted with permission from Salem M.S. et al., *J. Appl. Phys.* 100(8), 083520. Copyright 2006, American Institute of Physics.)

It should be noted that there are many technical solutions that realize the above-mentioned approaches. For example, Ruminski et al. (2008), to reduce the zero-point drift caused by the influence of humidity, suggested the use of PSi double-layer. One layer is hydrophobic-functionalized while the other layer is hydrophilic-functionalized. By comparing the response of both layers relative to one another, the interfering effect of RH can be subtracted out.

Finally, another method of avoiding interferents in the optical path is to reduce the segment of the optical path open to air. Rather than directing light from a bifurcated fiber through air, nitrogen, or a biological medium to the sensor surface, films of rugate PSi may be directly attached to optical fibers, greatly reducing the effects of any transient absorbing species in the optical path. In particular, such an approach was proposed by Karacali et al. (2013) and discussed in the previous section.

3.8 CONCLUSIONS AND FUTURE TRENDS

PSi nanostructured optical sensors show some advantages that are in some cases unique: their production is low cost; most of the devices are really easy-to-use and reusable; for many of them, high sensitivity and detection limits of ppm, or hundreds of ppb, have been reported; and sensing systems can span from simple structures on wafer to sophisticated transducers integrated with lab-on-chip. PSi photonic can be exploited in not only gas and vapor monitoring but also mostly as a useful platform for testing physical and chemical phenomena: absorption-desorption kinetic, dew point studies, effective medium model analysis, and so on. Even remote gas sensing as been proved to be enhanced by using PSi-modified materials (Schmedake et al. 2002; Ruminski et al. 2011). By changing PSi surface chemistry, some interactions can be enhanced and others partially suppressed.

Despite all the pros regarding PSi as optical transducer material, some critical considerations should be underlined, especially from the sensor theory point of view: passivation of PSi surface is mandatory because samples as produced undergo spontaneous aging, so that stability is not assured; also, thermal conditioning is required because capillary condensation depends on temperature and calibration curves, and could suffer strong thermal drift. The sensing mechanism is intrinsically specific, that is, it generates a different sensor response for different substances, but it is definitely not selective, that is, it does not recognize the element of a gas mixture. No one of these cons is overwhelming and as far as the technology ability will increase, the realization of a selective optical instrument based on a PSi transducer will be not so impractical.

An all-PSi-based microsystem composed by a chromatographic serpentine made of a porous channel could separate the gases injected in the microsystem, which once arrived in a test chamber equipped with a PSi optical transducer should be recognized as pure substances: an integrated approach of this kind is not just an idea, it could be the argument of our next publication.

REFERENCES

Adamson, A.W. (1990). *Physical Chemistry of Surfaces*. 5th ed. John Wiley & Sons, New York, pp. 58–59.

Albert, K.J., Lewis, N.S., Schaue, C.L. et al. (2000). Cross-reactive chemical sensor arrays. *Chem. Rev.* **100**(7), 2595–2626.

Allcock, P. and Snow, P.A. (2001). Time-resolved sensing of organic vapors in low modulating porous silicon dielectric mirrors. *J. Appl. Phys.* **90**, 5052–5057.

Alvarez, J., Kumar, N., Bettotti, P., Hill, D., and Martınez-Pastor, J. (2012). Phase-sensitive detection for optical sensing with porous silicon. *IEEE Phot. J.* **4**(3), 986–995.

Anderson, M.A., Tinsley-Brown, A., Allcock, P. et al. (2003). Sensitivity of the optical properties of porous silicon layers to the refractive index of liquid in the pores. *Phys. Stat. Sol. (a)* **197**, 528–533.

Arrand, H.F., Benson, T.M., Loni, A. et al. (1998). Novel liquid sensor based on porous silicon optical wave-guides. *IEEE Phot. Techn. Lett.* **10**, 1467–1469.

Arwin, H., Wang, G., and Jansson, R. (2003). Gas sensing based on ellipsometric measurement on porous silicon. *Phys. Stat. Sol. (a)* **197**, 518.

Baratto, C., Faglia, G., Sberveglieri, G. et al. (2002). Multiparametric porous silicon sensors. *Sensors* **2**, 121–126.

Buriak, J.M. (1999). Organometallic chemistry on silicon surfaces: Formation of functional monolayers bound through Si-C bonds. *Chem. Commun.* **12**, 1051–1060.

Buriak, J.M. (2002). Organometallic chemistry on silicon and germanium surfaces. *Chem. Rev.* **102**(5), 1272–1308.

Canham, L.T. (1990). Silicon quantum wire array fabrication by electrochemical and chimica dissolution of wafers. *Appl. Phys. Lett.* **57**, 1046–1048.

Casanova, F., Chiang, C.E., Li, C.P., and Schuller, I.K. (2007). Direct observation of cooperative effects in capillary condensation: The hysteretic origin. *Appl. Phys. Lett.* **91**, 243103.

Chazalviel, J.-N. (1987). Surface methoxylation as the key factor for the good performance of *n*-Si/methanol photoelectrochemical cells. *J. Electroanal. Chem.* **233**, 37–48.

Chow, E., Grot, A., Mirkarimi, L.W., Sigalas, M., and Girolami, G. (2004). Ultracompact biochemical sensor built with two-dimensional photonic crystal microcavity. *Opt. Lett.* **29**, 1093–1095.

Collins, B.E., Dancil, K.-P., Abbi, G., and Sailor, M.J. (2002). Determining protein size using an electrochemically machined pore gradient in silicon. *Adv. Funct. Mater.* **12**(3), 187–191.

Dancil, K.P.S., Greiner, D.P., and Sailor, M.J. (1999). A porous silicon optical biosensor: Detection of reversible binding of IgG to a protein A-modified surface. *J. Am. Chem. Soc.* **121**, 7925–7930.

De Stefano, L., Rendina, I., Moretti, L. et al. (2003). Optical sensing of flammable substances using porous silicon microcavities. *Mater. Sci. Eng. B* **100**, 271–274.

De Stefano, L., Moretti, L., Rossi, A.M. et al. (2004a). Optical sensors for vapors, liquids, and biological molecules based on porous silicon technology. *IEEE Trans. Nanotech.* **3**, 49–54.

De Stefano, L., Moretti, L., Rendina, I., Rossi, A.M., and Tundo, S. (2004b). Smart optical sensors for chemical substances based on porous silicon technology. *Appl. Opt.* **43**, 167–172.

De Stefano, L., Moretti, L., Rendina, I., and Rossi, A.M. (2004c). Time-resolved sensing of chemical species in porous silicon optical microcavity. *Sens. Actuators B* **100**, 168–172.

De Stefano, L., Malecki, K., Rossi, A.M. et al. (2006). Integrated silicon-glass opto-chemical sensors for lab-on-chip applications. *Sens. Actuators B* **114**, 625–630.

De Stefano, L., Rotiroti, L., Rea, I., Iodice, M., and Rendina, I. (2007). Optical microsystems based on a nano-material technology. *J. Phys.: Condens. Matter* **19**, 395008.

Dorvee, J. and Sailor, M.J. (2005). A low-power sensor for volatile organic compounds based on porous silicon photonic crystals. *Phys. Stat. Sol. (a)* **202**(8), 1619–1623.

Gao, J., Gao, T., and Sailor, M.J. (2000). A porous silicon vapor sensor based on laser interferometry. *Appl. Phys. Lett.* **77**(6), 901–903.

Gao, J., Gao, T., Li, Y.Y., and Sailor, M.J. (2002a). Vapor sensors based on optical interferometry from oxidized microporous silicon films. *Langmuir* **18**, 2229–2233.

Gao, J., Gao, T., and Sailor, M.J. (2002b). Tuning the response and stability of thin film mesoporous silicon vapor sensors by surface modification. *Langmuir* **18**(25), 9953–9957.

Geobaldo, F., Onida, B., Rivolo, P. et al. (2001). IR detection of NO_2 using $p+$ porous silicon as a high sensitivity sensor. *Chem. Commun.* 2196–2197.

Grattan, K.T.V. and Meggitt, B.T. (eds.) (2000). *Optical Fiber Sensor Technology.* Kluwer Academic Publishers, Boston.

Gurtner, C., Wun, A.W., and Sailor, M.J. (1999). Surface modification of porous silicon by electrochemical reduction of organo halides. *Angew. Chem. Int. Ed.* **38**, 1966–1968.

Heilig, A., Barsan, N., Weimar, U., Schweizer-Berberich, M., Gardner, J.W., and Göpel, W. (1997). Gas identification by modulating temperatures of SnO_2-based thick film sensors. *Sens. Actuators B* **43**, 45–51.

Huanca, D.R., Ramirez-Fernandez, F.J., and Salcedo, W.J. (2008). Porous silicon optical cavity structure applied to high sensitivity organic solvent sensor. *Microelectron. J.* **39**, 499.

Hutter, T. and Ruschin, S. (2010). Non-imaging optical method for multi-sensing of gases based on porous silicon. *IEEE Sens. J.* **10**, 97–103.

Imenes, A.G. and McKenzie, D.R. (2006). Flat-topped broadband rugate filters. *Appl. Optics* **45**(30), 7841–7850.

Israelachvili, J.N. (1992). *Intermolecular and Surface Forces.* 2nd ed. Academic Press, London, p. 331.

Jalkanen, T., Torres-Costa, V., Salonen, J. et al. (2009). Optical gas sensing properties of thermally hydro-carbonized porous silicon Bragg reflectors. *Opt. Expr.* **17**(7), 5446–5456.

Jang, S.H., Koh, Y.D., Kim, J. et al. (2008). Detection of organophosphates based on surface modified DBR porous silicon using LED light. *Mater. Lett.* **62**(3), 552–555.

Karacali, T., Hasar, U.C., Ozbek, I.Y., Oral, E.A., and Efeoglu, H. (2013). Novel design of porous silicon based sensor for reliable and feasible chemical gas vapor detection. *J. Lightwave Technol.* **31**(2), 295–305.

Kelly, T.L., Sega, A.G., and Sailor, M.J. (2011). Identification and quantification of organic vapors by time-resolved diffusion in stacked mesoporous photonic crystals. *Nano Lett* **11**(8), 3169–3173.

Kim, K. and Murphy, T.E. (2013). Porous silicon integrated Mach-Zehnder interferometer waveguide for biological and chemical sensing. *Opt. Exp.* **21**(17), 19488–19497.

King, B.H. (2010). Design and manipulation of 1-D rugate photonic crystals of porous silicon for chemical sensing applications. PhD Thesis. University of California, San Diego.

King, B.H., Wong, T., and Sailor, M.J. (2011). Detection of pure chemical vapors in a thermally cycled porous silica photonic crystal. *Langmuir* **27**, 8576–8585.

Kolasinski, K.W. (2005). Silicon nanostructures from electroless electrochemical etching. *Curr. Op. Sol. St. Mater. Sci.* **9**(1–2), 73–83.

Lambrecht, A., Hartwig, S., Schweizer, S.L. et al. (2007). Miniature infrared gas sensors using photonic crystals. *Proc. SPIE* 6480, *Photonic Crystal Materials and Devices VI*, 64800D. doi:10.1117/12.700792.

Lee, A.P. and Reedy, B.J. (1999). Temperature modulation in semiconductor gas sensing. *Sens. Actuators B* **60**, 35–42.

Lehman, V. (2002). *Electrochemistry of Silicon*. Wiley-VCH Verlag, Weinheim, pp. 17–20.

Letant, S.E. and Sailor, M.J. (2000). Detection of HF gas with a porous silicon interferometer. *Adv. Mater.* **12**(5), 355–359.

Letant, S.E., Content, S., Tan, T.T., Zenhausern, F., and Sailor, M.J. (2000). Integration of porous silicon chips in an electronic artificial nose. *Sens. Actuators B* **69**, 193–198.

Lewis, N.S. (1998). Progress in understanding electron-transfer reactions at semiconductor/liquid interfaces. *J. Phys. Chem. B* **102**(25), 4843–4855.

Li, S., Hu, D., Huang, J., and Cai, L. (2012). Optical sensing nanostructures for porous silicon rugate filters. *Nanoscale Res. Lett.* **7**, 79–84.

Lin, V.S.–Y., Motesharei, K., Dancil, K.-P.S., Sailor, M.J., and Ghadiri, M.R. (1997). A porous silicon based optical interferometric biosensor. *Science* **270**, 840–843.

Lin, H., Gao, T., Fantini, J., and Sailor, M.J. (2004). A porous silicon-palladium composite film for optical interferometric sensing of hydrogen. *Langmuir* **20**, 5104–5108.

Liu, N.H. (1997). Propagation of light waves in Thue-Morse dielectric multilayers. *Phys. Rev. B* **55**, 3543–3547.

Liu, R., Schmedake, T.A., Li, Y.Y., Sailor, M.J., and Fainman, Y. (2002). Novel porous silicon vapor sensor based on polarization interferometry. *Sens. Actuators B* **87**, 58–62.

Lorenzo, E., Oton, C.J., Capuj, N.E. et al. (2005). Porous silicon-based rugate filters. *Appl. Optics* **44**(26), 5415–5421.

Moretti, L., De Stefano, L., Rea, I., and Rendina, I. (2007). Periodic versus aperiodic: Enhancing the sensitivity of porous silicon based optical sensor. *Appl. Phys. Lett.* **90**, 191112–3.

Mulloni, V. and Pavesi, L. (2000). Porous silicon microcavities as optical chemical sensors. *Appl. Phys. Lett.* **76**, 2523–2525.

Mulloni, V., Gaburro, Z., and Pavesi, L. (2000). Porous silicon microcavities as optical and electrical chemical sensors. *Phys. Stat. Sol. (a)* **182**, 479–484.

Muriel, M.A. and Carballar, A. (1997). Internal field distributions in fiber Bragg gratings. *IEEE Photon. Technol. Lett.* **9**, 955–957.

Naderi, N., Hashim, M.R., and Amran, T.S.T. (2012). Enhanced physical properties of porous silicon for improved hydrogen gas sensing. *Superlatt. Microstruct.* **51**, 626–634.

Nakata, S., Nakasuji, M., Ojima, N., and Kitora, M. (1998). Characteristic nonlinear responses for gas species on the surface of different semiconductor gas sensors. *Appl. Surf. Sci.* **135**, 285–292.

Neimark, V. and Ravikovitch, P.I. (2001). Capillary condensation in MMS and pore structure characterization. *Microporous Mesoporous Mater.* **44–45**, 697–707.

Orosco, M.M., Pacholski, C., and Sailor, M.J. (2009). Real-time monitoring of enzyme activity in a mesoporous silicon double layer. *Nature Nanotech.* **4**, 255–258.

Oton, C.J., Pancheri, L., Gaburro, Z. et al. (2003). Multiparametric porous silicon gas sensors with improved quality and sensitivity. *Phys. Stat. Sol. (a)* **197**, 523–527.

Ouyang, H. and Fauchet, P. (2005). Biosensing using porous silicon photonic bandgap structures. *Proc. SPIE* **6005**, 600508–1.

Ouyang, H., Christophersen, M., Viard, R., Miller, B.L., and Fauchet, P.M. (2005). Macroporous silicon microcavities for macromolecule detection. *Adv. Funct. Mater.* **15**, 1851–1859.

Ouyang, H., Striemer, C., and Fauchet, P.M. (2006). Quantitative analysis of the sensitivity of porous silicon optical biosensors. *Appl. Phys. Lett.* **88**, 163108–3.

Pacholski, C. (2013). Photonic crystal sensors based on porous silicon. *Sensors* **13**(4), 4694–4713.

Pacholski, C. and Sailor, M.J. (2007). Sensing with porous silicon double layers: A general approach for background suppression. *Phys. Stat. Sol. (c)* **4**(6), 2088–2092.

Park, S.-H., Seo, D., Kim, Y.-Y., and Lee, K.-W. (2010). Organic vapor detection using a color-difference image technique for distributed Bragg reflector structured porous silicon. *Sens. Actuators B* **147**, 775–779.

Rickert, J., Weiss, T., and Gopel, W. (1996). Self-assembled monolayers for chemical sensors: Molecular recognition by immobilized supramolecular structures. *Sens. Actuators B* **31**, 45–50.

Rocchia, M.A., Garrone, E., Geobaldo, F., Boarino, L., and Sailor, M.J. (2003). Sensing CO_2 in a chemically modified porous silicon film. *Phys. Stat. Sol. (a)* **197**(2), 365–369.

Ruminski, A.M., Moore, M.M., and Sailor, M.J. (2008). Humidity-compensating sensor for volatile organic compounds using stacked porous silicon photonic crystals. *Adv. Funct. Mater.* **18**(21), 3418–3426.

Ruminski, A.M., King, B.H., Salonen, J., Snyder, J.L., and Sailor, M.J. (2010). Porous silicon-Based optical microsensors for volatile organic analytes: Effect of surface chemistry on stability and specificity. *Adv. Funct. Mater.* **20**(17), 2874–2883.

Ruminski, A.M., Barillaro, G., Chaffin, C., and Sailor, M.J. (2011). Internally referenced remote sensors for HF and Cl_2 using reactive porous silicon photonic crystals. *Adv. Funct. Mater.* **21**, 1511–1525.

Ruzicka, J. and Hansen, E.H. (1975). Flow injection analysis. Part I. A new concept of fast continuous flow analysis. *Anal. Chim. Acta* **78**, 145–157.

Ruzicka, J. and Hansen, E.H. (1998). *Flow Injection Analysis*, 2nd ed. John Wiley & Sons, New York.

Sailor, M.J. and Link, J.R. (2005). Smart dust: Nanostructured devices in a grain of sand. *Chem. Commun.* **2005**, 1375–1383.

Salem, M.S., Sailor, M.J., Harraz, F.A., Sakka, T., and Ogata, Y.H. (2006). Electrochemical stabilization of porous silicon multilayers for sensing various chemical compounds. *J. Appl. Phys.* **100**(8), 083520.

Salonen, J., Lehto, V.P., Björkqvist, M., Laine, E., and Niinistö, L. (2000). Studies of thermally-carbonize porous silicon surfaces. *Phys. Stat. Sol. (a)* **182**(1), 123–126.

Salonen, J., Bjorkqvist, M., Laine, E., and Niinisto, L. (2004). Stabilization of porous silicon surface by thermal decomposition of acetylene. *Appl. Surf. Sci.* **225**, 389–394.

Schmedake, T.A., Cunin, F., Link, J.R., and Sailor, M.J. (2002). Standoff detection of chemicals using porous silicon "smart dust" particles. *Adv. Mater.* **14**, 1270–1272.

Sciacca, B., Secret, E., Pace, S. et al. (2011). Chitosan-functionalized porous silicon optical transducer for the detection of carboxylic acid-containing drugs in water. *J. Mater. Chem.* **21**, 2294–2302.

Setzu, S., Letant, S., Solsona, P., Romenstain, R., and Vial, J.C. (1998). Improvement of the luminescence in *p*-type as-prepared or dye impregnated porous silicon microcavities. *J. Lumin.* **80**, 129–135.

Snow, P.A., Squire, E.K., Russell, P.S.J., and Canham, L.T. (1999). Vapor sensing using the optical properties of porous silicon Bragg mirrors. *J. Appl. Phys.* **86**, 1781–1784.

Song, J.H. and Sailor, M.J. (1999). Chemical modification of crystalline porous silicon surfaces. *Comments Inorg. Chem.* **21**(1–3), 69–84.

Soukoulis, C.M. and Economou, E.N. (1982). Localization in one-dimensional lattices in the presence of incommensurate potentials. *Phys. Rev. Lett.* **48**, 1043–1046.

Valcorcel, M. and Luque de Castro, M.D. (1987). *Flow-Injection Analysis. Principles and Applications.* Ellis Horwood Ltd., Chichester, UK.

Wolfbeis, O.S. (2006). Fiber-optic chemical sensors and biosensors. *Anal. Chem.* **78**(12), 3859–3874.

Zangooie, S., Bjorklund, R., and Arwin, H. (1997). Vapor sensitivity of thin porous silicon layers, *Sens. Actuators B* **43**, 168–175.

Zangooie, S., Jansson, R., Arwin, H. (1999). Ellipsometric characterization of anisotropic porous silicon Fabry-Pérot filters and investigation of temperature effects on capillary condensation efficiency. *J. Appl. Phys.* **86**, 850–858.

Zhao, Y., Zhang, Y.N., and Wang, Q. (2011). Research advances of photonic crystal gas and liquid sensors. *Sens. Actuators B* **160**, 1288–1297.

PSi-Based Electrochemical Sensors

4

Yang Zhou, Shaoyuan Li, Xiuhua Chen, and Wenhui Ma

CONTENTS

4.1 Introduction 96

4.2 Electrochemical Sensors 96

4.3 Preparation of Electrochemical Sensors 96

4.4 Electrochemical Sensors for Metal Ions 97

4.5 Electrochemical Sensors for pH Measurements 101

4.6 PSi-Based Electrochemical Gas and Vapor Sensors 102

4.7 Summary and Perspectives 105

References 105

4.1 INTRODUCTION

As it was shown in previous chapters, porous silicon (PSi), due to large specific surface area and high surface reactivity, is a promising material for designing various solid state and optical gas sensors. From general considerations, electrochemical sensors, aimed for operation mainly in a liquid environment for detection metal ions and dissolved toxic substances, also should be the area for PSi application. However, in reality, despite the advantages of electrochemical sensors such as easy preparation, low cost, high sensitivity, simplicity, and high functionality, there are very few works that have been done in this field of chemical sensors. All of these considerations make it particularly worthwhile to review the advantages of PSi-based electrochemical sensors.

4.2 ELECTROCHEMICAL SENSORS

Electrochemical sensors include those sensors that detect signal changes caused by an electric current being passed through electrodes that interact with chemicals. They can be categorized into three groups: (1) potentiometric (measurement of voltage), (2) amperometric (measurement of current), and (3) conductometric (measurement of conductivity). The fundamental requirement of the electrochemical method is a mobile electrolyte to maintain charge balance once an electron is removed or injected into the chemical being detected. The highly sensitive, reactive surface of PSi and the ability to record the changes in its electrical properties, along with its capacity to adsorb a wide range of different analytes are key factors that can help greatly in its utilization as electrical sensors.

4.3 PREPARATION OF ELECTROCHEMICAL SENSORS

PSi can be used for the realization of amperometric and voltammetric sensors, besides potentiometric ones. In this case, the immobilized probe species catalyses a redox reaction involving analytes oxidation/reduction, which produces a flux of electrons measured in terms of current intensity by the electrodes of the electrochemical cell. In these systems, PSi is usually one of the involved electrodes. In fact, due to its low conductivity properties, the PSi surface has to be modified using a conducting material such as gold (Ressine et al. 2010), platinum (Song et al. 2006, 2007), or a conductive polymer (Jin et al. 2010; Zhao and Jiang 2010). In that case, the main function of PSi is to act as a high surface area substrate to improve the sensitivity of sensing. Actually, it has been reported that porous electrodes can increase sensor sensitivity in comparison with flat surface electrodes. The electrochemical workstation and three-electrode test system are commonly needed for the PSi-based electrochemical sensor and the typical test configuration is shown in Figure 4.1. As we know, the electrical signal collection is the key step for the electrochemical sensor. The PSi with a low resistance is thus preferred for use in an electrochemical sensor. Moreover, the low resistance silicon substrate is easier to be anodized for fabricating uniform PSi with the size of 50~200 nm. PSi can be fabricated by various methods, of which

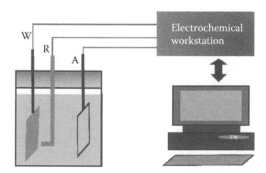

FIGURE 4.1 Typical three-electrode measurement system of PSi-based electrochemical sensor, the modified PSi is working-electrode (W), R is reference electrode, and A is auxiliary electrode.

electrochemical etching is often preferred to generate significant material with the uniform pore structure and convenience for controlling surface chemistry in electrochemical sensors.

PSi has high surface area and active surface chemical state, and thus a poor stability is predictable in the electrolyte, especially for the acidic electrolyte with fluorine ion. The blank PSi would not be used to fabricate an electrochemical sensor; the functionalization of PSi is not only helpful in specificity application, but also important in improving the stabilization.

In general, the functionalization of PSi for fabricating electrochemical sensors contains the following ways:

1. Silanization technique is the most popular way to modify the PSi electrochemical sensors. This method can generate the various end groups on the PSi via chemical surface modification.
2. Deposition of metal ions on the surface of PSi is considered a useful three-electrode electrochemical sensor.
3. Molecules' self-assembled monolayer on the PSi can generate the activated surface.
4. Other methods such as linkage group, physics adsorption, and ion sputtering are utilized to functionalize PSi.

4.4 ELECTROCHEMICAL SENSORS FOR METAL IONS

Electrochemical sensors based on modified PSi surfaces with recognition probes meet rapid, simple, sensitive, and low cost requirements for a fast and easy analysis of metal ions and they are likely to be miniaturized to allow the development of detection equipment capable of operating directly on site. Sam et al. (2010) developed covalent immobilization of amino acids on the PSi to detect complex Cu(II) ions from solution by cyclic voltammetry. They demonstrated that a dense monolayer of acid chains was first grafted via thermal hydrosilylation of undecylenic acid at the surface of hydrogenated PSi. The reaction takes place at the terminal C=C double bond and yields an organic monolayer covalently attached to the surface through Si–C bonds. Next, the acid terminal groups are transformed to succinimidyl ester terminations. The surface bearing an "activated ester" termination is subsequently made to react with the amino end of the amino acid, allowing for its covalent attachment through the formation of an amide bond. This feature makes the electrochemical detection of copper ions fast and easy and allows the detection to be achieved directly in the field. Shortly afterward, they further studied a Glycyl-Histidyl-Glycyl-Histidine (GlyHisGlyHis) peptide, which is covalently anchored to the PSi surface using a multistep reaction scheme compatible with the mild conditions required for preserving the probe activity (Sam et al. 2011). The property of peptides to form stable complexes with metal ions is exploited to achieve metal-ion recognition by the peptide-modified PSi-based biosensor. An electrochemical study of the GlyHisGlyHis-modified PSi electrode is achieved in the presence of copper ions. The recorded cyclic voltammograms showed a quasi-irreversible process corresponding to the Cu(II)/Cu(I) couple (shown in Figure 4.2). The kinetic factors (the heterogeneous rate constant and the transfer coefficient) and the stability constant of the complex formed on the PSi surface are determined. The molecular structure, the homogeneity of the layer, the surface density, bonds stability, and processes reproducibility are parameters that determine the performance of subsequent applications of those modified surfaces and, therefore, must be perfectly controlled. These two literatures mainly discussed the modification and immobilization organic macromolecules on the surface of PSi to form stable functionalized PSi electrodes.

Porosities, porous layer thicknesses, and pores morphologies of PSi and resistivities of bulk silicon are also factors of the influence on the reactivity and sensitivity of functionalized PSi electrodes. Ali et al. (1999) fabricated the large interfacial area of the PSi, which is oxidized by rapid thermal oxidation technique. The structure is then used as the working electrode of an electrochemical cell and sensor properties were investigated by capacitance measurements. This senor is an over-Nernstian response of the coated PSi to the sodium ion. Two years later, they further studied the dependence of the PSi sensitivity on the porosity of the samples, which were prepared from a lightly doped silicon substrate. Then, they presented a model to explain the mechanism of the ionic species adsorption at the electrolyte/SiO_2 interface, and to interpret the observed large

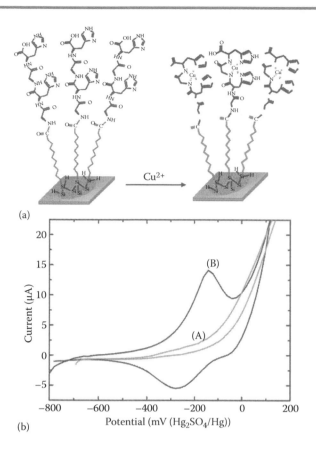

(a)

(b)

FIGURE 4.2 (a) Reaction scheme of the transition metal complexation on a PSi sensor modified with peptide. In this case, Gly-HisGly-His chelating Cu(II) cations. (b) Cyclic voltammetry of a GlyHisGlyHis-modified PSi surface, (A) before copper accumulation, (B) after copper accumulation. Scan rate = 0.5 V/s. (With kind permission from Springer Science+Business Media: *Nanoscale Res. Lett.* 6, 2011, 412, Sam S.S. et al.)

sensitivity against the different concentrations of the cations (Zairi et al. 2001a). They studied in detail the response of the PSi material against sodium ion concentrations with regard to its porosity, its pore and morphology, and its layer thickness. The dependence of the sensitivity on the porosity of the samples prepared from highly doped substrates has been studied. Maximal values of over-Nernstian sensitivities around 240 mV/p Na and ~92 mV/p Cu, corresponding to a PSi-layer porosity of about 65%, obtained, respectively, from p^- and p^+ silicon substrates have been registered. The responses of functionalized PSi samples toward sodium cation showed a large over-Nernstian behavior. It shows that this sensitivity enhancement as the PSi-layer porosity increases has been attributed to a crystallite quantum barrier effect. The best sensitivity that has been registered corresponds to a thin PSi layer and to a lightly doping level of the silicon material (Zairi et al. 2001b).

For nickel ion detection, Sakly et al. (2006) studied the electrochemical response of a field effect capacitor composed of a PSi/silicon dioxide structure as transducer's surface and p-tert-butylcalix[6] arene molecules as a recognizing agent toward nickel ions. They have estimated the active surface of the studied sensors (4.29 cm^2) deduced from C/V measurements and found the best potentiometric samples where SiO_2 growth has been accomplished by anodic oxidation in the KNO_3(1 M) solution than in H_2SO_4 (as shown in Figure 4.3). Before the immobilization of molecules on the surface of a PSi electrochemical sensor, pretreatment of PSi is usually necessary. Oxidizing the porous layers with KNO_3 and H_2SO_4 is a possible way but not the most efficient method to deal with it.

Our group proposed a novel voltammetric sensor based on 3-aminopropyltriethoxysilanes modified PSi electrode (APTES-PSE) to be used for determining Ag$^+$ in aqueous solution (Li et al. 2013). The schematic illustration of the preparation and measurement process of APTES-PSE is

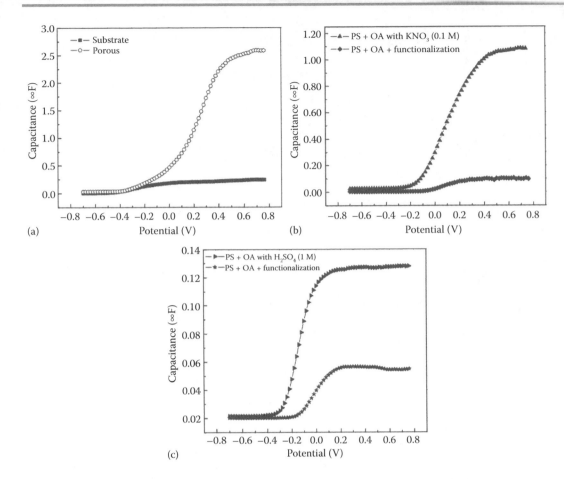

FIGURE 4.3 (a) Capacitance curves versus the applied potential of the substrate and the PSi samples. (b) Capacitance curves versus the applied potential of the oxidized PSi in KNO₃ (0.1 M) and the functionalized structure. (c) Capacitance curves versus the applied potential of the oxidized PSi in H₂SO₄ (1 M) and the functionalized structure. (Reprinted from *Mater. Sci. Eng. C* 26, Sakly H. et al., 232, Copyright 2006, with permission from Elsevier.)

shown in Figure 4.4. Under the optimal experimental conditions, the cathode peak current intensity ($|I\mathrm{pc}|$) of APTES-PSE linearly increases with logarithm of Ag^+ concentrations over a wide range from 1×10^{-3} mol/L to 1×10^{-7} mol/L. The calibration plot in Figure 4.5 highlights a linear response for logarithmic Ag^+ concentrations over the range from 1×10^{-3} mol/L to 1×10^{-7} mol/L. The linear equation is $|I\mathrm{pc}|$ (A/cm²) $= 118.28 + 15.15 \times \mathrm{lgCAg^+}$ (CAg⁺/mol/L) ($R^2 = 0.97173$).

Surface chemistry is important to the properties of PSi. In order to investigate the influence of modification on the surface of PSi on the detection characteristics, we fabricated cleavable

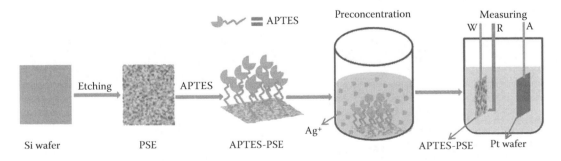

FIGURE 4.4 Schematic illustration of the preparation and measurement process of APTES-PSE. (From Li S.Y. et al., *Int. J. Electrochem. Sci.* 8, 1802, 2013. Published by ESG as open access. With permission.)

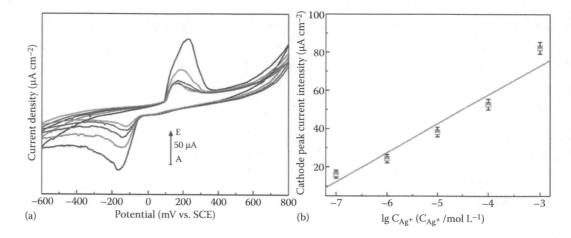

(a)   Potential (mV vs. SCE)

(b) $lg\ C_{Ag^+}\ (C_{Ag^+}\ /mol\ L^{-1})$

FIGURE 4.5   (a) Cyclic voltammograms of APTES-PSE in 0.1 mol/L KNO_3 after preconcentration of various concentrations: (A) 1×10^{-3} mol/L, (B) 1×10^{-4} mol/L, (C) 1×10^{-5} mol/L, (D) 1×10^{-6} mol/L, (E) 1×10^{-7} mol/L. Conditions: preconcentration time: 60 min, pH = 5, scan rate: 100 mV/s. (b) Calibration plot of cathode current intensity versus concentrations of Ag^+. (From Li S.Y. et al., *Int. J. Electrochem. Sci.* 8, 1802, 2013. Published by ESG as open access. With permission.)

PSi-based hybrid material by grafting benzimidazoledithio onto the surface of PSi using a multi-step covalent process (Li et al. 2012). The disulfide linkage of the benzimidazoledithio PSi can be dissociated successfully in the presence of reduced glutathione. The results demonstrated that the benzimidazoledithio PSi possessed a similar preferential adsorption trend for the studied metal ions (Cd > Cu ≫ Hg ~ Pb ~ Co) at different pH (from 2.0 to 6.0) (shown in Figure 4.6). At pH = 5.0, the benzimidazoledithio PSi showed the most efficient preenrichment for Cd ions. The concentration of Cd ions was increased 15.9 times and the recovery ratio was found to be 95.4% after the preenrichment. The silanization technique we utilized is the most popular way to modify the PSi layers. This method can generate the various end groups on the PSi via chemical surface modification and be easy to covalently bond to various probe molecules.

Recent research in our group demonstrated that the thiosemicarbazide derivative modified PSi could be used to detect Pb^{2+} in aqueous solution by using anodic stripping voltammetry. The results display that the anodic peak current density increases with the increasing Pb^{2+} concentration and

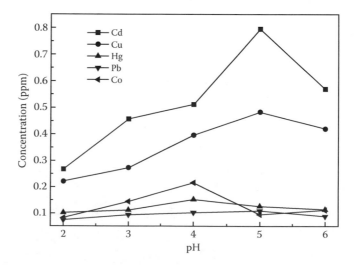

FIGURE 4.6   The effect of pH on the preenrichment for different metal ions. Initial metal ions concentration = 0.05 ppm, solution volume = 50 ml, enrichment time = 5 h, eluent = 3 ml GSH solution, desorption time = 12 h, temperature = 25°C. (Reprinted from *Appl. Surf. Sci.* 258, Li S.Y. et al., 5538, Copyright 2012, with permission from Elsevier.)

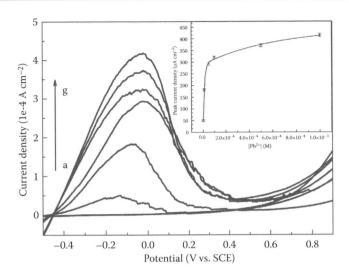

FIGURE 4.7 Anodic stripping voltammetric curves of the TSCD-PSi electrode after the accumulation of different concentrations of Pb^{2+} (a: 0 M, b: 1×10^{-6} M, c: 1×10^{-5} M, d: 5×10^{-5} M, e: 1×10^{-4} M, f: 5×10^{-4} M, g: 1×10^{-3} M). Inset: Calibration curve of anodic peak current density against copper concentration. Scanning rate: 100 mV/s; temperature: $25 \pm 2°C$.

has a good fitting with a second order exponential function, $I = -156\exp(-1562[Pb^{2+}]) - 260\exp(-10^5 [Pb^{2+}]) + 446$ ($R^2 = 0.998$) (shown in Figure 4.7).

4.5 ELECTROCHEMICAL SENSORS FOR pH MEASUREMENTS

Simonis et al. (2003) has developed a novel macroporous-type electrolyte–insulator–semiconductor sensor for the detection of (bio-)chemical substances in aqueous solutions. For this sensor, the SiO_2/ Si_3N_4 double layer insulator can be advantageously deposited by high-temperature processes to realize the porous pH sensor after the etching process. The results showed that the so-prepared pH sensors have an averaged sensitivity of 59 mV/pH (concentration range between pH 4 and 9) during a period of more than 1 year (shown in Figure 4.8). The measurement showed reproducible potentials for each pH value with short response times of smaller than 1 min. The hysteresis was less than 5 mV. The respective calibration curve demonstrated a nearly linear sensor response. Actually, Schoning et al. (1997)

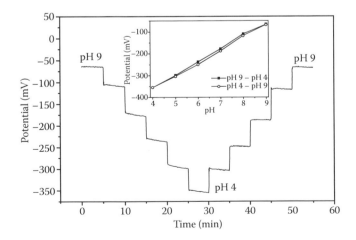

FIGURE 4.8 ConCap measurement of a porous pH sensor (stored for 86 days in Titrisol buffer), where the porous layer has been etched at a current density of 30 mA/cm². The sensitivity was calculated to 59.1 mV/pH (from pH 9 to 4) and 58.7 mV/pH (from pH 4 to 9). (Reprinted from *Sens. Actuators B* Simonis A., 91, 21, Copyright 2003, with permission from Elsevier.)

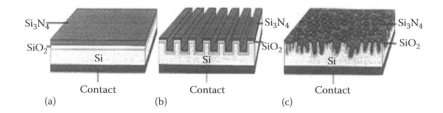

FIGURE 4.9 Schematic of the semiconductor/insulator capacitors: (a) planar, (b) structured, and (c) porous. (Reprinted from *Sens. Actuators B* 65, Schoning M.J. et al., 288–290, Copyright 2000, with permission from Elsevier.)

published a well-defined macroporous layer, which was formed on silicon by electrochemical etching and an SiO_2/Si_3N_4 sandwich was deposited as insulating a pH-sensitive layer for the first time early in 1997. Three years later, they alternated a concept based on structured and porous semiconductor/insulator capacitors. Different preparation methods have been investigated: anisotropic and anodic etching process of silicon and subsequent deposition of dielectric layers of SiO_2 and Si_3N_4 (Schoning et al. 2000). A schematic of the different capacitive electrolyte–insulator–semiconductor sensors for the pH determination is as shown in Figure 4.9. C/V (capacitance/voltage) measurements were performed to investigate the sensor properties.

Actually, a PSi-based electrochemical sensor for pH measurements is very sensitive to the circumstance of the detection media due to its surface structure and sensitive materials inside the pores. The reason utilizing porous instead of planar silicon to fabricate an electrochemical sensor is due to the enlarged sensor surface area yielding a higher capacitance and also a more stable anchoring and embedment of molecules inside the pores. Poor adhesion and fast leaching-out of the sensitive materials as well as electrochemical corrosion of the passivation layer are also the factors influencing the usage of sensor for pH measurements. Therefore, altering the surface structure of PSi and different dielectric layers might be an efficient method to overcome the drawback of these sensors.

4.6 PSi-BASED ELECTROCHEMICAL GAS AND VAPOR SENSORS

The gas and organics electrochemical sensing properties of PSi have been studied widely and different sensor designs and operation principles have been proposed (Boarino et al. 2000; Seals et al. 2002; Barillaro et al. 2005; Bjorkqvist et al. 2007; De Stefano et al. 2007; King et al. 2007). Besides PSi, actually, various materials (Korotcenkov et al. 2009; Korotcenkov 2014) such as membranes, electrolytes, electrodes, metal oxide-based nanocomposite, and so on are also usually utilized to fabricate the gas sensors. Humidity, organic solvents, CO, NOx, NH_3, O_2, H_2, HCl, SO_2, H_2S, and PH_3 have been detected using PSi gas sensors at or below ppm levels. These approaches consider the use of electrical contacts on the PSi layer made by metal/metal oxide deposition or covalent chemical surface modification to measure the changes in the electrical properties such as capacitance and conductance when an analyte is attached to the PSi layer.

Tebizi-Tighilt et al. (2013) have fabricated a gas sensor based on PSi and a polypyrrole obtained by covalent grafting. It was shown that Si/oxide PSi/PPy and Si/PSi/PPy structures could be used as a sensor for CO_2 gas. These sensors are extremely sensitive, have a good response and recovery time, and can operate at low voltages. Li et al. (2013a,b) published several papers on PSi with intermediate-sized pores for sensing NH_3 and NO_2 gases. The electrical contacts of Pt thin films were deposited onto the PSi surface by RF magnetron sputtering. Dynamic response curves of such PSi versus various concentrations of NH_3 and NO_2 gases were measured. The sensor showed high response, fast response–recovery characteristics, good reproducibility, and excellent repeatability.

Ensafi et al. (2014) reported a new electrochemical sensor based on PSi, Cu/PSi nanocomposite, which was synthesized by chemical etching of silicon microparticles followed by electrodeless deposition of copper nanoparticles on the etched silicon. Then, it is used for fabricating Cu/PSi nanocomposite-based carbon paste electrode. It has been found that this electrode has good electrocatalytic activity on the reduction of H_2O_2 in phosphate buffer solution at pH 7.0. The results implied that the detection limit and linear dynamic range of the modified electrode with Cu/PSi nanocomposite are better than are those recently obtained by other works (as shown in Figure 4.10).

A variety of analytical procedures for the assay of these organics has been proposed, most of them based on relatively complex spectrophotometric and fluorimetric methods. In these methods, extensive instrumentation is not suitable for point-of-care application or home-use testing (Chang et al. 2003). There was a report on the development of an electrochemical sensor assay using microelectro mechanical systems intended to measure levels of these organics (Song et al. 2009). The aim is to design a miniaturized measurement system using electrodes processed by PSi and array technology.

Song et al. (2006) reported that functionalized PSi could be used as sensors for the measurement of cholesterol concentrations. The results showed that the effective area of PSi-based sensors was 3.1-fold larger than that of PLS-based sensors and this caused the same level of increase in sensitivity in PSi-based cholesterol sensors compared with PLS-based sensors. The amperometric sensitivity of the PLS-based sensor was approximately 0.08567 A/mM (correlation coefficient $r = 0.975$) in the linear range of 10–100 mM, and that of the PSi-based sensor was approximately 0.2656 A/mM (correlation coefficient $r = 0.998$) in the linear range of 1–50 mM. The PSi electrode using the silanization technique appears to be an appropriate choice for the measurement of low concentrations of cholesterol, and is associated with low overall noise. Their group (Jin et al. 2003, 2006) also presented that urea sensors based on nano-porous silicon technology were fabricated. Stable and reproducible PSUEs and Ag/AgCl TFREs were fabricated on PSi layers that provided

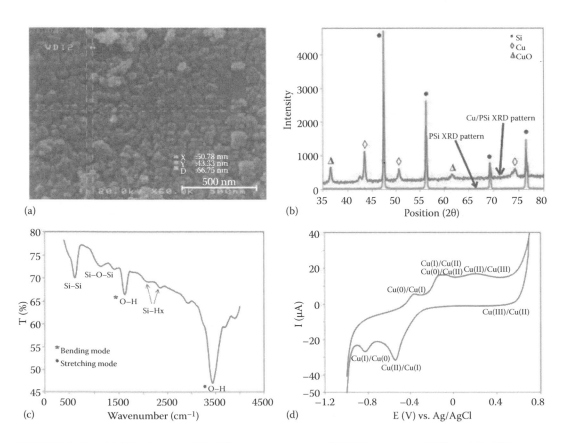

FIGURE 4.10 (a) SEM image of Cu/PSi nanocomposite; (b) XRD patterns of PSi and Cu/PSi nanocomposite; (c) FT-IR spectra of PSi; and (d) cyclic voltammograms of Cu/PSi-CPE in the 0.1 mol/L NaOH with scan rate of 50 mV/s. (Reprinted from *Sens. Actuators B* 196, Ensafi A.A., 398, Copyright 2014, with permission from Elsevier.)

better adhesive strength between thin-films and silicon-based electrodes to reduce the leaching out of TFRE components and to enhance the sensitivity of a sensing electrode. Limiting current flow dependence on urea concentration measured by referring to the TFRE was 191.5 A/decade cm^2. Amperometry for monitoring the urea concentrations caused by urease-catalyzed reactions is superior to a potentiometric method in that the amperometric urea sensor gives longer linear range, higher sensitivity, and shorter response time than the potentiometric urea sensor, especially at low urea concentrations.

Harraz (2014) has reviewed the use of PSi as a chemical sensor and biosensor for detection of many analytes. The surface properties and structural characteristics of the material are briefly described. The recent progress on utilization of such porous structures in chemical and biosensing applications is then addressed in the context of surface chemistry effects and nanostructures, measuring approaches, operating concepts, and device sensitivity and stability. Harraz et al. (2014) also reported PSi layer as a highly sensitive ethanol sensor. The PSi nanostructure that is generated in an electrochemical etching of crystalline silicon in HF-based solution was about 4.5 μm thick with an average pore size of 30 nm. The as-fabricated sensor exhibits highly sensitive, reversible response during the real-time measurements of capacitance and conductance. Excellent repeatability of the devise was obtained after six cyclic tests, demonstrating stability of the sensor. Long-term stability for the sensor performance was also observed after four week's storage (shown in Figure 4.11). They claim that the present PSi sensor shows a very sensitive and reversible response to liquid ethanol. The observed reversibility indicates no chemical reaction or surface modification has taken place during the sensing process.

The main factor response for changes in the electrical properties of the layers is the environmental composition present on the surface. For adsorption type gas and vapor detector, sensing elements including metal oxide semiconductor compounds of n-type conductivity, such as SnO_2, ZnO, NiO, Fe_2O_3, V_2O_5, and so on are often utilized. Moshnikov et al. (2012) studied the peculiarities of the formation of nanostructured materials based on the oxides of tin, iron, and nickel on PSi substrates and compared their sensory characteristics with the characteristics of metal oxide films produced on substrates of monocrystalline silicon and glass. They demonstrated that applying a variable frequency disturbance in a changing gas environment to a system of sensory structures can control their impedance response.

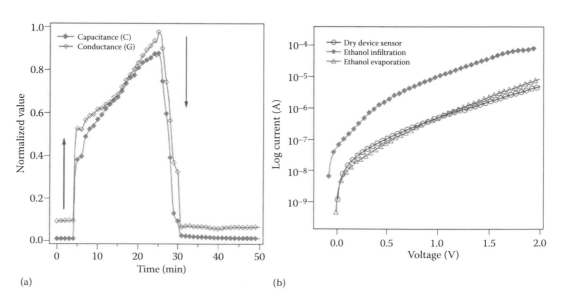

FIGURE 4.11   (a) Real-time capacitance (C) and conductance (G) of sensor measured at 100 kHz, 100 mV at room temperature after exposure to 10-μL ethanol. The vertical up-arrow indicates the injection of ethanol into the sensor device and the down-arrow indicates the evaporation of ethanol from the device. (b) Current-voltage curves measured for dry device, after exposure to ethanol and after ethanol evaporation. (From Harraz F.A. et al., *Int. J. Electrochem. Sci. 9*, 2149, 2014. Published by ESG as open access. With permission.)

4.7 SUMMARY AND PERSPECTIVES

The significant growth of interest in PSi sensors started from the huge internal surface area, surface reactivity, and its unique electrical properties. These advantages make the PSi-based sensing systems a very promising research area that would bring new discoveries, with real potential applications in related-sensing fields. The research field of PSi-related sensors is multidisciplinary; that is why various research skills on different backgrounds are required for tangible achievements on the subject. The ability to control the pore size, morphology, porosity, layer thickness, large surface area, and the ease to functionalize the surface will give researchers worldwide the possibilities for sensing various chemical species.

Chemical sensors are at the heart of the development of compact, self-contained devices for chemical analysis, which play an increasingly important role in our society. It seems that PSi is an ideal substrate for fabricating electrochemical sensors. The ability to control the pore size, porosity, and surface functionalization of PSi allows one to fabricate sensor devices capable of detecting very low analyte concentrations of chemical, gases, and biological molecules. Major developments in surface chemistry, however, are needed to further stabilize the PSi films particularly in aqueous solutions to help attach specific recognition molecules and to provide antifouling capabilities for the surface. Besides that, hybrid structures of PSi are also needed to be designed by incorporating other metal or metal oxide nanoparticles or conductive polymers inside the pores that allow two or more properties combined in a single material for developing future sensing applications.

Among these voltammetric, amperometric, and conductometric three-signal detection methods for PSi electrochemical sensors, the largest diversity of sensing approaches is observed with voltammetric sensors and detectors. Much of this richness is due to excellent past long-term fundamental advances, which found their way into modern chemical sensor research. Ion-transfer voltammetry, for example, was long a fundamental field of research but is now being used to fabricate useful, miniaturized devices that can be integrated into chips.

As long as the researchers are working to solve the above challenging tasks, "smart" PSi-based sensor devices are expected to emerge, which in turn would maximize the possibility for commercial use.

REFERENCES

Ali, M.B., Mlika, R., Ouada, H.B., Mghaieth, R., and Maaref, H. (1999). Porous silicon as substrate for ion sensors. *Sens. Actuators B* **74**, 123–125.
Barillaro, G., Diligenti, A., Marola, G., and Strambini, L.M. (2005). A silicon crystalline resistor with an adsorbing porous layer as gas sensor. *Sens. Actuators B* **105**, 278–282.
Bjorkqvist, M., Paski, J., Salonen, J., and Lehto, V. (2007). Studies on the hysteresis reduction in thermally carbonized porous silicon humidity sensor. *IEEE Sensors J.* **6**, 542–547.
Boarino, L., Baratto, C., Geobaldo, F. et al. (2000). NO$_2$ monitoring at room temperature by a porous silicon sensor. *Mater. Sci. Eng. B* **69–70**, 210–214.
Chang, K.S., Hsu, W.L., Chen, H.Y., Chang, C.K., and Chen, C.Y. (2003). Determination of glutamate pyruvate transaminase activity in clinical specimens using a biosensor composed of immobilized L-glutamate oxidase in a photo-crosslinkable polymer membrane on a palladium-deposited screen-printed carbon electrode. *Anal. Chim. Acta* **481**, 199–208.
De Stefano, L., Alfieri, D., Rea, I. et al. (2007). An integrated pressuredriven microsystem based on porous silicon for optical monitoring of gaseous and liquid substances. *Phys. Stat. Sol. (a)* **204**, 1459–1463.
Ensafi, A.A., Abarghoui, M.M., and Rezaei, B. (2014). Electrochemical determination of hydrogen peroxide using copper/porous silicon based non-enzymatic sensor. *Sens. Actuators B* **196**, 398–405.
Gamoudi, M. (2001). P-type porous-silicon transducer for cation detection: Effect of the porosity, pore morphology, temperature and ion valency on the sensor response and generalisation of the Nernst equation. *Appl. Phys. A* **73**, 585–593.
Harraz, F.A. (2014). Porous silicon chemical sensors and biosensors: A review. *Sens. Actuators B* **202**, 897–912.
Harraz, F.A., Ismail, A.A., Bouzid, H., Al-Sayari, S.A., Al-Hajry, A., and Al-Assiri, M.S. (2014). Mesoporous silicon layer as a highly sensitive ethanol sensor. *Int. J. Electrochem. Sci.* **9**, 2149–2157.
Jin, J.H., Paek, S.H., Lee, C.W., Min, N.K., and Hong, S.I. (2003). Fabrication of amperometric urea sensor based on nanoporous silicon technology. *J. Korean Phys. Soc.* **42**, S735–S738.

Jin, J.H., Min, N.K., and Hong, S.I. (2006). Poly(3-methylthiophene)-based porous silicon substrates as a urea-sensitive electrode. *Appl. Surf. Sci.* **252**, 7397–7406.

Jin, J.H., Alocilja, E., and Grooms, D. (2010). Fabrication and electroanalytical characterization of label-free DNA sensor based on direct electropolymerization of pyrrole on ptype porous silicon substrates. *J. Porous Mater.* **17**, 169–176.

King, H.B., Ruminski, A.M., Snyder, J.L., and Sailor, M.J. (2007). Optical fiber-mounted porous silicon photonic crystals for sensing organic vapor breakthrough in activated carbon. *Adv. Mater.* **19**, 4530–4534.

Korotcenkov, G. (2014). Metal oxide-based nanocomposites for conductometric gas sensors. In: Korotcenkov G. (Ed.) *Handbook of Gas Sensor Materials.* Springer, New York, pp. 197–207.

Korotcenkov, G., Han, S.D., and Stetter, J.R. (2009). Review of electrochemical hydrogen sensors. *Chem. Rev.* **109**, 1402–1433.

Li, S.Y., Ma, W.H., Zhou, Y., Wang, Y.F., Li, W., and Chen, X.H. (2012). Cleavable porous silicon based hybrid material for pre-enrichment of trace heavy metal ions. *Appl. Surf. Sci.* **258**, 5538–5542.

Li, M., Hu, M., Liu, Q., Ma, S., and Sun, P. (2013a). Microstructure characterization and NO2-sensing properties of porous silicon with intermediate pore size. *Appl. Surf. Sci.* **268**, 188–194.

Li, M., Hu, M., Zeng, P., Ma, S., Yan, W., and Qin, X. (2013b). Effect of etching current density on microstructure and NH3-sensing properties of porous silicon with intermediate-sized pores. *Electrochem. Acta* **108**, 167–174.

Li, S.Y., Ma, W.H, Zhou, Y. et al. (2013). 3-aminopropyltriethoxysilanes modified porous silicon as a voltammetric sensor for determination of silver ion. *Int. J. Electrochem. Sci.* **8**, 1802–1812.

Moshnikov, V.A., Gracheva, I., Lenshin, A.S. et al. (2012). Porous silicon with embedded metal oxides for gas sensing applications. *J. Non-Cryst. Solids* **358**, 590–595

Ressine, A., Vaz-Dominguez, C., Fernandez, V.M. et al. (2010). Bioelectrochemical studies of azurin and laccase confined in threedimensional chips based on gold-modified nano-/microstructured silicon. *Biosens. Bioelectron.* **25**, 1001–1007.

Sakly, H., Mlika, R., Chaabane, H., Beji, L., and Ben Ouada, H. (2006). Anodically oxidized porous silicon as a substrate for EIS sensors. *Mater. Sci. Eng. C* **26**, 232–235.

Sam, S., Chazalviel, J.N., Gouget-Laemmel, A.C. et al. (2010). Covalent immobilization of amino acids on the porous silicon surface. *Surf. Interface Anal.* **42**, 515–518.

Sam, S.S., Chazalviel, J.N., Gouget-Laemmel, A.C., Ozanam, F.F., Etcheberry, A.A., and Gabouze, N.E. (2011). Peptide immobilization on porous silicon surface for metal ions detection. *Nanoscale Res. Lett.* **6**, 412.

Schoning, M.J., Ronkel, F., Crott, M. et al. (1997). Miniaturization of potentiometric sensors using porous silicon microtechnology. *Electrochim. Acta* **42**, 3185–3193.

Schoning, M.J., Malkoc. U., Thust, M., Steffen, A., Kordos, P., and Luth, H. (2000). Novel electrochemical sensors with structured and porous semiconductor/insulator capacitors. *Sens. Actuators B* **65**, 288–290.

Seals, L., Tse, L.A., Hesketh, P.J., and Gole, J.L. (2002). Rapid, reversible, sensitive porous silicon gas sensor. *J. Appl. Phys.* **91**, 2519–2523.

Simonis, A., Ruge, C., Muller-Veggian, M., Luth, H., and Schoning, M.J. (2003). A long-term stable macroporous-type EIS structure for electrochemical sensor applications. *Sens. Actuators B* **91**, 21–25.

Song, M.J., Yun, D.H., Jin, J., Min, N., and Hong, S. (2006). Comparison of effective working electrode areas on planar and porous silicon substrates for cholesterol biosensor. *Jpn. J. Appl. Phy.* **45**, 7197–7202.

Song, M.J., Yun, D.H., Min, N., and Hong, S. (2007). Electrochemical biosensor array for liver diagnosis using silanization technique on nanoporous silicon electrode. *J. Biosci. Bioeng.* **103**, 32–37.

Song, M.J., Yun, D.H., and Hong, S.K. (2009). An electrochemical biosensor array for rapid detection of alanine aminotransferase and aspartate aminotransferase. *Biosci. Biotechnol. Biochem.* **73**, 474–478.

Tebizi-Tighilt, F.Z., Zane, F., Belhaneche-Bensemra, N., Belhousse, S., Sam, S., and Gabouze, N.E. (2013). Electrochemical gas sensors based on polypyrrole-porous silicon. *Appl. Surf. Sci.* **269**, 180–183.

Zairi, S., Martelet, C., Jaffrezic-Renault, N., Lamartine, R., Mgaieth, R., Maaref, H., Gamoudi, M., and Guillaud, G. (2001a). Porous silicon as a potentiometric transducer for ion detection: Effect of the porosity on the sensor response. *Appl. Surf. Sci.* **172**, 225–234.

Zairi, S., Martelet, C., Jaffrezic-Renault, N., Mgaieth, R., Maaref, H., and Lamartine, R. (2001b). Porous silicon a transducer material for a high-sensitive (bio)chemical sensor: Effect of porosity, pores morphologies and a large surface area on a sensitivity. *Thin Solid Films* **383**, 325–327.

Zhao, Z. and Jiang, H. (2010). Enzyme-based electrochemical biosensors. In: Serra P.A. (Ed.) *Biosensors.* InTech, Rijeka, pp. 1–22.

Porous Silicon-Based Biosensors

Some Features of Design, Functional Activity, and Practical Application

Nicolai F. Starodub

5

CONTENTS

5.1	Introduction: Biosensors, Their Importance, and Fields of Application	108
5.2	Preparation of the PSi Surface for Specific Biofunctionalization	110
5.3	General Aspects of PSi Practical Application	113
5.4	PSi-Based Immunobiosensors	114
5.4.1	Structural Analysis of the PSi Samples, Main Principles of the Specific Signal Registration, Construction of Working Devices, and Basic Algorithm of Analysis	114
5.5	PSi-Based Biosensor at the Control of Some Biochemical Quantities	119
5.5.1	Registration of Mb Concentration	119
5.5.2	Control of Level of Biological Substances in Air	121
5.5.3	Determination of the Level of Some Mycotoxins	122
5.5.4	Diagnostics of Retroviral Leucosis	122
5.5.5	Control of Other Types of Ag and Ab as Well as Some Peptides and Chemicals	124
5.5.6	Glucose and Appropriate Enzyme Determination	125
5.5.7	Detection of Viruses and Bacteria	126
5.5.8	DNA Sensors	127
5.5.9	Cells Analysis	128
5.6	Summary and Perspectives of PSi in Biosensorics	129
Acknowledgments		130
References		130

5.1 INTRODUCTION: BIOSENSORS, THEIR IMPORTANCE, AND FIELDS OF APPLICATION

Humanity has achieved significant progress in the development of an industry that is trying to go on the level of population growth in the world for providing its needs. Active and extensive human activity is the driving force of progress that leads to all the new wealth and values. Great intensity of life puts ordeals to a person and they often break morally and physically. Thus, trying to achieve a significant development of life, significantly a number of people cannot withstand overload and may be in a hospital bed. This leads to the urgent need to provide methodological basis and needed instrumental methods for the express diagnostics of some diseases and their successful treatment. However, it is necessary to pay attention to the fact that along with the creative activity of humans aimed at meeting needs, there is observed not only devastating effects on the body, but also a significant negative impact on the environment. Air pollution by the different biological active agents is part of the global problem of deterioration of the environment worldwide. In order to increase the social role of industry and minimize the harmful effects that they create because of the generation of additional pollutants, there is an urgent need for the presence of relevant sensitive, specific, and rapid methods for their registration. Thus, increasing the rate of progress on one hand leads to increasing material resources to ensure the needs of people, and on the other hand to the deterioration of their health and worsening environmental crisis. This makes finding a solution immediately and measures should be based primarily on the prevention of threatening situations using instrumental and methodological bases of a new generation that meet modern requirement's practices to ensure timely determination of diseases, their successful treatment, and the registration of number and type of harmful components, including chemical and pure biological contaminants in the environment.

The history of biochemical analytical methods includes at least three important steps: (1) involving biological components (enzymes, immune components, lectines, and others) as selective elements to provide specificity of the revealing analytes, (2) sharply increasing sensitivity of the immune analysis by transferring to registration of mono- and not polyvalent antibodies (Ab)-antigens (Ag) interactions and as result of which creation of a number of the immune-chemical approaches (Yalow and Berson 1971), and (3) achievement of having results online and in field conditions. The last was realized through the design and creation of different types of biosensors (Heinemann and Jensen 2006).

In general, the PSi-based biosensors in the dependence on the registration of the formed biospecific signal may be divided on the optical, electric, and electrochemical ones. In optical biosensors, different physical effects are used. Effects of biosensors are based on the changes in the optical spectral interference pattern. At the transfer of white light through PSi, the interference pattern is observed. This effect is called the Fabry–Perot fringe pattern, which changes may be quantified (Lin et al. 1997). Of course, this effect depends on the refractive index value of the analyzed solution and also on how it penetrates into the pores (De Stefano et al. 2004). It is possible to use mono- and double-layer films (Jane et al. 2009). To increase response, it is possible to build PSi biosensors with other complex optical structures as multilayer devices (Fauchet et al. 1995).

Photoluminescence (PhL) effects are very useful for the developing biosensors (Starodub et al. 2011a,b). In this case, the behavior was attributed to nonradiative recombination processes, but that this kind of biosensor is less accurate than its interferometric counterparts is not excluded (Jane et al. 2009). As a special variety, it can be considered fluorescence PSi-based biosensors. In that case, the intensity of a signal is measured from a fluorescence molecule using as a marker (e.g., fluorescein-5-maleimide) fixed at a PSi structure (Rossi et al. 2007).

Among optical transduction methods, fluorescence resonance energy transfer (FRET), surface enhanced Raman spectroscopy (SERS), and fluorescence spectroscopy are used for detection of biological samples. Spectroscopic techniques are used for detecting biological samples or events occurring in it because cells or tissues can absorb or emit light, thereby producing a signal or spectrum, which is a characteristic of that particular event. From this fingerprint spectrum, one can directly identify or quantify the sample or the event. FRET is a nonradiative energy transfer process from excited state donor molecules to an acceptor molecule, when appreciable overlap exists between the emission spectrum of the donor and the absorption spectrum of the acceptor.

This radiationless transfer of energy, when the excited state fluorophore and the second chromophore lie within a range of approximately 10 nm, provides vivid structural information about the donor-acceptor pair. This is a quantum mechanical process that does not require a collision and does not involve production of heat. When energy transfer occurs, the acceptor molecule quenches the donor molecule fluorescence, and if the acceptor is itself a fluorochrome, increased or sensitized fluorescence emission is observed.

The SERS is a surface sensitive technique that results in the enhancement of Raman scattering by molecules adsorbed on rough metal surfaces. The vibration modes of the adsorbates on the roughened surface are sometimes observed to have about 1 million times the intensity that would be predicted by comparison with their Raman spectra in the gaseous phase (Vo-Dinh 2004; Ghoshal et al. 2010).

The electrical biosensors consider the use of electrical contacts on the p-Si layer made by metal deposition to measure the changes in the electrical properties such as capacitance and conductance when an analyte is attached to the p-Si layer (Lenward et al. 2002). Another effective approach that is widely used at the development of PSi biosensors is the measurement of the electrochemical characteristics. In this case, generally, two main types of biosensors can be allocated according to the transduction principle: potentiometry and amperometry/voltammetry (Salis et al. 2011). In potentiometric biosensors, the main parameter is the potential difference between the cathode and the anode in an electrochemical cell. The highly sensitive surface of p-Si and the possibility to measure changes in its electrical properties added to its capacity to adsorb an enormous amount of different compounds, which can be used for electrical biosensor applications (Lugo et al. 2007).

Amperometric and voltammetric biosensors consider the redox reaction that takes place in the anodization cell. In this case, the analyte is immobilized and its oxidation/reduction process produces a flux of electrons measured, in terms of current intensity, across the electrodes of the electrochemical cell. These biosensors are too sensitive to pH modifications (Salis et al. 2011).

All modern practice requirements can be met only through biosensors. First, such instrumental analytical devices were established in 1957 as the enzymatic sensor based on oxygen electrode and intended for glucose determination. In general, biological sensors are primarily analytical instruments including a physical sensing element which is in close contact with the biological material and transforms physical-chemical signals generated by this material when interacting with the analyte in electric. Many different types of biosensors have been proposed, which differ both in physical transducer used and biological material, from haptens to high molecular nucleic acids. Based on the type of physical transducer, the biosensors can be classified into electrochemical, mechanical, calorimetric, magnetic, and optical. Electrochemical sensors include amperometric, potentiometric, conductometric, and capacitive devices. Potentiometer biosensors record changes in electrode potential of immobilized biological membranes that occur in the specific interaction of the membrane with the respective analyte. Amperometric biosensors measure current (proportional to the measured concentration dependent matter), which is the result of oxidation-reduction of electroactive substances on the electrode coated with biological matrix. As a rule, they are always enzyme or immune-enzyme type. Conductometric biosensors are based on the registration of changes in conductivity between a pair of electrodes through the electrolyte. These changes appear because of biochemical reactions followed by consumption or production of electrolyte. When using capacitive biosensors, the changes in the capacity of the system under the conditions of a specific interaction between biological molecules were determined. Mechanical biosensors are based on the piezoelectric materials, which are capable of fluctuations in the electric field. The interaction of a substance with its target molecule on the surface of a crystal of such material leads to changes in the surface mass, which in turn causes a shift in the resonant frequency of its oscillations. The effect of sound vibrations is based on the use of the piezoelectric crystals, but in combination with acoustic waves. Passage of a specific interaction between the molecules on the surface causes the change of acoustic propagation velocity of sound waves. The general principle of the calorimetric biosensors is to measure the heat released because of the interaction of, as a rule, enzyme and substrate or Ab and Ag. Using magnetic biosensors, the changes in the magnetic permeability arising from the use of immune components consisting of ferromagnetic labels are registered. In optical biosensors, the specific molecular interaction is determined due to the presence of appropriate elements sensitive to changes in photoluminescence (PhL) intensity, fluorescence, chemiluminescence (ChL), absorption, or scattering of light. Today, a number of approaches that use a variety of optical effects such as optical fibers, damped electromagnetic waves, nonirradiative energy transfer, surface plasmon resonance (SPR), and integrated optics have been proposed. Among the existing varieties of biosensors, they are distinguished based on direct and indirect (without using any additional component or with the application of one) methods for detecting interactions between molecules (Starodub and Starodub 2000, 2001).

Unfortunately, despite the significant development of science, most of the proposed biosensors are unable to meet the demands of practice. They are more complex, multistage, require the use of expensive reagents, and are not able to provide direct registration of specific interactions of biological molecules. At first, it concerns to the electrochemical biosensors based on the registration of signals with the help of the potentiometry and conductometry principles as well as some ones, in particular, which are using the piezocrystals as transducers. Therefore, the search for more attractive, sensitive, simple, and cheap transducers to create biosensors that would more fully meet all requirements of practice is continued. Taking into account that silicon is one of the most widely used elements in nature, the possibility of its using as a basis for creating simple, sensitive, and rapid in action biological sensors was proposed. Using simple methods of processing monocrystalline silicon (mcSi), it can be converted to PSi and the last may be used as a transducer, which will be able to registrate the biospecific interaction of components on the principles of visible PhL or electroconductivity. The results on the creation and the practical application of different types of biosensors based on PSi are analyzed next. This biosensorics area is rapidly developing. Each year, several hundred publications about new options for making and using PSi-based biosensors appear in the literature. At the same time, it is not possible to give a comprehensive description of all proposed options for such types of biosensors. That is why our chapter is focused on presenting the most common and significant achievements in this area.

5.2 PREPARATION OF THE PSi SURFACE FOR SPECIFIC BIOFUNCTIONALIZATION

For the functionalization of PSi by the selective biological molecules, a number of approaches were proposed, which are included (1) direct absorption through hydrophobic interaction and (2) preliminary formation of some intermediate layers on which the above-mentioned biomolecules could be attached through the formation of covalent bonds or electrostatistical adhesion.

In general, for the biofunctional characteristics of PSi (Gallach et al. 2010), two different molecular end-capping processes were principally allocated: antifouling polyethylene glycol (PEG) and polar binding APTS, which were evaluated by X-ray photoelectron spectroscopy (XPS). Both PEG and APTS binding to the particles could be confirmed from the analysis of Si 2p and C 1s XPS core level spectra. These modifications do not alter the properties of planar PSi such as PhL. Moreover, the process of obtaining PSi-molecule conjugates allows the reduction of fluorescence degradation with time in solution. APTS modified particles show notably stable luminescent properties, with no significant emission quenching or shift for more than 10 days. In addition, PEG termination allows getting spherical shapes in the micron scale after ultrasonic treatment of PSi. Arroyo-Hernandez et al. (2006) compared the efficiency of the application APTS at room temperature and at 100°C. The bioactivity of functionalized PS multilayers has been assessed by the detection of a fluorescent reagent that reacts with amine groups on the external surface of the device. They have paid attention that in both cases (room and high temperature) large surface biofunctionalization is observed. In the case of room temperature experiments, a quite homogeneous NH_2 functionalized surface was obtained, while at 100°C some evenly distributed colloidal aggregates were found leading to some surface inhomogeneity.

Early on, most of the methods for the functionalization by the selective biological structures were based on attaching the aliphatic or aromatic hydrocarbons to the PSi surface via routes as hydrosilylation of alkenes or alkynes. Unfortunately, such methods rarely provide opportunity for further modifications of the PSi for biomedical applications using standard organic chemistry strategies (Salonen and Lehto 2008). It was informed Boukherroub et al. (2002) have informed about a very simple and straightforward strategy with good stability of PSi surface in case of its hydrosilylation with the help of undecylenic acid. The fixed carboxyl groups could be easily bound with other biomolecules, for example, through an amine-reactive crosslinker (Faucheux et al. 2006; Schwartz et al. 2006). By using this approach, there is a possibility to bind bovine serum albumin (BSA) for the formation of its monolayer on the PSi surface (protein A), which is able for the interaction with specific sites on the Fc fragments of immunoglobulins to increase exposition of F(ab)$_2$ centers selective to appropriate Ag. The last may increase the sensitivity of the immunological analysis (Schwartz et al. 2007).

The PSi surface could be thermally treated in an ammonia-containing atmosphere to obtain partial amine termination or by using other wet chemical methods (Lees et al. 2003; Salonen and Lehto 2008). Another way was proposed based on the application of PSi reaction with liquid acetone at room temperature. It was shown that after reaction with liquid acetone, two infra-red absorption modes at 2253 and 2200 cm^{-1} developed and this was accompanied by loss of absorbance of the Si-H stretch modes at 2142, 2110, and 2089 cm^{-1}. Taking into account these results, it was postulated that oxidation of the surface silicon hydride species by acetone to form $(CH_3)_2HCO$-Si-H$_x$ and $[(CH_3)_2HCO]_2$-Si-H$_x$ species had occurred. This surface reaction may be utilized for passivation of the silicon surface (Rao and Yates 1993).

In the literature, a number of techniques and methods are described in what way there is possible to graft amines, oligonucleotides, and other biomolecules on the PSi surface. Surface-initiated atom-transfer radical polymerization of poly(ethylene glycol) monomethacrylate (PEGMA) was carried out on the hydrogen-terminated Si(100) substrates with surface-tethered alpha-bromoester initiator. Kinetic studies confirmed an approximately linear increase in polymer film thickness with reaction time, indicating that chain growth from the surface was a controlled "living" process. The "living" character of the surface-grafted PEGMA chains was further ascertained by the subsequent extension of these graft chains, and thus the graft layer. Well-defined polymer brushes of near 100 nm in thickness were grafted on the Si(100) surface in 8 h under ambient temperature in an aqueous medium. The hydroxyl end groups of the poly(ethylene glycol) (PEG) side chains of the grafted PEGMA polymer were derivatized into various functional groups, including chloride, amine, aldehyde, and carboxylic acid groups. The surface-functionalized silicon substrates were characterized by reflectance FT-IR spectroscopy and XPS. Covalent attachment and derivatization of the well-defined PEGMA polymer brushes can broaden considerably the functionality of single-crystal silicon surfaces (Xu et al. 2004).

It was informed (Hiraoui et al. 2011) that at the elaboration of PSi layers, with sufficient adjusted pore size and well defined porosity, the application of 3-aminopropyltriethoxysilane (APTES) molecules could provide using such bifunctional agents as glutaraldehyde (GL) for the next covalent binding BSA. The presence of each component on the PSi surface was registered by the FTIR and Raman spectroscopic analyses through the presence of the complementary spectra. These procedures were carried out to overcome the strongly hydrophobic PSi surface and to allow penetrating water solutions into pores. The fulfillment of the thermal oxidation and chemical adaptation were recommended, respectively, for the stabilization and functionalization of the internal surface of the PSi layer as well as for obtaining transparent porous silica for waveguide biosensor applications.

Zhang et al. (2010) have investigated efficiency of covalently bonding monolayers of two monofunctional aminosilanes (3-aminopropyldimethylethoxysilane, APDMES, and 3-amino-propyldiisopropylethoxysilane, APDIPES) and one trifunctional aminosilane (3-aminopropyl-triethoxysilane, APTES) deposited on dehydrated silicon substrates by chemical vapor deposition (CVD) at 150°C and low pressure (a few Torr) using reproducible equipment (Figure 5.1). Standard surface analytical techniques such as XPS, contact angle goniometry, spectroscopic ellipsometry, atomic force microscopy, and time-of-flight secondary ion mass spectroscopy (ToF-SIMS) have been employed to characterize the resulting films. These methods indicate that essentially constant surface coverage is obtained over a wide range of gas phase concentrations of

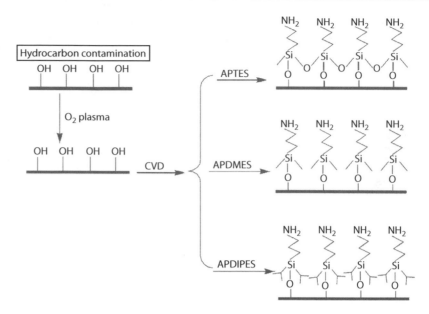

FIGURE 5.1 Detailed representation of surface cleaning followed by surface modification with APTES, APDMES, and APDIPES. (Reprinted with permission from Zhang F. et al., *Langmuir* 26(18), 14648. Copyright 2010, American Chemical Society.)

the aminosilanes. XPS data further indicate that the N1s/Si2p ratio is higher after CVD with the trifunctional silane (APTES) compared to the monofunctional ones, with a higher N1s/Si2p ratio for APDMES compared to that for APDIPES. AFM images show an average surface roughness of 0.12–0.15 nm among all three aminosilane films. Stability tests indicate that APDIPES films retain most of their integrity at pH 10 for several hours and are more stable than APTES or APDMES layers. The films also showed good stability against storage in the laboratory. ToF-SIMS of these samples showed expected peaks, such as CN-, as well as CNO-, which may arise from an interaction between monolayer amine groups and silanols. Optical absorption measurements on adsorbed cyanine dye at the surface of the aminosilane films show the formation of dimer aggregates on the surface. This is further supported by ellipsometry measurements. The concentration of dye on each surface appears to be consistent with the density of the amines.

For creation of the immune biosensors, state-of-the-art PSi films with good mechanical and optical properties have been effectively utilized for the biofunctionalization of the surface the silanization process using APTS as a precursor. The presence of reactive amino groups on the PSi surface along with GA as a linker aids in the covalent binding of the specific Ab (human IgG). Different Ag concentrations can be detected (Singh et al. 2009). Authors underline that the PhL, FTIR, and XPS indexes were characterized by the presence of amine groups for APTS-treated PS films. Moreover, they demonstrated that the PSi films prepared at an optimized Id ~50 mA/cm² had high PhL intensity, stable surface bond configurations, mechanically strong structure, and hydrogen-passivated surfaces. The immobilized Ab had native character. Such a method of biofunctionalization is a very easy and low cost technique that opens the possibility of using biofunctionalized PS in biosensing devices.

Misra and Angelucci (2001) have reported using PSi surfaces that were treated by vacuum depositing doped polyaniline in the form of thin film at the creation of the immune biosensor for the determination of *Escherichia coli*. The biosensor had high sensitivity level with a response time of about 5 sec.

Simultaneous with functionalization of the PSi surface, the problem of its durability is being solved. Han et al. (2011) showed that the durability of PSi may be improved by grafting a molecule, 2,4,6,8-tetramethyl-2,4,6,8-tetravinyl-1,3,5,7,2,4,6,8-tetraoxatetrasilocane (TE), with four terminal vinyl groups. The long-term durability of a native sample was compared to three modified PS samples: TE-, undec-10-enoic acid (UA)-, and TE/UA (TE first and followed by UA)-grafted PS, in a weak organic base of dimethyl sulfoxide, an aqueous mineral solution of CuBr₂, and phosphate buffered saline, respectively. Transmission FTIR was employed to compare the oxidation rate and degree by measuring the evolution of Si–O species against the incubation time. SEM was used to image their surface nanostructures. In all cases, TE- and TE/UA-grafted PSi samples had much less oxidation and their surface nanostructures remained without severe damaging. After screening the long-term durability of modified PSi samples, we fabricated a prototype protein microarray with the NTA-Ni$_2^+$/His-tag protein system. The attachment of His-tag Trx-urodilatin on microarrays was confirmed with MALDI-TOF-MS spectrometer and fluorescence scanning. Functionalization of PSi with some specific molecules and multiple components would be a closer step to practical applications of this structure in biosensing.

For the fabrication of an optical silicon-based label-free DNA sensor, n-type of crystalline silicon wafers have been electrochemically etched to form PSi layers. Then samples (0.25 cm^2) have been cut derivatized by using trimethoxy-3-bromoacetamidopropylsilane in order to link single strand DNA (ss-DNA). The efficiency of sensor work was characterized in terms of porosity, pore distribution, surface composition, and PhL. The last showed a 12% variation when the derivatized samples are exposed to 10-μM aqueous solution of the complementary strand-DNA while no effect is recorded with the noncomplementary strand-DNA at the same concentrations (Di Francia et al. 2005).

Several chemical and photochemical procedures of PSi surface functionalization at DNA-cDNA interaction detection were analyzed (De Stefano et al. 2006). The chemical functionalization was based on ethyl magnesium bromide (CH$_3$CH$_2$MgBr) as a nucleophilic agent that substitutes the Si-H bonds with the Si-C with the following washing surface by 1% solution of CF$_3$COOH in diethyl ether, deionized water, and pure diethyl ether. The photoactivated chemical modification of the PSi surface included the UV exposure of a solution of alkenes, which bring some carboxylic acid groups. The PSi chip was precleaned in an ultrasonic acetone bath for 10 min and then washed in deionized water. After being dried in an N$_2$ stream, it was immediately covered with 10% N-hydroxysuccinimide ester solution in CH$_2$Cl$_2$. Then the chip was washed in dichloromethane in an ultrasonic bath for 10 min and rinsed in acetone to remove any adsorbed alkene from the surface. At the etching process, organic acids (eptinoic and pentenoic acid with concentrations from 0.4 M to 3 M) were used in an electrochemical cell in the presence of a dilute HF solution (HF:EtOH=1:2). Photochemical passivation was chosen as the most promising method.

5.3 GENERAL ASPECTS OF PSi PRACTICAL APPLICATION

As it was already mentioned, the PSi characteristics are highly dependent on the method of its production and processing. One silicon sample may contain the crystalline filaments, spherical formation, and amorphous area. All this leads to the fact that the samples are not homogenous and this in turn causes difficulty in the study of their properties. However, despite these shortcomings, the ease of obtaining a PSi, its low cost, high efficiency PhL at room temperature (1–10%), and its diversity led to considerable interest in the use of such structures in creating practical optical devices, effectively operating at room temperature as well as the preparation of the analytical devices in the form of the biological sensors suitable for detecting a wide range of substances. It should be noted that one of the important and promising applications of PSi is its use as a bioactive material in the form of the biochip that can join the bone and thus provide a direct link between the biological and crystal systems. Early it was foreseen (Canham 1997a,b) creation of biochip-based PSi that could be implanted in organisms to provide functional electronic stimulation of organs, programmable drug delivery, and so on.

Examples of optical devices that use PSi can bring light emitting diodes (Namavar et al. 1992; Mehra et al. 1997) and photovoltaic cells, which are quite promising in the creation of solar cells (Vikulov et al. 1997). With the ability to obtain samples with different levels of porosity and selected characteristics, examples of effective refractive index in a wide area—from 1.1 to 2.6—can be seen. Due to this, it appears to be suitable material for making multilayer optical filters in the visible and infrared light (Mattei et al. 1998). In this case, we restrict ourselves to examples of the optical devices and proceed to consider the possibilities of using PSi to create biological sensors. Several biosensor devices based on PSi and intended for the analysis of a number of substances have been proposed. One of the easiest options for the PSi application is its use in the integrated silicon enzyme reactors (Laurell et al. 1996). The work of this enzymatic biosensor is based on the measuring changes in capacitance PSi with immobilized ionofore membrane. The ability to develop such sensor devices was shown on the example of the determination of Na$^+$ ions (Ben Ali et al. 1999). PSi was proposed as a conductor of light to create optical sensors as well as an effective thermal insulator for calorimetric biosensor (Lysenko et al. 2000). Thermal conductivity of PSi decreases with the increasing current density and time of anodyzation during the production of sensor. Because of this, PSi showed the best insulating properties at the creation of calorimetric biosensors. PSi was proposed as the basis of sensors to measure humidity (Rittersma and Benecke 1999). The work of these sensors is based on measuring small changes in capacitance resulting from the condensation of water on the surface of the PSi located on thermoelectric coolers. The same principle, namely measuring changes in capacitance, was used to develop sensors for determining the alcohol vapor in the air (Kim et al. 1999). It was reported (Thust et al. 1996; Bogue 1997) about the development of potentiometric biosensors for the determination of penicillin by measuring the capacitance-voltage curves. This biosensor showed sensitivity to penicillin at 0.1 mmol/L, but the increasing sensitivity may be achieved using the surface of a porous Si_3N_4. The changes in the effective optical thickness, which are the result of a specific interaction of biomolecules on the surface, are a parameter that is measured by the biosensor. The device uses a touch sensitive enough to register biological molecules such as biotin (sensitivity 10^{-12} mol/L), digoxihenin (10^{-6} mol/L), DNA oligomers containing 16 nucleotide bases (10^{-15} mol/L), and proteins—streptavidin (10^{-14} mol/L) and IgG (10^{-8} mol/L) (Dancil et al. 1998). The effectiveness of this biosensor was checked at registration interaction between biotin and streptavidin on the surface. The improved version interferometry method was demonstrated using the surface of the PSi activated ozone and then modifying its hydrophilic protein, which covalently bonded protein A, and the latter could register IgG (Dancil et al. 1999). Such surface modification is not only possible to carry out an effective immobilization of the immune component, but also helped to increase the stability of the PSi. Applying the method of ellipsometry to record interactions between biological molecules, the possibility of using a surface computer to create a biosensor device was tested in determining the streptavidin-biotin complex (Van Noort and Mandenius 2000). The sensitivity of this method with respect to biotin was at 40 mmol/L. Based on this method, a way to use PSi for recording various vapors and liquids (e.g., acetone, methanol, water) in the air was proposed (Zangooie et al. 1997). The minimum concentration of acetone, which could define this method, was found to be 12 mg/mL. Based on the properties of SO_2 to extinguish PhL PSi, an optical biosensor for determination of environmentally harmful gas (Kelly and Bocarsly 1998) was proposed. The possibility of using PC for development of gas sensors was also reported (Foucaran et al. 2000), where the electrical characteristics (dependence of current on the applied voltage) structure of gold-PSi in the presence of various gases was studied.

There was a detailed investigation of the application of chemical sensors based on PSi for control of acetone, ethanol, iso-propanol, and xylene (Rea 2008).

Thus, today PSi is a sufficiently promising material in developing not only optical devices such as light-emitting diodes, photovoltaic cells, and optical filters, but also in the development of a biochip that can be administered and guide treatment and diagnosis as well as the creation of biological sensors suitable for the determination of a number of substances.

5.4 PSi-BASED IMMUNOBIOSENSORS

In our investigations, we have developed the immune biosensors for the registration of myoglobin (Mb) as in model solutions and in real serum blood, a number of microbial biological substances in environmental air, control of some of the mycotoxins, and diagnostics of retroviral bovine infection (Starodub et al. 1996, 1998, 1999a,b, 2011, 2012; Starodub 1999; Starodub and Starodub 1999; Starodub and Slishek 2013). The specific biosensor signal was registered through measurements of PhL or electroconductivity.

5.4.1 STRUCTURAL ANALYSIS OF THE PSi SAMPLES, MAIN PRINCIPLES OF THE SPECIFIC SIGNAL REGISTRATION, CONSTRUCTION OF WORKING DEVICES, AND BASIC ALGORITHM OF ANALYSIS

In our investigations (Starodub et al. 1996, 1998, 1999a,b, 2011a,b, 2012, 2014a,b; Starodub 1999; Starodub and Starodub 1999, 2005; Starodub and Slishek 2013), the PSi plates were formed by the monocrystalline silicon modification chemically etched in a solution of hydrofluoric and nitric acid. Original plates were characterized by district p-type conductivity (boron doped), resistivity of 1 Ohm/cm, (100) crystallographic orientation, thickness of 350 µm. The thickness of PSi ranged from 3 to 60 nm. The overall scheme of the PSi preparation and the registration of its signal are given in Figure 5.2. This parameter was controlled during the process of the cultivation of a chemically modified silicon layer. Auger electron spectroscopy (AES), scanning tunneling microscopy (STM), and secondary ion mass spectroscopy measurements (SIMS) were used for the characterization of original silicon and PSi structures. The images of the original surface of the Si substrates for solar cells obtained by scanning electron microscopy (SEM) as well as the typical STM image of the surface of a target Si wafer are presented in Figure 5.3. The surface has a pronounced microrelief, which is formed by cutting the crystal into wafers. The vertical dimensions of the structural elements do not exceed 200–300 nm having a lateral size from 0.5 to 2 µm.

The chemical etching (during 2.5 min) leads to modification of the surface and formation of smaller surface agglomerates. These structures cannot be found in the SEM images due to the small size of the pores or with their continuous oxidation during chemical etching. The STM image of the PSi clearly shows the nanoscale properties of the modified surface (Figure 5.4). They correspond to vertices of silicon crystallites and deepening between them—a narrow exit to the surface of nanometer pores. This nanostructure has been formed on the entire surface of a target substrate, that is, the substrate's surface modulated by a nanoporous layer.

Analysis of STM images of the surface at different magnifications and the relevant sections suggests that the resulting PSi has a surface of fractal nature (Figure 5.3). The crystallites had

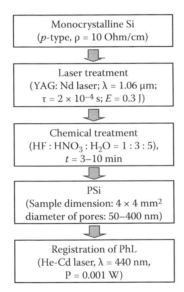

FIGURE 5.2 Overall scheme of PSi preparation and PhL signal registration. (Reprinted from *Sens. Actuators B* 35, Starodub et al., 44, Copyright 1996, with permission from Elsevier.)

(a)   (b)

FIGURE 5.3 View of the original Si sample obtained by SEM (a) and by STM (b). (From Starodub N.F. and Slishek N.F., *Adv. Biosens. Bioelectron.* 2(2), 7, 2013. Published by Science and Engineering Publishing Company as open access.)

(a)   (b)

FIGURE 5.4   Overall view of PSi obtained by SEM (a) and by STM (b) at the scanning field 1 × 1 μm and thickness of layer 20 nm. (From Starodub N.F. and Slishek N.F., *Adv. Biosens. Bioelectron.* 2(2), 7, 2013. Published by Science and Engineering Publishing Company as open access.)

lateral sizes of several nanometers and heights of 20 nm. The regular distribution of pores was found. The size of individual pores has a large fluctuation and their diameter varies in depth. With further increase of etching time (> 5 min), the developed microsurface flattened. In 10 min, the surface has become smoother and a large number of channels have been formed. The length and width of the channels varied in the range of 0.3–25 μm and 2.5–5 μm, respectively (Figure. 5.5). The channels increased with increase of etching time because of side channel growth. Surface roughness gradually smoothed during the etching for 20 min. The channel area is increased so that the terraces fade between channels (Figure 5.6). In this case, the channels become an oval-shaped form 5–50 μm in length and 6–16 μm in width. All received macrostructures demonstrated the pores of micron size (0.5–1.5 μm), which formed because of changes in surface topology. Thus, increasing the etching time results in changes of the original surface structure of silicon wafer and forming of micron-size pores on the surface.

It was shown that at the room temperature all studied PSi films have demonstrated PhL emission bands with high intensity centered at 640–650 nm (Figure 5.7). Chemical composition of PSi films versus technological conditions has been studied by AES and SIMS methods. It is shown that a process of chemical composition change has two stages.

In the first phase, the increase of surface content of SiO_x and oxygen was observed. Opposite to that, the content of C on the surface decreases with increasing time of etching. At the end of the first phase, the SiO_x surface concentration saturates, which corresponds to the maximum thickness of the nanostructured silicon film. The data of AES analysis is in a good agreement with results obtained by SIMS. Profile distributions of elements C, O, and Si obtained by two methods are similar. The most intense peaks in the mass spectra of secondary ions of the PSi in

FIGURE 5.5   Tree dimension of the chemical modified surface of monocrystalline silicon obtained by STM at the scanning field 0.5 × 0.5 μm. (From Starodub N.F. and Slishek N.F., *Adv. Biosens. Bioelectron.* 2(2), 7, 2013. Published by Science and Engineering Publishing Company as open access.)

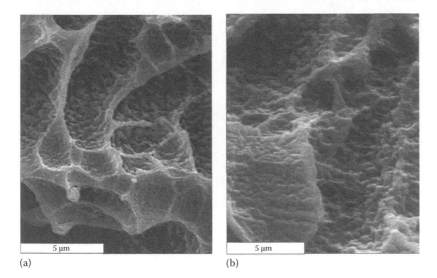

(a)   (b)

FIGURE 5.6   STM view of PSi with the etching during (a) 10 and (b) 20 min. (From Starodub N.F. and Slishek N.F., *Adv. Biosens. Bioelectron.* 2(2), 7, 2013. Published by Science and Engineering Publishing Company as open access.)

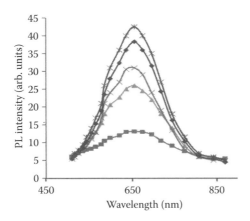

FIGURE 5.7   PhL spectra of PSi layers for different samples. Excitation wavelength in PhL measurements is 320 nm. (From Starodub N.F. and Slishek N.F., *Adv. Biosens. Bioelectron.* 2(2), 7, 2013. Published by Science and Engineering Publishing Company as open access.)

comparison with the initial substrate were found for H^+, C_2^+, SiH^+, $C_2H_6^+$, K^+, SiO^+, $SiOH^+$, SiF^+, and Si_2O^+ for positive ions and H^-, C^-, CH^-, O^-, F^-, C_2^-, and $C_2H_2^-$ for negative ions. SIMS results show that the observed ions (H^+, H^-, OH^-, O^-, F^-) are associated with traces of the chemical etching agents on the surface. The latter can be polisylany, siloxen with different content of oxygen, hydroxyl groups, and hydrogen and oxides. They can be present as SiH, SiO, $SiOH$, SiF, or Si_2O.

Thus, the results of AES and SIMS show that all formed PSi are nanocomposites with nanoscale silicon-oxide phase surrounded with SiO_x, hydrogen, oxygen, and carbon. The obtained PSi samples demonstrated high uniformity of the chemical composition and good reproducibility of structural parameters. Our studies of time stability of PhL of PSi obtained by a chemical method showed no drift of optical properties during five years. It allows creating and developing a model of the immune biosensor.

The registration of the specific signal was provided by two ways: the measurement of the level of PhL and photoresistivity of PSi. In the first one, the measurement setup includes ultraviolet (UV) light source with a wavelength of 350 nm, the two photodetectors (PhD) on the basis of single-crystal silicon, located at an angle of 20–250 with respect to the plane of the plate with a layer of the PSi and PhD to measure the incident UV light (Figure 5.8). A special optical quartz glass was used as a lid to seal the measurement chamber and to provide an excitation of PhL

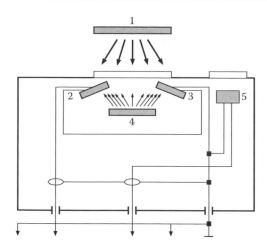

FIGURE 5.8   Construction of device for measuring PhL of PSi, where: (1) source of UV; (2, 3) PhD of the measuring channel; (4) PSi; (5) additional PhD for calibration of UVL level. (From Starodub N.F. and Slishek N.F., *Adv. Biosens. Bioelectron.* 2(2), 7, 2013. Published by Science and Engineering Publishing Company as open access.)

by UV. After adsorption of biomolecules deteriorating transmission quartz glass decreases, the value of the incident flux of UV decreases the value of the output voltage and PhL PhD that are connected in series. Using two PhD to record red emission increases sensitivity. To take in account the possible changes in incident UV, the additional PhD was used and the output signal was registered with other PD relatively from the output signal of additional ones. The last are diode type with n-p-$p+$ structures that work in the photogenerative mode. Such construction is of differential type. A scheme of photoresistance signal registration of the optical biosensors based on the PSi is given in Figure 5.9.

At the beginning of the measurement, 1 μl of the specific Ab was placed on the photoresistor surface between the contacts. Then this solution was evaporated at room temperature or at air stream. The bias (5 V) was applied between ohmic contacts and a dark current was measured by the digital voltmeter. A white light source was used to excite optical transitions on the sample surface (source A, illumination of 7000 lux) and current value of the excite sample was measured (light current). The photocurrent value was calculated as the difference between the light and dark currents. At the drawing of Ag layer on the sensitive plate and after its drying, the measurements of the dark and light current were repeated. These measurements were made after the immune complex formation, too. The control of reaching the sensor initial state was done according to the reduction of the dark current value after washing the sensitive surface by the buffer solution. The time of the single analysis was 5–10 min only.

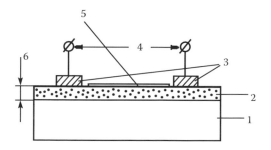

FIGURE 5.9   Scheme of the photoresistor structure based on the PSi and intended for the analysis of the interactions between biological structures. (1) the crystalline silicon; (2) the PSi; (3) the electrical contacts (Al with the thickness of ~3 μm); (4) the applied voltage; (5) the biological object; (6) the thickness of the PSi of 10–40 nm. (From Starodub N.F. and Slishek N.F., *Adv. Biosens. Bioelectron.* 2(2), 7, 2013. Published by Science and Engineering Publishing Company as open access.)

5.5 PSi-BASED BIOSENSOR AT THE CONTROL OF SOME BIOCHEMICAL QUANTITIES

Different types of chemical and biological sensors started from the determination of pure chemical substances, for example, polar and nonpolar organic solvents to control the number of biochemical and microbiological quantities, namely, bacteria, viruses, DNA or separate polynucleotides, some antigens (Ag), antibodies (Ab), enzymes, their substrates, and others analytes (Weiss et al. 2009; Harraz 2014; Harraz et al. 2014). The possibility with application of the PSi to detect the refractive index changes on the order of 10^{-4} for various water/ethanol mixtures was demonstrated (Barrios et al. 2007).

5.5.1 REGISTRATION OF Mb CONCENTRATION

Today it was clearly recognized that Mb might serve as a marker of myocardial infarction (Starodub et al. 1992, 1996). Moreover, it was found that this marker is more sensitive than other biochemical parameters, namely, with the activity of some serum enzymes (creatine kinase and its isoenzymes, as well as lactate dehydrogenase, aspartataminotransferase, and alaninaminotransferase) at identifying the disease and its treatment, too. Advantages definition Mb at the myocardial infarction compared with others biomarkers of serum are related primarily to the fact that Mb has smaller molecular weight, allowing to it to more easily pass through the membrane of damaged cells and continuing to appear in the circulation. Earlier leaching Mb of infarct zone and significant changes in its concentration in the bloodstream suggest the level of this protein is an important diagnostic test. Clinical studies have shown that the number of Mb in serum increases during the first hour of myocardial infarction, and in 2–6 h after feeling pain in the chest, it increases by 85–95% and reaches its maximum after 12 h. In general, myocardial infarction Mb concentration in serum exceeds the rate of 10–15 times and acute forms of heart attack a few dozen times. Dynamics of increase in the blood concentration of Mb are important not only in identifying infarct states, but also in their treatment using thrombolytic agents such as tissue plasminogen activator.

The determination of serum Mb in recent years was based on the principle of traditional immunochemical analysis. Now to fulfill all practice demands, efforts were directed to the development of alternative methods, including those based on biosensor approaches. Unfortunately, they are expensive, time-consuming as well as most of them are not sensitive enough and are not able to be automated. Moreover, they are based on indirect register specific immune complexes formed between Mb and specific Ab. This leads to a search for new approaches for the direct determination of Mb at the rapid diagnosis of myocardial infarction and its effective treatment. We have developed a number of immune biosensors for Mb determination, namely, based on SPR, ISFET, and PSi structures (Starodub et al. 1998, 1999a–d, 2010a,b; Starodub and Starodub 1999, 2001, 2005).

In the PSi-based immune biosensor, the PhL was registered. At the conditions of immune complex formation on the surface, the laser beam excitation falls sharply. The monoclonal anti-mouse Ab to Mb was immobilized on the surface by physical adsorption. The Mb solution was taken at a concentration of 100 mg/L. The optimal time for the Ab immobilization is 60 min (Figure 5.10). It turned out that the optimum amount of Ab to Mb in solution was at a concentration equal to 100 mg/L. The calibration curves for the determination of Mb in the model solutions and in the blood serum are presented in Figure 5.11. The detailed characteristics of process immobilization was shown by the atom force microscopy (Figure 5.12).

The lowest concentration of Mb, which can be registered using this immune biosensor, is 10 mg/L and the linear segment graph of PL characteristics of the quantitative content Mb in solution is in the range of 0.01–1 mg/L (Figure 5.13). The observed differences between the calibration curves of the above-mentioned samples may be stipulated by the possibility of the original presence of some Mb quantity in the analyzed serum blood.

To appreciate the efficiency of the developed immune biosensor based on PSi, data obtained by this method and others were compared. It was made in the comparison of results of determination of Mb by the PSi-based immune biosensor and the ordinary ELISA-method. The results of comparison investigations are presented in Table 5.1. To analyze the content, Mb sera used three

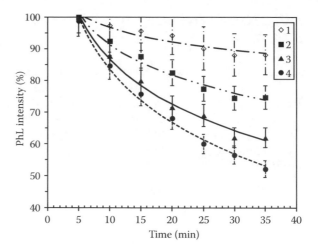

FIGURE 5.10 Changes of the intensity of PhL of PSi at the formation of the immune complex at the different time of the immobilization of Ab (100 mg/L) to Mb on the transducer surface. Curves: 1–4 respond to 15, 30, 60, and 90 min. (Reprinted from *Sens. Actuators B* 58, Starodub V.M. et al., 409, Copyright 1999, with permission from Elsevier.)

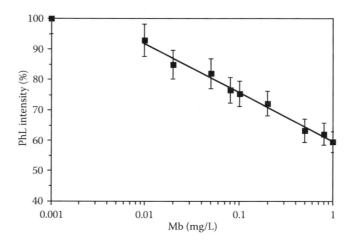

FIGURE 5.11 Changes of the PhL intensity of PSi with the immobilized specific Ab at the interaction with the Mb at different concentrations during 15 min.

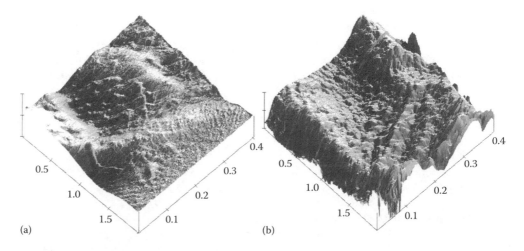

FIGURE 5.12 The view of PSi surface before (a) the Mb is immobilized and (b) after its interaction with the specific Ab (the time of the immobilization: 60 min, concentration of Mb: 100 mg/L).

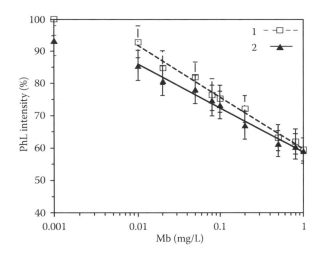

FIGURE 5.13 Changes of PhL of the PSi with the immobilized specific Ab at the interaction with the solutions containing different concentrations of Mb (curve 1) and with the serum blood dissolved by buffer to 1:10 (curve 2).

TABLE 5.1   The Determination of Mb Concentration (mg/L) in the Serum Blood of Healthy Persons and in Patients with Myocardial Infarct with Different Steps of Development

Type of Serum Blood	1	2	3	4	5
		ELISA Method			
I group	14.3 ± 0.58	19.7 ± 1.53	22.0 ± 1.43	22.6 ± 1.52	29.3 ± 1.27
II group	21.1 ± 1.13	15.8 ± 0.93	17.5 ± 0.82	17.9 ± 0.92	22.7 ± 1.33
III group	61.2 ± 2.55	42.4 ± 2.04	51.8 ± 2.32	68.9 ± 2.93	45.6 ± 1.86
		Immune Biosensor Based on the PSi			
I group	15.3 ± 1.82	18.6 ± 1.67	21.4 ± 2.07	23.3 ± 2.39	31.2 ± 3.28
II group	20.6 ± 1.87	16.3 ± 1.57	18.3 ± 1.67	17.3 ± 1.82	21.8 ± 2.24
III group	58.7 ± 5.82	40.6 ± 3.87	54.3 ± 5.16	71.6 ± 6.79	43.3 ± 4.41

types of it: from healthy subjects (group I), persons with primary form of myocardial infarction (group II), and patients with symptoms of acute infarct state (group III). A total of 15 people tested—5 from each group. Serum samples of patients with myocardial infarction PBS diluted 1:10, and healthy people were left undiluted. All samples were subjected to analysis using the developed optical sensor immune and standard method of ELISA. For each of the serum samples, Mb quantitative value was determined three times. The obtained results testified that the developed immune biosensor had the same characteristics as the ELISA-method but it provided much simpler and faster (up to 10 order) fulfillment of analysis.

5.5.2 CONTROL OF LEVEL OF BIOLOGICAL SUBSTANCES IN AIR

Today in the course of the intensive scientific-technical progress, we often have situations when different biologically active substances are appearing in the environment. We have similar situations in the case of working plants that produced nonsubstituted amino acids, such as lysine. Often around those plants, the air may contain biological components of lysine producers, which have very strong allergic effects. To prevent nondesirable situations, it is necessary to provide constant and express control of air around such plants. It is possible with the instrumental analytical devises of new generation and, in particular, based on the principles of biosensorics. To solve this problem, we have developed immune biosensors based on PSi (Starodub et al. 1998; Starodub 1999; Starodub and Starodub 1999, 2005). At the determining the biological components as air pollutants around biotechnology plants produced lysine the PSi-based immune biosensor showed sensitivity on the level of 100 mg/L of solution that corresponds to 0.33 µg/m³. The total time of the analysis was 60 min. It also can be reduced by reducing the time to immune complex formation and measurement of a kinetic mode. Highest level of air pollution among the surveyed areas of the plant was found in the packing of finished products. The nature of the distribution and storage of biological contaminants in the air surrounding the plant area is heavily dependent on climatic conditions and especially the availability of wind force and its direction.

The optical fiber immune biosensor based on enhanced chemiluminescence has a higher sensitivity compared to that based on PSi, namely, on the level of 20 mg/L or 66 ng/m³ and overall time of analysis during 45 min. However, manufacture of the first biosensor is more complete in comparison with the second one. In general, both biosensors had the sensitivity no less as ELISA-method, but they provide an analysis for considerably (almost 5 times) shorter time.

5.5.3 DETERMINATION OF THE LEVEL OF SOME MYCOTOXINS

The immune biosensor analysis was fulfilled in a "direct" way when specific Ab were immobilized on the PSi surface and then reacted with appropriate mycotoxin in solution. To examine the efficiency of the proposed immune biosensor in the case of the analysis of real products, oatmeal as well as tomato and pomegranate juices were selected as model objects taking into account that fruits and vegetables are predominantly affected by mycotoxins and, especially, by patulin. The extraction procedure is performed as follows. To tomato paste, 1 g of oatmeal, or 5 mL juice (pomegranate or tomato) soaked in the solvent in a ratio of 1:1 were added mycotoxin solution with careful stirring and its overall final concentration was kept at 200 ng/mL. After 2 h exposure of samples in sealed conditions, another half solvent was added and kept for 2 h. Then aliquot was taken by passing the mixture through a filter paper and at last the content of mycotoxin was determined. It is shown that the sensitivity of the proposed immune biosensor for both ways of the specific signal registration (by the ChL and electroconducticity) allows determining T-2 mycotoxin and patulin at a concentration of 10 ng/mL during several minutes (Starodub and Slishek 2013). To appreciate the efficiency of the proposed biosensor, its sensitivity has been compared with other similar approaches including the ELISA-method as well (Table 5.2).

The maximal sensitivity of the PSi-based immune biosensor is on the level of permissible concentration of mycotoxins in foodstuffs. For the verification of the results of analysis, namely to have more accurate determination of their level, it is necessary to find a way to increase this index for the proposed biosensor, or to apply another analytical approach.

5.5.4 DIAGNOSTICS OF RETROVIRAL LEUCOSIS

The deposition of the retroviral proteins on the PSi slightly increases the PhL level but its level decreases when the formation of the specific immune complex occurs. Moreover, the level of the change of PhL depends on the concentration of the specific Ab in the blood (Figure 5.14). If the

TABLE 5.2 Sensitivity of the Different Immune Biosensors in Case of T2 Control

N	Immune Biosensor Based on	Sensitivity	Ref.
1	Total internal reflection ellipsometry	0.15 ng/mL	Nabok et al. 2007a
2	SPR and thermistors ("direct" analysis)	~ 1.0 µg/mL	Starodub et al. 2010a
3	SPR ("to saturated" analysis)	~5.0–10.0 ng/m	Starodub et al. 2010a
4	Piesocrystal ("competitive" analysis)	1.5 ng/mL	Nabok et al. 2007b
5	sNPS, ChL ("direct" analysis)	20 ng/mL	Starodub and Slishek 2013
6	sNPS, photocurrent, ("direct" analysis)	~10 ng/mL	Starodub and Slishek 2013
7	ELISA-method ("competitive" analysis)	~10 ng/mL	Starodub and Slishek 2013

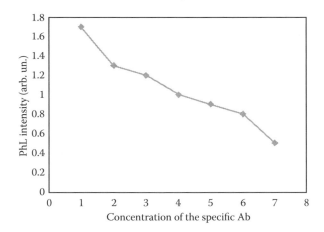

FIGURE 5.14 Dependence of the PhL intensity on the concentration of the specific Ab in the solution to be analyzed (serum blood of ill animals). Abscissa 1: immobilized Ag; 2–7: dilution of blood serum: 1:5000; 1:1000; 1:100; 1:50; 1:10; 1:1, respectively.

nonspecific Ab or the serum bovine albumin as Ag is used, the level of the PhL does not change (Starodub et al. 2011a, 2012).

We assume that PhL of PSi could be connected with the tunnel mechanism of the recombination of the charge bearers at the excitation of them in the nanocrystallites of oxide or interface. However, we do not exclude the hydrogen role for the generation of PhL extinguishing.

The photocurrent of the PSi decreased slightly after the immobilization of Ag (crude sample of the retroviral proteins) but at the addition of Ab (serum blood of ill cows) in the dilution of 1:5000 and, particular, in 1:1000 it decreased sharply. Unfortunately, at the lower level of blood dilution (from 1:100 to 1:1) the photosensitivity starts to decrease up to initial levels. It could be connected with the increase of the density of the solution to be analyzed or with other mechanisms of the electronic exchanges between the immune complex and the PSi surface.

If the blood serum from healthy cows is taken, the level of the photocurrent did not change in comparison with the initial one. The same situation was observed if bovine serum albumin was used instead of the crude samples of the retroviral proteins (Figure 5.15).

Unfortunately, there is no clear understanding of influence of the high concentration of the additional proteins into blood serum to the recombination process in the PSi. It will be our scientific task in the near future. Nevertheless, it is necessary to say that the overall time of the analysis is several dozen minutes instead of several hours in the case of the traditional ELISA-method or several days for the realization of the immune diffusion test.

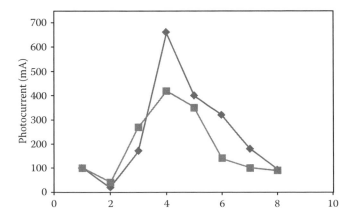

FIGURE 5.15 The level of the photocurrent before (1) and after deposition of Ag (2) and Ab in the different concentration (3–8 = 1:5000, 1:1000, 1:100, 1:50, 1:10, and 1:1, respectively). Line A and B are to two different samples of blood serum.

5.5.5 CONTROL OF OTHER TYPES OF Ag AND Ab AS WELL AS SOME PEPTIDES AND CHEMICALS

Practice is needed in control of a very big set of the different Ab or Ag based on the principle of biosensorics. For the detection of the rabbit IgG level, Meskini et al. (2007) proposed a multilayer immune biosensor structure of which was fabricated following the successive steps: APTS self-assembled monolayer, glutar aldehyde linker, and anti-rabbit IgG. Each step was controlled by high sensitive capacitance–voltage measurements. It was shown that the detection limit of rabbit IgG was on the 10 ng/mL levels.

The PSi photonic band gap structures were used for the detection of IgG though functionalization of the surface by the biotin and streptavidin and it was used to the effect that the first reagent characterizes much less mass than the second one due to it penetrating into silicon pores (Ouyang and Fauchet 2005). To control of the level of goat anti-mouse IgG, Betty et al. (2004) created the multilayer sensor structure (gold-silicon-PSi-SiO_2) which was aminosilaned with the next treatment by glutaraldehyde and mouse IgG. The content of a goat anti-mouse IgG was monitored via fluorescence microscopy and by change in the measured capacitance.

According to the data of Latterich and Corbeil (2008), typical PSi interferometry experiment shows that human IgG/anti-human IgG can be detected at 100 ng/mL, while in their investigation on an SPR instrument the detection is slightly less sensitive at 1 µg/mL.

The molecular interactions between the glutamine binding-protein (GlnBP) from *E. coli* and its main ligands, the L-glutamine (Gln) and the gliadin, a toxic peptide containing three Gln sequences is detected by means of an optical biosensor based on PSi technology. The binding effect was optically transduced in the wavelength shifts of the PSi reflectivity spectrum. If the GlnBPs are covalently bonded to the PSi, they are able to selectively recognize the gliadin in micromolar concentration (De Stefano et al. 2007).

The PSi functionalized by silane-glutaraldehyde was used for the investigation of biotin-streptavidin that was conjugated with the fluorescent label. Enhancement of the fluorescence emission of confined fluorescent biomolecules in the active layer of PSi microcavities was observed for a nonlabeled protein with a natural green fluorescence, the glucose oxidase enzyme (GOX). It was concluded that the use of smart silicon devices, enabling enhancement of fluorescence emission of biomolecules, offers easy to use biosensing, based on the luminescence response of the molecules to be detected (Palestino et al. 2008a).

Liang et al. (2012) developed multiwalled carbon nanotubes/silica nanoparticles array composite by the galvanostatic deposition technique. In this process, this composite were mixed with ammonium fluorosilicate to form a doped precursory sol solution. The formed electrochemically generated hydroxyl ions at negative potentials promote the hydrolysis of ammonium fluorosilicate and the simultaneously generated hydrogen bubbles assist the formation of three-dimensional PSi matrix. The construction of array with uniform distribution was confirmed by SEM. The formed matrix was used as a three-dimensional porous electrode and it served as an ideal platform for immobilization of alpha-fetoprotein (AFP). The proposed approach allowed achieving ultrasensitive detection of AFP antigen with ferricyanide as a probe. The linear response range was from 0.1 to 30.0 ng/mL AFP antigen with a lower detection limit of 0.018 ng/mL.

It was created the enzyme biosensor for the control of penicillin level with the help of the potentiometric capacitance-voltage measurements with PSi structures on which surface penicillinase was bound by physical adsorption (Thust et al. 1996). A linear potentiometric response for penicillin G has been detected from 0.5 to 20 mM with a sensitivity of about 40 mV. No significant degradation of the sensor output signal and good reversibility were observed for up to 50 days' storage. First investigations on porous penicillin sensors based on n-type silicon indicate even higher sensitivities of about 50 mV. In the next report of the authors (Schoning et al. 2000), it was demonstrated that the calibration curve possessed a wide linear range from 0.01 M up to 1 M with an average penicillin sensitivity of about 90 mV/mM.

A simple and low-cost optical biosensor based on PSi nanotechnology has been proposed for the control of levels of such pesticides as atrazine in water and humic acid solutions (Rotiroti et al. 2005). In both cases, it was registered up to 1 ppm. The authors explained this phenomenon as a result of the capillary infiltration of liquid into the pores, which changes the average refractive index of the structure.

The PSi-based microcantilever for biosensing triglycerides was proposed (Fernandez et al. 2008). In this case, lipase from *Pseudomonas cepacia* was covalently immobilized on the surface of PSi.

5.5.6 GLUCOSE AND APPROPRIATE ENZYME DETERMINATION

Some efforts were made in the direction of the application of PSi devices for the determination of GOX, in particular of its low concentration (Palestino et al. 2008b). For this purpose, the stable PSi microcavities (PSiMcs) were used. The method is based on the direct silanization of the PSiMc structures. This approach is very fast, occurs at room temperature, ambient pressure, and avoids preliminary thermal oxidation. The direct infiltration of silane molecules into the pores of the as-etched PSiMc structures produces only slight changes in the quality factor, contrary to the thermal oxidation that degrades the quality factor to half of its initial value. Therefore, it was demonstrated that the use of direct silanization method is a good option toward achieving high detection resolution for the low concentrations of organic species. When silanized-PSi was directly exposed to the GOX solution, the release of the adsorbed organic material and/or the oxidation of the PSi structure due to incomplete silane coverage was observed. When nonspecific (positively charged) protein binding was performed into the (negatively charged) PSiMc surfaces, the enzyme could be removed from the PSiMc structures by simply exposing them to buffer. It was demonstrated that activation of aminosilane by glutaraldehyde is needed to really bind GOX into the porous scaffold. In this case, penetration of organic molecules along the entire PSiMc structure was demonstrated by the presence of C, P, O, and Na in the energy dispersive X-ray (EDX) profile along the PSiMc structure. The secondary structure of the adsorbed GOX has been resolved and found to contain a majority (80%) of α-helix, β-sheet, and β-turn conformations, indicating that GOX retains in native form within the PSi structures. A detection limit of 25 nM protein has been achieved, thus demonstrating the high quality and potential of our functionalized PSi microcavity structures.

In another communication (Drottyx et al. 1997), it was informed about PSi as the carrier matrix in microstructured enzyme reactors yielding high enzyme activities. This paper describes a method to accomplish a highly efficient silicon microstructure reactor. It was fabricated as flow-through cells comprising 32 channels, 50 µm wide, spaced 50 µm apart and 250 µm deep micromachined in (110) oriented silicon, p-type (20–70 Ω·cm), by anisotropic wet etching. The overall dimension of the microreactors was 13:1 × 3:15 mm. To make the PSi layer, the reactor structures were anodized in a solution of hydrofluoric acid and ethanol. In order to evaluate the surface enlarging effect of different pore morphologies, the anodization was performed at three different current densities, 10, 50, and 100 mA/cm^2.

Glucose oxidase was immobilized onto the three porous microreactors and a nonporous reference reactor. The enzyme activity of the reactors was monitored following a colorimetric assay. To evaluate the glucose monitoring capabilities, the reactor anodized at 50 mA cm^{-2} was connected to a system for glucose monitoring. The system displayed a linear response of glucose up to 15 mM using an injection volume of 0.5 µl. An increase in enzyme activity by a factor of 100, compared to the nonporous reference, was achieved for the reactor anodized at 50 mA/cm^2. The obtained results of glucose turnover rate clearly demonstrate the potential of PSi as a surface-enlarging matrix for microenzyme reactors.

A group of authors (Lopez-Garcia et al. 2007) has fabricated PSi-based devices with the structure Al/PSi/Si/Al, which shows a typical rectifying behavior for both longitudinal and transversal *I–V* measurements. Glucose biomolecules were immobilized onto the surface of PSi from solutions of glucose in distilled water using different concentrations. Electrical measurements (current–voltage characteristics) were performed for both longitudinal and transversal conduction through the PSi-based structures. It was found that for increasing glucose concentration, there is a reduction of both longitudinal and transversal current for a given voltage. The experimental results open the way to the development of PS sensors based on changes in the electrical behavior.

A high-performance electrochemical biosensor for the detection of glucose was described using a Pt-dispersed polyanyline nanorod electrode with a hierarchical porous surface (Soug and Hwang 2009). The nanorodes were synthesized directly on an Au-coated PSi electrode. The high-density Pt nano-particles were dispersed on the nanorodes by the electrochemical deposition. The electrode surface's structure exhibited order on multiple length scales from ~30 nm Pt nano-particles to ~200 nm diameter nanorods to micrometer pore sizes for the PSi. The sensitivity of such hierarchical electrode was 35.64 μA/mM, which corresponds to 22.1-fold larger than a planar one. The linear response was in the range from 9×10^{-6} to 5×10^{-3} M with a short time (<10 s). The stability of the biosensor retained 80% of the initial value after a 45-day storage.

D'Auria et al. (2006) published a review with the analysis of the existed experimental results about application of PSi as an effective transducer for the determination of not glucose only and a number of other analytes including L-glutamine, ammonium nitrate, and other low molecular substrates. In all cases, the specific recognition elements such as protein receptors and enzymes were immobilized on hydrogenated PSi wafers and used as probes in optical sensing systems. The binding events were optically transduced as wavelength shifts of the PSi reflectivity spectrum or were monitored via changes of the fluorescence emission.

5.5.7 DETECTION OF VIRUSES AND BACTERIA

Taylor et al. (2008) summarized the main principles that are realized at the detection of bacteria on the basis of biosensorics. Thin films of nano-PSi that were prepared using pulsed anodic etching in an HF electrolyte solution were used (Rossi et al. 2007) at the development of a new method for improving the sensitivity for detection of the bacteriophage virus MS2. Bioconjugation of the PSi is accomplished by replacing the hydrogen with a functional organic group that can be bound to the desired protein molecule using a crosslinker. Two procedures were used to make a covalent bond between the rabbit MS2 antibody and the PSi surface. One of them was realized through carbodiimide and the other using TDBA-OSu (a commercially available photoactive aryldiazirine cross-linker). The amounts of Ab and virus bound to the PSi surface were quantified by measuring the dye-labeled fluorescence intensity at the emission maximum. The reported sensor allows measurement of viral concentrations ranging from 2×10^7 to 2×10^{10}. The system exhibits sensitivity and dynamic range similar to the Luminex liquid array-based assay while outperforming protein microarray methods.

For the detection of *E. coli*, the structure Au-NiCr/nano-PSi/Au-NiCr was used (Recio-Sánchez et al. 2010). Layers of nano-PSi were grown by electrochemical etching of boron-doped (*p*-type) silicon wafers (orientation: <100>, resistivity: 0.1–0.5 Ω·cm). Low resistivity ohmic contacts were formed by coating the backside of the Si wafers with Al and subsequently annealing at 400°C for 5 min. The wafers were cut into 1×1 cm² pieces, which were mounted into a sample holder with an exposed area to the electrochemical etching solution of 0.64 cm². The electrolyte consisted of HF (48 wt %):ethanol (98 wt %) solutions with different HF-to-ethanol ratios. The wafers were galvanostatically etched using etching current densities ranging from 10 mA/cm² to 100 mA/cm², and etching times from 20 to 180 s. Illumination from a 100-W halogen lamp was used to promote the generation of electron-hole pairs. Tthis contributed to reducing the typical crystallite size. Previously, PSi was functionalized by the use of solutions of aminopropyltriethoxysilane in toluene. These solutions were found to be highly adapted for the subsequent immobilization of *E. coli* Ab through the high density of amino groups was induced on the surface of nano-PSi. After Ab immobilization, fragments of *E. coli*, which are specifically recognized by the Ab, were present on the surface of the nano-PSi. It was stated that the nano-PSi-based devices showed a strong dependence of their electrical behavior (*I–V* curves) on the presence and concentration of *E. coli* fragments. It is necessary to underline that the authors have shown the efficiency of such construction of biosensors at the determination of glucose in the model solutions.

An optical label-free biosensing platform for the direct (without cell lyses) bacteria detection (*E. coli* K12 as a model system) based on nanostructured oxidized PSi was described (Massad-Ivanir et al. 2011). The optical reflectivity spectrum of the nanostructure displays Fabry-Perot fringes characteristic of thin-film interference, enabling direct, real-time observation of bacteria attachment within minutes. The biofunctionalizing of the PSi was made as shown in Figure 5.16.

Because of the investigation, it was shown that the limit of the sensitivity of bacteria analysis was on the level of 10^4 cells/mL and a response time of several minutes. Several approaches are currently being explored to improve the detection limit and sensitivity of the biosensor, including

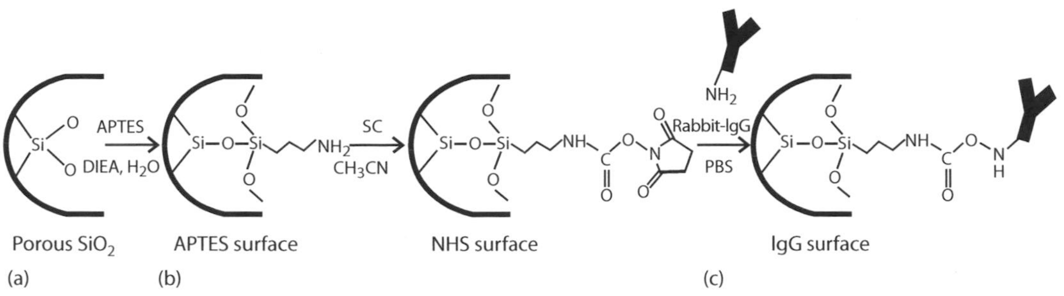

FIGURE 5.16 Schematic representation of the synthesis steps followed to biofunctionalize the PSiO$_2$ surface with IgG. (a) The PSiO$_2$ surface is initially reacted with APTES and is catalyzed by an organic base to create an APTES surface. (b) The APTES surface terminal amine groups react with the NHS activated ester of SC to create an NHS surface. (c) Amine-terminated IgG is attached to the NHS surface to create an IgG surface. (Reprinted with permission from Massad-Ivanir N. et al., *Anal. Chem.* 83, 3282. Copyright 2011, American Chemical Society.)

improving optimization of the antibody concentration and orientation, enhancement of the coupling chemistry, and incorporation of a hydrogel into the PSi nanostructure. For comparison, the detection limit of current state-of-the-art SPR biosensors is in the range of 10^2–10^4 cells/mL (Taylor et al. 2008).

It was reported (Mace and Miller 2008) that the application of glycine methyl ester as a spacer molecule at the functionalization of silicon surface gives the possibility to create a lipid A biosensor that is able to distinguish gram-negative and gram-positive bacteria. Therefore, at the exposure biosensor to lysates of *Esherichia coli* or *Salmonella minnesota* (both gram-negative), the red-shift in PhL spectra was observed. At the same time, in the case of contact of a biosensor with the lysates of *Bacillus subtilis* or *Lactobacillus acidophilus* (gram-positive bacteria) any signals were registered. The authors concluded that this biosensor is an analog of a gram stain as a core procedure of bacteriology.

5.5.8 DNA SENSORS

For the analysis of DNA molecules, the PSi surface was functionalized with 3-aminopropyltriethoxysilane (3-APTES) and sulfosuccinimidyl 4–[N-maleimidomethyl] cyclohexane 1–carboxylate (Sulfo-SMCC), which also served as intermediate small molecules (Rodriguez et al. 2013).

The biosensor is an electrochemical device that transduces the hybridization of DNA into a chemical oxidation of guanine by Ru(bpy)$_3^{+2}$ the reduced form of which is then detected electrochemically (Lugo et al. 2007). DNA hybridization has been detected voltammetrically. In order to prevent the DNA probe from reacting with the guanosine 5'-monophosphate, it has been replaced in the DNA probe sequence with inosine 5'-monophosphate, which is much less reactive to oxidization. Ruthenium bipyridine showed a catalytic effect on the anodic peak current which is related to the concentration of target DNA in the hybridization reaction. The anodic peak current of Ru(bpy)$_3^{+2}$ was linearly related to the target DNA sequence in the range 0.5×10^{-10} to 500×10^{-10} M with a detection limit of 0.5×10^{-10} M. In addition the ruthenium bipyridine, the indicator was able to selectively discriminate against different DNA sequences; a necessary property for sensing applications. The authors concluded that the obtained results confirmed that the DNA sensing method using them is a quick and convenient way for the specific and quantitative detection of DNA.

Chan et al. (2001) developed the PSi-based DNA biosensor for the direct detection of cDNA. In this case, the complementary DNA olyconucleotides were attached to a silanized surface. A large differential signal is obtained before and after formation of complementary interaction. A 7-nm red-shift is observed indicating a change in optical thickness that is consistent with an increase in refractive index by 0.03. When a noncomplementary strand of DNA is exposed to the DNA sequence on the biochip, no shifting of the luminescence peaks is observed and the differential signal is negligible. It is necessary to take into account that at exposure times of less than 7 min, no PhL shifts were detected. This time is insufficient for the DNA molecules to seek and find their complementary pair. On longer exposure time (30 min), a ~5-nm red-shift in photoluminescence is observed. Saturation of the PhL red-shift occurs when the complementary DNA exposure time is greater than 60 min. Therefore, in all subsequent experiments, the complementary strand of DNA is allowed to bind to the DNA-immobilized PSi biochip for at least 1 h. The appropriate 30-base pair complementary DNA segment was immobilized into the silanized porous matrix. After exposure to the full-length viral DNA, the biochip is heated to 89°C for 3 min to denature the phage lambda DNA into two separate strands. The segmented and immobilized cDNA on the PSi biochip is then allowed to hybridize with the full-length DNA strands at 37°C for 1 h. The recognition and binding of bacteriophage lambda to a partial cDNA sequence immobilized on the PSi surface was registered through PhL red-shifts which achieved 12 nm.

Vamvakaki and Chaniotakis (2008) have formed PSi by electrochemically etching and employed it as a substrate for the entrapment of oligonucleotides and the subsequent development of stable DNA biosensors. The controlled potential anodic etching of p-type silicon wafers was optimized in order to obtain a surface layer with pore diameters that are close to those of the adsorbed DNA helix. The stabilization and hybridization of DNA inside the PSi layer is confirmed using ATR-FTIR. The hybridization was verified by the large and reproducible impedance changes at the interface layer. It was concluded that the PSi DNA sensor paves the way for the label-free detection of oligonucleotide sequences in DNA microarrays.

Rong et al. (2008) consider too that the PSi is an excellent material for biosensing due to its large surface area and its capability for molecular size selectivity. They have developed the label-free nanoscale PSi resonant waveguide biosensor with the average diameter of pores in 20 nm. DNA was attached inside the pores through successive application of standard amino-silane and GA. This biosensor can selectively discriminate between complementary and noncomplementary DNA. It had high-level sensitivity with a detection limit on the level of 50 nM DNA concentration or equivalently in 5 pg/mm^2.

A number procedures needed for the functionalization of PSi surfaces at the creation of DNA biosensors are detailed in the review of Rodriguez et al. (2014).

There is information (Kerman et al. 2004) that at last three major electrochemical DNA chips produced by GeneOhm Sciences Inc., Toshiba Corp., and Motorola Life Sciences Inc. should be at the molecular diagnosis market. Electrochemical DNA biosensors with their cost-effectiveness and suitability for microfabrication can be expected to become increasingly popular in the near future. The main target in all the DNA chip technologies has been to eliminate the role of the polymerase chain reaction (PCR) from their protocols. Unfortunately, this goal has not yet been reached in commercialized electrochemical technologies. Some efforts are now being made toward decreasing the time needed for PCR.

5.5.9 CELLS ANALYSIS

Flavel et al. (2011) described the use of conventional photolithographic techniques to create arrays of PSi by electrochemical anodization without the use of ceramic or metal assistive layers thereby dramatically reducing the complexity of the fabrication process. The obtained PSi layers were modified by hydrosilylation with the following immobilization of the dye of lissamine and the cell adhesion on the attached arginine-glycine-aspartic acidserine peptide. Such surface facilitated attachment of human epithelial cells. This opens new vistas toward the development of optical sensory components of cell culture systems, which may become viable alternatives to conventional intrusive cell-based assays.

The cell adhesion and its viability during cultivation on the PSi surface were investigated with the rat hepatocytes. It was determined by measuring protein production from the cells, which were coated with collagen on the ozone oxidated PSi (Chin et al. 2001). Human iris epithelium cells and rat pheochromocytoma cells had more high adhesion to the collagen-coated and amino-silanized PSi. This process was poor in the case of ozone-oxidized and polyethylene glycol silanized surfaces (Low et al. 2009). It was demonstrated (Low et al. 2009) that human neuroblastoma cells prefer the large (1–3 mm) to small (100–300 nm) pores by comparison. Based on the investigations of the possible PSi role at the cell analysis, conception of a so-called "smart Petri dish," which enables monitoring of the health of cell cultures by means of light scattering from the surface of a rugate filter, was proposed (Alvarez et al. 2007). The surface of a PSi rugate filter is illuminated with white light at an oblique angle, and a charge coupled device spectrometer was positioned perpendicular to the surface. Changes in the morphology of the cells on the surface indicate typically on a decrease in viability that reflected in change in the scattered light. It is an important advance in using PSi for cell biosensing and in cell chips although exact biochemical information was lacking up to now (Gupta et al. 2007).

5.6 SUMMARY AND PERSPECTIVES OF PSi IN BIOSENSORICS

Previously it was mentioned that silicon is very widespread. Now it is used intensively in electronics and according to a number of given demonstrations, PSi is a perspective transducer in different types of biosensors. It turns out that the construction of PSi-based biosensors is very simple; the created devices may be applied in field condition and may provide analysis in an online regime. As a rule, the sensitivity of such types of biosensors respond to practice demands for the screening observation.

The next step in the development of PSi-based biosensors is connected, according to our opinion, with the formation of the integrated devices in the form of lab on a chip. Today, the investigations in this direction are actively growing. There is information about the creation of a chemical sensor in such a form for simultaneous detection of the number of organic solvents (Rea 2008).

We have created prototypes of such integrated devices, too (Karpiuk et al. 2014; Starodub et al. 2014). It is able for simultaneous control of several samples and for analysis of a number of biochemical quantities. The last concerns a special type of group of metabolic disorders. One among them is diabetes characterized by abnormal fuel metabolism, which results most notably in hyperglycemia and dyslipidemia, due to defects in insulin secretion, insulin action, or both. The specific pathophysiology of type 1 diabetes is the attacks of the immune system and destroying the insulin producing beta cells of the pancreas. To overcome this disorder, it is necessary to constantly control the glucose level and the content of the anti-insulin antibody. Today, a number of instrumental approaches for the control of the first parameter are proposed and among them, some types of amperometric enzymatic biosensors are the most dispersed. Unfortunately, the similar devices for express control of the anti-insulin antibodies in the blood of patients are absent.

We have realized the possibility to create such a prototype as lab on a chip with the application of PSi as transducers. This prototype is able to provide not only simultaneous analysis of glucose and anti-insulin antibodies but also control of other biochemical quantities, which are diagnostically significant in diabetes. The manufacturing of such lab on a chip biosensor is fully compatible with silicon planar technology, which is realized at the production of semiconductor devices. In this case, the registration of specific sensor signals in the form of a change of such parameters as photoluminescence or photocurrent of PSi surface may be done with the help of simple and portable devices. The model of a new type of lab on a chip immune biosensor based on PSi for simultaneous detection of the levels of insulin and anti-insulin antibodies in the sample to be analyzed is presented in Figure 5.17. In this case, the possibility to register not only specific interaction of the immune components and the enzyme reaction, in particular, but also catalytically the depletion of glucose by the glucose oxidase at the control level of sugar in the blood of patients was realized. Therefore, we have the possibility to do simultaneous control glucose and anti-insulin specific antibodies during treatment of patients by constantly introducing this hormone.

To test the proposed lab on a chip in other diagnostics the interaction of immune complex (antibody–antigen in determining retroviral leukemia) with the surface of silicon was studied. On the surface of a photoresistor (in determining leukemia) with the already immobilized antigen solution, positive serum blood (containing specific antibodies—Ab$^+$) was applied and on the surface of another one the solution serum blood without selective Ab$^-$ was applied. Then the formed complexes were washed and dried in air. The analysis of the photocurrent makes it possible to assert that when applying a positive serum, the formation of immune complexes takes place (there is a sharp increase in the photocurrent). When applying the negative serum, virtually nothing happens.

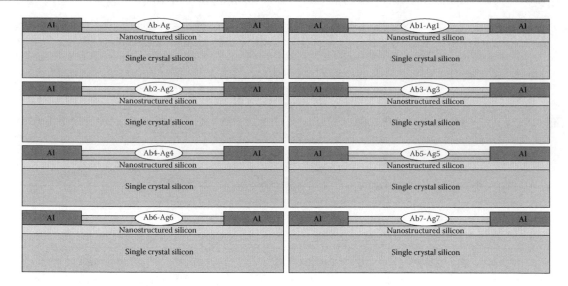

FIGURE 5.17 General view of lab on a chip prototype based on PSi.

The proposed sensor meets all modern requirements such as ease of use, inexpensive, high precision, selectivity, speed of analysis, miniaturization, and it can be connected to a phone or tablet with built-in software. The very big perspectives of the practical application of such a type of biosensor are not in the field of medicine only and in solving problems of ecology, namely, at the control environment state as well as at the estimation of quality food and feeds, for example, at the simultaneous control of a number of samples or the determination of several analytes (different mycotoxins or some toxic substances). Therefore, we developed the detailed protocol of the analysis of a number of types of toxins (first of all mycotoxins, such as T2, aflatoxins, patulin, and others) among different environmental objects. It was stated that the proposed biosensors having all of the previously mentioned advantages allow wide screening of these toxins simultaneously in many objects and for the different types of analytes in full agreement with the modern practice demands.

ACKNOWLEDGMENTS

The work was partially supported by NATO project "Optical Bio-Sensors for Bio-Toxins" NUKR. SFPP 984637 in the Science for Peace and Security (SPS) Programme and EU funded BIOSENSORS-AGRICULT project (Development of Nanotechnology Based Biosensors for Agriculture FP7-PEOPLE-2012-IRSES, contract no. 318520) as well as Ministry of Education of Ukraine, project in fundamental program and National University of Life and Environmental Sciences of Ukraine, project no. 110/476-Appl. We especially thank Prof. G. Korotcenkov (Gwangju Institute of Science and Technology, Gwangju, Republic of Korea) for his very productive help in the preparation of our material for publication as this chapter.

REFERENCES

Alvarez, S.D., Schwartz, M.P., Migliori, B., Rang, C.U., Chao, L., and Sailor, M. (2007). Using a porous silicon photonic crystal for bacterial cell-based biosensing. *Phys. Stat. Sol. (a)* 204, 1439–1443.

Arroyo-Hernandez, M., Martin-Palma, R.J., Torres-Costa, V., and Martinez Duart, J.M. (2006). Porous silicon optical filters for biosensing applications. *J. Non-Crystall. Solids* 352, 2457–2460.

Barrios, C.A., Banuls, M.J., Gonzalez-Pedro, V. et al. (2007). Label-free optical biosensing with slot-waveguides. *Opt. Lett.* 33(7), 708–710.

Ben Ali, M., Mlika, R., Ben Ouada, H., M'ghaeth, R., and Maaref, H. (1999). Porous silicon as substrate for ion sensors. *Sens. Actuators A* 74, 123–125.

Betty, C.A., Lal, R., Sharma, D.K., Yakhmi, J.V., and Mittal, J.P. (2004). Macroporous silicon based capacitive affinity sensor-fabrication and electrochemical studies. *Sens. Actuators B* 97, 334–343.

Bogue, R.W. (1997). Novel porous silicon biosensor. *Biosens. Bioelectron.* 12(1), xxvii–xxix.

Boukherroub, R., Wojtyk, J.T.C., Wayner, D.D.M., and Lockwood, D.J. (2002). Thermal hydrosilylation of undecylenic acid with porous silicon. *J. Electrochem. Soc.* 149, 59.

Canham, L.T. (1997a). Porous semiconductors: A tutorial review. *Proc. Mat. Res. Soc. Symp.* 452, 29–42.

Canham, L.T. (1997b). Biomedical applications of porous silicon. In: Canham L.T. (Ed.) *Properties of Porous Silicon.* EMIS Data Rev. Series 18, London, pp. 371–376.

Chan, S., Li, Y., Rothberg, L.J., Miller, B.L., and Fauchet, Ph.M. (2001). Nanoscale silicon microcavities for biosensing. *Mater. Sci. Eng. C* 15, 277–282.

Chin, V., Collins, B.E., Sailor, M.J., and Bhatia, S.N. (2001). Compatibility of primary hepatocytes with oxidized nanoporous silicon. *Adv. Mat.* 13(24), 1877–1880.

Dancil, K.-P.S., Douglas, P.G., Gurtner, C., and Sailor, M.J. (1998). Development of a porous silicon based biosensor. *Proc. Mat. Res. Soc. Symp.* 536, F3.3.

Dancil, K.-P.S., Greiner, D.P., and Sailor, M.J. (1999). A porous silicon optical biosensor: Detection of reversible binding of IgG to a protein A-modified surface. *J. Am. Chem. Soc.* 121, 7925–7930.

D'Auria, S., de Champdor, M., Aurilia, V. et al. (2006). Nanostructured silicon-based biosensors for the selective identification of analytes of social interest. *J. Phys.: Condens. Matter* 18, 2019–2028.

De Stefano, L., Ivo, R., Moretti, L. et al. (2004). Smart optical sensors for chemical substances based on porous silicon technology. *Appl. Opt.* 43, 167–172.

De Stefano, L., Rotiroti, L., Rea, I. et al. (2006). Porous silicon-based optical biochips. *J. Opt. A: Pure Appl. Opt.* 8, 540–544.

De Stefano, L., Rendina, I., Rossi, A. M., Rossi, M., Rotiroti, L., and D'Auria, S. (2007). Biochips at work: Porous silicon microbiosensor for proteomic diagnostic. *J. Phys. Condens. Matter.* 19, 395007.

Di Francia, G., La Ferrara, V., Manzo, S., and Chiavarini, S. (2005). Towards a label-free optical porous silicon DNA sensor. *Biosens. Bioelectron.* 21, 661–665.

Drottyx, J., Lindstromy, K., Rosengrenz, L., and Laurelly, T. (1997). Porous silicon as the carrier matrix in microstructured enzyme reactors yielding high enzyme activities. *J. Micromech. Microeng.* 7, 14–23.

Fauchet, P. M., Tsybeskov, L., Peng, C. et al. (1995). Light-emitting porous silicon: Materials science, properties, and device applications. *IEEE J. Selected Topics Quant. Electron.* 1, 1126–1139.

Faucheux, A., Gouget-Laemmel, A.C., Henry de Villeneuve, C. et al. (2006). Well-defined carboxyl-terminated alkyl monolayers grafted onto H-Si(111): Packing density from a combined AFM and quantitative IR study. *Langmuir* 22, 153–162.

Fernandez, R.E., Bhattacharya, E., and Chadha, A. (2008). Covalent immobilization of Pseudomonas cepacia lipase on semiconducting materials. *Appl. Surf. Sci.* 254(15), 45124519.

Flavel, B.S., Sweetman, M.J., Shearer, C.J., Shapter, J.G., and Voelcker, N.H. (2011). Micropatterned arrays of porous silicon: Toward sensory biointerfaces. *Appl. Mater. and Interfaces* 3, 2463–2471.

Foucaran, A., Sorli, B., Garcia, M. et al. (2000). Porous silicon layer coupled with thermoelectric cooler: A humidity sensor. *Sens. Actuators A* 79, 189–193.

Gallach, D., Sancheza, G.R., Noval, A.M. et al. (2010). Functionality of porous silicon particles: Surface modification for biomedical applications. *Mater. Sci. Eng. B* 169, 123–127.

Ghoshal, S., Mitra, D., Roy, S., and Dutta Majumder, D. (2010). Biosensors and biochips for nanomedical applications: A review. *Sens. Transducers J.* 113(2), 1–17.

Gupta, B., Zhu, Y., Guan, B, Reece, P.J., and Gooding, J.J. (2007). Functionalised porous silicon as a biosensor: Emphasis on monitoring cells in vivo and in vitro. *Phys. Stat. Sol. (a)* 204, 1439–1443.

Han, H.-M., Li, H.-F., and Xiao, S.-J. (2011). Improvement of the durability of porous silicon through functionalisation for biomedical applications. *Thin Solid Films* 519, 3325–3333.

Harraz, F.A. (2014). Porous silicon chemical sensors and biosensors: A review. *Sens. Actuators B* 202, 897–912.

Harraz, F.A., Ismail, A.A., Bouzida H., Al-Sayari S.A., Al-Hajry A., and Al-Assiri M.S. (2014). A capacitive chemical sensor based on porous silicon for detection of polar and non-polar organic solvents. *Appl. Surf. Sci.* 307, 704–711.

Heinemann, W.R. and Jensen, W.B. (2006). Leland C. Clark Jr. (1918–2005). *Biosens. Bioelectron.* 21(8), 1403–1404.

Hiraoui, M., Guendouz, M., Lorrain, N. et al. (2011). Spectroscopy studies of functionalized oxidized porous silicon surface for biosensing applications. *Mater. Chem. Phys.* 128, 151–156.

Jane, A., Dronov, R., Hodges, A., and Voelcker, N. H. (2009). Porous silicon biosensors in the advance. *Trends Biotech. Rev.* 27, 230–240.

Karpiuk, A.D., Starodub, N.F., Luchenko, A.I., and Melnichenko, M.M. (2014). Nano-structured silicon based lab on a chip for diagnostics. In: *Proceedings of Int. Conf. of ELANO*, Kiev, Ukraine (in press).

Kelly, M.T. and Bocarsly, A.B. (1998). Mechanisms of photoluminescent quenching of oxidized porous silicon. Applications to chemical sensing. *Coord. Chem. Rev.* 171, 251–259.

Kerman, K., Kobayashi, M., and Tamiya, E. (2004). Recent trends in electrochemical DNA biosensor technology. *Measurements Sci. Technol.* 15, R1–R11.

Kim, S.-J., Jeon, B.-H., and Choi, K.-S. (1999). Improvement of the sensitivity by UV light in alcohol sensors using porous silicon layer. In: *Proceedings of Int. Semiconductor Conference*, Sinaia Romania, Vol. 2, pp. 475–478.

Latterich, M. and Corbeil, J. (2008). Label-free detection of biomolecular interactions in real time with a nano-porous silicon-based detection method. *Proteome Sci.* 6(31), 1–11.

Laurell, T., Drott, J., Rosengren, L., and Lindstrom, K. (1996). Enhanced enzyme activity in silicon integrated enzyme reactors utilizing porous silicon as the coupling matrix. *Sens. Actuators B* 31(3), 161–168.

Lees, I.N., Lui, H., Canaria, C.A. et al. (2003). Chemical stability of porous silicon surfaces electrochemically modified with functional alkyl species. *Langmuir* 19(23), 9812–9817.

Lenward, S., James, L., Gole, L. et al. (2002). Rapid, reversible, sensitive porous silicon gas sensor. *J. Appl. Phys.* 91–94, 2519.

Liang, R.-P., Wang, Zh.-X., Zhang, L., and Qiu, J.-D. (2012). A label-free amperometric immunosensor for alpha-fetoprotein determination based on highly ordered porous multi-walled carbon nanotubes/silica nanoparticles array platform. *Sens. Actuators B* 166–167, 569–575.

Lin, V.S.Y., Motesharei, K.K., Dancil, P.S. et al. (1997). A porous silicon-based optical interferometric biosensor. *Science* 278, 840–843.

Lopez-Garcia, J., Martin-Palma, R.J., Manso, M., and Martinez-Duart, J.M. (2007). Porous silicon based structures for the electrical biosensing of glucose. *Sens. Actuators B* 126(1), 82–85.

Low, S.P., Voelcker, N.H., Canham, L.T., and Williams, K.A. (2009). The compatibility of porous silicon in tissues of the eye. *Biomaterials* 30(15), 2873–2880.

Lugo, J.E., Ocampo, M., Kirk, A.G. et al. (2007). Electrochemical sensing of DNA with porous silicon layers. *J. New Mater. Electrochem. Syst.* 10, 113–116.

Lysenko, V., Roussel, P., Remaki, B. et al. (2000). Study of nano-porous silicon with low thermal conductivity as thermal insulating material. *J. Porous Mater.* 7, 177–182.

Mace, C.R. and Miller, B.L. (2008). Porous and planar silicon sensors. In: Zourab M. et al. (Eds.) *Principles of Bacterial Detection: Biosensors, Recognition Receptors and Microsystems.* Springer Science+Business Media LLC, pp. 231–253.

Massad-Ivanir, N., Stenberg, G., Tzur, A., Krepker, M.A., and Segal, E. (2011). Engineering nanostructured porous SiO surfaces for bacteria detection via "Direct Cell Capture." *Anal. Chem.* 83, 3282–3289.

Mattei, G., Marucci, A., and Yakovlev, V.A. (1998). Splitting of porous silicon microcavity mode due to the interaction with Si-H vibrations. *Mater. Sci. Eng. B* 51, 158–161.

Mehra, R.M., Agarwal, V., and Mathur, P.C. (1997). Development and characterization of porous silicon: A review. *Solid State Phenom.* 55, 71–76.

Meskini, O., Abdelghani, A., Tlili, A., Mgaieth, R., Jaffrezic-Renault, N., and Martelet, C. (2007). Porous silicon as functionalized material for immunosensor application. *Talanta* 71, 1430–1433.

Misra, S.C.K. and Angelucci, R. (2001). Polyamine thin film-porous silicon sensors for detection of microorgamizm. *Indian. J. Pure Appl. Phys.* 39, 726–730.

Nabok, A., Tsargorodskaya, A., Holloway, A. et al. (2007a). Specific binding of large aggregates of amphiphilic molecules to the respective antibodies. *Langmuir* 23, 8485–8490.

Nabok, A.V., Tsargorodskaya, A., Holloway, A. et al. (2007b). Registration of T-2 mycotoxin with total internal reflection ellipsometry and QCM impedance methods. *Biosens. Bioelectron.* 22, 885–890.

Namavar, F., Maruska, H.P., and Kalkhoran, N.M. (1992). Visible electroluminescence from porous silicon np heterojunction diodes. *Appl. Phys. Lett.* 60(20), 2514–2516.

Ouyang, H. and Fauchet, Ph. M. (2005). Biosensing using porous silicon photonic bandgap structures. In: *Proceedings of SPIE Optics East 2005, Proc. SPIE.* 6005, 31–45 (doi:10.1117/12.629961).

Palestino, G., Agarwal, V., Aulombard, R., Perez, E., and Gergely, C. (2008a). Biosensing and protein fluorescence enhancement by functionalized porous silicon devices. *Langmuir* 24, 13765–13771.

Palestino, G., Legros, R., Agarwal, V., Pérez, E., and Gergely, C. (2008b). Functionalization of nanostructured porous silicon microcavities for glucose oxidase detection. *Sens. Actuators B* 135(1), 27–34.

Rao, L.-F. and Yates, J.T. (1993). Surface chemistry of hydrogen-passivated porous silicon—Oxidation of surface Si-H groups by acetone. *Technical Report of Naval Research*, AD-A264655.

Rea, I. (2008). Porous silicon based optical devices for biochemical sensing. PhD Thesis, University of Naples "Federico II", Naples, Italy, p. 105.

Recio-Sánchez, G., Torres-Costa, V., Manso, M., Gallach, D., López-García, J., and Martín-Palma, R.J. (2010). Towards the development of electrical biosensors based on nano-structured porous silicon. *Materials* 3(2), 755–763.

Rittersma, Z.M. and Benecke, W. (1999). A novel capacitive porous silicon humidity sensor with integrated thermo- and refresh resistors gas sensors. In: *Proc. of 13th Eur. Conf. Solid-State Transducers "EUROSENSORS XIII,"* Sept. 12–15, the Netherlands, pp. 371–374.

Rodriguez, G., Ryckman, J., Jiao, Y. et al. (2013). Real-time detection of small and large molecules using a porous silicon grating-coupled Bloch surface wave label-free biosensor. *Proc. SPIE* 8570, *Frontiers in Biological Detection: From Nanosensors to Systems V,* 857004, March 5 (doi:10.1117/12.2004455).

Rodriguez, G.A., Lawrie, J.L., and Weiss, S.M. (2014). Nanoporous silicon biosensors for DNA sensing. In: Santos H.A. (Ed.) *Porous Silicon for Biomedical Applications.* Woodhead Publ., Cambridge, UK, pp. 304–329.

Rong, G., Najmaie, A., Sipe, J.E., and Weiss, Sh.M. (2008). Nanoscale porous silicon waveguide for label-free DNA sensing. *Biosens. Bioelectron.* 23(10), 1572–1576.

Rossi, A.M., Wang, L., Reipa, V., and Murphy, Th.E. (2007). Porous silicon biosensor for detection of viruses. *Biosens. Bioelectron.* 23(5), 741–745.

Rotiroti, L., De Stefan, L., Rendin, I. et al. (2005). Optical microsensors for pesticides identification based on porous silicon technology. *Biosens. Bioelectron.* 20, 2136–2139.

Salis, A., Setzu, S., Monduzzi, M., and Mula, G. (2011). Porous silicon-based electrochemical biosensors. In: Serra P.A. (Ed.) *Biosensors-Emerging Materials and Applications*, InTech., Rijeka, Croatia, pp. 334–352.

Salonen, J. and Lehto, V.-P. (2008). Fabrication and chemical surface modification of mesoporous silicon for biomedical applications. *Chem. Eng. J.* 137, 162–172.

Schoning, M.J., Kurowski, A., Thust, M., Kordos, P., Schultze, J.W., and Luth, H. (2000). Capacitive microsensors for biochemical sensing based on porous silicon technology. *Sens. Actuators B* 64, 59–64.

Schwartz, M.P., Derfus, A.M., Alvarez, S.D. et al. (2006). The smart petri dish: A nanostructured photonic crystal for real-time monitoring of living cells. *Langmuir* 22, 7084–7090.

Schwartz, M.P., Alvares, S.D., and Sailor, M.J. (2007). Porous SiO_2 Interferometric biosensor for quantitative determination of protein interaction. Binding of protein A to immunoglobulins derived from different species. *Anal. Chem.* 79, 327–334.

Singh, S., Sharma, S.N., Govind, Shivarpasad, S.M., Lal, M., and Khan, M.A. (2009). Nanostructured porous silicon as functionalized material for biosensor application. *J. Mater. Sci.: Mater. Med.* 20, 181–187.

Soug, M.-J. and Hwang, J.W. (2009). Amperometric glucose biosensor based on a Pt-dispersed hierarchically porous electrode. *J. Korean. Phys. Soc.* 54(4), 1612–1618.

Starodub, V.M. (1999). Optical immune sensors for the express monitoring of the air contamination by biological substances. In: *Proc. NATO ASI Human Monitoring after Environmental and Occupational Exposure to Chemical and Physical Agents*, Sept. 23–Oct. 3, Turkey, pp. 81–82.

Starodub, V.M. and Starodub, N.F. (1999). Optical immune sensors for the monitoring protein substances in the air. In: *Book of Abstracts of 13th European Conf. on Solid-State Transducers*, Sept. 12–15, the Netherlands, pp. 87–88.

Starodub, N.F. and Starodub, V. M. (2000). Immune sensors: Original, achievements and perspectives. *Ukr. Biochem. J.* 4–5, 147–163.

Starodub, V.M. and Starodub, N.F. (2001). Electrochemical and optical biosensors: Origin of development, achievements and perspectives of practical application. In: Bozoglu F., Deak F., and Ray B. (Eds.) *Novel Processes and Control Technologies in the Food Industry*, NATO series. IOS Press, Amsterdam, pp. 63–94.

Starodub, N.F. and Starodub, V.M. (2005). Biosensors based on the photoluminescence of porous silicon. Application for the monitoring of environment. *Sens. Electron. Microsyst. Technol.* 1, 63–71.

Starodub, N.F. and Slishek, N.F. (2013). Nano-porous silicon based immune biosensor for the control of level of mycotoxins. *Adv. Biosens. Bioelectron* 2(2), 7–15.

Starodub, N.F., Korobov, V.M., and Nazarenko, V.I. (1992). *Myoglobin: Structure, Properties, Biological Role.* Naukova Dumka, Kiev.

Starodub, N.F., Fedorenko, L.L., Starodub, V.M. et al. (1996). Use of the silicon crystals photoluminescence to control immunocomplex formation. *Sens. Actuators B* 35, 44–47.

Starodub, V.M., Fedorenko, L.L., Torbich, W. et al. (1998). Optical sensors for environmental monitoring in field: The measurement of the biological pollutant level of the air. In: *Proc. of NATO ARW New Trends in Biosensor Development*, Ukraine, July 6–9, pp. 51–52.

Starodub, V.M., Fedorenko, L.L., Sisetsky, A.P., and Kurskij, M.D. (1999a). Immune sensor for the determination of myoglobin level. *Ukr. Biochem. J.* 71(3), 68–72.

Starodub, V.M., Fedorenko, L.L., Sisetsky, A.P., and Starodub, N.F. (1999b). Control of myoglobin level in a solution by an immune sensor based on the photoluminescence of porous silicon. *Sens. Actuators B* 58(1–3), 409–414.

Starodub, V.M., Dibrova, T.L., Kostjukevich, K. et al. (1999c). Optical sensors for medical diagnostics and environmental monitoring. In: Bergveld P. and Torbicz W. (Eds.) *Lecture Notes of the ICB Seminars*, Warsaw, April 1998. ICB, pp. 192–205.

Starodub, N.F., Dibrova, T.L., Shirshov, Yu.M., and Kostjukevich, K.V. (1999d). Development of myoglobin sensor based on the surface plasmon resonance. *Ukr. Biochem. J.* 71(2), 33–37.

Starodub, N.F., Pylypenko, I.V., Pylypenko, L.N. et al. (2010a). Biosensors for the determination of mycotoxins: Development, efficiency at the analysis of model samples and in case of the practical applications. In: *Lecture Notes of the 86 ICB Seminars.* ICB, pp. 81–101.

Starodub, N.F., Sitnik, J.A., Melnichenko, M.M., and Shmyryeva, O.M. (2010b). Biosensors based on the nanostructured silicon and intended for the determination of number of biochemical quantities. In: *Abstract Book of 4th International Scientific and Technical Conference on Sensors electronics and Microsystems Technology*, (SEMST-4), Ukraine, June 28–July 2, pp. 32–33.

Starodub, N.F., Shulyak, L.M., Shmyryeva, O.M. et al. (2011a). Nanostructured silicon and its application as the state of the art in biosensors—General aspects transducer in immune biosensors. In: Mikhalovsky S. and Khajibaev A. (Eds.) *Biodefence: Advanced Materials and Methods for Health Protection*. NATO Science for Peace and Security Series A: Chemistry and Biology. Springer Science+Business Media B, Vol. 2, pp. 87–98.

Starodub, N.F., Sitnik, J.A., Melnichenko, M.M., and Shmyryeva, O.M. (2011b). Optical immune biosensors based on the nanostructured silicon and intended the diagnostics of retroviral bovine leucosis. In: *Proceedings of International Conference The SENSOR+TEST 2011*, Nurenberg, pp. 127–132.

Starodub, N.F., Ogorodniichuk, Y.A., Sitnik, Y.A., and Slishik, N.F. (2012). Biosensors for the control of some toxins, viral and microbial infections to prevent actions of bioterrorists. In: Nikolelis D.P. (Ed.), *Portable Chemical Sensors: Weapons Against Bioterrorism*, NATO Science for Peace and Security Series A: Chemistry and Biology. Springer Science+Business Media B, pp. 95–117.

Starodub, N.F., Slyshyk, N.F., Shavanova, K.E. et al. (2014a). Experimental demonstration and theoretical explanation of the efficiency of the nano-structured silicon as the transducer for optical immune biosensors. In: Spigulis J. (Ed.), *Proceedings of 8th International Conference on Advanced Optical Materials and Devices (AOMD-8), Proc. of SPIE*, 9421, 94210F1-11 (doi:10.1117/12.2081163).

Starodub, N.F., Prilutskij, M., and Mel'nichenko, M.M. (2014b). Application of biosenors as effective approaches for the overcoming hard situation at the development of some endocrine disorders. In: *Abstract Book of International Scientific and Technical Conference Sensor Electronics and Microsystem Technologies SEMST-6*, Ukraine, September 29–October 3, pp. 29–30.

Taylor, A.D., Ladd, J., Homola, J., and Jiang, S. (2008). *Principles of Bacterial Detection: Biosensors, Recognition Receptors and Microsystems.* Springer, New York, pp. 83–108.

Thust, M., Schoening, M.J., Frohnhoff, S., Arens-Fischer, R., Kordos, P., and Luth, H. (1996). Porous silicon as a substrate material for potentiometric biosensors. *Meas. Sci. Technol.* 7, 26–29.

Vamvakaki, V. and Chaniotakis, N.A. (2008). DNA stabilization and hybridization detection on porous silicon surface by EIS and total reflection FT-IR spectroscopy. *Electroanal.* 20(17), 1845–1850.

Van Noort, D. and Mandenius, C.-F. (2000). Porous gold surfaces for biosensor applications. *Biosens. Bioelectron.* 15, 203–209.

Vikulov, V., Verba, A., and Kirichenco, Y. (1997) Silicon solar cells with porous silicon layers. *Proceedings of MRS Fall Meeting Thin-Film Structures for Photovoltaics*, Dec. 2–5, *Mat. Res. Soc. Symp. Proc.* 485, G2.2.

Vo-Dinh, T. (2004). Biosensors, nanosensors, and biochips: Frontiers in environmental and medical diagnosis. In: *Proceedings of the 1st International Symposium on Micro and Nano-Technology*, Honolulu, HI, pp. 14–17.

Weiss, S.M., Rong, G., and Lawrie, J.L. (2009). Current status and outlook for silicon-based optical biosensors. *Physics E* 41, 1071–1075.

Xu, D., Yu, W.H, Kang, E.T., and Neoh, K.G. (2004). Functionalization of hydrogen-terminated silicon via surface-initiated atom-transfer radical polymerization and derivatization of the polymer brushes. *J. Colloid Interface Sci.* 279(1), 78–87.

Yalow, R.S. and Berson, S.A. (1971). Size heterogeneity of immunoreactive muman ACTH in plasma and in extracts of pituitary glands and ACTH-producing thymoma. *Biochem. Biophys. Res. Commun.* 44, 439–445.

Zangooie, S., Bjorklund, R., and Arwin, H. (1997). Vapor sensitivity of thin porous silicon layers. *Sens. Actuators B* 43, 168–174.

Zhang, F., Sautter, K., Larsen, A.M. et al. (2010). Chemical vapor deposition of three aminosilanes on silicon dioxide: Surface characterization, stability, effects of silane concentration, and cyanine dye adsorption. *Langmuir* 26(18), 14648–14654.

MEMS-Based Pressure Sensors

Ghenadii Korotcenkov and Beongki Cho

6

CONTENTS

6.1 Introduction: Silicon-Based Pressure Sensors 136

6.2 Technology of Porous Silicon in Pressure Sensor Fabrication 138

    6.2.1 SOI Structure Pressure Sensors 138

    6.2.2 Pressure Sensors with Monocrystalline Si-Membranes Fabricated Using Porous Silicon 139

    6.2.3 Pressure Sensors with Porous Membrane and Porous Piezoresistors 142

    6.2.4 Field-Emission Array Pressure Sensor 144

Acknowledgments 144

References 144

6.1 INTRODUCTION: SILICON-BASED PRESSURE SENSORS

Pressure sensors such as absolute pressure sensor, gauge pressure sensor, vacuum pressure sensor, differential pressure sensor, and sealed pressure sensor are devices that convert pressure to physical movement with subsequent transformation into electrical signal or other output. Thus, a standardized pressure transmitter consists of three basic components: a pressure transducer, its power supply, and a signal conditioner/retransmitter that converts the transducer signal into a standardized output (Eaton and Smith 1997). Pressure sensors are now widely used for aerospace, biomedical, automobile, defense, and consumer applications (Tandeske 1991; Puers 1993; Judy 2001; Ripka and Tipek 2007).

At present, there are many different types of pressure sensors. However, most of them, especially silicon microelectromechanical system (MEMS) devices, are based on diaphragms. In diaphragm-based sensors, the pressure is determined by the deflection of the diaphragms due to applied pressure. Figure 6.1a illustrates a schematic cross-section of a typical pressure sensor diaphragm. The reference pressure can be a sealed chamber or a pressure port so that absolute or gauge pressures are measured, respectively. The shape of the diaphragm as viewed from the top is arbitrary, but generally takes the form of a square or circle.

Considerable interest in recent years to the MEMS-based silicon pressure sensors is conditioned not only by the excitement naturally associated with a nascent technology, but also because of the great promise of increased miniaturization and performance of MEMS devices over conventional devices (Eaton and Smith 1997; Bhat 2007; Lindroos et al. 2010; Eswaran and Malarvizhi 2013; Kumar and Pant 2014). First, silicon is an ideal material for mechanical sensors because of its excellent mechanical properties required for reproducible elastic deformations under identical mechanical load (Petersen 1982; Lindroos et al. 2010). Silicon is also free from hysteresis and creep. Moreover, the material properties and fabrication procedure are well defined and easily available for silicon material. Second, micromechanical devices and systems are inherently smaller, lighter, and faster than their macroscopic counterparts and are often more precise. Therefore, recently MEMS-based pressure sensors in many cases replaced the bulkier version of their traditional electrical and mechanical counterparts. Most of the advantages of

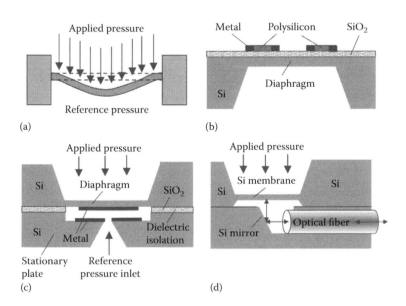

FIGURE 6.1 (a) A schematic cross-section of a typical pressure sensor diaphragm. Dotted lines represent the undeflected diaphragm. (b) The schematic cross-section of polysilicon piezoresistive pressure sensor. (c) A cross-section schematic diagram of a bulk-micromachined, capacitive pressure sensor. (Idea from Clark S.K. and Wise K.D. *IEEE Trans. Electron. Dev.* ED-26, 1887–1896, 1979.) (d) A cross-sectional view of the optical pressure sensor. The sensor consists of two micromachined silicon wafers which are fusion bonded together to form an optical cavity. (Reprinted from *Sens. Actuators A* 43, Chan M.A. et al., 196, Copyright 1994, with permission from Elsevier.)

micromachined pressure sensors stem from the fact that many of the manufacturing processes and tools are borrowed from the integrated circuit (IC) industry, such as photolithography, oxidation and diffusion, wet cleaning and etching, thin-film deposition, metallization, ion implantation, and others. Many of these processes are used directly, while others have been modified or developed to meet the specific needs of micromachined devices (Kloeck and de Rooij 1994). The use of well-developed technologies of microelectronic industry reduces development costs. As a result, MEMS pressure sensors currently dominate the market for pressure sensors. These sensors can be fabricated using both the bulk micromachining and the surface micromachining (Eaton and Smith 1997; Bustillo et al. 1998; Bhat 2007; Müller et al. 2010). In Chapter 11 of this book it will be shown that porous silicon (PSi) technology can compete with traditional bulk and surface micromachining technologies in the manufacturing of various types of diaphragms (membranes) and cantilever beam structures. This means that the development of pressure sensors is also the field of applying technology of silicon porosification.

Micromachined pressure sensors can be classified by transduction mechanisms. From all variety of sensors, it is necessary to allocate capacitive and piezoresistive sensors because they are the most common (Zhou et al. 2005), and optical and resonant devices, which have good working characteristics, are used much less (Eaton and Smith 1997; Bhat 2007; Eswaran and Malarvizhi 2013). Piezoresistive pressure sensors have piezoresistors mounted on or in a diaphragm (see Figure 6.1b). The basic structure of a piezoresistive pressure sensor consists of four sense elements connected in a Wheatstone bridge configuration. These sensors make use of the change in the resistance of conductors due to the change in their physical dimensions when subjected to strain. Thus, when assembled on a membrane they give information about the strain experienced by the membrane when it is subjected to stress by the applied pressure. For thin diaphragms and small deflections, the resistance change is linear with applied pressure. Usually p-type piezoresistors are being formed on the surface of an n-type single crystalline diaphragm using diffusion or ion implantation. However, the leakage current of a p–n junction increases exponentially with temperature. As a result, the isolation between the resistors of the conventional piezoresistive pressure sensors becomes poor at temperatures in excess of about 100°C (Li et al. 2012). Therefore, in many pressure sensors designed for applications such as aircraft gas turbine combustion control, chemical processing, automotive industry, aviation engineering, oil fields, and industrial measurements, polysilicon piezoresistors are used (Guo et al. 2009; Kähler et al. 2012). Polysilicon resistors are laid out on oxide grown on the diaphragm region and hence are isolated from each other by the oxide layer (see Figure 6.1b). Therefore, isolation is maintained even at temperatures exceeding 300°C (Chung 1993).

Capacitive sensors are based on parallel plate capacitors. A typical bulk-micromachined capacitive pressure sensor is shown in Figure 6.1c. This means that the capacitive pressure sensor detects the deflection of the membrane as a variation in capacitance between two plates. The principal advantages of capacitive pressure sensors over piezoresistive pressure sensors are increased pressure sensitivity and decreased temperature sensitivity (Clark and Wise 1979; Lee and Wise 1982; Eaton and Smith 1997; Kovacs 1998). However, the output of a capacitive pressure sensor is not linear with respect to the applied pressure. In addition, excessive signal loss from parasitic capacitance is a serious disadvantage, which hindered the development of miniaturized capacitive sensors. However, studies have shown that monolithic integration helps in reducing the parasitic capacitance and increasing reliability compared to separate fabrication and subsequent assembly (Judy 2001).

Many diaphragm-based optical sensors have also been reported (Dzuiban et al. 1992; Chan et al. 1994; Wagner et al. 1994). Figure 6.1d shows one such pressure sensor. Optical transduction has the major advantage that it is completely passive, that is, no at-site electronic device or circuitry is necessary. Therefore, optical sensors can be quite accurate, but often suffer from temperature sensitivity problems (Bartelt and Unzeitig 1993). Furthermore, aligning the optics and calibrating the sensors can be challenging and expensive.

The resonant pressure sensors, designed during the last decades, operate by monitoring the resonant frequency of a beam or diaphragm, which act as a sensitive strain gauge (Zook et al. 1992; Burns et al. 1995). As the stress state of the diaphragm changes, the tension in the embedded structures changes and so does the resonant frequency. A digital counter circuit detects the shift. Because this change in frequency can be detected quite precisely, this type of transducer can be used for low differential pressure applications as well as to detect absolute and gauge pressures. The stability is determined only by the mechanical properties of the resonator material, which is generally very stable. The most significant advantage of the resonant pressure transducer is that it generates an inherently digital signal, and therefore can be sent directly to a stable crystal clock in a microprocessor. Resonant pressure sensors have lower temperature sensitivity than pure piezoresistive sensors. However, the sensitivity to temperature variation also presents in these sensors. Other limitations include a nonlinear output signal and some sensitivity to shock and vibration. These limitations typically are minimized by using a microprocessor to compensate for nonlinearities as well as ambient and process temperature variations.

Piezoelectric pressure sensors can also be developed. However, silicon is not piezoelectric. Therefore, in order to translate the deformation or strain into an electrical signal with this approach, it is necessary to externally glue a piezoelectric pellet on the membrane. This technique does not form part of the conventional integrated circuit technology. Furthermore, the piezoelectric pressure sensors cannot be used for static pressure sensing purposes due to charge leakage.

As we have shown before, each pressure sensor type has advantages and disadvantages. Moreover, each pressure sensor has unique requirements and, therefore, a pressure sensor with a particular specification and application must be custom designed (Kumar and Pant 2014). A better design will ensure better linearity and sensitivity without putting load on the postprocessing circuits. The various parameters such as diaphragm dimensions, the technology to be used for fabrication, the nominal resistance of the piezoresistors, and their placement on the diaphragm must all be carefully chosen based on the target application. For example, when the pressure sensors are used for biomedical applications such as intracranial or cardiovascular pressure measurement, the size of the sensor must be small and the sensor must be biocompatible because the sensor would be implanted within the body. However, such constraints may not be applicable to applications such as a tire pressure monitoring system, where the primary consideration may be to reduce the sensor cost. In this case, the use of expensive deep reactive ion etching (DRIE) for the diaphragm fabrication should be limited. Some applications may require the sensor to work reliably in a harsh environment with high temperatures and therefore silicon-on-insulator (SOI)-based sensors may be preferred. However, in low temperature applications, this may not be necessary. In addition, the thickness chosen for the diaphragm will be dependent on the application. Low-pressure applications generally require thinner diaphragms and high-pressure applications require thicker diaphragms. Considering some of the above arguments, one can observe that there are a large number of design principles that a pressure sensor designer must keep in mind before designing a pressure sensor.

Among the various micromachined pressure sensors mentioned previously, the piezoresitive pressure sensors are the simplest to fabricate and provide electrical output directly. As it will be seen in subsequent sections, they can be easily integrated with microelectronic circuits because the fabrication processes are compatible with the integrated circuit fabrication techniques. In addition, piezoresistive pressure sensors have mainly been studied and commercialized because of high yield and wide dynamic range. As a result, piezoresistive pressure sensors are some of the most reported and developed micromachined devices (Esashi et al. 1998). However, the piezoresistive pressure sensor suffers from the low sensitivity and large size as compared with the capacitive pressure sensor. Therefore, many of the piezoresistive pressure sensor studies are trying to improve their low sensitivity using various materials such as high-doped silicon (Toriyama et al. 2002) and PSi (Armbruster et al. 2003).

6.2 TECHNOLOGY OF POROUS SILICON IN PRESSURE SENSOR FABRICATION

6.2.1 SOI STRUCTURE PRESSURE SENSORS

In previous section it was shown that using polysilicon piezoresistiors, formed on the insulator structures, one can design a pressure sensor capable of operating at elevated temperatures. At the same time, it is known that silicon porosification is one of the approaches to the implementation of SOI technology, allowing formation of the single crystalline SOI structures with perfect electrical

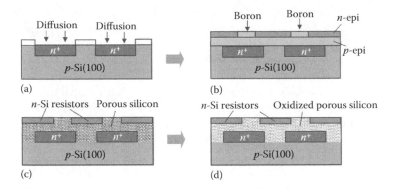

FIGURE 6.2 Key process steps for fabrication of an SOI structure designed for application in pressure sensors: (a) Patterns on an SiO_2 film about 200 nm thick are formed as n^+ buried layer diffusion masks on a p-type silicon substrate (5–6 $\Omega \cdot$cm). Then antimony is diffused into the silicon substrate to form an n^+ barrier layer. (b) A p-type 2-μm thick epitaxial layer with resistivity of 1–1.5 $\Omega \cdot$cm is deposited on the substrate. A 2-μm n-type epitaxial layer with resistivity of 0.8 $\Omega \cdot$cm is then grown on the p-type epitaxy layer. This upper layer is to be used as material for the piezoresistive SOI structure pressure transducer. Then, boron diffusion is performed to form isolated n-resistors. (c) Anodization of the wafer with hydrofluoric acid (42%). The anodization ($J = 70$ mA/cm^2) is carried out in the dark. In this anodic reaction, only p-type silicon around an n-island is changed to PSi. The pressure of the n^+ barrier layer forces the current of the anodic reaction to flow laterally rather than down into the substrate. Formed layers had a porosity of ~45%. (d) The PSi is thermally oxidized at 700°C for 2 h in wet O_2 ambient. Since the oxidation rate of PSi is 10–20 times greater than that of single crystal silicon, fully isolated n-type single crystalline silicon islands can be obtained. (Reprinted from *Sens. Actuators A 21–23*, Zhao G. et al., 840, Copyright 1990, with permission from Elsevier.)

isolation properties. This means that the SOI technology should attract a special interest for the pressure sensor's design. Research carried out by Zhao et al. (1990) has shown that PSi-based SOI technology can be used in the pressure sensor's manufacturing. The key processing steps used for the formation of the SOI structure are shown in Figure 6.2. The output characteristics of this SOI structure pressure transducer, fabricated by Zhao et al. (1990), showed high sensitivity like the single crystal silicon transducer, high operating temperatures up to 350°C, and good stability.

6.2.2 PRESSURE SENSORS WITH MONOCRYSTALLINE Si-MEMBRANES FABRICATED USING POROUS SILICON

It is known that single crystal membranes have better mechanical properties and can withstand higher mechanical loads. In addition, the sensitivity of the polysilicon piezoresistive pressure sensor is always lower compared to the single crystalline piezoresistive pressure sensor because of the lower gauge factor of polysilicon than that of single crystalline silicon (Schafer et al. 1989). Therefore, monocrystalline silicon membranes are more advantageous than polycrystalline for pressure sensor design. Such membranes conventionally are fabricated using either bulk or surface micromachining (Eaton and Smith 1997; Bustillo et al. 1998; Bhat 2007). In particular, besides producing monocrystalline silicon membranes by anisotropic etching with electrochemical etch stop (Kress et al. 1991), such membranes can be fabricated using underetching of epitaxial layers grown on expensive silicon-on-sapphire (SOS) wafers (Suzuki et al. 1997) or underetching of epitaxial layers grown on PSi (Seefeld et al. 1999). Both methods require sacrificial layer etching and additionally a subsequent sealing of etch holes. However, the necessary subsequent sealing of the cavity typically done with dielectric layers is a possible weak spot regarding long-term stability and leak tightness of pressure sensors. Local plasma etching of a cavity into a first substrate and subsequent wafer bonding of a second silicon substrate also can be used (Parameswaran et al. 1995). Afterward, the bonded wafer is thinned using grinding and electrochemical etching in KOH to form the monocrystalline membrane.

Armbruster et al. (2003) from Bosh GmbH proposed another more cheap and efficient approach to fabrication of monocrystalline Si-membranes. A schematic overview of the membrane process flow, proposed by Armbruster et al. (2003), is shown in Figure 6.3. The core of this process was (1) anodic etching of PSi in concentrated hydrofluoric acid (HF) forming PSi double layers, (2) sintering of PSi, and (3) epitaxial growth of the silicon membrane. Since PSi is etched from the monocrystalline bulk, the atoms inside the remaining pore walls retain their original monocrystalline order. Thus, it is possible to use PSi as a substrate for the deposition of monocrystalline silicon by epitaxy, for example, high-temperature chemical vapor deposition (CVD). The appearance of the cavity in the formed structure is because the rearrangement, which leads to a solidification of the low porosity layer and a dissolution of the high porosity layer, takes place in the PSi double layers during high-temperature annealing (Ott et al. 2003). At the interface of PSi and vacuum (equal to 100% porosity), the same effect leads to the sealing of surface pores. A detailed description of PSi sintering can be found in Chapter 12 in *Porous Silicon: Formation and Properites* and of epitaxial growth on the PSi layer in Chapter 7 in *Porous Silicon: Optoelectronics, Microelectronics and Energy Technology Applications*.

Armbruster et al. (2003) established that the cavity, formed during the proposed process, was hermetically sealed. As it is known, this is a prerequisite for successful operation of pressure sensors. In addition, the pressure sensitive membrane has shown excellent long-term stability. For demonstration of such properties, the sensor has passed 1.5×10^5 pressure cycles between 200 mbar and 2 bar at a temperature of 150°C. Highly successful and commercialized Bosch-BMP085 barometric piezoresistive pressure sensors fabricated by Bosh GmbH using the indicated technology exhibited an excellent linearity of the output signal between 200 mbar and 1000 mbar in the entire temperature range from 40°C to 125°C.

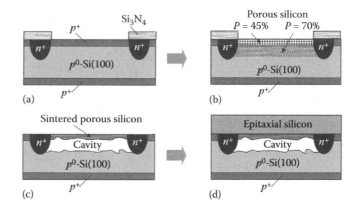

FIGURE 6.3   Schematic steps of monocrystalline Si-membrane fabrication. (a) Substrate before anodic etching. Before the anodization, several implantations have been performed, including a p^+ rear doping of the substrate for sufficient electrical contact during the porosification, and front side local ion implanted with a shallow p+ and a deep n+ doping. A silicon nitride layer was deposited onto the wafer, leaving unprotected only the membrane regions. Silicon nitride is rather HF resistant and therefore acts as a protection of the substrate surface during the anodization. (b) Substrate after anodic etching. The anodic etching was carried out in HF. The anodization process was held in two steps. First, the implanted p+ silicon was etched, using a low current density. This resulted in an approximately 1-μm thick mesoporous layer with a porosity of about 45%. Second, the p^o-doped substrate was anodized using a higher current density. This led to forming of a buried nanoporous silicon layer with a thickness of approximately 4 μm and a porosity of about 70%. (c) Substrate after sintering in hydrogen atmosphere. The annealing was carried out in a CVD reaction chamber at temperatures between 900°C and 1100°C in hydrogen atmosphere during several minutes. The hydrogen diffuses into the porous network and reduces the native oxide covering the PSi surface. Once the native oxide has been removed, the PSi starts to rearrange. The surface pores in the mesoporous layer are sealed and at the same time the pores in the buried nanoporous layer start to merge into one large cavity. (d) Substrate after epitaxial growth. epitaxial growth was conducted in the same CVD chamber. Immediately after annealing in hydrogen, the temperature was raised above 1100°C for epitaxial growth. (Idea from Armbruster S., Schäfer F., Lammel G. et al. In: *Proceedings of 12th International Conference on Solid-State Sensors, Actuators, and Microsystems*, June 8–12, 2003, Boston, Vol. 1, pp. 246–249.)

Subsequently, it was shown that the technology of membrane fabrication, which uses PSi and the previously mentioned APSM process, can be integrated in standard semiconductor processes, suitable for high volume production in an IC-fab (Lammel et al. 2005). According to Lammel et al. (2005), only two mask layers and one electrochemical etching step were inserted at the beginning of a standard IC-process to transform the epitaxial silicon layer from the electronic process into a monocrystalline membrane with a vacuum cavity under it. The piezoresistive pressure sensor fabricated using the proposed technology looked like a standard IC chip. Lammel et al. (2005) believe that the easy packaging makes it suitable for new consumer applications and tire pressure measurement systems, aside from the known automotive applications of manifold air pressure and barometric pressure sensor.

Knese et al. (2008, 2009) have shown that the technology of monocrystalline silicon membrane fabrication also can be used for development of capacitive absolute pressure sensors. Knese et al. (2009) believe that expanding this technology to capacitive transduction allows for a greater flexibility in tailoring the sensor properties to specific applications. This expansion is implemented by adding a poly-Si counterelectrode layer on top of the membrane in a surface micromachining step. Since only front side processing on standard silicon substrates is being used, this method is very cost-efficient and fully CMOS compatible, enabling monolithic integration of circuitry. A schematic view of the sensor concept is displayed in Figure 6.4d. A thin silicon oxide layer separates and electrically isolates the electrodes. To define the lateral confinement of the counterelectrode and thereby reduce the parasitic capacitance of the sensor, an additional trench is etched around the membrane region. The vent holes allow the applied pressure to penetrate the poly-silicon electrode and deflect the single crystal membrane. With a proper design of the vent holes, the perforated poly-silicon electrode is not affected by the pressure. Therefore, the deflection results in a variation of the electrode gap. This can be detected electrically as a change of capacitance AC. The process of electrode forming on the surface of monocrystalline silicon membranes is shown in Figure 6.4a–c.

According to Knese et al. (2008, 2009), the novel concept of a capacitive pressure sensor has the following benefits: First, it provides a continuous single crystal pressure-sensitive membrane taking advantage of its unique properties like high strength, high reliability, and no mechanical hysteresis. Second, this membrane as well as the polycrystalline counterelectrode added on top of it is formed

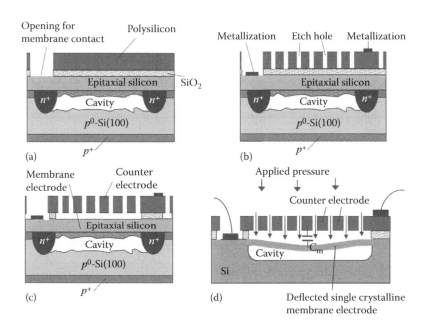

FIGURE 6.4 (a–c) Schematic illustration of the process of electrode fabrication in capacitive pressure sensors: (a) deposition and structuring of sacrificial layer (SiO₂) and deposition of poly-Si (counterelectrode), (b) metallization and perforation of poly-Si by DRIE, (c) removal of sacrificial layer by HF vapor etch and thereby releasing membrane from counterelectrode, and (d) functional principle of capacitive pressure sensor with single crystalline membrane electrode. (Idea from Knese K., Armbruster S., Weber H. et al. In: *IEEE 22nd International Conference on Micro Electro Mechanical Systems, MEMS 2009*, January 25–29, 2009, Sorrento, pp. 697–700.)

by only using front side and fully CMOS-compatible process steps. Third, since the dimensions of the electrodes are confined by an etched trench instead of a *p-n*-junction and the electrodes are isolated by a dielectric layer, the sensor has decreased sensitivity to temperature variations.

One should note that the advanced porous silicon membrane (APSM) process, discussed in this section, offers a high degree of freedom in designing different membrane geometries and therefore can be used for fabricating other sensors as well. For example, recently Etter et al. (2012) reported about integrated microbolometer fabrication using this technology.

6.2.3 PRESSURE SENSORS WITH POROUS MEMBRANE AND POROUS PIEZORESISTORS

It is known that the pressure sensitivity of bulk MEMS silicon is rather low (~1 µV/V/mbar) due to low piezoresistive coefficient of monoscrystaline silicon (Guckel 1991). As a result, it becomes difficult to use bulk silicon for very low pressure sensing applications such as biomedical and other applications (Pramanik and Saha 2006a). The sensitivity of the pressure sensor can generally be improved by reducing the membrane thickness. However, very thin membranes exhibit high non-linearity and are difficult to handle. On the other hand, PSi, owing to its nanocrystalline nature, shows higher piezoresistive coefficient and hence much better sensitivity to pressure (~0.02 mV/V/mbar) (Pramanik and Saha 2006b). Taking this feature of PSi into account, it is clear that the use of PSi can provide the improvement of the parameters of MEMS-based pressure sensors and assist in an expansion of the field of their application.

Studies have confirmed these assumptions. In particular, Gangopadhyay et al. (2005) have shown that the use of PSi/Si sandwich membrane (see Figure 6.5b) for pressure sensor fabrication gave an improvement of sensitivity by more than three times in comparison with conventional bulk silicon membrane (Merlos et al. 2000). At that, Si/micro PSi membranes exhibit higher sensitivity than Si/macro PSi ones (Sujatha and Bhattacharya 2009a). Sandwich contact geometry was chosen because the current lines should pass through PSI in order to observe its electrical response on application of pressure. In the case of lateral contacts, the current lines pass almost entirely through the underneath bulk silicon because PSi is much more resistive compared to bulk silicon and hence the response of PSi will be masked.

Figure 6.6 shows the effect of the thickness and the porosity of the PSi layer on the pressure sensor performances. It is seen that the sensitivity increases with porosity. However, there is an optimum porosity required. In this case, for a 100-µm-thick diaphragm the optimum porosity is ~60%. With further increase of porosity to 70%, the sensitivity reduces significantly. The increased sensitivity with increasing porosity up to 60% can be explained by the increase of the piezoresistive coefficient of the PSi sensors with increasing porosity (Pramanik and Saha 2006b).

Sujatha and Bhattacharya (2007, 2009b) used another approach to sensitivity improvement. It is known that the sensitivity of pressure sensors can be improved by fabricating the membrane with a material having the Young's modulus lower than monocrystalline silicon. PSi has a very low Young's modulus and it drastically reduces with increase in porosity (Populaire et al. 2003). This means that when a part of the silicon membrane is converted into PSi, the effective Young's modulus of the membrane is reduced, which should give more deformation with the application of pressure. Since the deformation is greater, the strain is also greater compared to single crystalline silicon and hence the sensitivity should be higher. For realization of this approach, Sujatha and Bhattacharya (2007, 2009b) proposed using the configuration of pressure sensors

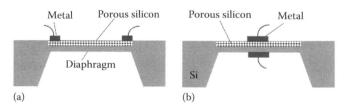

FIGURE 6.5   (a) Contacts are taken from the top surface of the PSi membrane (lateral structure). (b) Contacts are taken from the top surface of the PSi and the bottom surface of bulk silicon (sandwich structure). (Idea from Pramanik C. and Saha H. *IEEE Sensors J.* 6(2), 301–309, 2006b.)

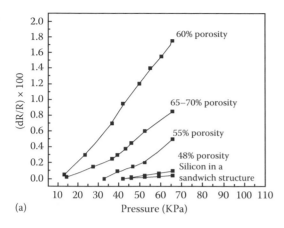 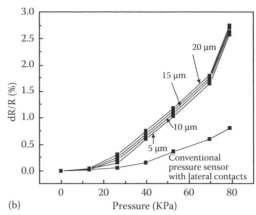

FIGURE 6.6   (a) Variation of fractional change in resistance of PSi layer of thickness 20 μm and various porosities. (b) Fractional change in resistance for different thicknesses of PSi and a fixed porosity of 60% with affixed diaphragm of 100 μm. (From Gangopadhyay U. et al. *J. Korean Phys. Soc.* 47, S450, 2005. Copyright 2005: The Korean Physical Society. With permission.)

shown in Figure 6.7. The advantage of this approach avoids the difficulty in handling thin silicon diaphragms compared to composite membranes giving the same high sensitivity especially for low-pressure applications. Sujatha and Bhattacharya (2007, 2009b) established that the Si/PSi composite membranes showed higher deformation compared to silicon membranes, but Si/PSi composite membranes show poor hysteresis at higher pressure ranges and they lose the linearity earlier. This is not seen in the Si membranes and could be due to the low elastic constants of PSi, which loses its elasticity at a higher pressure range. Hence, Sujatha and Bhattacharya (2007, 2009b) believe that the Si/PSi composite membrane is a viable option to measure low pressure.

Sujatha and Bhattacharya (2009b) have reported that the enhancement in sensitivity of pressure sensors can be also achieved by converting the polysilicon piezoresistors on the membrane into porous polysilicon (PPSi). Fabricated pressure sensors with PPSi piezoresistors have shown modest improvement in sensitivity without increasing the offset voltage. Pressure sensors with composite membrane and PPSi piezoresistors showed the highest sensitivity and hence are a viable option for pressure sensors.

However, the same studies found that PSi sensors had high temperature sensitivity, 60 mV/V/°C in reverse bias mode and 8 mV/V/°C in forward bias mode (Futane et al. 2011), while bulk silicon had a temperature sensitivity of <0.1 mV/V/°C (Lisec et al. 1996). This means that as PSi exhibits a stronger dependence on ambient temperature, the need for a signal conditioning circuit to compensate for the temperature-drift in the output response of PSi piezoresistive pressure sensor therefore becomes pertinent (Pramanik et al. 2006; Futane et al. 2011). Prominent nonlinearity of a sensor signal (of the order of 20%), as compared to that of a conventional silicon piezoresistive pressure sensor, is another limitation of the PSi pressure sensor (Pramanik et al. 2006). The increased nonlinearity in comparison with the conventional silicon pressure sensor may arise from porous microstructure, which undergoes more prominent deformation under pressure. In

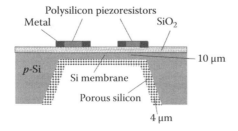

FIGURE 6.7   Cross-sectional diagram of the pressure sensor with an Si/PSi composite membrane. (Adapted from Sujatha L. and Bhattacharya E., *J. Micromech. Microeng.* 17, 1605, 2007. Copyright 2007: IOP. With permission.)

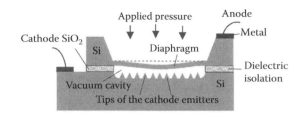

FIGURE 6.8 Schematic diagram of a field-emission emitter array pressure sensor.

addition, PSi formation produces stress in the membrane raising the offset voltage in the pressure sensors. However, Sujatha and Bhattacharya (2009b) believe that the higher offset voltages can be nullified by external circuit or by optimizing the process parameters after PSi formation.

6.2.4 FIELD-EMISSION ARRAY PRESSURE SENSOR

As it was shown earlier in the Chapter 1 of this book, field ionization can be used for designing efficient gas sensors. During experimental studies, it was established that field emission pressure sensors could be also developed (Lee and Huang 1991; Busta and Pogemiller 1993; Nicolaescu 1995; Xia and Liu 1998; Liao et al. 2000). A schematic diagram of a field-emission emitter array pressure sensor is presented in Figure 6.8. The field-emission array pressure sensor is made up of an anode collector, cathode emitter array, sensible film-receiving exterior pressure, and the vacuum microchamber between the cathode and the anode. The design principle of the sensor is based on the field emission of electrons from the tips of the cathode emitters, which is described by Fowler-Nordheim field emission theory. Such a field emission current is sensitive to the distance change between the cathode and the anode. Therefore, the designed sensor device can monitor the exterior pressure that causes the change in distance through measuring the emission current. It was found that such pressure sensors have higher sensitivity, better endurance at high and low temperature, radiation resistance, and smaller volume than other pressure sensors (e.g., piezoresistance and capacitance types), and are capable of being used in various civil and military engineering fields for measuring pressure, especially in the tactile system of intelligent robotics in a severe operating environment.

Liao et al. (2003), for improving emission efficiency of the cathode cones in pressure sensors, proposed to coat the surface of the cathode cones by the nanocrystalline silicon film. However, in Chapter 1 in this book and Chapter 8 in *Porous Silicon: Optoelectronics, Microelectronics and Energy Technology Applications* it was shown that the porosification of Si tip also helps to increase emission and improve the stability of the emission properties of cathodes. Moreover, the application of the technology of PSi gives possibility to designing planar field emission devices without using any pyramidal Si tips. This means that we can expect that in the near future, a field-emission array pressure sensor with advanced parameters, which will be developed using the technology of PSi, may appear.

ACKNOWLEDGMENTS

This work was supported by the Ministry of Science, ICT, and Future Planning (MSIP) of the Republic of Korea, and partly by the National Research Foundation (NRF) grants funded by the Korean government (No. 2011-0028736 and No. 2013-K000315).

REFERENCES

Armbruster, S., Schäfer, F., Lammel, G. et al. (2003). A novel micromachining process for the fabrication of monocrystalline Si-membranes using porous silicon. In: *Proceedings of 12th International Conference on Solid-State Sensors, Actuators, and Microsystems*, June 8–12, Boston, Vol. 1, pp. 246–249.
Bartelt, H. and Unzeitig, H. (1993). Design and investigation of micromechanical bridge structures for an optical pressure sensor with temperature compensation. *Sens. Actuators A* **37–38**, 167–170.
Bhat, K.N. (2007). Silicon micromachined pressure sensors. *J. Indian Inst. Sci.* **87**(1), 115–131.

Burns, D.W., Zook, J.D., Horning, R.D., Herb, W.R., and Guckel, H. (1995). Sealed-cavity resonant micro-beam pressure sensor. *Sens. Actuators A* **48**, 179–186.

Busta, H.H. and Pogemiller, J.E. (1993). The field-emitter triode as a displacement/pressure sensor. *J. Micromech. Microeng.* **3**, 49–56.

Bustillo, J.M., Howe, R.T., and Muller, R.S. (1998). Surface micromachining for microelectromechanical systems. *Proc. IEEE* **86**(8), 1552–1574.

Chan, M.A., Collins, S.D., and Smith, R.L. (1994). A micromachined pressure sensor with fiber-optic inter-ferometric readout. *Sens. Actuators A* **43**, 196–201.

Chung, G.-S. (1993). Thin SOI structures for sensing and integrated circuit applications. *Sens. Actuators A* **39**, 241–251.

Clark, S.K. and Wise, K.D. (1979). Pressure sensitivity in anisotropically etched thin-diaphragm pressure sensors. *IEEE Trans. Electron. Dev.* **ED-26**, 1887–1896.

Dzuiban, J.A., Gorecka-Drzazga, A., and Lipowics, U. (1992). Silicon optical pressure sensor. *Sens. Actuators A* **32**, 628–631.

Eaton, W.P. and Smith, J.H. (1997). Micromachined pressure sensors: Review and recent developments. *Smart Mater. Struct.* **6**, 530–539.

Esashi, M., Sugiyama, S., Ikeda, K., Wang, Y., and Miyashita, H. (1998). Vacuum-sealed silicon microma-chined pressure sensors. *Proc. IEEE* **86**, 1627–1639.

Eswaran, P. and Malarvizhi, S. (2013). MEMS capacitive pressure sensors: A review on recent development and prospective. *Int. J. Eng. Technol. (IJET)* **5**(3), 2734–2746.

Etter, D.B., Zimmermann, M., Ferwana, S., Hutter, F.X., and JBurghartz, J.N. (2012). Low-cost CMOS compatible sintered porous silicon technique for microbolometer manufacturing. In: *Proceedings of IEEE International Conference on Micro Electro Mechanical Systems—MEMS*, Jan. 29–Feb. 2, Paris, pp. 273–276.

Futane, N.P., RoyChowdhury, S., RoyChaudhuri, C., and Saha, H. (2011). Analog ASIC for improved tem-perature drift compensation of a high sensitive porous silicon pressure sensor. *Analog Integr. Circ. Sig. Process* **67**, 383–393.

Gangopadhyay, U., Pramanik, C., Saha, H., Kim, K., and Yi, J. (2005). Porous silicon as pressure sensing material. *J. Korean Phys. Soc.* **47**, S450–S453.

Guckel, H. (1991). Surface micromachined pressure transducers. *Sens. Actuators A* **28**, 133–146.

Guo, S., Eriksen, H., Childress, K., Fink, A., and Hoffman, M. (2009). High temperature smart-cut SOI pres-sure sensor. *Sens. Actuators A* **154**, 255–260.

Judy, J.W. (2001). Microelectromechanical systems (MEMS): Fabrication, design and applications. *IOP Smart Mater. Struct.* **10**, 1115–1134.

Kähler, J., Stranz, A., Doering, L. et al. (2012). Fabrication, packaging, and characterization of p-SOI Wheatstone bridges for harsh environments. *Microsyst. Technol.* **18**, 869–878.

Kloeck, B. and de Rooij, N.F. (1994). Mechanical sensors. In: Sze S.M. (Ed.) *Semiconductor Sensors*. Wiley-Interscience, New York, pp. 153–199.

Knese, K., Armbruster, S., Benzel, H., and Seidel, H. (2008). Novel surface micromachining technology for fabrication of capacitive pressure sensors based on porous silicon. In: *Microsystems Technology in Germany 2008*. Trias Consult, Berlin, pp. 46–47.

Knese, K., Armbruster, S., Weber, H. et al. (2009). Novel technology for capacitive pressure sensors with monocrystalline silicon membranes. In: *IEEE 22nd International Conference on Micro Electro Mechanical Systems, MEMS 2009*, January 25–29, Sorrento, pp. 697–700.

Kovacs, G.T.A. (1998). *Micromachined Transducers Sourcebook*. McGraw-Hill, New York.

Kress, H.-J., Bantien, F., Marek, J., and Willmann, M. (1991). Silicon pressure sensor with integrated CMOS signal conditioning circuit and compensation of temperature coefficient. *Sens. Actuators A* **25**, 21–26.

Kumar, S.S. and Pant, B.D. (2014). Design principles and considerations for the 'ideal' silicon piezoresistive pressure sensor: A focused review. *Microsyst. Technol.* **20**, 1213–1247.

Lammel, G., Armbruster, S., Schelling, C. et al. (2005). Next generation pressure sensors in surface micro-machining technology. In: *Proceedings of the 13th International Conference on Solid-State Sensors, Actuators and Microsystems, Transducers '05*, June 5–9, Seoul, Korea, pp. 35–36.

Lee, Y.S. and Wise, K.D. (1982). A batch fabricated silicon capacitive pressure transducer with low tempera-ture sensitivity. *IEEE Trans. Electron. Dev.* **29**(1), 42–49.

Lee, H.C. and Huang, R.S. (1991). A novel field emission array pressure sensor. In: *Proceedings of the 6th Intl. Conf. on Solid State Sensors and Actuators, Transducers 91*, June 24–27, San Francisco, pp. 241–244.

Li, X., Liu, Q., Pang, S., Xu, K., Tang, H., and Sun, C. (2012). High-temperature piezoresistive pressure sen-sor based on implantation of oxygen into silicon wafer. *Sens. Actuators A* **179**, 277–282.

Liao, B., Lin, H.Y., and Wans, Y. (2000). Lately progress in field-emission pressure sensor. *J. Vac. Sci. Technol.* **20**(6), 413–417.

Liao, B., Chen, M., Kong, D., Zhang, D., and Li, T. (2003). Application of nano-crystalline silicon film in the fabrication of field-emission pressure sensor. *Sci. China E* **46**(4), 418–422.

Lindroos, V., Tilli, M., Lehto, A., and Motooka, T. (eds.) (2010). *Handbook of Silicon Based MEMS Materials and Technologies*. Elsevier, Oxford.

Lisec, T., Kreutzer, M., and Wagner, B. (1996). Surface micromachined piezoresistive pressure sensors with step-type bent and flat membrane structures, *IEEE Trans. Electron Dev.* **43**(9), 1547–1552.

Merlos, A., Santander, J., Alvarz, M.D., and Campabadal, F. (2000). Optimized technology for the fabrication of piezoresistive pressure sensors *J. Micromech. Microeng.* **10**, 204–208.

Müller, G., Friedberger, A., and Knese, K. (2010). Porous silicon based MEMS. In: Lindroos V., Tilli M., Lehto A., and Motooka T. (Eds.) *Handbook of Silicon Based MEMS Materials and Technologies.* Elsevier, Oxford, pp. 409–431.

Nicolaescu, D. (1995). Modeling of the field emitter triode as a displacement/pressure sensor. *Appl. Surf. Sci.* **87–88**, 61–68.

Ott, N., Nerding, M., Muller, G., Brendel, R., and Strunk, H.P. (2003). Structural changes in porous silicon during annealing. *Phys. Stat. Sol. (a)* **197**(1), 93–97.

Parameswaran, L., Mirsa, A., Chan, W.K., and Schmidt, M.A. (1995). Silicon pressure sensors using a wafer-bonded sealed cavity process. In: *Proceedings of the 8th International Conference on Solid-State Sensors and Actuators, and Eurosensors IX,* June 25–29, Stockholm, Sweden, pp. 582–585.

Petersen, K.E. (1982). Silicon as a mechanical material. *Proc. IEEE* **70**, 420–457.

Populaire, Ch., Remaki, B., Lysenko, V., Barbier, D., Atrmann, H., and Pannek, T. (2003). On mechanical properties of nanostructured meso-porous silicon. *Appl. Phys. Lett.* **83**, 1370–1372.

Pramanik, C. and Saha, H. (2006a). Low pressure piezoresistive sensors for medical applications. *Mater. Manufactur. Process.* **21**, 233–238.

Pramanik, C. and Saha, H. (2006b). Piezoresistive pressure sensing by porous silicon membrane. *IEEE Sensors J.* **6**(2), 301–309.

Pramanik, C., Saha, H., and Ganguli, U. (2006). An integrated pressure and temperature sensor based on nanocrystaline porous silicon. *J. Micromech. Microeng.* **16**, 1340–1348.

Puers, R. (1993). Capacitive sensors: When and how to use them. *Sens. Actuators A* **37–38**, 93–108.

Ripka, P. and Tipek, A. (2007). *Modern Sensors Handbook.* Wiley, Wiltshire.

Schafer, H., Greeter, V., and Kobs, R. (1989). Temperature–independent pressure sensors using polycrystalline silicon strain gauges. *Sens. Actuators A* **17**, 521–527.

Seefeld, J., Gianchandani, Y., Mattes, M., and Reimer, L. (1999). Porous silicon process for encapsulated single crystal surface micromachined microstructures. *Proc. SPIE* **3874**, 258–268.

Sujatha, L. and Bhattacharya, E. (2007). Enhancement in sensitivity of pressure sensors with composite Si/porous silicon membrane. *J. Micromech. Microeng.* **17**, 1605–1610.

Sujatha, L. and Bhattacharya, E. (2009a). Composite Si/PS membrane pressure sensors with micro and macro-porous silicon. *Sadhana* **34**(4), 643–650.

Sujatha, L. and Bhattacharya, E. (2009b). Porous silicon/polysilicon for improved sensitivity pressure sensors. *Phys. Stat. Sol. (c)* **6**(7), 1759–1762.

Suzuki, Y., Kudo, T., and Lkeda, K. (1997). Accurate, cost effective absolute pressure sensor. In: *Digest of Technical Papers of the International Conference on Solid-State Sensors and Actuators, Transducers '97,* June 16–19, Chicago, pp. 1493–1496.

Tandeske, D. (1991). *Pressure Sensors: Selection and Application.* Marcel Dekker, New York.

Toriyama, T., Tanimoto, Y., and Sugiyama, S. (2002). Single crystal silicon nano-wire piezoresistors for mechanical sensors. *J. Microelectromech. Syst.* **11**, 605–611.

Wagner, D., Frankenberger, J., and Deimel, P. (1994). Optical pressure sensor using two Mach–Zehnder interferometers for the TE and TM polarizations. *J. Micromech. Microeng.* **4**, 35–39.

Xia, S.H. and Liu, J. (1998). Vacuum-microelectronic pressure sensor with novel stepped or curved cathode. *J. Vac. Sci. Technol. B* **16**(3), 1226–1232.

Zhao, G., Huang, Y., and Bao, M. (1990). SOI structure pressure transducer formed by oxidized porous silicon. *Sens. Actuators A* **21–23**, 840–843.

Zhou, M.X., Huang, Q.A., Qin, M., and Zhou, W. (2005). A novel capacitive pressure sensor based on sandwich structures. *J. Microelectromech. Syst.* **14**, 1272–1282.

Zook, J.D., Burns, D.W., Guckel, H., Sniegowski, J.J., Engelstad, R.L., and Feng, Z. (1992). Characteristics of polysilicon resonant microbeams. *Sens. Actuators A* **35**, 51–59.

Micromachined Mechanical Sensors

Ghenadii Korotcenkov and Beongki Cho

7

CONTENTS

7.1   Introduction   148

7.2   Silicon Micromachined Accelerometers   149

7.3   Micromachined Gyroscopes   156

7.4   Technology of Micromachined Mechanical Sensors Fabrication   158

Acknowledgments   164

References   164

7.1 INTRODUCTION

Micromachining mechanical sensors such as accelerometers, gyroscopes, tactile, angle, and force sensors are other types of devices where the technology of microporous and macroporous silicon formation can be applied (Howe 1988; Anderson et al. 1994; Bell et al. 1996; Kaienburg and Schellin 1999; Gardner et al. 2001; Yang et al. 2002; Ádám et al. 2004, 2008; French and Sarro 2012). As a rule, all these devices are based on the membrane and cantilever principles (see Figure 7.1), and technology of their fabrication involves a deep etch to produce viable devices. As it was shown before in *Porous Silicon: Formation and Properties* (Chapter 12) and in the Chapter 11 of this book, silicon porosification is efficient technology for such applications. In particular, technology based on silicon porosification eliminates the need of bulk micromachining (Bean 1978), which sometimes causes low yield and high cost. In addition, the use of silicon porosification allows increasing the sensitivity of mechanical sensors in comparison with sensors fabricated using conventional surface micromachining (Howe 1988). The resulting large separation between the beams and the substrate on porous silicon (PSi) removal helps also to minimize parasitic capacitance in the devices. This large separation distance, as well as the roughness of the surface after PSi removal, helps to reduce sticking between beams and the substrate (Bell et al. 1996). For example, Mohammad et al. (2011) have found that with the 10-μm cavity, even 1-mm long and slender structures never get stuck to the substrate.

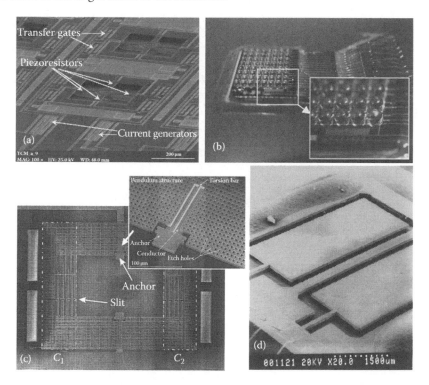

FIGURE 7.1 (a) Scanning electron micrograph of the two sensing elements of piezoresistive tactile sensor. (b) Micrograph of a mounted 8 × 8 element tactile sensor chip covered with periodic silicon rubber bumps protruding 300 μm from the surface. The single crystalline sensing elements consist of a central plate, suspended by four bridges over an etched cavity. Each of the four bridges includes p^+-piezoresistor, acting as an independent strain gauge, providing the signals for resolving the vector components of the load. (c) Top view of sensing element (size: 950 = 750 μm^2) of angle sensor with scanning electron microscope photo of one torsion bar. The sensing element consists of a movable polysilicon structure suspended by two torsion bars forming a micromachined pendulum structure. (d) SEM picture of micromachined vibratory gyroscope fabricated by combining anisotropic etching and deep reactive ion etching (DRIE) process. ([a, b] Reprinted from *Sens. Actuators A* 142, Ádám M. et al. 192, Copyright 2008, with permission from Elsevier; [c] Reprinted from *Sens. Actuators A* 73, Kaienburg J.R. and Schellin R., 68, Copyright 1999, with permission from Elsevier.); [d] Reprinted from *Sens. Actuators A* 96, Yang H. et al. 145, Copyright 2002, with permission from Elsevier.)

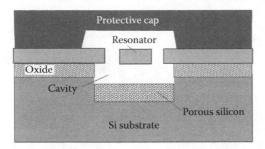

FIGURE 7.2 Schematic diagram of vacuum-encapsulated MEMS resonator with PSi layer as getter material. (Idea from Mohammad W. et al., In: *Proceedings of Joint Conference of the IEEE on Frequency Control and the European Frequency and Time Forum (FCS),* May 2–5, San Francisco, 2011.)

Studies have shown that the use of PSi can also be useful in the manufacture of vacuum-encapsulated microelectromechanical systems (MEMS) resonators. It is known that the vacuum encapsulation improves the resonators designed for gyroscopes. Vacuum-encapsulated MEMS resonators are required for high-precision frequency control. Mohammad et al. (2011) have shown that PSi may act as a getter material in the cavity formed around a resonator (see Figure 7.2). In particular, Mohammad et al. (2011) established that due to large internal surface area, PSi is highly reactive and can absorb gases such as oxygen and other gas molecules. This helps in maintaining low pressures in the cavity of the bonded MEMS resonators. As a result, the devices with PSi getter reported two times better quality factor than the nongetter devices. PSi in the cavity was formed by electrochemical etching in diluted HF solution (10% HF, current density of 3.0 mA/cm^2 for 20 min). According to Mohammad et al. (2011), the advantages of PSi getter along with cavity are: (1) The cavity and porous layer reduce the parasitic capacitance and air damping that helps resonators perform better, (2) the porous layer needs no mask and is self aligned, and (3) the porous getter material needs no external heat to be activated.

As it was indicated before, there are many types of mechanical sensors that can be fabricated using micromachining technology based on silicon porosification. However, in this chapter we will consider only accelerometers and gyroscope, which found the widest application. Moreover, as a rule, all these devices are manufactured using the same technological approaches. Descriptions of some of them can be found in Chapters 6, 8, 11, and 15 of this book.

7.2 SILICON MICROMACHINED ACCELEROMETERS

Accelerometer is one of the most important types of mechanical microsensors (Rudolf et al. 1990; Kuehnel and Sherman 1994; Nemirovsky et al. 1996; Yazdi et al. 1998; Yazdi and Najafi 2000; Xia et al. 2014). This device is inertial and measures linear acceleration. Accelerometers are required in a diverse range of activities in everyday life and industries and the need for cost-effective reliable sensors with reduced dimensions is ever increasing. In particular, there are numerous applications and opportunities in transport and avionics for GPS-augmented navigation and guidance systems, microgravity measurements and platform stabilization in space, in industrial control and automation, seismometry for oil-exploration and earthquake prediction, underwater acoustic measurements, entertainment products (toys), and even in biomedicine (see Figure 7.3). However, a reliable and cost-effective accelerometer for automotive applications, for example, in automatic braking system (ABS) and suspension systems (0 to 2 g) and airbag systems (up to 50 g) (Kuehnel and Sherman 1994; Zimmermann et al. 1995), is currently the most established need. In the case of a central airbag system, the electronic circuit unit (ECU) is placed in the middle of the car on its transmission tunnel. The ECU contains at least one acceleration sensor, which measures the acceleration and deceleration of the car. In the event of a crash, the acceleration sensor supplies a microcontroller with an electrical signal, which corresponds to the progress of deceleration. Algorithms in the controller decide whether to fire the airbag, and, if required, calculate the firing time point. They have to be small in their geometrical dimensions and must be available in high production volumes. Although the quality requirements are very high, costs

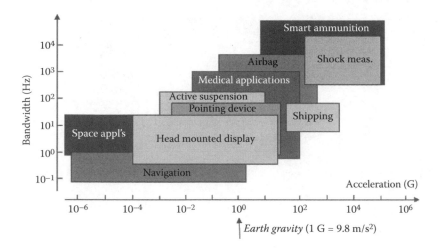

FIGURE 7.3   Overview about applications and required performances for micromachined accelerometers. (Data from Kraft M., *Meas. Control* 33, 164, 2000.)

must be extremely low. One should to note that microaccelerometers are now produced in million units with sophisticated damping and overload protection.

As a rule, accelerometers are based on the cantilever principle in which an end mass (or shuttle) displaces under an inertial force. Thus, the dynamics can be described in simple terms by the second-order system of a mass-spring damper (see Figure 7.4). Since the resolution of an accelerometer with a given seismic mass is proportional to the square of the natural frequency or, equivalently, the stiffness of the sensor structure along the sensing direction, the natural frequency must be lowered to improve the resolution, requiring the use of a bigger mass and longer beam springs for the sensor structure (Park et al. 1998). Therefore, to increase the accelerometer sensitivity, additional seismic masses can be added, for example, via electroplating of gold on the sensor plates. However, according to Rudolf et al. (1990) these added solid masses induce thermal stress problems.

Experiment has shown accelerometers can be classified into the following categories according to the forms of transduction mechanism (Yazdi et al. 1998): piezoresistive (Roylance and Angell 1979; Sim et al. 1998a,b), capacitive (Matsumoto et al. 1996), based on tunneling effect (Waltman and Kaiser 1989; Rockstad et al. 1996), piezoelectric (Nemirovsky et al. 1996; Yu and Lan 2001), thermal (Dauderstadt et al. 1995), and optical (Abbaspour-Sani et al. 1995) devices. Piezoresistive, capacitive, and piezoelectrical-based transducers are most popular in micromachined silicon accelerometers. Whereas piezoresistive, capacitive, and tunneling devices are fabricated in silicon, piezoelectric sensors are fabricated in ceramics. These methods of sensing mechanical forces and hence acceleration are compared in Ristic (1994), KIoeck and de Rooij (1994), and Yazdi et al. (1998). Each sensing method has inherent advantages as well as drawbacks. The basic structures of the electromechanical part of the accelerometers are shown in Figure 7.5.

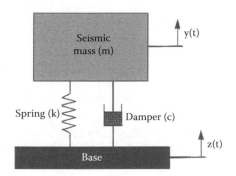

FIGURE 7.4   Dynamic model of accelerometer. (Idea from Park K.-Y. et al., *Sens. Actuators A* 73, 109, 1999.)

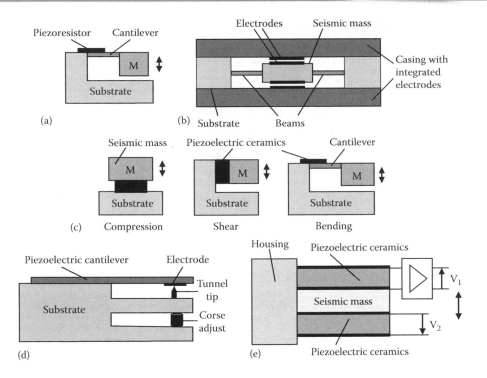

FIGURE 7.5   Basic structures of microaccelerometers: (a) piezoresistive; (b) capacitive; (c) piezoelectrical; (d) tunneling; and (e) PZE cell structure. (With kind permission from Springer Science+Business Media: *Microsystem Technology and Microrobots*, 1997, Fatikow S. and Rembold U. Idea from Yu J.-C. and Lan C.-B., *Sens. Actuators A* 88, 178, 2001; Waltman S.B. and Kaiser W.J., *Sens. Actuators* 19, 201, 1989; and Spineanu A. et al., *Sens. Actuators A* 60, 127, 1997.)

Among micromachining accelerometers, the piezoresistive accelerometer (see Figure 7.5a) has been the most successful in meeting the requirements for high reliability, low cost, and the potential for mass production. The piezoresistive effect describes change in the electrical resistivity of a semiconductor or metal when mechanical strain is applied. In contrast to the piezoelectric effect, the piezoresistive effect only causes a change in electrical resistance, not in electric potential. The main advantage of piezoresistive accelerometers is the simplicity of their structure, fabrication process, and read-out circuitry. One of the first commercialized microaccelerometers was piezoresistive. Typically, the basic element of a bulk micromachined piezoresistive accelerometer is a mass spring system consisting of a detectable cantilever supporting a seismic mass that was first developed by Roylance and Angell in 1979. The beams carry piezoresistors, which, in order to achieve a maximum output signal, are placed at the edges of the beams where the maximum bending strain occurs (see Figure 7.5a). The distance change in these devices is converted to a voltage change. Piezoresistors can be fabricated using a wide variety of piezoresistive materials. However, the most common are the diffused silicon piezoresistors. To control damping, the sensors can be placed in oil-filled housings. Another possibility for controlling damping is to build up narrow air gaps surrounding the seismic mass in the direction of deflection. In this case, damping is caused by streaming resistances of the air in the gaps.

An accelerometer with single cantilever suspension has the advantage of a high sensitivity; however, it has the disadvantages of a very fragile mechanical structure and a high off-axis sensitivity because the seismic mass is geometrically asymmetric in relation to the X-Y plane due to the fabrication process. Since a single cantilever suspension is more sensitive to cross-acceleration, to overcome these problems, bridge-type accelerometers have been designed with two (Crazzolara et al. 1993), four (Allen et al. 1989), six (Sim et al. 1998a), or eight beams (Sim et al. 1998b) to support the mass (Figure 7.6). The outputs of these sensors, achieved through an appropriate resistor arrangement, are only sensitive to normal accelerations in the device plane and not to lateral accelerations. Furthermore, Sim et al. (1998a) have shown that the off-axis sensitivity of the accelerometer can be strongly minimized using a summing circuit. It was established

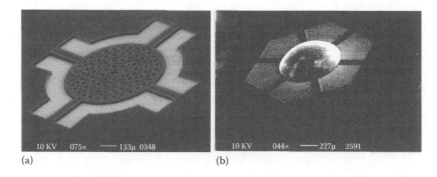

10 KV   075×   ———133µ 0348   10 KV   044×   ———227µ 2591

(a)   (b)

FIGURE 7.6 SEM micrographs of the (a) four-beam and (b) six-beam accelerometers. (From Sim J.-H. et al., *J. Korean Phys. Soc.* 33, S427, 1998: Copyright 1998: Korean Physical Society; and Sim J.-H. and Lee J.-H., *Jpn. J. Appl. Phys.* 38, 1915, 1999: Copyright 1999: The Japan Society of Applied Physics. With permission.)

that piezoresistive accelerometers show high temperature dependence and a great influence of mounting stress. Piezoresistive devices are quite sensitive to mechanical strain induced by the fabrication process or by packaging.

In capacitive accelerometers (see Figure 7.5b), a movable plate acting as inertial mass and capacitor plate is suspended by one, two, or four flexure bars. The displacement of the plate due to acceleration is measured by the associated capacitance change between the movable plate and fixed electrodes on either side. This means that the seismic mass is simultaneously used as one electrode of a variable capacitance, moving with respect to a counter-electrode situated on a second plate. Thus, a capacitive sensor measures a variable distance due to the deflection of the movable plate. Suitable damping is achieved by adjusting the gas pressure surrounding the plate. The device can be operated either in a plate displacement mode, where the capacitance difference is proportional to acceleration, or in a force-balancing mode, where the capacitance changes are used in a servo loop to adjust electrostatic forces to maintain the plate in the middle position. A suitably packaged complete accelerometer is composed of an electromechanical silicon chip and adequate measurement electronics delivering an output voltage proportional to acceleration. Several examples of seismic mass used in capacitive accelerometers are shown in Figure 7.7.

One should note that modern capacitive micromachined sensors have the most complex silicon structures (see Figure 7.7b). Advantageously, they are designed as differential capacitors because this guarantees an excellent linearity and provides the possibility to integrate an electromechanical feedback loop (Kuehnel and Sherman 1994). In particular, the accelerometer designed by Park et al. (1999) consists of a planar proof mass with branched comb-fingers, differential capacitive type electrodes to sense the relative displacement between the seismic mass and the electrodes themselves, tuning electrodes to reduce the stiffness of the structure by generating electrostatic force (Zimmermann et al. 1995), and four folded type beam springs. The black squares are the anchors that fix the sensor structure to the substrate. The comb-finger type of capacitive sensing element that is commonly used in surface-micromachined accelerometers is shown in Figure 7.8a, where the variations of the capacitances C_1 and C_2 are reversed in sign. The performance of the differential capacitive sensing type electrode can be improved by reducing the gap between the seismic mass and electrode, and by increasing the sensing area. On the other hand, in this electrode, the electrostatic stability deteriorates with better sensitivity especially in the accelerometer with a high resolution. The probability of clash between mass and electrode increases because of its low stiffness. Exactly to overcome such clash problems, branched comb-fingers (see Figure 7.8b) are introduced in the construction of accelerometers (Park et al. 1999). In this case, the gap along the sensing direction d_{bo} of a branched comb-finger can be larger than that of a straight comb-finger, while maintaining the capacitance variation and occupied wafer area unchanged. The sticking problem can then be resolved with a larger gap. The gap g_c along the perpendicular axis to the sensing direction can be made much smaller than d_{bo} because the stiffness of folded spring along the perpendicular axis to the sensing direction is much higher than that of the sensing direction. Actually, the minimum gap g_c is limited normally by the performance of etching.

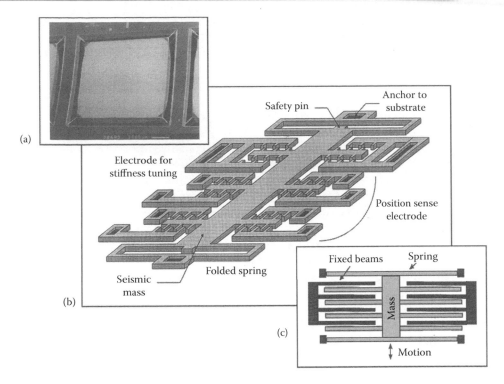

FIGURE 7.7 (a) Scanning electron micrograph of the seismic mass of first accelerometer with four-beams; (b) schematics of modern capacitive accelerometers; and (c) planar view of a lateral accelerometer. (Reprinted from *Sens. Actuators A* 21–23, Seidel H. et al., 312, Copyright 1990, with permission from Elsevier; Reprinted from *Sens. Actuators A* 73, Park K.-Y. et al., 109, Copyright 1999, with permission from Elsevier; and Bell T.E. et al., *J. Micromech. Microeng.* 6, 361, 1996. Copyright 1996: IOP. With permission.)

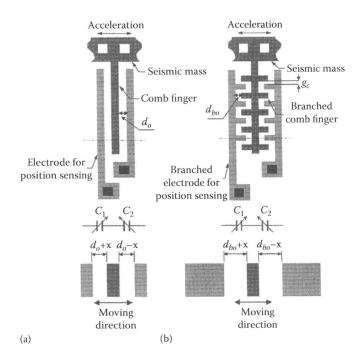

FIGURE 7.8 Electrodes for position sensing: (a) straight comb-finger type and (b) branched comb-finger type. (Reprinted from *Sens. Actuators A* 73, Park K.-Y. et al., 109, Copyright 1999, with permission from Elsevier.)

Currently, capacitive sensing is considered to possess several advantages over piezoresistive sensing: smaller size of device, higher sensitivity, wider operating temperature range, lower temperature coefficients, and better overall system linearity and compatibility with complementary metal oxide semiconductor (CMOS) signal processing. In addition, capacitive accelerometers have good noise performance, low temperature sensitivity, and low drift (Yu and Lan 2001). Therefore, the capacitive silicon micromachined accelerometers are probably the most prevalent devices in the market. Furthermore, self-testing and force-balancing techniques to deflect the seismic mass and verify operation and calibration of the device are possible by utilizing the electrostatic force (Matsumoto et al. 1996). This feature is very important in high-volume production (Nemirovsky et al. 1996). However, capacitive accelerometers have problems with electromagnetic interference.

Bulk piezoelectric ceramic accelerometer sensors (see Figure 7.5c) are already in production (Waanders 1991). However, experiment has shown that thin film sensors are more promising for the sensor market. The most widely used piezoelectric ceramics are $PbZrTiO_6$ solid solutions (PZT). The fundamental advantage of piezoelectric thin films compared to the bulk material is that they can be integrated with silicon wafers to obtain a monolithic system where the sensor elements and the electronics are implemented on the same substrate. This approach provides higher reliability, lower cost, lighter systems, smaller dimensions, and higher frequency response. It also exhibits a low cut-off frequency of about 1 Hz. Furthermore, with the thin film and proposed design, it is possible to implement controlled damping, temperature compensation, and self-test features (Nemirovsky et al. 1996).

There are three basic modes of piezoelectric transducers: compression mode, shear mode, and bending mode (or cantilever mode) (see Figure 7.5c). Bending mode transducers have a better sensitivity and manufacturability for microaccelerometers using piezoelectric thin films (Yu and Lan 2001). The structure of the piezoelectric measuring cell, called the "PZE cell," in the piezoelectric thin-film accelerometer designed by Spineanu et al. (1997) with sigma-delta servo technique is shown in Figure 7.5e. The device is operated in a force-balancing mode as follows: the PZE cell realizes the acceleration sensing and the servo loop summer. It is composed of two piezoelectric thin plate electrodes on their faces, called the sensor and actuator, sandwiched with a seismic mass in the housing of the accelerometer. The mass, with only one degree of freedom, is the movable electrode. When acceleration is applied to the system, the mass under inertial force will follow the motion of the housing. During acceleration in the axial direction, the seismic mass exerted a force on the piezoelectric layer, resulting in the generation of a charge due to the piezoelectric effect.

Figure 7.9 shows four basic suspension configurations of bending mode piezoelectric microaccelerometers (Yu and Lan 2001). When some acceleration is acting on the structure, the seismic masses using suspension of one and two beams are subject to rotations along the x-axis and the y-axis in addition to the translation along the z-axis. On the other hand, the centered seismic mass suspended by four symmetric beams, as shown in Figure 7.9d, will only vibrate along the z-axis with negligible motions in other directions. This configuration will greatly simplify the modeling process. Although tension stresses are present in addition to bending stresses of the suspension beam, which might affect the linearity of accelerometer, the effect is negligible for small displacements.

The main advantages of piezoelectric sensors are wide operating temperature range (up to 300°C) and high operating frequency range (up to 100 kHz). In addition, piezoelectric accelerometers have the advantages of easy integration in existing measuring systems. Due to their excellent dynamic performance and linearity, they have been widely used in condition monitoring systems to measure machinery vibration (Scheeper et al. 1996). The main limitations are related to depolarization due to shock and temperature variation as well as to low-frequency operation. Under a constant force, the charges due to strain leak away and hence DC operation is not possible and the low-frequency operation is limited by the electrical properties of the sensor and the electronic readout. Nemirovsky et al. (1996) concluded also that self-test features are difficult to implement in these devices and that damping is not available. In the field of vibration measurement, which requires low sensitivity and large bandwidth, the piezoelectric approach is optimum (Spineanu et al. 1997).

The first tunnel accelerometer described by Waltman and Kaiser (1989) comprised a single suspended mass component, the proof mass (see Figure 7.5e). Electron tunneling took place between a tip on the proof mass and a fixed counter-electrode on the case. As it is known, electron tunneling appeared only in tunnel devices where a thin solid (0.5–3 nm) insulator barrier separated two conducting electrodes. Operation of this accelerometer uses electronic feedback circuitry to

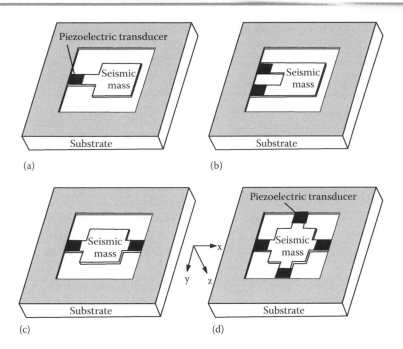

FIGURE 7.9 Basic suspension configurations of piezoelectric microaccelerometers. (a) Cantilever beam; (b) two cantilever beams; (c) two-symmetric-beam suspension; and (d) four-symmetric-beam suspension. (Reprinted from *Sens. Actuators A* 88, Yu J.-C. and Lan C.-B., 178, Copyright 2001, with permission from Elsevier.)

maintain a constant tip to-counter electrode spacing, as in scanning tunneling microscopy, by controlling an electrostatic deflection voltage on the suspended mass (Waltman and Kaiser 1989; Kenny et al. 1991). For these purposes, the tunnel accelerometer employs a piezoelectric actuator. When the sensor is accelerated, the proof mass experiences an inertial force, which causes its motion to lag that of the sensor. The feedback circuit generates an electrostatic rebalance force, which causes the proof mass to follow the motion of the sensor. In order to achieve stable operation, the bandwidth of the feedback circuit is restricted to frequencies below the natural frequency of the proof mass, 200 Hz in this case. Electron tunneling displacement transducers offer extremely high resolutions, which can be utilized for sensitive microinstruments. One example of a tunneling-based acceleration sensor is shown in Figure 7.10.

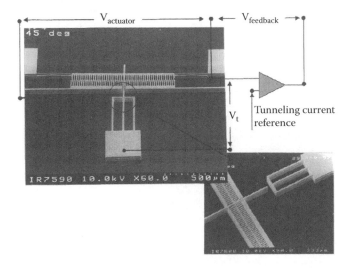

FIGURE 7.10 Tunneling-based acceleration sensor with electrostatic driven comb actuator for tip distance approach. (Reprinted from *Vacuum* 62, Rangelow I.W., 279, Copyright 2001, with permission from Elsevier.)

7.3 MICROMACHINED GYROSCOPES

The second type of inertial mechanical sensors is the gyroscope, which measures the change in orientation of an object (Grieff et al. 1991; Ayazi and Najafi 1998, 2001; Geiger et al. 1998; Yazdi et al. 1998; Xia et al. 2014). In particular, gyroscopes are used to measure angular rate or attitude angle. Compared to traditional gyroscopes, micromachined gyroscopes have many advantages such as small size, light weight, low cost, high precision, and easy integration. As a result, there is considerable interest in micromachined gyroscopes in the defense industries for controlling missiles, but low-cost commercial devices for nonmilitary applications are now appearing. It was found that gyroscopes could be widely applied in many fields, including automotive applications for ride stabilization and rollover detection, some consumer electronic applications, such as video-camera stabilization, virtual reality, and inertial mice for computers, robotics applications, and so on (see Figure 7.11).

The basic principle of a gyroscope is a transfer of energy from one rotating system to another because of the Coriolis force (Gardner et al. 2001; Xia et al. 2014). In other words, a gyroscope responds to the Coriolis motion and converts to an electrical signal. The Coriolis force (Thomson 1986) is a noninertial quantity that arises from the velocity of a particle in the rotating coordinate system. A conventional spinning wheel offers an excellent performance and is accepted for high-cost applications. It should be noted that reducing size on a silicon wafer, such as a spinning wheel, is not easy to fabricate because of the wear of the bearing and the lifetime. On the contrary, vibrating structures are easier to fabricate because they have no rotating parts that require bearings; they just need to be suspended on the silicon substrate with well-designed flexures. This means that they can be easily miniaturized. Hence, the vibratory gyroscope has been the main subject in the micromachining technology (An et al. 1999). As a rule, these devices have been fabricated based on coupled resonators. When the suspended proof mass is vibrating, angular rate induces the Coriolis force. The Coriolis force is proportional to the proof mass, the vibration velocity, and the angular rate. Micromachined structures have a small mass and a small sensing area, which degrades the signal-to-noise ratio. For minimum electrical noise, the high velocity of vibration can increase the signal. To resonate in the large linear range, flexures should be designed to release the stress from the fabrication imperfection and the large deflection. Folded springs have larger linear range than simple springs. With linear actuation of electrostatic comb force (Tang et al. 1989), the resonators are subjected to a harmonic motion. Due to the small electrostatic force and the large air damping, a high amplitude of vibration is achieved by increasing the Q-factor in a high vacuum state where air damping is not the dominant variable.

Figure 7.12 shows several types of vibratory gyroscopes (An et al. 1999). One of the simplest resonating types is the one-proof-mass supported by flexures. The mass vibrates parallel to the substrate and detects an angular rate orthogonal to the first vibration. Consequently, this

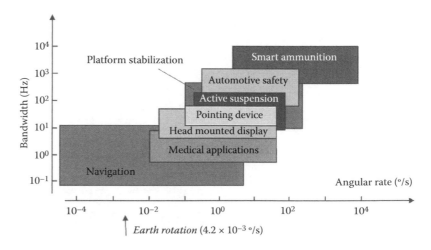

FIGURE 7.11   Overview about applications and required performances for micromachined gyroscopes. (Data from Kraft M., *Meas. Control* 33, 164, 2000.)

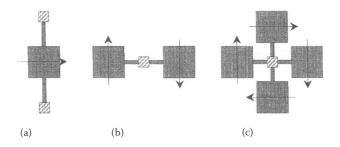

(a)   (b)   (c)

FIGURE 7.12 Types of vibratory microgyroscopes. (a) One-plate; (b) two-plate; and (c) four-plate. (Reprinted from *Sens. Actuators A* 73, An S. et al., 1, Copyright 1999, with permission from Elsevier.)

one-proof-mass type has acceleration sensitivity. To decrease it, two-plate vibration is a choice. Two plates are vibrated in opposite directions and the Coriolis force moves the vibrating plates into each opposite sensing direction. The electrodes beneath the plates make the differential position sensing, which cancels the acceleration force. With two plates, a tuning fork is achieved with the proper flexure designs. In addition, by adding the two plates into the orthogonal direction, the vibrating elements work as a dual-axis gyro where each plate responds to the corresponding input rate in each axis. With differential sensing, the accelerations are canceled. There have also been reports of a performance of a two-input axis gyro with one circular plate where each quadrature has a sensing electrode beneath (Juneau et al. 1997). Furthermore, Juneau et al. (1997) reported that this structure has a force balancing capability with a simple rebalancing tilting torque.

The first silicon coupled resonator gyrometer was developed by Draper Laboratory in the early 1990s (Grieff et al. 1991). The device is bulk micromachined and supported by torsional beams with micromass made from doped (p^{++}) single-crystal silicon. The outer gimbal was driven electrostatically at the constant amplitude and the inner gimbal motion was sensed. The rate resolution was only 4 deg/s and bandwidth was just 1 Hz. More advanced gyroscopes have been fabricated using surface micromachining of poly silicon. There are a number of examples of coupled resonator gyroscopes such as the MARS-RR gyroscope reported by Geiger et al. (1998, 1999). A schematic diagram of this device is shown in Figure 7.13. In these devices, comb-drives are used to force an oscillation of the mechanical structure. By turning the device, Coriolis forces generate a second oscillation perpendicular to the first one with the amplitude proportional to the angular rate. Respectively, these oscillations or modes are referred to as the primary and the secondary motion. Similar configurations have a silicon micromachining gyroscope designed by An et al. (1999) (see Figure 7.14). The main sources of errors of comb driven gyroscopes are mechanical and electromechanical coupling effects and the small deflections, which are to be measured. The

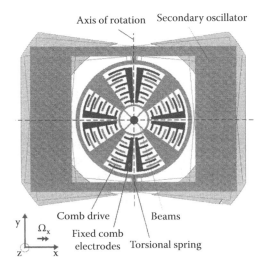

FIGURE 7.13 Top view of the angular rate sensor MARS-RR. (Reprinted from *Sens. Actuators A* 73, Geiger W. et al., 45, Copyright 1999, with permission from Elsevier.)

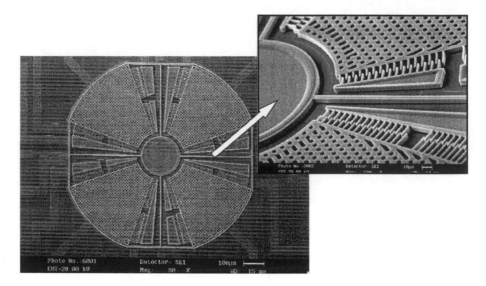

FIGURE 7.14 SEM photo of the top view of a silicon gyroscope designed by An et al. (1999). (Reprinted from *Sens. Actuators A* 73, An S. et al., 1, Copyright 1999, with permission from Elsevier.)

last mentioned problem has to be solved by using a detection capacitance as large as possible. The coupling effects include purely mechanical crosstalk and at least two electromechanical effects. First, due to the substrate supporting the comb drives, the electric field is unsymmetric and electrostatic forces arise, which pull the mechanical structure upward. This electromechanical coupling effect is called levitation (Tang et al. 1992). Second, if the comb drives are not decoupled from the secondary oscillation, the overlap of a pair of combs changes in dependence of the input rate. Thus, the driving forces change and a nonlinear behavior results. MARS-RR gyroscope was designed in a way that a large detection capacitance allows very sensitive measurement and that the mentioned coupling effects are reduced largely. As pointed out by Geiger et al. (1998), the use of eight pairs of combs and their proper, highly symmetrical arrangement compensates the levitation forces up to the second order. Terms of higher order are effectively suppressed by the decoupling and the high stiffness of the suspension beams in the z-direction. There are reports of a number of other types of devices to measure precision rates (Xia et al. 2014). For example, Ayazi and Najafi (1998) designed the ring gyroscope that again works by the Coriolis force transferring energy from one mode into another at 45°.

7.4 TECHNOLOGY OF MICROMACHINED MECHANICAL SENSORS FABRICATION

As we wrote before, silicon beam, cantilever, or membrane are the main elements of mechanical micromachining sensors. The technology acceptable for these elements forming using silicon porosification will be described in Chapter 11 in this book. Therefore, in this section we will discuss technologies of accelerometers and gyroscopes fabrication.

There are several approaches to accelerometers and gyroscopes fabrication. However, the central idea of all technological approaches to design of mechanical micromachining sensors is combining surface and bulk micromachining to fabricate devices with high sensitivity, low noise floor, and controllable damping on a single silicon wafer (Yazdi and Najafi 2000). At that, as a rule for resolving this problem all these approaches (1) use the entire wafer thickness to attain a large proof mass, (2) use a sacrificial film to form a uniform and conformal gap over a large area, and (3) use deep trench etching to create mechanically stiff electrodes on both sides of the proof mass circumventing the need for multiple wafer bonding. As it is known, in order to increase the sensitivity and reduce the noise floor, the proof mass is desired to be as large as possible. The noise and the damping factor are reduced also by forming damping holes in the electrodes. Air gap also should be maximal because the accelerometer sensitivity inversely increases with the air gap squared. The lower limit of the air gap is determined by the device release time and stiction. The air gap is defined by the thickness of a sacrificial layer.

The first approach to accelerometers and gyroscopes fabrication is based on traditional electrochemical etching of a selectively diffused (111)- or (100)-oriented $n/n^+/n$ silicon wafer in hydrouoric acid (HF) at room temperature (Bell et al. 1996; Sim et al. 1998a,b; Artmann and Frey 1999; Sim and Lee 1999). It is well known that PSi formation is highly selective with respect to different dopant types and concentrations. In particular, since the anodic reaction does not proceed to the lightly doped n-type substrate, the reaction stops automatically after the complete conversion of the n^+ region to PSi. This means that the resultant microstructure will be well defined and uniform without a cusp or any evidence of the undercutting phenomenon. These features of electrochemical etching were analyzed in detail in *Porous Silicon: Formation and Properties*. Utilizing this property, an electrochemical etch-stop, using the electrochemical etching of an n^+-diffused layer and stopping on an n-epitaxial layer, was applied to the manufacture of the free-standing microstructure (see Figure 7.15). In particular, in microstructure fabricated by Sim and Lee (1999), the air gap from the microstructure to the substrate was 20 μm. The fabrication process of the silicon piezoresistive accelerometer designed by Sim et al. (1998b) is shown in Figure 7.16. The beam thickness and air gap height exactly matched the respective thicknesses of the 5-μm epitaxial n-Si layer and the 20-μm diffused n^+-layer, indicating precise controllability for determining the microstructures of devices. The full details of the fabrication process for the accelerometer have been presented in Sim et al. (1998b). The anodic reaction was performed in 10% aqueous HF solution for 2.5 min at room temperature by applying a constant current density of 15 mA/cm² to the wafer. The PSi layer was then etched away in 5 wt.% NaOH solution. After the Au/Ni-Cr masking layer was removed, contacts for the piezoresistors were made by etching nitride and oxide through contact windows. A new Au/Ni-Cr layer was deposited and patterned to form both the resistor pads and the mass paddle. These wafers were then annealed at 330°C in nitrogen for 30 min to reduce the contact resistance of the piezoresistor and to provide a good adhesion to the surface of the silicon paddle.

It is known that the sensitivity of accelerometers and gyroscopes depends on the thickness of gaps between oscillating structures and substrate. In particular, due to limited thickness of conventional sacrificial layers as SiO_2 or TEOS, resulting in small gaps between oscillating structures and substrate, surface-micromachined vibrational structures such as gyroscopes require low-pressure environment to ensure low air damping for proper operational conditions. Earlier we have shown that the use of PSi allows resolving this problem. Artmann and Frey (1999) found that the combination of silicon porosification technology with electrochemical etching gives the possibility to realize even larger substrate gaps of surface-micromachined structures and as result to reduce air damping at a given ambient pressure. For electropolishing after PSi removal, the same

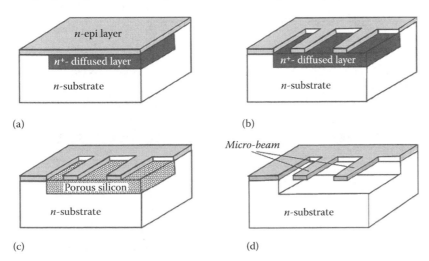

FIGURE 7.15   Concept of the microstructure fabrication process using PSi. (a) $n/n^+/n$ structure; (b) microstructure define; (c) PSi layer formation; and (d) PSi layer etching. (Reprinted from *Sens. Actuators A* 66, Sim J.H. et al., 273, Copyright 1998, with permission from Elsevier.)

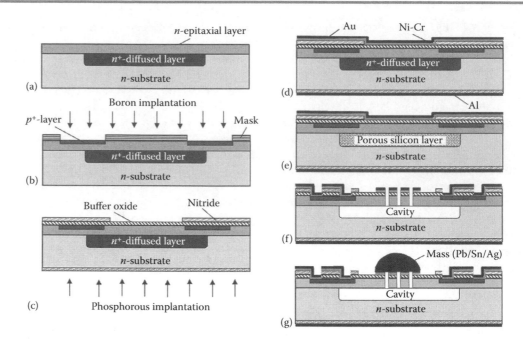

FIGURE 7.16 Fabrication process of the silicon piezoresistive accelerometer. This process includes: (a) implantation and epitaxial layer growth; (b) formation of piezoresistors; (c) surface passivation and backside implantation; (d) metallization; (e) Si porosification by anodic etching; (f) PSi layer removal; and (g) seismic mass forming by soldering. (Reprinted from *Sens. Actuators A* 66, Sim J.H. et al., 273, Copyright 1998, with permission from Elsevier.)

apparatus and sample preparation as for PSi generation was used. It is reached at low HF concentrations (~5%, HF:H_2O:ethanol = 1:9:10) and current densities (J = 10 mA/cm^2). The use of lower HF concentrations led to a reduced attack of the contact. Higher current densities (J = 25–75 mA/cm^2) led to a strong concave bow profile under the perforated mass. Figure 7.17 shows measured profiles obtained for different conditions of etching. Artmann and Frey (1999) established that with this increased substrate gap, the air damping of interdigital oscillating structures was reduced and led to a Q factor increase of around 100% compared to the conventional gap of 2 μm.

Other approaches based on electrochemical etching designed for the macroporous silicon forming was proposed by Ohji et al. (1999, 2000). The process flow for electrochemical etching is shown in Figure 7.18. Starting material was *p*- or *n*-type (100) silicon. (100) silicon wafer is good for vertical etching. Silicon nitride or silicon dioxide was deposited on the top of the surface and patterned to define initial pit by KOH. After making these initial pits, the sample was etched electrochemically. Electrolyte used for electrochemical etching was prepared by mixing dimethylformamide (DMF), deionized water, and 50 wt% hydrofluoric acid. All electrolytes, when mixed, contained 5 wt% hydrofluoric acid. Tetrabutylammonium perchlorate was added in enhanced conduction. After the desired depth was obtained, current density was increased by increasing the light intensity. The width of the trenches increased and trenches were connected under the structures. Thus, freestanding structures made of single crystal silicon were fabricated with only one mask.

We have to note that despite the progress achieved in the development of accelerometers fabricated using electrochemical etching technology, the sensor market is dominated by devices manufactured using dry etching technology (Park et al. 1999; Rangelow 2001, 2003; Tsuchiya and Funabashi 2004). Description of this method can be found in Chapter 10 in *Porous Silicon: Formation and Properties*. Experiment has shown that dry etching technology has great potential in MEMS devices fabrication, which is caused by achieving high vertical aspect ratios and extending the limits of deep etching down to the 40–300 μm range (Aachboun and Ranson 1999; Blauw et al. 2002; Abdolvand and Ayazi 2008). For these purposes, usually two approaches are used. The first approach (the Bosch process) is based on using SF$_6$ plasma for silicon etching and CH$_4$/CHF$_3$ or C$_4$F$_8$ for sidewall passivation (Volland et al. 1999; Rhee et al. 2008, 2009).

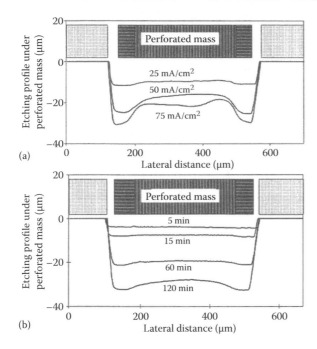

FIGURE 7.17   Bow profile under the perforated mass with the schematic view of the mass: (a) electropolishing with different current densities in 5% HF for 30 min; (b) electropolishing with 10 mA/cm² in 5% HF for different etching times. (Reprinted from *Sens. Actuators A* 74, Artmann H. and Frey W., 104, Copyright 1999, with permission from Elsevier.)

The Bosch process, also known as pulsed or time-multiplexed etching, alternates repeatedly between two modes to achieve nearly vertical structures. The depth of etching depends on the number of cycles of the Bosch process: etching–deposition–etching. Etching with gas chopping provides higher etch rate and selectivity. However, the sidewalls of the etched structures still bear the ripples resulting from individual isotropic etch and deposition steps. It was found that

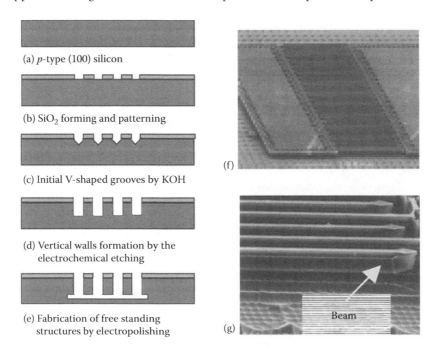

FIGURE 7.18   (a–e) Process flow for fabrication of freestanding beam structures in *p*-type silicon wafer with epitaxial layer; (f) freestanding beam structures; and (g) close-up view of beams. (Reprinted from *Sens. Actuators A* 73, Ohji H. et al., 95, Copyright 1999, with permission from Elsevier.)

the balance between etching and deposition steps controls the anisotropy. At the same time, the selectivity also can be improved by the selective high deposition rate of the passivating layer on the resist mask surface caused due to the high sticking coefficient. These techniques also can be applied to conventional reactive-ion etching (RIE) equipment with Cl-, F-, and Br-based plasma chemistry for etching of semiconductors, dielectric materials, or metals. With these techniques, open stencil masks for the charged particle lithography with 100-nm openings have been realized.

The second approach is based on "cryogenic" etching (Tachi et al. 1988; Isakovic et al. 2008; Henry et al. 2009). It was found that an inductively coupled plasma source (DPS) Oxford deep trench etcher with hardware for cryogenic substrate temperatures using a conventional SF_6/O-based process shows capability for etching sub-0.5 μm feature size deep trenches with high etch rate (http://www.oxford-instruments.com). Anisotropy in the reactive ion etching of silicon essentially results from the probability of halogen reaction at the sidewall and the surface coverage (sticking coefficient) of passivation films. High anisotropy of "cryogenic" etching is achieved by keeping the substrate temperature in the range of −40 to −150°C, where the reaction probability (halogen reactivity) of fluorine radicals decreases and the sticking coefficient of the passivation precursor increases. In other words, the low temperature slows down the chemical reaction that produces isotropic etching. However, ions continue to bombard upward-facing surfaces and etch them away. As a result, this process under "cryogenic" conditions provides the high aspect ratio etching and produces trenches with highly vertical sidewalls. Moreover, along with the high etch rate, the inductively coupled plasma (ICP) deep trench etcher demonstrates superior process performance with minimum mask oxide facet, excellent uniformity, and good profile and critical dimension control. In addition, this conventional SF_6/O_2-based process results in minimum sidewall deposition during etching leading to the elimination of a dry clean with enhanced MEMS productivity due to increased throughput. The disadvantages of this process include cracking the standard masks on the substrates under the extreme cold, and etch by-products have a tendency to deposit on the nearest cold surface, that is, the substrate or electrode.

One of technological routes of accelerometers' fabrication using methods of dry etching is shown in Figure 7.19. A z-axis capacitive accelerometer was fabricated using vertical comb electrodes. The fabrication process of the sensor used backside polished SOI wafers, double-layer masks, and trench RIE. Two-layer masks of plasma CVD SiO_2 films were fabricated after aluminum-pad fabrication. A time-controlled trench RIE was processed twice to make one side of the comb electrodes thinner than the other. The etching depth of each RIE process was half the thickness of the SOI layer. The first mask layer was removed between the two trench RIE processes. After the front-side passivation by plasma CVD SiO_2, the substrate under the device structure was anisotropically etched from the backside by using TMAH solution. Finally, the SiO_2 films for the trench mask and the passivation and sacrificial layers were removed by BHF.

The same approach was used for gyroscope fabrication (An et al. 1999; Geiger et al. 1999). For example MARS-RR gyroscope discussed in the previous section was produced within the Bosch Foundry process, which is a process similar to conventional surface micromachining with an additional protective silicon cap (Figure 7.20). This process features a polycrystalline silicon layer with a thickness of 10.3 μm for the freestanding structure. The large thickness was achieved by using an epitaxial deposition of polysilicon ("epipoly-Si"). A special trench technique allowed the formation of vertical sidewalls with high aspect ratio. In addition to the functional polysilicon layer, a second thin poly-Si-layer ("buried poly") was provided underneath, which served as interconnect, shield, or counterelectrode. The buried poly layer was isolated from the "epipoly-Si" by the sacrificial oxide and from the substrate by a lower oxide.

However, we have to note that recently there have been reports that the electrochemical etching may also provide parameters of micromachining devices that previously may be achieved only by dry etching. Bassu et al. (2012) have shown that the fabrication of high-complexity silicon microstructures and microsystems with submicrometer accuracy at aspect-ratio (AR) values (about 100) can be achieved through dynamic control of the etching anisotropy and the use of high-aspect-ratio functional and sacrificial structures. As an example of the potential of ECM technology, a few experimental results concerning the fabrication of silicon microstructures and microsystems are shown in Figure 7.21. Figure 7.21a shows a SEM image of a MEMS structure consisting of an inertial, free-standing mass (340 μm × 340 μm × 130 μm) equipped with

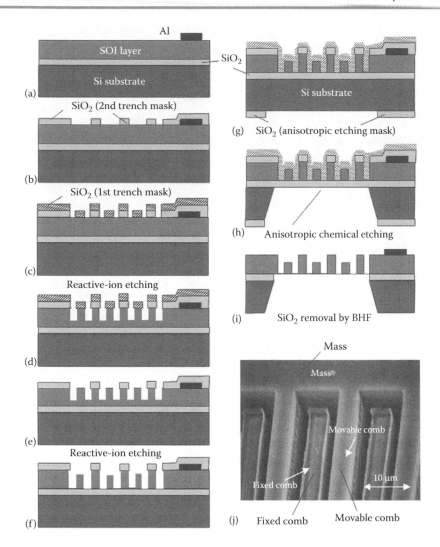

FIGURE 7.19 (a–i) Fabrication process and (j) close-up view of vertical comb electrodes. (Reprinted from *Sens. Actuators A* 116, Tsuchiya T. and Funabashi H., 378, Copyright 2004, with permission from Elsevier.)

high-aspect-ratio comb-fingers and suspended by high-aspect-ratio folded springs (AR of about 100 for both the fingers and the springs) that are fixed to an anchor structure composed of a 2D array of square holes. Figure 7.21d shows a mechanical microgripper. Figure 7.21b,c,f show details of the above-mentioned structures. Bassu et al. (2012) believe that electrochemical micromachining (ECM) technology should obtain worldwide dissemination for silicon microstructuring, due to its low cost and high flexibility.

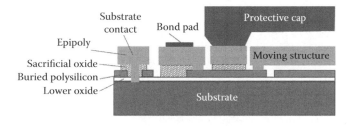

FIGURE 7.20 Layer structure of the Bosch foundry process. (Reprinted from *Sens. Actuators A* 73, Geiger W. et al., 45, Copyright 1999, with permission from Elsevier.)

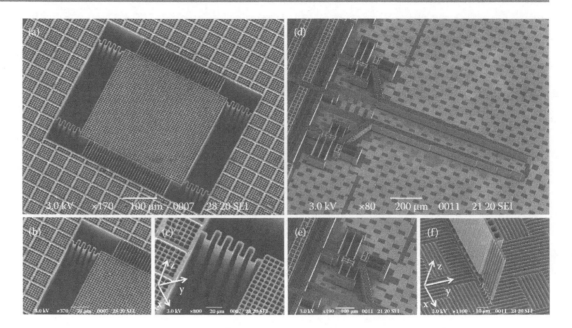

FIGURE 7.21   (a–f) ECM-fabricated microstructures and microsystems: SEM image of a MEMS structure consisting of an inertial, freestanding mass (340 μm × 340 μm × 130 μm) equipped with high-aspect-ratio comb-fingers and suspended by high-aspect-ratio folded springs (*AR* about 100 for both fingers and springs) that are fixed to an anchor structure composed of a 2D array of square holes (a); details at two different magnifications of the MEMS structure in panel (a) are shown in (b, c); one of the folded springs of the structure in panel (a), which clearly highlights the remarkable accuracy in microfabrication by ECM at such high aspect-ratio values (c); SEM image of a mechanical microgripper consisting of two fingers (length of 1 mm and thickness of 70 μm), electrically actuable by means of comb-finger batteries driving a spring system, which allows rotation of the two fingers to be performed symmetrically (d); detail of the spring system exploited to allow symmetric rotation of the two microgripper fingers (e); and detail of one of the microgripper finger's tip provided with a sub-micrometer artificial surface ripple with the aim of increasing the grasping capability of the gripper (f). (Bassu M. et al.: *Adv. Funct. Mater.* 2012. 22. 1222. Copyright Wiley-VCH Verlag GmbH & Co. KGaA. Reproduced with permission.)

ACKNOWLEDGMENTS

This work was supported by the Ministry of Science, ICT, and Future Planning (MSIP) of the Republic of Korea, and partly by the National Research Foundation (NRF) grants funded by the Korean government (No. 2011-0028736 and No. 2013-K000315).

REFERENCES

Aachboun, S. and Ranson, P. (1999). Deep anisotropic etching of silicon. *J. Vac. Sci. Technol. A* 17, 2270–2273.
Abbaspour-Sani, E., Huang, R.S., and Kwok, C.Y. (1995). A wide-range linear optical accelerometer. *Sens. Actuators A* 49, 149–154.
Abdolvand, R. and Ayazi, F. (2008). An advanced reactive ion etching process for very high aspect-ratio submicron wide trenches in silicon. *Sens. Actuators A* 144, 109–116.
Ádám, M., Vázsonyi, É., Bársony, I., Vésárhelyi, G., and Dücsö, C. (2004). Three dimensional single crystalline force sensor by porous Si micromachining. In: *Proceedings of the IEEE Conference Sensors 2004*, October 24–27, 2004 Vienna, Austria, IEEE, 1, 501–504.
Ádám, M., Mohácsy, T., Jónás, P., Dücső, C., Vázsonyi, É., and Bársony, I. (2008). CMOS integrated tactile sensor array by porous Si bulk micromachining. *Sens. Actuators A* 142, 192–195.
Allen, H.W., Terry, S.C., and De Bruin, D.W. (1989). Accelerometer systems with self-testable features. *Sens. Actuators A* 20, 153–161.

An, S., Oh, Y.S., Park, K.Y., Lee, S.S., and Song, C.M. (1999). Dual-axis microgyroscope with closed-loop detection. *Sens. Actuators A* 73, 1–6.

Anderson, R.C., Muller, R.S., and Tobias, C.W. (1994). Porous polycrystalline silicon: A new material for MEMS. *J. Microelectromech. Syst.* 3, 10–7.

Artmann, H. and Frey, W. (1999). Porous silicon technique for realization of surface micromachined silicon structures with large gaps. *Sens. Actuators A* 74, 104–108.

Ayazi, F. and Najafi, K. (1998). Design and fabrication of high-performance polysilicon vibrating ring gyroscope. In: *Proc. of the 11th Int. Workshop on MEMS*, Jan. 25–29, Heidelberg, Germany, pp. 621–626.

Ayazi, F. and Najafi, K. (2001). A HARPSS polysilicon vibrating ring gyroscope. *J. Microelectromech. Syst.* 10, 169–79.

Bassu, M., Surdo, S., Strambini, L.M., and Barillaro, G. (2012). Electrochemical micromachining as an enabling technology for advanced silicon microstructuring. *Adv. Funct. Mater.* 22, 1222–1228.

Bean, K.E. (1978). Anisotropic etching in silicon. *IEEE Trans. Elec. Devices* ED-25, 1185–1193.

Bell, T.E., Gennissen, P.T.J., DeMunter, D., and Kuhl, M. (1996). Porous silicon as a sacrificial material. *J. Micromech. Microeng.* 6, 361–369.

Blauw, M.A., Craciun, G., Sloof, W.G., French, P.J., and van der Drift, E. (2002). Advanced time-multiplexed plasma etching of high aspect ratio silicon structures. *J. Vac. Sci. Technol. B* 20, 3106–3110.

Crazzolara, H., Flach, G., and Von Munch, W. (1993). Piezoresistive accelerometer with overload protection and low cross-sensitivity. *Sens. Actuators A* 39, 201–207.

Dauderstadt, U.A., de Vries, P.H.S., Hiratsuka, R., and Sarro, P.M. (1995). Silicon accelerometer based on thermalpiles. *Sens. Actuators A* 46/47, 201–204.

Fatikow, S. and Rembold, U. (1997). *Microsystem Technology and Microrobots.* Springer, Berlin.

French, P.J. and Sarro, P.M. (2012). Integrated MEMS: Opportunities and challenges. In: Kahrizi M. (Ed.) *Micromachining Techniques for Fabrication of Micro and Nano Stru*ctures. InTech, Rijeka, Croatia, pp. 253–276.

Gardner, J.W., Varadan, V.K., and Awadelkarim, O.O. (2001). *Microsensors, MEMS, and Smart Devices.* John Wiley & Sons, New York.

Geiger, W., Folkmer, B., Sobe, U., Sandmaier, H., and Lang, W. (1998). New design of micromachined vibrating rate gyroscope with decoupled oscillation modes. *Sens. Actuators A* 66, 118–124.

Geiger, W., Folkmer, B., Merz, J., Sandmaier, H., and Lang, W. (1999). A new silicon rate gyroscope. *Sens. Actuators A* 73, 45–51.

Grieff, P., Boxenhorn, B., King, T., and Niles, L. (1991). Silicon monolithic micromechanical gyroscope. In: *Proceedings of Transducers '91,* June 24–28, San Francisco, pp. 966–969.

Henry, M.D., Welch, C., and Scherer, A. (2009). Techniques of cryogenic reactive ion etching in silicon for fabrication of sensors. *J. Vac. Sci. Technol. A* 27, 1211–1217.

Howe, R.T. (1988). Surface micromachining for microsensors and microactuators. *J. Vac. Sci. Technol. B* 6(6), 1809–1813.

Isakovic, A.F., Evans-Lutterodt, K., Elliott, D., Stein, A., and Warren, J.B. (2008). Cyclic, cryogenic, highly anisotropic plasma etching of silicon using SF_6/O_2. *J. Vac. Sci. Technol. A* 26, 1182–1187.

Juneau, T., Pisano, A.P., and Smith, J.H. (1997). Dual axis operation of a micromachined rate gyroscope. In: *Proceedings of Transducer 97,* June 16–19, Chicago, pp. 883–886.

Kaienburg, J.R. and Schellin, R. (1999). A novel silicon surface micromachining angle sensor. *Sens. Actuators A* 73, 68–73.

Kenny, T.W., Waltman, S.B., Reynolds, J.K., and Kaiser, W.J. (1991). Micromachined silicon tunnel sensor for motion detection. *Appl. Phys. Lett.* 58, 100–102.

KIoeck, B. and de Rooij, N.F. (1994). Mechanical sensors. In: Sze S.M. (Ed.), *Semiconductor Sensor.* Wiley, New York, pp. 153–204.

Kraft, M. (2000). Micromachined inertial sensors: The state-of-the-art and a look into the future. *Meas. Control* 33, 164–168.

Kuehnel, W. and Sherman, S. (1994). A surface micromachined silicon accelerometer with on-chip detection circuitry. *Sens. Actuators A* 45, 7–16.

Matsumoto, Y., Iwakiri, M., Tanaka, H., Ishida, M., and Nakamura, T. (1996). A capacitive accelerometer using SDB-SOI structure. *Sens. Actuators A* 53, 267–272.

Mohammad, W., Wilson, C., and Kaajakari, V. (2011). Introducing porous silicon as a getter using the self aligned maskless process to enhance the quality factor of packaged MEMS resonators. In: *Proceedings of Joint Conference of the IEEE on Frequency Control and the European Frequency and Time Forum (FCS),* May 2–5, San Francisco. IEEE (doi: 10.1109/FCS.2011.5977875).

Nemirovsky, Y., Nemirovsky, A., Muralt, P., and Setter, N. (1996). Design of a novel thin-film piezoelectric accelerometer. *Sens. Actuators A* 56, 239–249.

Ohji, H., Trimp, P.J., and French, P.J. (1999). Fabrication of free standing structure using single step electrochemical etching in hydrofluoric acid. *Sens. Actuators A* 73, 95–100.

Ohji, H., French, P.J., and Tsutsumi, K. (2000). Fabrication of mechanical structures in p-type silicon using electrochemical etching. *Sens. Actuators A* 82, 254–258.

Park, K.Y., Lee, C.W., Jang, H.S., Oh, Y.S., and Ha, B.J. (1998). Capacitive sensing type surface-micromachined silicon accelerometer with a stiffness tuning capability. In: *Digest of IEEErASME MEMS Workshop*, Feb., Heidelberg, Germany, pp. 637–642.

Park, K.-Y., Lee, C.-W., Jang, H.-S., Oh, Y., and Ha, B. (1999). Capacitive type surface-micromachined silicon accelerometer with stiffness tuning capability. *Sens. Actuators A* 73, 109–116.

Rangelow, I.W. (2001). Dry etching-based silicon micro-machining for MEMS. *Vacuum* 62, 279–291.

Rangelow, I.W. (2003). Critical tasks in high aspect ratio silicon dry etching for microelectromechanical system. *J. Vac. Sci. Technol. A* 21, 1550–1562.

Rhee, H., Kwon, H., Kim, C.-K., Kim, H.J., Yoo J., and Kim, Y.W. (2008). Comparison of deep silicon etching using SF_6/C_4F_8 and SF_6/C_4F_6 plasmas in the Bosch process. *J. Vac. Sci. Technol. B* 26, 576–581.

Rhee, H., Lee, H.M., Namkoung, Y.M., Kim, C.-K., Chae, H., and Kim, Y.W. (2009). Dependence of etch rates of silicon substrates on the use of C_4F_8 and C_4F_6 plasmas in the deposition step of the Bosch process. *J. Vac. Sci. Technol. B* 27, 33–40.

Ristic, L. (1994). *Sensor Technology and Devices.* Artech House, Boston, pp. 377–455.

Rockstad, H.K., Tang, T.K., Reynolds, J.K., Kenny, T.W., Kaiser, W.J., and Gabrielson, T.B. (1996). A miniature, high-sensitivity, electron tunneling accelerometer. *Sens. Actuators A* 53, 227–231.

Roylance, L.M. and Angell, J.B. (1979). A batch-fabricated silicon accelerometer. *IEEE Trans. Elec. Devices* ED-26, 1911–1917.

Rudolf, F., Jornod, A., Bergqvist, J., and Leuthold, H. (1990). Precision accelerometers with μg resolution. *Sens. Actuators A* 21–23, 297–302.

Scheeper, P., Gullov, J.O., and Kofoed, L.M. (1996). A piezoelectric triaxial accelerometer. *J. Micromech. Microeng.* 6, 131–133.

Seidel, H., Riedel, H., Kolbeck, R., Muck, G., Kupre, W., and Koniger, M. (1990). Capacitive silicon accelerometer with highly symmetrical design. *Sens. Actuators A* 21–23, 312–315.

Sim, J.-H. and Lee, J.-H. (1999). A piezoresistive silicon accelerometer using porous silicon micromachining and flip-chip bonding. *Jpn. J. Appl. Phys.* 38, 1915–1918.

Sim, J.-H., Yu, I.-S., and Lee, J.-H. (1998a). Simulation and characterization of a piezoresistive accelerometer with self-canceling off-axis sensitivity. *J. Korean Phys. Soc.* 33, S427–S435.

Sim, J.H., Cho, C.S., Kim, J.S., and Lee, J.H. (1998b). Eight-beam piezoresistive accelerometer fabricated by using a selective porous-silicon etching method. *Sens. Actuators A* 66, 273–278.

Spineanu, A., Benabes, P., and Kielbasa, R. (1997). A digital piezoelectric accelerometer with sigma-delta servo technique. *Sens. Actuators A* 60, 127–133.

Tachi, S., Tsujimoto, K., and Sadayuki, O. (1988). Low temperature reactive ion etching and microwave plasma etching of silicon. *Appl. Phys. Lett.* 52, 616–618.

Tang, W.C., Nguyen, T.-C.H., and Howe, R.T. (1989). Laterally driven polysilicon resonant microstructure. *Sens. Actuators A* 20, 25–32.

Tang, W.C., Lim, M.G., and Howe, R.T. (1992). Electrostatic comb drive levitation and control method. *J. Microelectromech. Syst.* 1, 170–178.

Thomson, W.T. (1986). *Introduction to Space Dynamics*, Dover Publications, New York, Chaps. 2–6.

Tsuchiya, T. and Funabashi, H. (2004). A z-axis differential capacitive SOI accelerometer with vertical comb electrodes. *Sens. Actuators A* 116, 378–383.

Volland, B., Shi, F., Hudek, P., Heerlein, H., and Rangelow, I.W. (1999). Dry etching with gas chopping without rippled sidewalls. *J. Vac. Sci. Technol. B* 17(6), 2768–2771.

Waanders, J.W. (1991). *Piezoelectric Ceramics.* Philips Components, Eindhoven.

Waltman, S.B. and Kaiser, W.J. (1989). An electron tunneling sensor. *Sens. Actuators* 19, 201–210.

Xia, D., Yu, C., and Kong, L. (2014). The development of micromachined gyroscope structure and circuitry technology. *Sensors* 14, 1394–1473.

Yang, H., Bao, M., Yin, H., Shen, S. (2002). A novel bulk micromachined gyroscope based on a rectangular beam-mass structure. *Sens. Actuators A* 96, 145–151.

Yazdi, N. and Najafi, K. (2000). An all-silicon single-wafer micro-g accelerometer with a combined surface and bulk micromachining process. *J. Microelectromech. Syst.* 9(4), 544–550.

Yazdi, N., Ayazi, F., and Najafi, K. (1998). Micromachined inertial sensors. *Proc. IEEE* 86(8), 1640–1659.

Yu, J.-C. and Lan, C.-B. (2001). System modeling of microaccelerometer using piezoelectric thin films. *Sens. Actuators A* 88, 178–186.

Zimmermann, L., Ebersohl, J.Ph., Le Hung, F. et al. (1995). Airbag application: A microsystem including a silicon capacitive accelerometer, CMOS switched capacitor electronics and true self-test capability. *Sens. Actuators A* 46, 190–195.

Porous Silicon as a Material for Thermal Insulation in MEMS

Pascal J. Newby

CONTENTS

8.1	Why Is Thermal Insulation Necessary In MEMS?		168
	8.1.1	Gas Sensors	168
	8.1.2	Gas Chromatography Preconcentrators	168
	8.1.3	Microfluidics and Biological Applications	168
	8.1.4	Bolometers	169
	8.1.5	Summary	169
8.2	Overview of Common Thermal Insulation Solutions		169
	8.2.1	What Makes a Good Thermal Insulation?	170
	8.2.2	Insulating Substrate	170
	8.2.3	Thick Layers	170
		8.2.3.1 Silicon Oxide	170
		8.2.3.2 Polymers	171
		8.2.3.3 Aerogels	171
	8.2.4	Suspended Membranes	171
8.3	Porous Silicon as a Thermal Insulation Material		172
	8.3.1	Porous Silicon Morphology	172
	8.3.2	Integration Strategies	172
		8.3.2.1 Pretreatments	172
		8.3.2.2 Geometry	173
		8.3.2.3 Indirect Uses of Porous Silicon for Thermal Insulation	174

8.4 Review of Devices Fabricated with Porous Silicon as a Thermal Insulation Material 175

    8.4.1 Flow Sensors 175

    8.4.2 Gas Sensors 176

    8.4.3 Bolometers 177

    8.4.4 Micro-Hotplates 178

    8.4.5 Thermoacoustic Sound Sources 178

    8.4.6 Heat Flux Sensor and Thermoelectric Generator 179

8.5 Discussion and Conclusions 180

References 180

8.1 WHY IS THERMAL INSULATION NECESSARY IN MEMS?

Several types of microelectromechanical systems (MEMS) are based on the use of microheaters, or are used to measure temperature variations. In these devices, effective thermal insulation is essential to achieve good performance. We will review some examples of common thermal MEMS that require thermal insulation.

8.1.1 GAS SENSORS

There are several types of gas sensors; one of the most common is the resistive sensor. In these devices, a layer of sensing material (often SnO_2) reacts with the surrounding gasses, which leads to a variation of its resistivity. In order to work, these sensors must be heated to 200–400°C (Simon et al. 2001; Sharma and Madou 2012). Macroscale gas sensors of this type are already available commercially, but several groups are working on microfabricating these sensors in order to integrate them into wireless sensor networks and portable applications (Simon et al. 2001; Ali et al. 2008; Sharma and Madou 2012), which implies a low power consumption. Gas sensors require heaters to reach their target temperature. To reach this temperature using as little power and as rapidly as possible, these heaters must be thermally isolated.

8.1.2 GAS CHROMATOGRAPHY PRECONCENTRATORS

Gas-phase chromatography is used to separate and identify volatile chemical species. Work is underway to microfabricate and miniaturize these systems in order to use them as portable analysis systems (Voiculescu et al. 2008). Smaller devices mean the volume being analysed is also smaller, which limits the detection sensitivity (Yeom et al. 2008). A solution to this problem is to use a preconcentrator, which uses an adsorbing surface to accumulate the analyte prior to the measurement. The adsorbed analyte is then released by heating the preconcentrator. This can improve detection sensitivity by up to three orders of magnitude (Tian and Pang 2003). As these systems are used in portable devices, their power consumption must be as low as possible. However, in a gas-phase chromatography system, preconcentrators are the subsystem that uses the most power (Cook and Sastry 2005); therefore, effective thermal insulation is essential to reduce power consumption. Porous silicon can also be used as the functional material for preconcentrators, which is discussed in Chapter 12 of this book.

8.1.3 MICROFLUIDICS AND BIOLOGICAL APPLICATIONS

Many devices have been developed in the field of microfluidics and "lab on chip" systems. We will look at two examples: polymerase chain reaction (PCR) devices and microreactors. A PCR system is used to "amplify," or replicate, ADN molecules by heating them at three different successive temperatures (94°C, 45–55°C, and 72°C), and the first temperature must be less than a degree from the target temperature (Sadler et al. 2003; Jain and Goodson 2011). This can be achieved in a microfabricated device either by keeping the ADN in the same area and changing the temperature, or by moving it through three zones with different temperatures (Jain and Goodson 2011). In both configurations, thermal insulation is necessary to reduce power consumption while heating the ADN, and in the latter configuration, thermal insulation is essential to separate the different temperature zones. Several examples of microfluidic devices fabricated using porous silicon (PSi) technology are described in Chapter 16 of this book.

Another application is the fabrication of nanoparticles in microreactors. Synthesizing nanoparticles in a micro-reactor reduces fabrication costs, and also allows better control of the process. The reaction temperature depends on the product being synthesized, but in the case of semiconductor nanoparticles (in particular CdSe quantum dots), it can be higher than 300°C (Jensen 2006; Marre and Jensen 2010). Again, good insulation reduces power consumption and insulates the heated area from the rest of the chip. This latter point is very important for a "lab on chip" system, as the integration of several different features on a single chip is one of the aims of these systems. PSi can also be used in the fabrication of microreactors, as discussed in Chapter 14 in *Porous Silicon: Optoelectronics, Microelectronics, and Energy Technology Applications.*

8.1.4 BOLOMETERS

Bolometers are used to detect infrared light, and can be used in devices such as thermal cameras or night-vision systems. Uncooled microfabricated bolometers are typically composed of an infrared-absorbing material, whose temperature increases when it absorbs infrared radiation, and a temperature-sensing layer, whose resistance changes with the temperature increase. The most common materials for this latter layer are amorphous silicon or vanadium oxide (VOx), which are used, among other reasons, for their high temperature coefficient of resistance (TCR) (typically a few %) (Niklaus et al. 2008).

To ensure high sensitivity, the temperature increase should be as high as possible for a given quantity of incident infrared radiation. To achieve this, the sensing materials must be thermally isolated from the substrate. These devices are often integrated in cameras, which must acquire several images per second. The time constant of such a system is $\tau = RC$, where R is the thermal resistance of the system, and C is the thermal capacity of the system. So, while a low thermal conductivity is desirable to obtain a good sensitivity, a compromise must be found to ensure an acceptable time constant. However, the time constant can also be decreased by lowering the thermal capacity of the system (Niklaus et al. 2008), in other words, by reducing its mass.

8.1.5 SUMMARY

These examples of devices have shown that thermal insulation is indeed essential to ensure good performance of thermal MEMS. Most of the examples shown are based around a heating element, sometimes referred to as a micro-hotplate. Good insulation of the heater means that reaching its target power will require less power, and the target temperature will be reached faster. However, in some cases, rapid cooling between cycles is required, so low thermal conductivity should be balanced by low thermal capacity.

However, the most common material for MEMS fabrication is crystalline silicon, which has a high thermal conductivity of 156 W/(m·K) (Glassbrenner and Slack 1964). Therefore, silicon cannot ensure the required thermal insulation for the devices described above, so other solutions or materials must be used.

8.2 OVERVIEW OF COMMON THERMAL INSULATION SOLUTIONS

Several solutions for thermal insulation of microfabricated devices have already been developed. In the following section, we will present the different materials, as well as the geometries and configurations in which they are used.

8.2.1 WHAT MAKES A GOOD THERMAL INSULATION?

The main material property that determines if a material is a good thermal insulator is its thermal conductivity, given in W/(m·K). Copper (400 W/[m·K]) and silicon (156 W/[m·K]) are excellent thermal conductors, whereas glass (~1 W/[m·K], depending on the type of glass) is a good insulator. However, the geometry of the material is also important, which is expressed by its thermal resistance.

Thermal resistance R gives the heat flow $\dot{Q}$ through a material for a given temperature gradient ΔT, such as: $\Delta T = R\dot{Q}$. Thermal resistance is defined by analogy with electrical resistance, and for a material with a thermal conductivity λ, and of constant section A and length l, is given by $R = l/(\lambda A)$. The obvious conclusion of this is that increasing the thickness of a thermal insulation material is just as effective as reducing its thermal conductivity.

8.2.2 INSULATING SUBSTRATE

The most conceptually simple approach for thermal insulation is to replace the silicon substrate by a material with a low thermal conductivity, such as glass or ceramic. Low temperature cofired ceramics (LTCC) are an example, and have been used to fabricate PCR devices (Sadler et al. 2003), microreactors (Martínez-Cisneros et al. 2012), and microfluidic devices (Wilcox et al. 2011). These ceramic materials have a thermal conductivity of 3 W/(m·K) (Martínez-Cisneros et al. 2012), and are typically patterned using screen-printing, with 25 μm being the minimum feature size (Wilcox et al. 2011). Glass substrates such as quartz or pyrex can also be used. Devices using these substrates include a glucose biosensor (Xie et al. 1994) and a microreactor (Jensen 2006; Marre and Jensen 2010). Glass has the advantage of being transparent, which means optical *in situ* characterization techniques can be used (Marre and Jensen 2010).

The main disadvantage of these materials is that they are not compatible with the majority of existing microfabrication techniques, and specific fabrication techniques must be developed. Furthermore, the minimum feature size is generally much larger than what can be achieved using silicon-based microfabrication. For some applications, this may not be a problem; however. the use of such substrates excludes the fabrication of an integrated circuit on the same chip. Work is underway to remove these obstacles, and recent progress includes the fabrication of vias through glass substrates (Haque and Wise 2013). However, fabrication on silicon for the moment remains much simpler than on glass substrates.

8.2.3 THICK LAYERS

Another approach for thermal insulation in MEMS is to integrate a layer of insulating material onto a silicon substrate. This layer should be as thick as possible in order to ensure a high thermal resistance. Several materials can be used; we will review here silicon oxide, polymers, and aerogel.

8.2.3.1 SILICON OXIDE

Silicon oxide is a commonly used material for microfabrication, has a low thermal conductivity (1.4 W/[m·K]) (Cahill 1990), and can be fabricated using several different methods. The first method is thermal oxidation of silicon; however, this is not a realistic option for fabricating thick layers, as it is too slow. Indeed, making a 10-μm thick layer would take 100 h by wet oxidation and 2000 h by dry oxidation (Deal and Grove 1965). SiO_2 can be deposited at higher speeds by low-pressure chemical vapor deposition (LPCVD) or plasma-enhanced chemical vapor deposition (PECVD). LPCVD is carried out at temperatures between 300 and 900°C (Franssila 2004), which can lead to high stress caused by different coefficients of thermal expansion (CTE) between the SiO_2 film and the silicon substrate. PECVD is carried out at lower temperatures, but films are generally of inferior quality due to hydrogen incorporation (Franssila 2004). PECVD films can be densified by annealing, but this will also cause high stress and even breaking of the substrates (Chen et al. 2002). In any case, films deposited by CVD methods are still limited to a thickness of about 10 μm, which is generally not enough to provide good thermal insulation. A final method

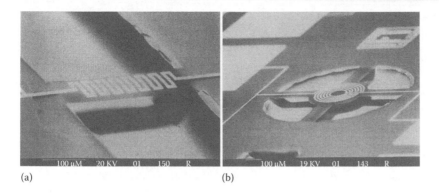

(a)   (b)

FIGURE 8.1   (a) A freestanding bridge and (b) a suspended circular bolometer structure made by sacrificial layer technology using PSi. (Reprinted from *Sens. Actuators A* 43, Lang W. et al., 185, Copyright 1994, with permission from Elsevier.)

for fabricating thick SiO_2 layers is to etch trenches in a silicon substrate by deep reactive ion etching (DRIE), and then refill the trenches by thermal oxidation (Zhang and Najafi 2004). In this manner, 100-μm thick layers have been obtained. However, these layers are also subject to high stress and the surface state is bad due to traces of the initial trench patterning.

8.2.3.2 POLYMERS

Several different polymers have been used for thermal insulation in microfabricated devices, including polyimide (Briand et al. 2006), parylene (Lei et al. 2009), benzocyclobutene or BCB (Modafe et al. 2005), and SU-8 (Vilares et al. 2010). While these materials have a very low thermal conductivity, between 0.08 and 0.3 W/(m·K) for the polymers mentioned previously, their integration in thick layers is not simple, they have low Young's modulus values, and most importantly they cannot withstand high temperatures. Parylene and SU-8 have a glass transition temperature T_g of, respectively, 150°C (Harder et al. 2002) and 200°C,* whereas T_g = 350°C for BCB and polyimide (Modafe et al. 2005; Briand et al. 2006). This limits the microfabrication processes that can be used, and excludes their application in environments at higher temperatures.

8.2.3.3 AEROGELS

Aerogels are made by evaporating the liquid phase of a colloidal gel in order to keep only the solid phase. The solid obtained in this manner is highly porous and has a sponge-like morphology. The most common aerogel is silica aerogel. Silica aerogels with extremely high porosities of 99.99% have been made (Bauer et al. 2011), and thermal conductivity values as low as 0.006 W/(m·K) have been measured (Zeng et al. 1996). However, aerogels with such high porosity do not have good mechanical resistance, and indeed Hrubesh and Poco (1995) noted that aerogels with a porosity of 95% and higher were destroyed simply by applying tape with the aim of carrying out a "tape test" to characterize adhesion. Aerogel can be fabricated on silicon substrates; however, there is a risk that the aerogel's porous structure may collapse during evaporation, and generally, films have a thickness on the order of 1 μm (Hrubesh and Poco 1995; Hu et al. 2000; Bauer et al. 2011).

8.2.4 SUSPENDED MEMBRANES

The most commonly used method for thermal insulation is to fabricate the device to be insulated on a thin suspended membrane (see Figure 8.1). This is an effective solution, as the low thickness (generally ~1 μm) of the membrane as well as the length and reduced width of the legs suspending the membrane ensure a high thermal resistance. Indeed, these structures have been used in

---

* http://www.microchem.com/pdf/SU-8-table-of-properties.pdf.

commercial devices, such as gas sensors and bolometers (Wood 1993). However, these structures are fragile and their fabrication is complex, with the release process being particularly delicate. Stress in the membrane should also be tuned to avoid its buckling. Finally, integrating suspended structures into a device is a considerable design constraint, and is not suitable for certain, more complex, devices such as the Rankine-cycle microturbine designed by Liamini et al. (2011).

8.3 POROUS SILICON AS A THERMAL INSULATION MATERIAL

As we have seen, several thermal insulation techniques exist. While they may be suitable for certain applications, they all have certain drawbacks. PSi has also been proposed as a material for thermal insulation (Lysenko et al. 2002), and it has several advantages, which make it a potentially interesting material for this application.

As shown in Chapter 9 in *Porous Silicon: Formation and Properties*, the thermal conductivity of PSi is 2–3 orders of magnitude lower than that of bulk crystalline silicon, which has a value of 156 W/(m·K). In other words, by making silicon porous, its thermal conductivity can be reduced to levels comparable to or lower than that of silicon oxide (approximately 1.4 W/[m·K]). This means PSi can be used for thermal insulation in MEMS (Lysenko et al. 2002).

PSi has a few major advantages over the thermal insulation techniques presented earlier. It is a low-cost technique, it is more robust and mechanically resistant than thin suspended membranes, and 100-µm-thick layers are relatively easy to fabricate, which provides a high thermal resistance. However, its main advantage resides in the fact that it is a silicon-based technology. Using a mask (Defforge et al. 2012), the geometry of the PSi areas can be controlled and fabricated only where thermal insulation is required. In this way, the rest of the substrate can be kept crystalline and standard silicon microfabrication processes can be used, which greatly facilitates its integration into devices.

8.3.1 POROUS SILICON MORPHOLOGY

PSi can be fabricated in a wide variety of morphologies and pore sizes (Lehmann et al. 2000), and depending on the latter parameter, can be classified as nano-, meso-, or macroporous. The characteristic dimension of nano- and mesoporous silicon is on the order of ~10 nm, which is lower than the phonon mean free path in silicon (49 nm) (Chen 1996). This leads to phonon scattering effects, which explains the strong reduction in thermal conductivity for these morphologies. In contrast, the structures in macroporous silicon are larger and there are no phonon scattering effects, so its thermal conductivity is higher.

Although nanoporous silicon has the lowest thermal conductivity (Lysenko et al. 2000b), it is not necessarily the best material for thermal insulation in a device. Indeed, for use in a device, mechanical resistance during fabrication and throughout the device's lifetime must also be considered. The mechanical resistance of nanoporous silicon is poor due to the small size of its crystallites. When both thermal and mechanical properties are taken into account, mesoporous silicon is a good choice, as it offers a good compromise, having better mechanical resistance than nanoporous silicon, while offering a significantly lower thermal conductivity than macroporous silicon (Lysenko et al. 1999). Nevertheless, all three morphologies have been used in devices, as we will see in the following paragraphs.

8.3.2 INTEGRATION STRATEGIES

8.3.2.1 PRETREATMENTS

As shown in Chapter 9 in *Porous Silicon: Formation and Properties*, certain pretreatments can be used to further reduce the thermal conductivity of PSi. The most common treatment is low-temperature partial oxidation, generally at 300°C for 1 h, which creates a very thin layer of silicon oxide on the surface of the PSi nanostructures. This reduces the thermal conductivity of PSi (Lysenko et al. 1999), and has the added advantage of stabilizing its structure so that it can withstand temperatures up to 800°C in inert environments (Herino et al. 1984). A similar treatment is nitridation (Maccagnani et al. 1999; Amato et al. 2000). Heavy ion irradiation has also been used to reduce the thermal conductivity of PSi, by a factor of three (Newby et al. 2013).

8.3.2.2 GEOMETRY

PSi can be integrated in a wafer using a variety of geometries. The most basic is simply to porosify the whole surface of the silicon chip or substrate (Roussel et al. 1999), but this prevents the monolithic integration of electronic circuitry on the same chip, and makes ulterior microfabrication or wafer bonding more complicated.

To increase freedom of design and fabrication on the areas of the chip where thermal insulation is not required, the best approach is to fabricate the PSi layer using a mask to define the geometry of the porous areas (Kaltsas and Nassiopoulou 1999). Many masking techniques and materials have been described in the literature, including a poly-silicon/SiO_2 bilayer, silicon-rich nonstoechiometric silicon nitride, doping (Kaltsas and Nassiopoulou 1999), or fluoropolymer (Defforge et al. 2012). This leads to the formation of a PSi island, embedded within the crystalline silicon substrate. This approach is ideal, as it uses the same starting material as the majority of microelectronic and MEMS devices, and by porosifying the substrate where required, it can be transformed from a material with a high thermal conductivity to a good thermal insulator.

As shown in Section 8.2.1, the effectiveness of thermal insulation provided by a PSi layer increases with its thickness. Indeed, the ideal geometry from this point of view would be to porosify the whole thickness of a substrate (Roussel et al. 1999). This is difficult, however, as PSi layers become increasingly nonuniform as their thickness increases (Thönissen et al. 1996), but can be achieved using adequate etch-stops to allow electrolyte regeneration (Billat et al. 1997). This can be combined with double-sided porosification to further increase uniformity of the PSi layers (Lysenko et al. 2000a). In this approach, silicon wafers are porosified alternately on their front and back side. While this significantly improves the uniformity of the PSi layers, a thin crystalline silicon layer remains at the center of the wafer, at the interface between the two PSi layers (Lysenko et al. 2000a; Populaire 2005; Lucklum et al. 2014a). This remaining layer acts as a thermal short-circuit, and is detrimental to the performance of the thermal insulation layer (Lucklum et al. 2014b). Recently, the process has been optimized, and 350-µm-thick PSi layers have been obtained with no residual crystalline silicon (Lucklum et al. 2014b). In any case, layers of mesoporous silicon with a thickness of about 100 µm are easily fabricated (Populaire et al. 2003), and numerical simulations should be conducted in order to determine the minimum thickness required for a given application.

In Section 8.2.4, we showed the advantages of suspended membranes as thermal insulation structures, as well as the difficulties regarding their fabrication. Some authors have successfully fabricated suspended membranes constituted of PSi. While the problems encountered when fabricating membranes are the same for PSi membranes, thanks to its low thermal conductivity and the fact that thick porous layers can easily be formed, thicker membranes can be used (several tens of micrometers), rather than the thin (~1 µm) membranes traditionally used with Si_3N_4 or SiO_2, thus improving mechanical performance. Therefore, if this type of structure is compatible with the device, it can be a good solution as it provides very good thermal insulation performance. Various membrane structures have been fabricated in this way, notably 25- to 30-µm-thick nitrided mesoporous closed membranes (Maccagnani et al. 1999), 4-µm-thick suspended membranes (Tsamis et al. 2003b), illustrated in Figure 8.2a, and a membrane formed over a cavity created by electropolishing (Pagonis et al. 2004b) (Figure 8.2b).

In the geometries described thus far, the elements requiring insulation are fabricated directly on top of the insulating layer. However, the insulating layer can also be formed around the device, thus allowing greater freedom in fabricating the device, and avoiding the problems sometimes caused by PSi, such as swelling or worsened surface state. However, to avoid a thermal short-circuit with the rest of the substrate, the porous layer should be through-wafer. As explained two paragraphs earlier, this implies a very thick layer, or the use of a membrane structure. The latter solution has the additional advantage of reducing the volume of crystalline silicon within the insulating ring, and therefore the thermal inertia of the device. This type of geometry has been used with macroporous silicon (Splinter et al. 2002) (see Figure 8.3), as well as fully oxidized PSi (Dominguez et al. 1993; Ye et al. 2005).

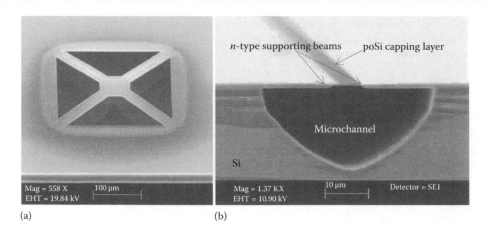

(a)   (b)

FIGURE 8.2 Suspended PSi membranes: (a) A 4-μm-thick PSi suspended membrane, and (b) a membrane over a cavity. ([a] Tsamis C. et al.: *Phys. Stat. Sol.* (a). 2003. 197(2). 539. Copyright Wiley-VCH Verlag GmbH & Co. KGaA. Reproduced with permission. [b] Pagonis D.N.N. et al.: *Phys. Stat. Sol.* (a). 2007. 204(5). 1474–1479. Copyright Wiley-VCH Verlag GmbH & Co. KGaA. Reproduced with permission.)

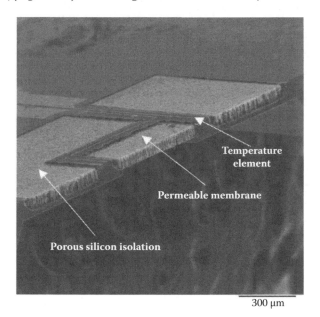

FIGURE 8.3 Heater and macroporous silicon membrane, surrounded by macroporous-based thermal insulation. (Reprinted from *Sens. Actuators B* 83, Splinter A. et al., 169, Copyright 2002, with permission from Elsevier.)

8.3.2.3 INDIRECT USES OF POROUS SILICON FOR THERMAL INSULATION

This chapter is mainly focused on the use of PSi as the thermal insulation material. However, PSi is used as an intermediate step in two other thermal insulation strategies, whose popularity justify their inclusion in this chapter.

The first is the use of PSi as a sacrificial layer for bulk micromachining of silicon. Nano- and mesoporous silicon layers can easily and rapidly be dissolved in alkaline solutions, such as KOH (Lai et al. 2011). This is a low-cost alternative to deep reactive ion etching (DRIE), and using the appropriate doping can even be used to fabricate suspended crystalline silicon membranes (Zeitschel et al. 1999). This technique has been applied to fabricated thermally isolated suspended structures, such as micro-hotplates (Friedberger et al. 2001; Fürjes et al. 2004), gas sensors (Bársony et al. 2004), and bolometers (Lang et al. 1994). The use of PSi for the fabrication of suspended membranes is discussed in detail in Chapter 11 in this book.

The second technique is the oxidation of PSi layers to form thick SiO_2 layers, sometimes referred to as full isolation by porous oxidized silicon (FIPOS). Due to its high specific surface and small structure size, PSi can be completely oxidized relatively easily; for instance, a 51% porosity layer can be completely oxidized after 50 min of wet oxidation at 800°C (Yon et al. 1987). This is generally followed by a densification anneal at temperatures higher than 1000°C (Yon et al. 1987). In this way, silicon oxide layers up to 100 µm thick can be obtained, which is significantly higher than the standard oxidation methods described earlier. This technique has been used with nano- and mesoporous silicon, as well as macroporous silicon (Barillaro et al. 2003). Stress remains an issue with this technique (Imai and Unno 1984), and when macroporous silicon is the starting material, the oxide surface is not flat and retains traces of the initial macropores (Barillaro et al. 2003).

SiO_2 layers obtained through the oxidation of PSi have been used for thermal insulation in numerous devices, including flow sensors (Tabata 1986; Tabata et al. 1987; Dominguez et al. 1993), gas sensors (Maccagnani et al. 1998), microreactors (Ye et al. 2005), and microfluidic devices (Sukas et al. 2013).

8.4 REVIEW OF DEVICES FABRICATED WITH POROUS SILICON AS A THERMAL INSULATION MATERIAL

PSi has been used as a material for thermal insulation in several devices, which we will now review. As there are several examples for the most common types of thermal MEMS, we have sorted these by type of device. The operation principles for some of these devices have already been explained in Section 8.1, so they will not be repeated here.

8.4.1 FLOW SENSORS

Among the devices fabricated using PSi as the thermal insulation material, thermal flow sensors are the most common. In its most basic form, a thermal flow sensor requires a heating element and a sensing element. Ideally, these two elements should be thermally isolated, so that no heat conduction occurs between the two, and the only heat transfer is via the fluid being measured. In this way, the temperature change at the sensing element can be related to fluid velocity (Kuo et al. 2012).

Nassiopoulou's group has fabricated several thermal flow sensors based on PSi, using different architectures and configurations for the thermal insulation. Their first device used a heater and two thermopiles, one upstream and the other downstream. The hot contacts of the thermopiles and the heater were fabricated on a PSi layer, embedded within a crystalline Si substrate, and the cold contacts were placed on the silicon substrate. In this way, the PSi was used to isolate the cold contact from the heater, and to ensure a thermal gradient between the hot and cold contacts of the thermopile. Figure 8.4a shows a top view of this device. The porous layers for this device were 40 µm thick, with 75% porosity, the initial substrates were p-type, and the PSi was partly oxidized (Kaltsas and Nassiopoulou 1999; Nassiopoulou and Kaltsas 2000). Extensive characterization of this device was carried out (Kaltsas et al. 2002), and demonstrated good performance of the device, with notably a very fast response time of 1.5 ms.

This sensor was used for several different applications. Specific packages were designed to use the flow sensor in medical equipment for respiratory control (Kaltsas and Nassiopoulou 2004) and as an airflow meter for automobile engines (Hourdakis and Nassiopoulou 2014). In the latter case, the sensor was fully packaged and tested *in situ* in the inlet of a truck engine. During testing, a commercial sensor was used as a reference, and the PSi sensor displayed the same signal variations as the reference, as well as linear behavior over a wide range of flow values.

The authors also used their flow sensor as a thermal accelerometer. The most common type of accelerometer uses an inertial mass suspended by springs. In contrast, in a thermal accelerometer, a fluid is in contact with the sensor's heater. Acceleration causes displacement of the fluid, and convection causes variation in the heat exchange between the heater and the hot contact of the thermopile (Goustouridis et al. 2007). The sensor can be used in air or can be packaged with oil, and the fluid can be sealed or with a free surface. The type of packaging, as well as the

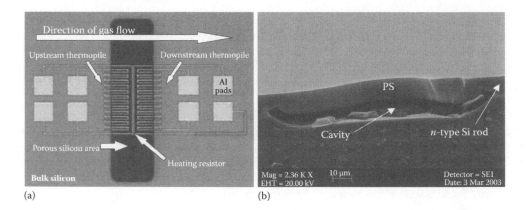

(a) (b)

FIGURE 8.4 Flow sensors using PSi as a thermal insulation material. (a) A top view of a thermal sensor on top of a 40-µm-thick PSi island and (b) a PSi membrane over a cavity destined to be used in a flow sensor. ([a] Nassiopoulou A.G. and Kaltsas G.: *Phys. Stat. Sol.* (a). 2000. 182(1), 307 Copyright Wiley-VCH Verlag GmbH & Co. KGaA. Reproduced with permission. [b] From Pagonis D.N. et al., *J. Micromech. Microeng.* 14(6), 793, 2004. Copyright 2004: IOP Publishing. With permission.)

properties of the fluid (viscosity, thermal conductivity, and heat transfer coefficient) can have a huge effect on the sensitivity, linearity, and frequency response of the device (Kaltsas et al. 2006). The accelerometer performs well at low frequency, with a cut-off frequency between 12 and 70 Hz depending on the fluid used, and the fluid can also be chosen in order to obtain a linear response versus acceleration (Goustouridis et al. 2007). The main advantage of this type of sensor, compared to spring-mass accelerometers, is that they have no moving parts, which makes them more resistant to shocks and reduces aging.

The second flow sensor developed by Nassiopoulou's group uses the same architecture for the flow sensor heater and two thermopiles, but uses an improved thermal insulation structure. This structure is a 10-µm-thick PSi membrane (75% porosity) suspended over a 10-µm air gap (Pagonis et al. 2004a). This structure is fabricated by carrying out an electropolishing step just after the porosification process. In order to prevent the membrane from detaching itself from the substrate, the authors used a p-type substrate and fabricated an n-type doped ring around the membrane. This n-type ring was not etched during the electropolishing step, and formed a crystalline silicon support for the porous membrane (Pagonis et al. 2004b). Figure 8.4b shows a cross-section of such a structure. This structure is more efficient than the one presented in the previous paragraphs, as it requires 27% less power to achieve the same sensitivity (Pagonis et al. 2004a).

The authors used the same PSi membrane-over-cavity structure to fabricate microchannels in silicon substrates, and subsequently fabricated a flow sensor on top of the structure to measure flow in the channel. The authors have published extensive details on the fabrication of the channels and sensor and have simulated its performance (Kaltsas et al. 2003; Pagonis et al. 2007), but no results showing the working device have been published to our knowledge.

A different group worked on a similar type of thermal flow sensor, but used in biological applications for measuring blood flow. Two types of sensors were studied. The first was a small size sensor, which was designed to be integrated into a needle and used as an implantable sensor. This sensor was designed (Lysenko et al. 1998) and two successive iterations were fabricated (Roussel et al. 1999; Périchon et al. 2000; Périchon 2001), using a 100-µm-thick layer of 56% porosity PSi. A larger non-invasive surface-sensor was also fabricated (Périchon 2001), again using 100-µm layers of PSi, but two porosities were tested (56% and 70%). For both types of sensors, reference structures were fabricated directly on crystalline silicon, with no PSi thermal insulation, and in both cases an increased efficiency and sensitivity due to the PSi isolation was demonstrated (Périchon 2001).

8.4.2 GAS SENSORS

PSi has received much attention as a material for gas sensors (Ozdemir and Gole 2007; Korotcenkov and Cho 2010). However, most of this interest has been focused on using PSi as the actual sensing material. Indeed, several of its properties change when exposed to certain gases (e.g., its capacitance changes when exposed to humidity), and its very high specific surface is a very useful property for increasing the sensitivity of the sensor. This application is analyzed in detail in Chapter 1 of this book. As mentioned in Section 8.1.1, certain types of gas sensors function at relatively high temperatures, and thermal insulation of the heater reduces the power required to reach the target temperature. However, very little work has been done on using the low thermal conductivity of PSi for the thermal insulation of gas sensors.

Nevertheless, a few articles have been published on the subject. Maccagnani et al. (1999) developed a process for fabricating suspended membranes made of PSi, which had been passivated using a nitridation process. These structures were intended to be used for the fabrication of thermally insulated gas sensors. While 20- to 30-µm-thick membranes with porosities of 55% and 75% were successfully fabricated, the actual devices were not fabricated on the membranes.

The same group published work on a gas sensor using macroporous silicon permeated with Sn-V mixed oxides (Angelucci et al. 1997, 1999). In this case, the PSi did not serve as the sensing material, as the Sn-V oxide was used for this, but was used instead as a support for the sensing material, so the sensor still benefitted from the high specific surface of PSi. The device used 30-µm-thick membranes formed of partially oxidized macroporous silicon. The final device was supposed to have both a sensing element and a heater on the membrane, but the implementation presented used an external heater, so the authors were not able to evaluate the gain in power consumption provided by the PSi insulation. Nevertheless, the sensor achieved sub-ppm sensitivity to benzene.

Fürjes (2003) used 20-µm-thick porous layers both as a sensing material for a humidity sensor and for thermally insulating the heating element, but did not study the performance gain due to the PSi insulation.

Bársony et al. (2004) directly compared the performance of thermal insulation using a PSi island and an airgap to isolate a heating element used in a pellistor-type gas sensor. No information is given on the dimensions of the airgap and the PSi layer, but they are the same size, as the airgap is fabricated simply by using the porous island as a sacrificial layer. The silicon nitride layer under the heater has a size of $100 \times 100 \times 1\ \mu m^3$. The authors show that the suspended membrane has significantly better performance; at a constant heating power of 18 mW, the heater on a suspended membrane reaches a maximum temperature of 507°C, whereas the heater on PSi only reaches a temperature of 226°C. The device insulated by PSi also has inferior dynamic properties, with a time constant of 1.8 ms instead of 1.1 ms for the airgap insulation.

Splinter et al. (2002) developed a gas precombustor to be used as a preprocessing stage for gas sensors. The aim of this device is to convert a detectable gas into a nondetectable gas (e.g., CO into CO_2) in order to improve the sensor's sensitivity to other gases and to help with gas differentiation. The authors designed their device around a macroporous silicon membrane. The membrane had two roles: it served as insulation for the heater necessary to the reaction, and was impregnated with a catalytic material (in this case palladium). A SEM picture of this structure is shown in Figure 8.3.

8.4.3 BOLOMETERS

Bolometer pixels were fabricated by depositing and patterning 50-nm niobium films on nanoporous silicon layers (Boarino et al. 1999; Monticone et al. 1999). Two types of PSi layers were used: 65% porosity/48-µm thickness, and 70% porosity/8-µm thickness, but no other information is provided on the device geometry. Commercial microbolometers generally provide noise equivalent temperature difference (NETD) as their performance specification, whereas the authors measure the responsivity and noise equivalent power (NEP) of their devices,[*] which makes any comparison difficult. However, they measure a time constant of 4 ms for their PSi devices, which compares favorably with commercial devices.

---

[*] http://www.ulis-ir.com/index.php?infrared-detector=products; http://www.flir.com/cvs/cores/view/?id=64979.

8.4.4 MICRO-HOTPLATES

Micro-hotplates are simply micro-heaters, and are a generic building block that can be used in any type of thermal MEMS that require a heater. This is the case for many of the devices mentioned at the beginning of this chapter. Micro-hotplates specifically applied to gas sensors are covered in Chapter 11 of this book.

Tsamis et al. (2003a) made microheaters on suspended PSi membranes (4 μm thick, 60 μm^2 area). They developed a novel fabrication process, where the PSi is partially oxidized. This passivates the PSi so that it can withstand the subsequent SF$_6$ plasma etch used to etch the crystalline silicon underneath it and release the membrane (Tsamis et al. 2003b). An example of this structure is shown in Figure 8.5. The authors estimate that the hotplates can reach temperatures of almost 600°C, for an input power of about 30 mW.

Lucklum et al. (2014b) fabricated hot plates using a 10-mm^2 molybdenum heater on 350-μm thick mesoporous silicon layers with 50% porosity (see Figure 8.6). The initial substrates were 550 μm thick, and were thinned by electropolishing prior to porosification. This process enabled the fabrication of thick PSi layers, with no residual crystalline silicon in the center of the wafer. The authors also fabricated heaters on 300-μm-thick silica glass substrates and 550-μm silicon wafers in order to compare their heating efficiency. The power efficiencies measured for PSi, glass, and silicon were, respectively, 0.4 K/mW, 0.37 K/mW, and 0.1 K/mW. Apart from improved efficiency, the authors demonstrate two other advantages of PSi compared to silica glass. PSi can be operated reversibly up to 400°C, while the silica's glass transition limits it to temperatures of 350°C. Furthermore, although the PSi and glass substrates have similar thickness, the PSi has a lower heat capacity due to its porosity. Therefore, the heater on PSi has a faster response than on glass, with a 10 s rise time (10–90%) compared to 13.1 s for glass.

8.4.5 THERMOACOUSTIC SOUND SOURCES

PSi has been used to make sound sources based on the thermoacoustic effect. Applying an AC current to a conductor will induce time-dependent temperature variations, through Joule heating.

FIGURE 8.5 SEM image of a micro-hotplate: a heater is fabricated on a 4-μm-thick porous membrane suspended over a cavity. (Reprinted from *Sens. Actuators B* 95, Tsamis C. et al., 78, Copyright 2003, with permission from Elsevier.)

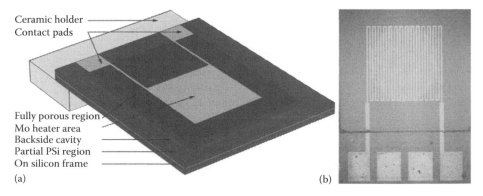

Ceramic holder
Contact pads

Fully porous region
Mo heater area
Backside cavity
Partial PSi region
On silicon frame

(a)　　　　　　　　　　　　　　　　　　　(b)

FIGURE 8.6 Design of a microheater (a) and micrograph of the fabricated heater (b). (Reprinted from Lucklum F. et al., *Sens. Actuators A* 213, 35, Copyright 2014, with permission from Elsevier.)

If this conductor is exposed to air, then some of the heat will be transferred to the surrounding air, in which there will also be temperature oscillations. These will in turn cause pressure oscillations or, in other words, sound waves (Niskanen et al. 2009; Venkatasubramanian 2010).

The idea for using the thermoacoustic effect to make a sound source, or "thermophone," was proposed almost 100 years ago (Arnold and Crandall 1917). However, the implementation of this device was prevented by the lack of adequate materials. For such a device to produce sound efficiently, the conductor must have a low heat capacity and, most importantly, the heat produced must be predominantly transferred to the surrounding air (Niskanen et al. 2009; Venkatasubramanian 2010).

The low thermal conductivity of PSi was used by Shinoda et al. (1999) to make a working thermophone. The authors fabricated the device by patterning 30-nm-thick aluminum layers on top of 70% porosity nanoporous silicon islands embedded in crystalline silicon substrates. The required thickness of the porous layer depends on the frequencies that will be used; a lower frequency has a higher diffusion length in the PSi layer, so thicknesses up to 95 μm were used to reach frequencies as low as 100 Hz (Koshida et al. 2013).

The PSi-based thermophone is able to produce both audible sound and ultrasounds, and can span frequencies from 300 Hz to 40 kHz (Koshida et al. 2013). It is particularly suited to producing ultrasounds, and these devices have been applied to biological research to reproduce the calls of mouse-pups (Kihara et al. 2006). Further detail on PSi-based ultrasound emitters can be found in Chapter 10 of this book.

8.4.6 HEAT FLUX SENSOR AND THERMOELECTRIC GENERATOR

These two devices have been fabricated using PSi, and although they are used for different applications, their designs and operation principles are essentially the same.

Ziouche et al. (2010) used periodic mesoporous silicon trenches in a crystalline substrate to fabricate heat-flux sensors. The heat fluxes crossing the device are preferentially deviated into the higher conductivity areas, that is, the crystalline silicon. This generates a periodic temperature difference at the surface of the substrate, which is measured using microfabricated thermopiles. The mesoporous layers were 130 μm thick and the PSi was partially oxidized. The sensor had a sensitivity of 6.6 μV/(W/m^2) and a time constant between 8 and 12 ms. This is comparable to the specifications of several commercial sensors cited by the authors (Ziouche et al. 2010).

Hourdakis and Nassiopoulou (2013) use the same architecture, with periodic PSi trenches and thermopiles, to fabricate a thermoelectric generator. The device is very similar to the heat flux sensor presented in the previous paragraph, and the main difference is that the hot contacts are on the PSi. Photoresist is used to insulate the cold contacts and a thick aluminum layer is used to direct heat to the hot contacts. An image of the device, without the photoresist and final aluminum layer is shown in Figure 8.7. The authors use 60% porosity mesoporous silicon, and two different thicknesses are used (25 and 50 μm). As expected, the thicker layer provides better

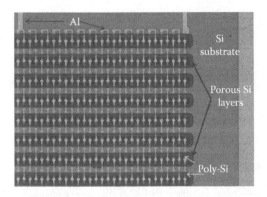

FIGURE 8.7 Top view of a thermoelectric generator, based on thermopiles patterned on periodic PSi trenches embedded in a crystalline silicon substrate. (From Hourdakis E. and Nassiopoulou A.G., *Sensors* 13(10), 13596, 2013. Published by MDPI as open access. With permission.)

performance, with an output of 0.39 $\mu W/cm^2$ for a 10 K temperature difference. While this is not very high, the device fabrication is extremely simple, and thus low-cost.

8.5 DISCUSSION AND CONCLUSIONS

Effective solutions for thermal insulation are essential in thermal MEMS to decrease power consumption, but also to reduce heating time and separate hot areas from areas that need to be kept cool. Numerous solutions have been developed based on different materials, but these have various disadvantages. Among these, the method that provides the best thermal insulation performance is suspended membranes. However, these structures have lower mechanical resistance, fabrication of these structures is complex, with potentially lower yield, and they limit design flexibility.

PSi is an interesting alternative. Its nanoscale structures cause phonon scattering, which can lower its thermal conductivity up to 2–3 orders of magnitude compared to crystalline silicon, which makes it a good thermal insulator. In addition, it has the major advantage of being a silicon-based technology, and therefore compatible with standard microfabrication processes. Furthermore, by using a mask during porosification, the geometry of the porous areas can be controlled. Finally, it is generally used in thick layers, so it is more robust than thin suspended membranes. PSi is effectively a method for locally transforming crystalline silicon into a thermal insulator. Indeed, PSi has successfully been integrated into several devices, using several different geometries and pretreatments. The devices where PSi has been integrated include thermal flow sensors, gas sensors, and thermoacoustic sound sources.

While a couple of papers compare the performance of their device insulated with PSi to other thermal insulation solutions, a rigorous and systematic comparison is lacking. The two papers that do carry out this comparison find that a fully PSi substrate offers better insulation than a silica glass substrate of similar thickness, as well as a faster response time thanks to the lower heat capacity of PSi. As expected, they also show that a heater on a membrane suspended over an air-gap has a better heating efficiency than the same heater on top of a porous layer with a thickness of several tens of micrometers. However, these comparisons are carried out only for two specific cases, and the thermal performance of a device is highly dependent on its dimensions and the geometry of the insulation layer.

Even if PSi does not provide as effective thermal insulation as a suspended membrane, it has other advantages. It is a very simple, low-cost process, easy to integrate, and PSi layers are more robust than suspended membranes. These properties mean that the trade-off on thermal conductivity, compared to suspended membranes, can definitely be worthwhile for certain applications.

REFERENCES

Ali, S.Z., Udrea, F., Milne, W.I., and Gardner, J.W. (2008). Tungsten-based SOI microhotplates for smart gas sensors. *J. Microelectromech. Syst.* **17**(6), 1408–1417.

Amato, G., Angelucci, R., Benedetto, G. et al. (2000). Thermal characterisation of porous silicon membranes. *J. Porous Mater.* **7**(1), 183–186.

Angelucci, R., Poggi, A., Dori, L. et al. (1997). Porous silicon layer permeated with Sn–V mixed oxides for hydrocarbon sensor fabrication. *Thin Solid Films* **297**(1–2), 43–47.

Angelucci, R., Poggi, A., Dori, L. et al. (1999). Permeated porous silicon for hydrocarbon sensor fabrication. *Sens. Actuators A* **74**(1–3), 95–99.

Arnold, H.D. and Crandall I.B. (1917). The thermophone as a precision source of sound. *Phys. Rev.* **10**(1), 22–38.

Barillaro, G., Diligenti, A., Nannini, A., and Pennelli, G. (2003). A thick silicon dioxide fabrication process based on electrochemical trenching of silicon. *Sens. Actuators A* **107**(3), 279–284.

Bársony, I., Fürjes, P., Ádám, M. et al. (2004). Thermal response of microfilament heaters in gas sensing. *Sens. Actuators B* **103**(1–2), 442–447.

Bauer, M.L., Bauer, C.M., Fish, M.C. et al. (2011). Thin-film aerogel thermal conductivity measurements via 3ω. *J. Non. Cryst. Solids* **357**(15), 2960–2965.

Billat, S., Thönissen, M., Arens-Fischer, R., Berger, M.G., Krüger, M., and Lüth, H. (1997). Influence of etch stops on the microstructure of porous silicon layers. *Thin Solid Films* **297**, 22–25.

Boarino, L., Monticone, E., Amato, G. et al. (1999). Design and fabrication of metal bolometers on high porosity silicon layers. *Microelectron. J.* **30**(11), 1149–1154.

Briand, D., Colin, S., Gangadharaiah, A. et al. (2006). Micro-hotplates on polyimide for sensors and actuators. *Sens. Actuators A* **132**(1), 317–324.

Cahill, D.G. (1990). Thermal conductivity measurement from 30 to 750 K: The 3ω method. *Rev. Sci. Instrum.* **61**(2), 802–808.

Chen, G. (1996). Nonlocal and nonequilibrium heat conduction in the vicinity of nanoparticles. *J. Heat Transfer* **118**, 539–545.

Chen, K.-S., Chen, J.-Y., and Lin, S.-Y. (2002). Fracture analysis of thick plasma-enhanced chemical vapor deposited oxide films for improving the structural integrity of power MEMS. *J. Micromech. Microeng.* **12**(5), 714–722.

Cook, K.A. and Sastry, A.M. (2005). Influence of scaling effects on designing for power efficiency of a micropreconcentrator. *J. Vac. Sci. Technol. B* **23**(2), 599–611.

Deal, B.E. and Grove, A.S. (1965). General relationship for the thermal oxidation of silicon. *J. Appl. Phys.* **36**(12), 3770–3778.

Defforge, T., Capelle, M., Tran-Van, F., and Gautier, G. (2012). Plasma-deposited fluoropolymer film mask for local porous silicon formation. *Nanoscale Res. Lett.* **7**(1), 344.

Dominguez, D., Bonvalot, B., Chau, M.T., and Suski, J. (1993). Fabrication and characterization of a thermal flow sensor based on porous silicon technology. *J. Micromech. Microeng.* **3**, 247–249.

Franssila, S. (2004). *Introduction to Microfabrication*. John Wiley & Sons, Chichester.

Friedberger, A., Kreisl, P., Müller, G., and Kassing, R. (2001). A versatile and modularizable micromachining process for the fabrication of thermal microsensors and microactuators. *J. Micromech. Microeng.* **11**, 623–629.

Fürjes, P. (2003). Porous silicon-based humidity sensor with interdigital electrodes and internal heaters. *Sens. Actuators B* **95**(1–3), 140–144.

Fürjes, P., Dücső, C., Ádám, M., Zettner, J., and Bársony, I. (2004). Thermal characterisation of micro-hotplates used in sensor structures. *Superlattices Microstruct.* **35**(3–6), 455–464.

Glassbrenner, C.J. and Slack, G.A. (1964). Thermal conductivity of silicon and germanium from 3 K to the melting point. *Phys. Rev.* **134**(4A), A1058–A1069.

Goustouridis, D., Kaltsas, G., and Nassiopoulou, A.G. (2007). A silicon thermal accelerometer without solid proof mass using porous silicon thermal isolation. *IEEE Sens. J.* **7**(7), 983–989.

Haque, R.M. and Wise, K.D. (2013). A glass-in-silicon reflow process for three-dimensional microsystems. *J. Microelectromech. Syst.* **22**(6), 1470–1477.

Harder, T.A., Yao, T.J., He, Q., Shih, C.Y., and Tai, Y.C. (2002). Residual stress in thin-film parylene-C. In: *Proceedings of the Fifteenth IEEE International Conference on Micro Electro Mechanical Systems*, pp. 435–438.

Herino, R., Perio, A., Barla, K., and Bomchil, G. (1984). Microstructure of porous silicon and its evolution with temperature. *Mater. Lett.* **2**(6), 519–523.

Hourdakis, E. and Nassiopoulou, A.G. (2013). A thermoelectric generator using porous Si thermal isolation. *Sensors* **13**(10), 13596–13608.

Hourdakis, E. and Nassiopoulou, A.G. (2014). Single photoresist masking for local porous Si formation. *J. Micromech. Microeng.* **24**(11), 117002.

Hrubesh, L.W. and Poco, J.F. (1995). Thin aerogel films for optical, thermal, acoustic and electronic applications. *J. Non. Cryst. Solids* **188**(1–2), 46–53.

Hu, C., Morgen, M., Ho P.S. et al. (2000). Thermal conductivity study of porous low-k dielectric materials. *Appl. Phys. Lett.* **77**(1), 145–147.

Imai, K. and Unno, H. (1984). FIPOS (full isolation by porous oxidized silicon) technology and its application to LSI's. *IEEE Trans. Electron Devices* **31**(3), 297–302.

Jain, A. and Goodson, K.E. (2011). Thermal microdevices for biological and biomedical applications. *J. Therm. Biol.* **36**(4), 209–218.

Jensen, K.F. (2006). Silicon-based microchemical systems: Characteristics and applications. *MRS Bull.* **31**(02), 101–107.

Kaltsas, G. and Nassiopoulou, A.G. (1999). Novel C-MOS compatible monolithic silicon gas flow sensor with porous silicon thermal isolation. *Sens. Actuators A* **76**(1–3), 133–138.

Kaltsas, G. and Nassiopoulou, A.G. (2004). Gas flow meter for application in medical equipment for respiratory control: Study of the housing. *Sens. Actuators A* **110**(1–3), 413–422.

Kaltsas, G., Nassiopoulos, A.A., and Nassiopoulou, A.G. (2002). Characterization of a silicon thermal gas-flow sensor with porous silicon thermal isolation. *IEEE Sens. J.* **2**(5), 463–475.

Kaltsas, G., Pagonis, D.N., and Nassiopoulou, G.G. (2003). Planar CMOS compatible process for the fabrication of buried microchannels in silicon, using porous-silicon technology. *J. Microelectromech. Syst.* **12**(6), 863–872.

Kaltsas, G., Goustouridis, D., and Nassiopoulou, A.G. (2006). A thermal convective accelerometer system based on a silicon sensor-study and packaging. *Sens. Actuators A* **132**(1), 147–153.

Kihara, T., Harada, T., Kato, M. et al. (2006). Reproduction of mouse-pup ultrasonic vocalizations by nano-crystalline silicon thermoacoustic emitter. *Appl. Phys. Lett.* **88**(4), 043902.

Korotcenkov, G. and Cho, B.K. (2010). Porous semiconductors: Advanced material for gas sensor applications. *Crit. Rev. Solid State Mater. Sci.* **35**(1), 1–37.

Koshida, N., Hippo, D., Mori, M., Yanazawa, H., Shinoda, H., and Shimada, T. (2013). Characteristics of thermally induced acoustic emission from nanoporous silicon device under full digital operation. *Appl. Phys. Lett.* **102**(12), 123504.

Kuo, J.T.W., Yu, L., and Meng, E. (2012). Micromachined thermal flow sensors—A review. *Micromachines* **3**(4), 550–573.

Lai, M., Parish, G., Liu, Y., Dell, J.M., and Keating, A.J. (2011). Development of an alkaline-compatible porous-silicon photolithographic process. *J. Microelectromech. Syst.* **20**(2), 418–423.

Lang, W., Steiner, P., Schaber, U., and Richter, A. (1994). A thin film bolometer using porous silicon technology. *Sens. Actuators A* **43**, 185–187.

Lehmann, V., Stengl, R., and Luigart, A. (2000). On the morphology and the electrochemical formation mechanism of mesoporous silicon. *Mater. Sci. Eng. B* **69–70**, 11–22.

Lei, Y., Wang, W., Yu, H. et al. (2009). A parylene-filled-trench technique for thermal isolation in silicon-based microdevices. *J. Micromech. Microeng.* **19**(3), 035013.

Liamini, M., Shahriar, H., Vengallatore, S., and Fréchette L, .G. (2011). Design methodology for a Rankine micro-turbine: Thermomechanical analysis and material selection. *J. Microelectromech. Syst.* **20**(1), 339–351.

Lucklum, F., Schwaiger, A., and Jakoby, B. (2014a). Development and investigation of thermal devices on fully porous silicon substrates. *IEEE Sens. J.* **14**(4), 992–997.

Lucklum, F., Schwaiger, A., and Jakoby, B. (2014b). Highly insulating, fully porous silicon substrates for high temperature micro-hotplates. *Sens. Actuators A* **213**, 35–42.

Lysenko, V., Roussel, P., Delhomme, G. et al. (1998). Oxidized porous silicon: A new approach in support thermal isolation of thermopile-based biosensors. *Sens. Actuators A* **67**(1–3), 205–210.

Lysenko, V., Perichon, S., Remaki, B., Barbier, D., and Champagnon, B. (1999). Thermal conductivity of thick meso-porous silicon layers by micro-Raman scattering. *J. Appl. Phys.* **86**(12), 6841–6846.

Lysenko, V., Remaki, B., and Barbier, D. (2000a). Double-sided mesoporous silicon formation for thermal insulating applications. *Adv. Mater.* **12**(7), 516–519.

Lysenko, V., Roussel P., Remaki, B., Delhomme, G., Dittmar, A., and Barbier, D. (2000b). Study of nano-porous silicon with low thermal conductivity as thermal insulating material. *J. Porous Mater.* **7**(1–3), 177–182.

Lysenko, V., Perichon, S., Remaki, B., and Barbier, D. (2002). Thermal isolation in microsystems with porous silicon. *Sens. Actuators A* **99**(1–2), 13–24.

Maccagnani, P., Angelucci, R., Pozzi, P. et al. (1998). Thick oxidised porous silicon layer as a thermo-insulating membrane for high-temperature operating thin-and thick-film gas sensors. *Sens. Actuators B* **49**(1–2), 22–29.

Maccagnani, P., Angelucci, R., Pozzi, P. et al. (1999). Thick porous silicon thermo-insulating membranes. *Sensors Mater.* **11**(3), 131–147.

Marre, S. and Jensen, K.F. (2010). Synthesis of micro and nanostructures in microfluidic systems. *Chem. Soc. Rev.* **39**(3), 1183–1202.

Martínez-Cisneros, C.S., Gómez-de Pedro, S., Puyol, M., García-García, J., and Alonso-Chamarro, J. (2012). Design, fabrication and characterization of microreactors for high temperature syntheses. *Chem. Eng. J.* **211–212**, 432–441.

Modafe, A., Ghalichechian,, N., Powers, M., Khbeis, M., and Ghodssi, R. (2005). Embedded benzocyclobutene in silicon: An integrated fabrication process for electrical and thermal isolation in MEMS. *Microelectron. Eng.* **82**(2), 154–167.

Monticone, E., Boarino, L., Lerondel, G., Steni, R., Amato, G., and Lacquaniti, V. (1999). Properties of metal bolometers fabricated on porous silicon. *Appl. Surf. Sci.* **142**(1–4), 267–271.

Nassiopoulou, A.G. and Kaltsas, G. (2000). Porous silicon as an effective material for thermal isolation on bulk crystalline silicon. *Phys. Stat. Sol. (a)* **182**(1), 307–311.

Newby, P.J., Canut, B., Bluet, J.-M. et al. (2013). Amorphization and reduction of thermal conductivity in porous silicon by irradiation with swift heavy ions. *J. Appl. Phys.* **114**(1), 014903.

Niklaus,, F., Vieider, C., and Jakobsen, H. (2008). MEMS-based uncooled infrared bolometer arrays—A review. In: *MEMS/MOEMS Technologies and Applications III.* **6836**, 68360D.

Niskanen, A.O., Hassel, J., Tikander, M., Maijala P., Grönberg, L., and Helistö, P. (2009). Suspended metal wire array as a thermoacoustic sound source. *Appl. Phys. Lett.* **95**(16), 163102.

Ozdemir, S. and Gole, J.L. (2007). The potential of porous silicon gas sensors. *Curr. Opin. Solid State Mater. Sci.* **11**(5–6), 92–100.

Pagonis, D.N., Kaltsas, G., and Nassiopoulou, A.G. (2004a). Fabrication and testing of an integrated thermal flow sensor employing thermal isolation by a porous silicon membrane over an air cavity. *J. Micromech. Microeng.* **14**(6), 793–797.

Pagonis, D.N.N., Nassiopoulou, A.G.G., and Kaltsas, G. (2004b). Porous silicon membranes over cavity for efficient local thermal isolation in Si thermal sensors. *J. Electrochem. Soc.* **151**(8), H174–H179.

Pagonis, D.N.N., Petropoulos, A., Kaltsas, G., Nassiopoulou, A.G.G., and Tserepi, A. (2007). Novel micro-fluidic flow sensor based on a microchannel capped by porous silicon. *Phys. Stat. Sol. (a)* **204**(5), 1474–1479.

Périchon, S. (2001). Technologie et propriétés de transport dans les couches épaisses de silicium poreux: Applications aux microsystèmes thermiques. PhD Thesis, Institut National des Sciences Appliquées de Lyon, INSA, France.

Périchon, S., Roussel, P., Lysenko, V. et al. (2000). Micro-blood flow measurement using thermal conductivity micro-needles: A new CMOS compatible manufacturing process onto porous silicon. In: *Proceedings of 1st Annual International IEEE-EMBS Special Topic Conference on Microtechnologies in Medicine and Biology*, pp. 184–187.

Populaire, C. (2005). Propriétés physiques du silicium poreux: Traitements et applications aux micro-systèmes. PhD Thesis, Institut National des Sciences Appliquées de Lyon, INSA, France.

Populaire, C., Remaki, B., Lysenko, V., Barbier, D., Artmann, H., and Pannek, T. (2003). On mechanical properties of nanostructured meso-porous silicon. *Appl. Phys. Lett.* **83**(7), 1370–1372.

Roussel, P., Lysenko, V., Remaki, B., Delhomme, G., Dittmar, A., and Barbier, D. (1999). Thick oxidised porous silicon layers for the design of a biomedical thermal conductivity microsensor. *Sens. Actuators A* **74**(1–3), 100–103.

Sadler, D.J., Changrani, R., Roberts, P., Chou, C.-F., and Zenhausern, F. (2003). Thermal management of BioMEMS: Temperature control for ceramic-based PCR and DNA detection devices. *IEEE Trans. Components Packag. Technol.* **26**(2), 309–316.

Sharma, S. and Madou, M. (2012). A new approach to gas sensing with nanotechnology. *Philos. Trans. R. Soc. A* **370**(1967), 2448–2473.

Shinoda, H., Nakajima, T., Ueno, K., and Koshida, N. (1999). Thermally induced ultrasonic emission from porous silicon. *Nature* **400** (August), 853–855.

Simon, I., Bârsan, N., Bauer, M., and Weimar, U. (2001). Micromachined metal oxide gas sensors: Opportunities to improve sensor performance. *Sens. Actuators B* **73**(1), 1–26.

Splinter, A., Stürmann, J., Bartels, O., and Benecke, W. (2002). Micro membrane reactor: A flow-through membrane for gas pre-combustion. *Sens. Actuators B* **83**(1–3), 169–174.

Sukas, S., Tiggelaar, R.M., Desmet, G., and Gardeniers, H.J.G.E. (2013). Fabrication of integrated porous glass for microfluidic applications. *Lab Chip* **13**(15), 3061–3069.

Tabata O. (1986). Fast-response silicon flow sensor with an on-chip fluid temperature sensing element. *IEEE Trans. Electron Dev.* **33**(3), 361–365.

Tabata, O., Inagaki, H., and Igarashi, I. (1987). Monolithic pressure-flow sensor. *IEEE Trans. Electron Dev.* **34**(12), 2456–2462.

Thönissen, M., Billat, S., Krüger, M. et al. (1996). Depth inhomogeneity of porous silicon layers. *J. Appl. Phys.* **80**(5), 2990–2993.

Tian, W.-C. and Pang, S.W. (2003). Thick and thermally isolated Si microheaters for microfabricated pre-concentrators. *J. Vac. Sci. Technol. B* **21**(1), 274–279.

Tsamis, C., Nassiopoulou, A.G., and Tserepi, A. (2003a). Thermal properties of suspended porous silicon micro-hotplates for sensor applications. *Sens. Actuators B* **95**(1–3), 78–82.

Tsamis, C., Tserepi, A., and Nassiopoulou, A.G.G. (2003b). Fabrication of suspended porous silicon micro-hotplates for thermal sensor applications. *Phys. Stat. Sol. (a)* **197**(2), 539–543.

Venkatasubramanian, R. (2010). Nanothermal trumpets. *Nature* **463**(7281), 619.

Vilares, R., Hunter, C., Ugarte, I. et al. (2010). Fabrication and testing of a SU-8 thermal flow sensor. *Sens. Actuators B* **147**(2), 411–417.

Voiculescu, I., Zaghloul, M., and Narasimhan, N. (2008). Microfabricated chemical preconcentrators for gas-phase microanalytical detection systems. *Trends Anal. Chem.* **27**(4), 327–343.

Wilcox, D.L., Burdon, J.W., Changrani, R. et al. (2011). Add ceramic "MEMS" to the pallet of microsystems technologies. *MRS Proc.* **687**, B7.1.

Wood, R.A. (1993). High-performance infrared thermal imaging with monolithic silicon focal planes operating at room temperature. In: *Proceedings of Electron Devices Meeting, IEDM '93*, pp. 175–177.

Xie, B., Mecklenburg, M., Danielsson, B., Öhman, O., and Winquist, F. (1994). Microbiosensor based on an integrated thermopile. *Anal. Chim. Acta* **299**(2), 165–170.

Ye, S.-Y., Tanaka, S., Esashi, M., Hamakawa, S., Hanaoka, T., and Mizukami, F. (2005). Thin palladium membrane microreactors with oxidized porous silicon support and their application. *J. Micromech. Microeng.* **15**(11), 2011–2018.

Yeom, J., Field, C.R., Bae, B., Masel, R.I., and Shannon, M.A. (2008). The design, fabrication and character-ization of a silicon microheater for an integrated MEMS gas preconcentrator. *J. Micromech. Microeng.* **18**(12), 125001.

Yon, J.J., Barla, K., Herino, R., and Bomchil, G. (1987). The kinetics and mechanism of oxide layer formation from porous silicon formed on p-Si substrates. *J. Appl. Phys.* **62**(3), 1042–1048.

Zeitschel, A., Friedberger, A., Welser, W., and Müller, G. (1999). Breaking the isotropy of porous silicon formation by means of current focusing. *Sens. Actuators A* **74**(1–3), 113–117.

Zeng, J.S.Q., Stevens, P.C., Hunt, A.J., Grief, R., and Lee, D. (1996). Thin-film-heater thermal conductivity apparatus and measurement of thermal conductivity of silica aerogel. *Int. J. Heat Mass Transf.* **39**(11), 2311–2317.

Zhang, C. and Najafi, K. (2004). Fabrication of thick silicon dioxide layers for thermal isolation. *J. Micromech. Microeng.* **14**(6), 769–774.

Ziouche, K., Godts, P., Bougrioua, Z., Sion, C., Lasri, T., and Leclercq, D. (2010). Quasi-monolithic heat flux microsensor based on porous silicon boxes. *Sens. Actuators A* **164**(1–2), 35–40.

PSi-Based Microwave Detection

Jonas Gradauskas and Jolanta Stupakova

9

CONTENTS

9.1   Introduction   186

9.2   Basic Principles of Microwave Detection   186

9.3   Microwave Detection on Porous Silicon Layers   187

9.4   Summary   193

References   193

9.1 INTRODUCTION

From the whole electromagnetic spectrum, extending from DC to gamma rays, microwaves cover frequencies ranging from 300 MHz to 300 GHz, with equivalent wavelengths ranging from 1 m to 1 mm (Sorrentino and Bianchi 2010). High frequencies and short wavelengths of the microwaves provide unique opportunities for their application in various life sectors including, among others:

- Wireless networking and communication
- Radio-astronomy
- Weather forecasting and remote sensing
- Automotive aids and control
- Civil and military surveillance systems
- Healthcare and biomedicine
- Microwave imaging, test of materials, objects, or constructions
- Food processing
- Basic and applied research
- Application in the field of energy

9.2 BASIC PRINCIPLES OF MICROWAVE DETECTION

A vast spectrum of application possibilities requires suitable microwave (MW) sources and a variety of detectors having features good for each specific sector of application. The process of detection of radiation involves two requirements to be fulfilled. First, radiation has to be able to invoke any changes (physical, chemical, etc.) in a detector and, second, the latter should produce any sensible signal, preferably electrical or optical. In practice, MW radiation is usually detected using substitution techniques based on its heating effect or by rectification.

A *calorimeter* is a directly heat-measuring device: it consists of a load directly heated by the absorbed MW radiation, and the rise of its temperature is taken into consideration. High accuracy of calorimeters has opened the doors for them to be used as primary as well as secondary standards of MW power.

The main element of a *bolometer* is a temperature sensitive resistor. The change in resistance resulting from the absorption of MW power is determined in a resistance-measuring circuit.

Thermoelectric elements develop a current, mainly due to the Seebeck effect (Fantom 1990), because of the absorption of heat from the incident microwave power. The base parts of the thermoelectric elements are made in different configurations, such as semiconductor (hot carrier) diode, wire (thermocouples), or thin film devices.

The operation of *diode*-type detectors is based on rectification of MW currents. A diode's variable resistance is used for rectification and to produce a DC voltage proportional to the power of incident MW signal. The classical *p-n* diodes are not suitable for high frequency application because they have relatively large junction capacitance. In contrast, semiconductor *lh* (n-n^+ or p-p^+) junctions containing diodes reveal the ability to rectify the MW currents (Ašmontas et al. 1982). The most popular currently in use diode-type devices are Schottky diodes and backward diodes (a variation of a tunnel diode). These both have low junction capacities and demonstrate high voltage-to-power responsivity (the ratio of the generated DC voltage across the diode to the incident MW power) values, but their operation is limited to low-level MW signals.

Other types of detectors employ physical phenomena such as pyroelectric effect, radiation pressure, magnetic effects, superconductivity, frequency mixing (heterodyne detection), and so on. More detailed descriptions of operation of all these kinds of MW detectors can be found in Fantom (1990) and Collier and Skinner (2007).

Special interest is worth to be paid to semiconductor detectors with operations based on hot carrier effects because MW power in a semiconductor is initially absorbed by the free charge carriers. The carriers are heated directly by the MW electric field while the crystal lattice itself remains cool. Fast response time is inherent for the detectors of this type because their speed of operation is fundamentally limited by the energy relaxation time, which in typical semiconductors is of an order of picoseconds (Dargys and Kundrotas 1994). Such semiconductor detectors demonstrate uniform frequency characteristics in a wide MW frequency range due to nonselective free carrier absorption of the radiation.

Energy increase of the carriers and the corresponding change in mobility is a measure of the incident power. This has made the basis of indium antimonide (Kinch and Rollin 1963; Sakai and Sakai 1976) and silicon (Dagys et al. 2001) hot electron bolometers.

Detectors of another type, the *hot carrier diodes*, make use of the thermoelectromotive force arising across a cat-whisker–type diode (Ašmontas et al. 1978; Fantom 1990) or across an asymmetrically narrowed planar diode structure (Ašmontas 1975; Sužiedėlis et al. 2003) because of the nonhomogeneous heating of the free carriers.

Hot carrier thermoelectromotive force is also induced across a semiconductor potential barrier exposed to MW radiation, whether it is a diode with the *lh* junction (Ašmontas et al. 1982), with the *p-n* junction (Veinger et al. 1975), or with the heterojunction (Ašmontas et al. 1999). In general, the hot carrier thermoemf arising across a potential barrier depends on the barrier height and on the heating level of the carriers (Ašmontas 1984):

$$(9.1) \qquad U_{themf} = V_C \left(\frac{T_n}{T} - 1 \right).$$

Here V_C stands for the contact potential difference (or qV_C would be the potential barrier height; q is elementary charge), T_n and T are the hot carrier and the lattice temperatures, respectively.

The pulsed action of MW radiation gives rise to the voltage signal across the semiconductor formation containing depleted space charge region. The hot carrier induced pulsed alteration of the space charge was the reason for the signal formation across a metal-oxide-semiconductor structure (Kalvėnas et al. 1975) or a closed *p-n* junction (Ašmontas and Olekas 1991).

Last, maximum MW power is transferred to the detector when the requirement to match load and source impedances is fulfilled (Collier and Skinner 2007). The better the matching, the more incident MW power is absorbed by the detector and thus higher values of voltage-to-power responsivity can be achieved.

9.3 MICROWAVE DETECTION ON POROUS SILICON LAYERS

The conductivity of PSi increases under the action of DC electric field (Ben-Chorin et al. 1994), as well as exposed to 1 Hz to 100 kHz (Ben-Chorin et al. 1995a) and GHz (Shatkovskis et al. 2007; Šatkovskis et al. 2008) frequencies. Activation nature was supposed to be responsible for the electric-field-enhanced conduction of PSi. Although the conductivity has a rather complicated, activation energy dependent, character of its dependence on the MW power (Shatkovskis et al. 2007), the PSi layer containing device operating in a photoconductive mode can be used for the MW detection, that is, the bolometric effect in PSi can be brought into play (Figure 9.1).

Summarizing the basic principles of microwave detection but the bolometric effects, one can conclude that MW radiation may be detected by employing a semiconductor region with structural, carrier concentration, or, under the illumination, temperature gradient. The samples containing PSi layers do have such inhomogeneities in the contacts PSi/metal, PSi/dielectric, PSi/crystalline silicon (c-Si) (Ben-Chorin et al. 1995b), and PSi/PSi of different porosities (Lehmann 2002). Any of these interlayers in principle are potential working elements of a MW detector.

Two terminal (diode-like) samples containing layers of PSi were fabricated by the usual technology of electrochemical etching (Shatkovskis et al. 2007). Boron-doped *p*-type conductivity (100) oriented c-Si wafers of 0.4 Ω cm resistivity were used for preparation of the PSi structures. Mix of fluoric acid (48%) with ethanol (96%) in the ratio 1:2 was used as an electrolyte. The density of current during the etching cycle was 10 and 80 mA/cm², and the etching lasted for 5 and

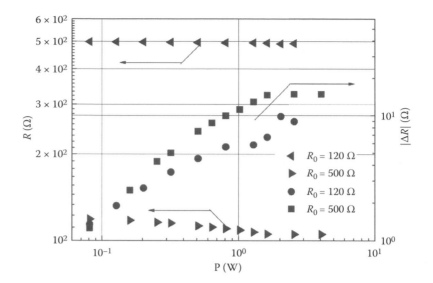

FIGURE 9.1   Resistance R under the pulsed action of 10 GHz MW radiation and absolute value of the resistance decrease $|\Delta R| = R - R_0$ of PSi structures versus MW power. R_0 is the initial resistance of the sample. (From Šatkovskis, E. et al., *Acta Phys. Pol. A* **113**(3), 993, 2008. Copyright 2008: Institute of Physics, Polish Academy of Sciences. With permission.)

10 min, respectively. As a result, two PSi layers of different porosity have been fabricated with the denser one situated at the surface. To get ohmic electrical contacts, boron diffusion was performed on both surfaces of the initial plate, and Al contacts (100 μm in diameter for PSi and entire for the bottom surface) have been thermally evaporated. Though the voltage–current characteristics of the samples were linear (Stupakova et al. 2006), actually, such structures contained five potential barriers (Figure 9.2).

When exposed to 10 GHz MW pulsed radiation, the voltage signal was generated between the aluminum contacts of the structure. The response signal was fast; it followed the shape of the MW-modulating 2-μs-long pulse. It almost linearly depended on the incident MW power in distinct power regions (Stupakova et al. 2006). Both these findings let the authors arrive at the conclusion that the voltage signal was caused by the heating of free holes. Change of the signal polarity at higher power values (see Figure 9.3, sample 2) was attributed to the competition of major two hot holes' thermoemfs of opposite polarities, the first one being generated across the p^+-PSiL1/p-PSiL2 contact, and the second one across the p-PSiL2/p-c-Si contact.

Such consideration agrees well with the theory of hot carrier thermoemf formation across a potential barrier according to Equation 9.1 because, first, at higher excitation levels there

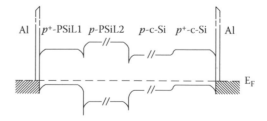

FIGURE 9.2   The energy band structure of the two PSi layers of different porosity containing diode-like samples. Indications: Al—aluminum contacts; p^+-PSiL1—denser PSi layer having undergone boron diffusion; p-PSiL2—PSi layer of higher porosity and thus of wider forbidden energy gap due to quantum confinement effect; p-c-Si—crystalline silicon layer; p^+-c-Si—crystalline silicon layer having undergone boron diffusion; E_F—Fermi level. (From Šatkovskis, E. et al., *Acta Phys. Pol. A* **113**(3), 993, 2008. Copyright 2008: Institute of Physics, Polish Academy of Sciences. With permission.)

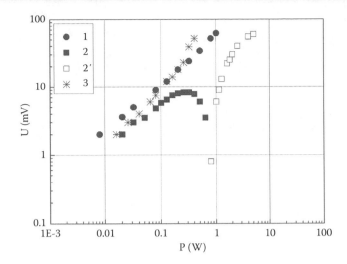

FIGURE 9.3 The dependence of the response signal across the structures on 10 GHz incident MW power. Sample resistance without illumination: 1—300 Ω, 2 and 2'—500 Ω, 3—42 Ω. Full marks indicate positive signal polarity, open squares (2') indicate negative signal polarity of sample 2. (From Stupakova, J. et al., *Acta Phys. Pol. A* **110** (6), 817, 2006. Copyright 2006: Institute of Physics, Polish Academy of Sciences. With permission.)

dominates the thermoemf generated across the contact with higher potential barrier height (the p-PSiL2/p-c-Si contact has higher inhomogeneity as compared to the p^+-PSiL1/p-PSiL2 one); and, second, the conversion of the signal polarity was observed only on the sample with the highest resistance, that is, with the highest porosity and thus stronger inhomogeneity. Nonuniform heating of the holes in the sample has been taken into consideration as well.

The crystalline silicon c-Si sample without porous layers has been fabricated in parallel following the same technological steps, including boron diffusion and metallization, except the formation of porous layers (see Figure 9.4a).

The responsivity of the samples containing PSi layers (Figure 9.4b) was found to be 3 to 5 orders higher than that of the c-Si sample (Figure 9.4c). To interpret the results, the hot carrier point-contact semiconductor detector model has been used. The presence of the quantum confinement effect, supported by the photoluminescence investigations, indicated that the cross-section of silicon stem in the p-PSiL2 layer should be in the nanometer range (Shatkovskis et al. 2011). Therefore, point-contact-like dimensionally nonuniform junctions should have been formed on both sides of the p-PSiL2 layer, as indicated in Figure 9.2, with a higher level of inhomogeneity on the p-PSiL2/c-Si contact side. The latter contact seems to dominate in the overall voltage signal formation, since the voltage responsivity of a point-contact MW detector is inversely proportional to the radius r of a point contact in the third degree (Guoga and Pozhela 1969)

(9.2)
$$S \propto \frac{1}{r^3}.$$

Much higher levels of spatial inhomogeneity in the PSi layers containing samples in comparison with c-Si samples, that is, nanometer size "point-contact" radius compared to 50-μm size Al island radius (see Figure 9.4a), qualitatively explains much higher responsivity values of the PSi samples according to Equation 9.2. Furthermore, the dependence of the responsivity magnitude of different PSi samples on their resistance (Figure 9.4c) supports the invoked model: higher values of voltage signal are detected across the samples of higher resistance, that is, the lower dimensions of the PSi stem cause higher levels of inhomogeneity which, in turn, causes higher responsivity of the detector.

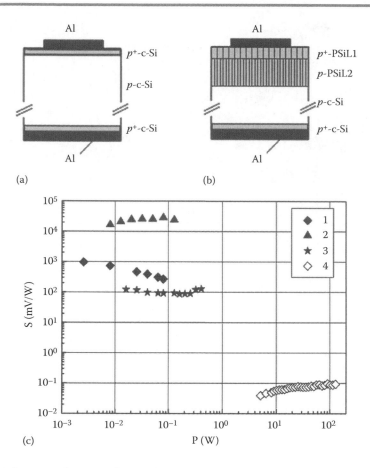

FIGURE 9.4 Schematic drawing of two-terminal crystalline c-Si structure without PSi layer (a) and of the structure with two PSi layers (b). (c) The responsivity of the structures versus incident microwave power: 1, 2, and 3 are PSi layers containing samples, 4 is the reference c-Si sample. Sample resistance: 1 – 10 kΩ, 2 – 4.2 kΩ, 3 – 500 Ω. (From Shatkovskis, E. et al., *Lith. J. Phys.* **51**(2), 143, 2011. Copyright 2011: Lithuanian Academy of Sciences. With permission.)

The samples similar to the ones shown in Figure 9.4b, but having undoped additionally the upper PSi layer, demonstrate asymmetrical voltage-current characteristics as compared to the linear one of the sample with both surfaces doped (Figure 9.5a).

The B samples obviously demonstrate higher responsivity to the MW radiation than the A samples (Figure 9.5b). Moreover, the polarities of the induced voltage signals are opposite for both samples (the B sample produced positive voltage sign on the PSi-side in respect to the bottom contact). In this case, the competition of two effects should be discussed considering the A samples. The first is the formation of the hot carrier thermoemf, the same process as described previously and which takes place in the B samples. The second, inspired by the asymmetrical voltage-current dependencies, is the rectification of the MW currents similar to that in the Schottky diode. Since the voltage signals caused by each of these two effects are of opposite polarities (Fantom 1990), their competition most probably results in the decrease of the total response signal magnitude and even in the change of its polarity.

Another possibility to detect pulsed MW radiation was demonstrated on *n*-type PSi structures (Gradauskas et al. 2014). The (100)-oriented Si wafer of 1 Ω cm resistivity was used to fabricate the porous structures by electrochemical etching in a mix of fluoric acid with ethanol and water in the ratio of 1:1:1. The PSi samples were fabricated at current density of 30 mA/cm² for 45 min etching time. After the etching process, rinsing in deionized water and desiccation, the upper nanoporous layer was cleaned out with 5% KOH aqueous solution. Electric wiring was contacted on both sides of the sample using silver epoxy (Figure 9.6a). For comparison, similar samples but only crystalline, c-Si, without porous layer were fabricated in parallel. MW radiation of 10 GHz frequency modulated variably in 10-µs-long pulses at 100 Hz repetition rate was used for excitation.

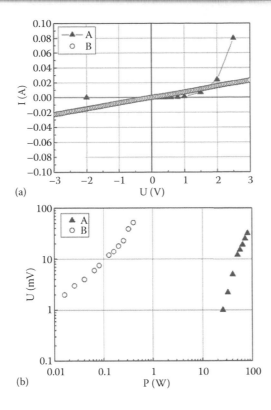

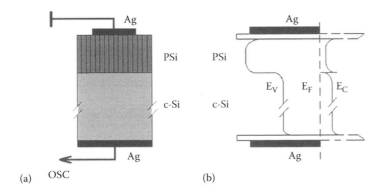

FIGURE 9.5 (a) Voltage-current characteristics of the A and B samples: "A" indicates the samples with undoped upper porous layer, "B" stands for the PSi sample with boron-doped both surfaces (the same as in Figure 9.4b). (b) Typical plots of the response voltage versus incident MW pulse power. (From Ašmontas, S. et al., *Tech. Phys. Lett.* **32**(7), 603, 2006. Copyright 2006: St. Petersburg publishing and bookselling firm "Nauka". With permission.)

FIGURE 9.6 (a) Schematic view of the n-PSi layer containing sample and measurement scheme. (b) Energy band diagram of the sample. (From Gradauskas, J. et al., *Eighth International Conference on Advanced Optical Materials and Devices [AOMD-8]*, Spigulis, J. (Ed.). Proceedings of SPIE, 9421, 9421–7, 2014. Copyright 2014: SPIE—the International Society for Optics and Photonics. With permission.)

A voltage signal was induced across the PSi sample under the action of the MW radiation. It linearly depended on the applied MW power (Figure 9.7a). The polarity of the signal (positive sign with respect to the grounded PSi layer) indicated on the radiation-inspired flow of electrons from the bulk c-Si side to the PSi side of the sample. Actually, three potential barriers could be distinguished in the sample. Two of them are within the metal-oxide-semiconductor (MOS) structures because of the naturally grown SiO_2 layer, the top M/O/PSi and the bottom M/O/c-Si. The third one can be treated as a PSi/c-Si heterojuction barrier, similarly as it was considered in the *p*-type PSi case above (see Figure 9.6b). Since in this case the porosity of the PSi layer was not high (at

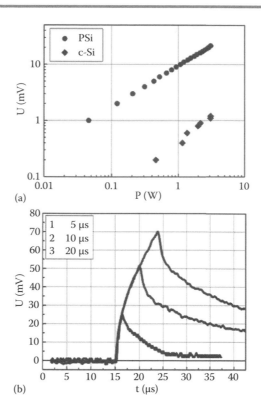

(a)

(b)

FIGURE 9.7   (a) The dependence of the detected voltage signal on microwave pulse power: circles—across the n-type PSi sample, diamonds—across the n-type c-Si sample. (b) Time dependence of the detected voltage signal across the PSi sample at various durations of modulating MW pulse as indicated in the legend. (From Gradauskas, J. et al., *Eighth International Conference on Advanced Optical Materials and Devices [AOMD-8]*, Spigulis J. (Ed.). Proceedings of SPIE, 9421, 9421–7, 2014. Copyright 2014: SPIE—the International Society for Optics and Photonics. With permission.)

most 30%), and respectively the potential barrier height of the PSi/c-Si heterojunction was low, thus, the hot carrier thermoemf across it could be considered negligible.

The formation of the voltage signal under the action of the MW radiation could be explained in the following way. As the carriers get heated, redistribution of charges in the depleted region of a MOS structure takes place (Kalvénas et al. 1975). As a result, an emf is induced with polarity opposite to that of the carrier generation-induced one. Thus, the top M/O/PSi junction and the bottom M/O/c-Si junction "work" in opposite directions. Two reasons can be pointed out why the first one surpasses the second one and determines the polarity of the total signal. Both of them are based on hot carrier point-contact MW diode philosophy and Equation 9.2. First, the radius of the top metal contact is much smaller than that of the bottom contact, and the free electron heating is stronger at the top. This consideration is proved by the signal of the same polarity detected across the similar sample with no porous layer (see Figure 9.7a). In addition, the second reason predicts that the edge of the porous layer of the sample can be treated as a set of micro- or even submicro-columns covered with metal, that is, as a set of point-contacts. Thus, a much higher value of the detected signal across the PSi layer containing the sample again can be explained considering the $1/r^3$ dependence. This is why the responsivity of the PSi sample is higher than that of the c-Si one (Figure 9.7a): a narrower radius of the contact causes stronger heating of the electrons and thus induces thermoemf of higher magnitude.

The porosity also influences the shape of the detected signal. The voltage signal across the c-Si sample was fast enough; it perfectly followed the rectangular MW pulse, while much slower response of the PSi sample (Figure 9.7b) most probably was due to increased RC time constant of the structure. The possibility of heat accumulation in the porous layer during the action of the MW pulse and slow heat flow from it after the pulse due to much lower heat conductivity of the PSi compared to the one of the crystalline Si (Amato et al. 1997) should not be disregarded as well. Nevertheless, the response to longer MW pulses reveals the hidden ability of the PSi samples to provide even higher magnitudes of the voltage-power responsivity.

9.4 SUMMARY

In conclusion, the above-surveyed results of investigation of MW radiation interaction with PSi suggest that promising operation of the MW detector containing PSi structures would be mostly based on the phenomenological theory of hot carriers and point-contact diodes. The PSi structures may have a potential for application in MW detection thanks to

- Simplicity of their fabrication
- Possibility to increase their responsivity due to strongly reduced active area of the detector down to nanometer size
- Ease of reaching impedance matching between the sample and the MW waveguide by choosing appropriate conditions of sample fabrication

With regard to shortcomings of the PSi-based MW detectors, the main problem seems to be the achievement of uniform and repeatable geometrical and electrical parameters of the detector's active elements during its fabrication process.

REFERENCES

Amato, G., Benedetto, G., Boarino, L., Brunetto, N., and Spagnolo, R. (1997). Photothermal and photo-acoustic characterization of porous silicon. *Opt. Eng.* **36**(2), 423–431.

Ašmontas, S. (1975). Investigation of electron heating in non-uniform electric field in Ge. *Lith. J. Phys.* **15**(5), 791–798 (in Russian).

Ašmontas, S. (1984). *Electrogradient Phenomena in Semiconductors.* Mokslas, Vilnius (in Russian).

Ašmontas, S. and Olekas, A. (1991). Investigation of the kinetics of the thermo e.m.f. of hot carriers in reverse biased *p-n*-junction. *Lith. J. Phys.* **31**(2), 213–219 (in Russian).

Ašmontas, S., Lapinskas, R., and Olekas, A. (1978). Investigation of electric properties of silicon n-n+ point contact diodes in the microvawe fields. *Lith. J. Phys.* **18**(6), 741–747 (in Russian).

Ašmontas, S., Vingelis, L., and Subačius, L. (1982). Thermoemf of hot electrons in silicon. *Fiz. Tekh. Poluprovodn. (Sov. Phys.-Semicond.)* **16**(12), 2110–2115 (in Russian).

Ašmontas, S., Gradauskas, J., Kundrotas, J., Sužiedėlis, A., Šilėnas, A., and Valušis, G. (1999). Influence of composition in GaAs/AlGaAs heterojunctions on microwave detection. *Mater. Sci. Forum* **297–298**, 319–322.

Ašmontas, S., Gradauskas, J., Zagadsky, V., Stupakova, J., Sužiedėlis, A., and Šatkovskis, E. (2006). Microwave detectors based on porous silicon. *Tech. Phys. Lett.* **32**(7), 603–605.

Ben-Chorin, M., Möller, F., and Koch, F. (1994). Nonlinear electrical transport in porous silicon. *Phys. Rev. B* **49**(4), 2981–2984.

Ben-Chorin, M., Möller, F. and Koch, F. (1995a). Hopping transport on a fractal: AC conductivity of porous silicon. *Phys. Rev. B* **51**(4), 2199–2213.

Ben-Chorin, M., Möller, F. and Koch, F. (1995b). Band alignment and carrier injection at the porous-silicon–crystalline-silicon interface. *J. Appl. Phys.* **77**(9), 4482–4488.

Collier, R.J. and Skinner, A.D. (eds) (2007). *Microwave Measurements*, 3rd ed., IET Electrical and Measurement Series, Vol. 12. The Institution of Engineering and Technology, London.

Dagys, M., Kancleris, Z., Simniskis, R., Schamiloglu, E., and Agee, F.J. (2001). The resistive sensor: A device for high-power microwave pulsed measurements. *IEEE Antennas Propag. Mag.*, **43**(5), 64–79.

Dargys, A. and Kundrotas, J. (1994). *Handbook of Physical Properties of Ge, Si GaAs and InP.* Science and Encyclopedia Publishers, Vilnius.

Fantom, A. (1990). *Radio Frequency Microwave Power Measurement.* P. Peregrinus on behalf of the Institution of Electrical Engineers, London.

Gradauskas, J., Stupakova, J., Sužiedėlis, A., and Samuolienė, N. (2014). Detection of microwave radiation on porous silicon nanostructures. In: *Eighth International Conference on Advanced Optical Materials and Devices (AOMD-8)*, Spigulis J. (Ed.). SPIE Proceedings **9421**, 942103-1–942103-4.

Guoga, V.I. and Pozhela, J.K. (1969). About the sensitivity of hot carrier microwave detector *J. Comm. Tech. El., (Sov. Radiotekh. Elektron.)* **14**(3), 565–566 (in Russian).

Kalvėnas, S.P., Juškevičienė, M.M., and Versockas, A.P. (1975). Surface thermoemf of hot electrons in n-Si. *Fiz. Tekh. Poluprovodn. (Sov. Phys.-Semicond.),* **9**(9), 1662–1667 (in Russian).

Kinch, M.A. and Rollin, B.V. (1963). Detection of millimetre and sub-millimetre wave radiation by free carrier absorption in a semiconductor. *Br. J. Appl. Phys.* **14**, 672–676.

Lehmann, V. (2002). *Electrochemistry of Silicon.* Willey-VCH, Weinheim.

Sakai, K. and Sakai, J. (1976). Characteristics of n-InSb hot electron submillimeter detector. *Jpn. J. Appl. Phys.* **15**, 1335–1341.

Šatkovskis, E., Gradauskas, J., Česnys, A., Stupakova, J., and Sužiedėlis, A. (2008). Charge carrier heating effect in porous silicon structures investigated by microwaves. *Acta. Phys. Pol. A* **113**(3), 993–996.

Shatkovskis, E., Gradauskas, J., Stupakova, J., Česnys, A., and Sužiedėlis, A. (2007). Effect of microwave radiation on conductivity of porous silicon nanostructures. *Lith. J. Phys.* **47**(2), 169–173.

Shatkovskis, E., Stupakova, J., Gradauskas, J., Sužiedėlis, A., and Mitkevičius R. (2011). Sensitivity improvement in porous silicon microwave detector. *Lith. J. Phys.* **51**(2), 143–146.

Sorrentino, R. and Bianchi, G. (2010). *Microwave and RF Engineering.* John Wiley & Sons, Chichester.

Stupakova, J., Ašmontas, S., Gradauskas, J., Zagadskij, V., Shatkovskis, E., and Sužiedėlis, A. (2006). Studies of response of metal-porous silicon structures to microwave radiation. *Acta. Phys. Pol. A* **110**(6), 817–822.

Sužiedėlis, A., Gradauskas, J., Ašmontas, S., Valušis, G., and Roskos, H.G. (2003). Giga- and terahertz frequency band detector based on an asymmetrically-necked n-n+-GaAs planar structure. *J. Appl. Phys.* **93**(5), 3034–3038.

Veinger, A.I., Paritskyi, L.G., Akopian, E.A., and Dadamirzaev, G. (1975). Thermoemf of hot carriers on p-n junction. *Fiz. Tekh. Poluprovodn. (Sov. Phys.-Semicond.)* **9**(2), 216–224 (in Russian).

Auxiliary Devices

PSi-Based Ultrasound Emitters (Acoustic Emission)

10

Ghenadii Korotcenkov and Vladimir Brinzari

CONTENTS

10.1   Introduction   198

10.2   Mechanism of Ultrasound Generation by PSi-Based Emitter   198

10.3   PSi-Based Ultrasonic Devices   201

Acknowledgments   205

References   205

10.1 INTRODUCTION

The most common mechanism for generating ultrasound in air is via a piezoelectric transducer, whereby an electrical signal is converted directly into a mechanical vibration (Manthey et al. 1992; Wang et al. 2005; Drinkwater and Wilcox 2006). However, the acoustic pressure so generated is usually limited to less than 10 Pa, the frequency bandwidth of most piezoelectric ceramics is narrow, and it is difficult to assemble such transducers into a fine-scale phase array with no crosstalk (Mo et al. 1992). An alternative strategy using micromachined electrostatic diaphragms is showing some promise (Haller and Khuri-Yakub 1998), but the high voltages required and the mechanical weakness of the diaphragms may prove problematic for applications. Shinoda et al. (1999) proposed to resolve these problems via porous silicon (PSi) as active material in ultrasound emitters. Koshida's team (Shinoda et al. 1999; Koshida et al. 2001; Asamura et al. 2002) studying PSi formed by conventional anodization technique has found that the experimental device shown in Figure 10.1 can operate as an efficient ultrasound emitter. Under a constant electrical AC input power, a significant acoustic signal was observed over a wide frequency range. This device in the present and following experiments was composed of a patterned, thin aluminum film electrode (30 nm thick), a microporous silicon layer (10–50 μm thick), and a p-type (100) single crystalline silicon (c-Si) wafer. The PSi layer (with porosity of ~70%) was formed in a solution of 55% HF:ethanol = 1:1 at a temperature of 20°C at a current density of 20 mA/cm² for 8–40 min. The effective area of the device was usually about 2 × 2 cm. The aluminum electrode was used to input a sinusoidal current into the PSi layer, the temperature of which was raised by Joule's heating. The emitted acoustic pressure was measured as a function of output frequency by a microphone placed at a position of 3.5–5 cm from the front surface. This configuration is regarded as an open space operation, in which the spacing between the emitter surface and the detector is larger than the sound wavelength of interest.

10.2 MECHANISM OF ULTRASOUND GENERATION BY PSi-BASED EMITTER

Shinoda et al. (1999) explained the effect of ultrasound generation by PSi, based on a theoretical analysis of thermal conduction phenomena (Holman 1963) in the PSi/air system. Later generalized theory of thermo-acoustic emission from nanocrystalline PSi was proposed by Hu et al. (2012) and Wang et al. (2014). One should note that the idea of a thermal sound generator (a "thermophone") was offered 80 years ago by Arnold and Crandall (1917). In their proposal, the acoustic element was a simple self-supporting thin metal film. Experimental and theoretical studies have shown that the thermo-acoustic effect observed in PSi is presumably because of complete carrier depletion—both the thermal conductivity and the heat capacity per unit volume of PSi are extremely low in comparison to those of c-Si. Ultrasound is emitted from the still PSi device as illustrated in Figure 10.2 by an AC electrical input and subsequent fast heat transfer to air. According to Shinoda et al. (1999), the temperature change induces an acoustic pressure $P(x, \omega)$

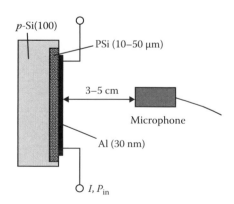

FIGURE 10.1   A schematic of the cross-sectional view of the fabricated nanocrystalline PSi (nc-PSi) device and experimental configuration for sound emission measurements. (From Migita T. et al., *Jpn. J. Appl. Phys.* 41, 2588, 2002. Copyright 2002: The Japan Society of Applied Physics. With permission.)

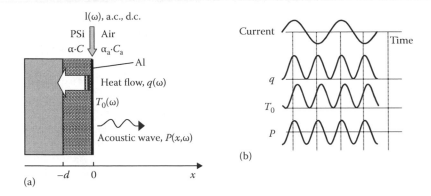

FIGURE 10.2 Device operation. (a) The air/thin-Al-film PSi/c-Si structure and the coordinate configuration. (b) Current I introduced into the top electrode induces the surface temperature change T_0 by Joule's heat q, which produces acoustic pressure P. Following an induced Joule heating, the temperature at the device surface fluctuates with a frequency two times higher than the input frequency. Because of an effectively quenched thermal conductivity in the nc-PSi layer, the change in the surface temperature is instantly and directly transferred to the expansion and compression of air in the region in proximity to the front surface. It directly induces sound pressure, while both the bulk and the surface of the PSi layer remain mechanically stable. (Reprinted by permission from Macmillan Publishers Ltd. Shinoda H. et al., 400, 853, copyright 1999.)

$\exp(j\omega t)$ through the alternating thermal expansion of the air. Using the fundamental equations of photoacoustic analysis (McDonald and Wetsel 1978), it was found that

$$(10.1) \qquad P(x,\omega)=A\frac{\exp(-j\kappa x)}{\sqrt{\alpha C}}q(\omega), \quad A=\sqrt{\frac{\gamma\alpha_a}{C_a}}\cdot\frac{P_a}{\upsilon T_a}$$

where α and C are the heat conductivity and the heat capacity per unit volume of the insulator, i.e., PSi, P_a is atmospheric pressure, T_a is room temperature, v is the sound velocity, $\gamma=\dfrac{C_p}{C_\gamma}=1.4$, k is the wavenumber of free-space sound, α is the thermal conductivity of air, and C_a is the heat capacity per constant unit volume of air. The assumptions in this analysis are as follows: $\kappa\ll\sqrt{\dfrac{\omega C_a}{\alpha_a}}$ (that is, the sound wavelength is much larger than thermal diffusion length), $\alpha C\gg\gamma\alpha_a C_a$ (that is, the heat flow into the device is much larger than that into the air), and the PSi itself does not

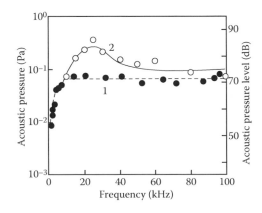

FIGURE 10.3 Observed frequency characteristics of thermally induced nc-PSi ultrasonic emitter. The electrical power input was 1 W/cm². A decrease in the sound pressure observed in the low frequencies is related to the situation in which the sound wavelength is larger than the device size. (Data extracted from Shinoda H. et al., *Nature* 400, 853, 1999; and Migita T. et al., *Jpn. J. Appl. Phys.* 41, 2588, 2002.)

vibrate. Too large a value of PSi layer thickness (d) causes a stationary temperature rise on the surface (which is useless for our purposes) and, in contrast, too small a d decreases the signal of the acoustic pressure amplitude itself. We note that the ratio $|P(\omega)|/|q(\omega)|$ is constant, and independent of ω. This means that an ideal, flat frequency response is expected from this device. Following experiments (Asamura et al. 2002; Migita et al. 2002; Watabe et al. 2006a) and theoretical simulations, Hu et al. (2012) and Wang et al. (2014) confirmed this model. In particular, Figure 10.3 shows flat frequency response of PSi-based emitters.

As it follows from Equation 10.1, the most important key parameter of PSi-based emitters is the product $\alpha \cdot C$ because the theoretical output sound pressure is inversely proportional to $(\alpha \cdot C)^{1/2}$ (Shinoda et al. 1999). This means that efficiency of PSi-based emitters should be controlled by the porosity of PSi layer because both α and C strongly depend on this parameter of PSi. As it is known, $\alpha \cdot C$ value rapidly decreases with increasing porosity. The results presented in Figure 10.4a are in full agreement with this statement. Some other characteristic features of the PSi emitter are summarized in Table 10.1 in comparison to those of the conventional ones.

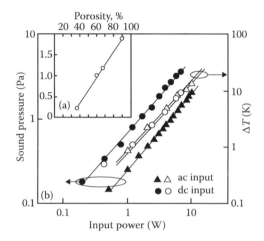

FIGURE 10.4   (a) Measured sound pressure amplitudes at 100 kHz as a function of the porosity. (b) Input power influences on the measured output sound pressures under AC-voltage driving mode (input frequency: 25 kHz) and DC superimposed driving mode (input frequency: 50 kHz). ([a] Data extracted from Tsubaki K. et al., *Proceedings of 206th Meeting of the Electrochemical Society*, Abs. 1028, 2004. [b] From from Tsubaki K. et al., *Jpn. J. Appl. Phys.* 45(4B), 3642, 2006. Copyright 2006: The Japan Society of Applied Physics. With permission.)

TABLE 10.1   Comparative Survey of the PSi Ultrasound Emitter and a Conventional Emitter

Item	PSi Emitter	Conventional Emitter
Principle	Thermo-acoustic	Electro-acoustic
Operation mode	Still	Vibration
Acoustic output	Proportional to P_{in}	Proportional to V_{in}
Efficiency	Pa/W	Pa/V
Frequency characteristics	Flat	Resonant
Impulse signal	Ideal	–
Scaling merit	Available	Nothing
Total harmonic distortion	<0.1%	–
Dynamic range at 10–100 kHz	>60 dB	–
Arrayed integration	Easy	Difficult

Source: Data extracted from Koshida N. et al., In: *Pits and Pores II: Formation, Properties, and Significance for Advanced Materials.* ECS, Pennington, NJ, 2001; and Kihara T. et al., *Jpn. J. Appl. Phys.* 43, 2973, 2004.

10.3 PSi-BASED ULTRASONIC DEVICES

According to Koshida and co-workers (Shinoda et al. 1999; Koshida et al. 2001; Asamura et al. 2002; Migita et al. 2002; Hirota et al. 2004; Watabe et al. 2006a), the PSi ultrasonic device has many advantageous features over the conventional ones in which electromechanical or electrostatic vibration mechanism of solid surface is used. They are the following:

1. The thermoacoustic operation can emit ultrasound without any mechanical vibration.
2. The observed acoustic output of the PSi emitter is independent of frequency. Flat frequency behavior is a very important performance as acoustic devices because an ideal acoustic impulse can be generated. It makes it possible to improve the accuracy for detecting a distance. In conventional ultrasound emitters, in contrast, the ultrasound operation is produced under a resonant mode and consequently the available frequency is limited to a considerably narrower band.
3. A wide range of operating frequency. For the appropriate device structure, we can estimate that a maximum frequency reaches 1 GHz. It is difficult to get such a high frequency by conventional ultrasound generators. Regarding the maximum operating frequency, there are the following two limiting factors. The first one is a difference in the frequency dependence between the acoustic wavelength and the air layer thickness d related to heat exchange. Another limiting factor is the heat capacity of the top electrode, although it is negligible for the electrode with a sufficiently small thickness.

   According to simulations carried out by Hu et al. (2012), lower-end frequency of thermo-acoustic (TA) emission depends on the thickness and properties of the TA sample, while its upper-end frequency is only related to the state and properties of gas as well as the distance from the sample surface. Its bandwidth shrinks with the increase in the distance. The flat frequency response of thermo-acoustic ultrasound could basically exist up to 10 MHz frequency and 100 m distance from the sample surface for typical samples at normal atmospheric pressure and room temperature. At that, the TA sample thickness should be larger than 2–3 times of thermal diffusion length.
4. Reliable response under impulse operation (Figure 10.4b).
5. The output pressure of the conventional emitters is proportional to the input voltage, whereas that of the PSi devices should increase in proportion to the input power supplied to the top Al electrode (see Figure 10.4b). To enhance the efficiency of the PSi emitter, it looks most promising to fabricate a PSi array because there is a universal scaling principle in the thermal conduction phenomena.
6. To enhance the efficiency defined as the ratio of output pressure to input power, it looks most promising to fabricate an integrated fine array because there is a universal scaling principle in the thermal conduction phenomena: when the emitting size is reduced by a factor of n, the efficiency should be increased by a factor of n. According to quantitative estimation, an intense output of 50 Pa or more would be obtained from an arrayed periodic structure with an electrode size of about 10 μm. This can be achieved by standard processing without difficulty.
7. Small total harmonic distortion. Moreover, the distortion of the output signal for the PSi emitter is extremely small even at large input power. Effectively no distortion is observed. One should note that the frequency-independent dynamic range is never obtainable in the conventional devices.
8. The observed wide-range linearity between the input power and output acoustic-amplitude is also sufficient for practical use.

Koshida and co-workers (Koshida et al. 2001; Tsubaki et al. 2006) believe that these merits of PSi-based emitters should be enhanced by a combination of the scaling effect and the monolithic integration. Possible improvement in the efficiency by scaling and arraying device structures would make it possible to develop novel ultrasonic technology. According to Koshida et al. (Koshida et al. 2001; Kihara et al. 2004; Tsubaki et al. 2006), the PSi emitter is directly applicable to various functional devices such as high-precision sensors, single chip sonar, functional speaker, a noncontact actuator with an intense beam source, and so on. For example, Tsubaki et al. (2005)

reported that the nc-PSi ultrasonic emitter combined with an arrayed condenser microphone is useful for three-dimensional (3D) image sensing in air. The 3D object was detected with a spatial resolution considerably higher than that in the case of conventional techniques based on piezo-electric transducers. It was also shown that the device was available for the detection of several objects with different acoustic reflectance coefficients. Gavrilchenko et al. (2009) have found that the acoustic phenomena in PSi/c-Si heterostructure are sensitive to ambient gas atmosphere. A 1.0 kHz shift of resonance peak to high frequency region has been observed in saturated alcohol vapors in comparison to air. This means that this effect can also be applied as new transducer for chemical sensors based on PSi.

It was demonstrated that a thermally induced ultrasound device based on the thermo-acoustic effect in nc-PSi is useful as a pulse acoustic emitter with tunable output directivity. Watabe et al. (2006b) have found that the nc-PSi ultrasound generator under the operation of pulse width modulation can control the directivity angle with nearly constant sound pressure amplitude. The measured directivity angle is decreased with decreasing pulse width (see Figure 10.5). This characteristic is extremely different from that of conventional ultrasound devices based on the electric-acoustic effect. By utilizing this advantage, the detection area for ultrasound sensing can be extended only by controlling the input pulse width. This feature will make it possible to develop practical 3D ultrasonic imaging sensors in air without any complicated system for signal processing.

Isozaki et al. (2007) have also shown that PSi-based emitters allow designing ultrasonic transmitters with a variable directional pattern. One should note that the ability to control the direction of ultrasound emission is an important advantage of PSi-based emitters. As is known, ultrasonic transmitters for proximity sensors need to induce an acoustic pressure field with high directivity and variable direction of the main lobe. The directional pattern in PSi-based emitters was realized by using an interference effect. An ultrasonic wave was emitted from electrodes on the nc-PSi layer according to the applied electrical current. Figure 10.6 illustrates features of such emitters. When two alternating currents (AC) are applied to the two emission areas, two ultrasonic waves are emitted. The two emitted waves interfere with each other and create the directional pattern. Ultrasonic pressure intensity at a certain point depends on the relative phase difference between the waves at that point. Figure 10.6a shows the directional pattern without interference effect (one emission area). Figure 10.6b shows the directional patterns with interference effect (two emission areas) with and without phase difference of the applied current, φ. The distance between two emission areas d decides the lobe number and the sharpness of the directional pattern. The phase difference decides the direction of the main lobe. It was demonstrated that the directional pattern of ultrasonic pressure intensity was able to be controlled by the distance between the two emission areas and the phase difference of applied AC currents.

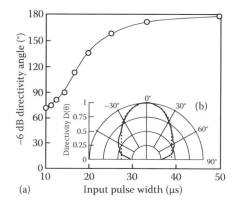

(a)

FIGURE 10.5   (a) Pulse width dependence of directivity angle corresponding to acoustic output level of −6 dB. (b) Directivity pattern of ultrasound emission from nc-PSi device driven under pulse input with a width of 20 ms. Experimental and theoretical results are indicated by the solid and dashed curves, respectively. (From Watabe Y. et al., *Jpn. J. Appl. Phys.* 45(9A), 7240, 2006. Copyright 2006: The Japan Society of Applied Physics. With permission.)

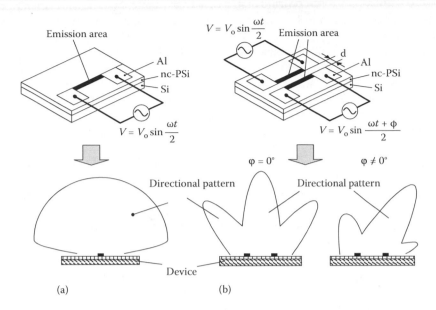

FIGURE 10.6   Schematic view on ultrasonic transmitters and types of directional patterns. (a) One-line pattern and (b) two-line pattern. (Idea from Isozaki A. et al., In: *Proceedings of the IEEE International Conference on Micro Electro Mechanical Systems (MEMS)*, January 21–25, 2007, Kobe, Japan, p. 55.)

In the following, Gelloz et al. (2008) confirmed the effectiveness of this approach to control the direction of ultrasound emission. For this purpose, a 3 × 3 ultrasonic emitter array was designed and fabricated (Figure 10.7a). Gelloz et al. (2008) established that by the use of phase-shift driving in a PSi ultrasonic emitter array, the acoustic output of the device could be significantly enhanced, and the emission directivity could also be controlled with sufficient resolution. An appropriate combination of the device unit size, its spatial array density, the operation frequency,

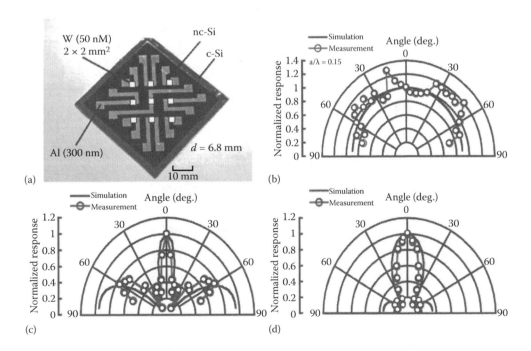

FIGURE 10.7   (a) Photograph of the fabricated device consisting of a 3 × 3 ultrasonic emitter array. (b–d) Simulated and measured two-dimensional intensity profiles of the relative acoustic output obtained from a single device operated at different modes. (From Gelloz B. et al., *Jpn. J. Appl. Phys.* 47(4), 3123, 2008. Copyright 2008: The Japan Society of Applied Physics. With permission.)

and the phase shift makes it possible to focus the emission intensity and tune the directivity (see Figure 10.7b–d).

Tsubaki et al. (2004) have also shown that the DC-superimposed driving mode is more useful for the efficient operation of a PSi-based thermally induced ultrasonic emitter than the conventional simple AC-voltage driving mode. In the DC-superimposed driving mode, the output frequency is the same as the input frequency and a stationary temperature rise is kept constant independent of input peak-to-peak voltage. This acoustic emission characteristic is very useful for speaker applications because no signal processing is required to use the original signal frequency. In addition, power efficiency significantly increases compared with that in the AC-voltage driving mode without affecting the temperature rise. Moreover, acoustic output has a linear relationship with input voltage. The present results ensure the technological potential of the nc-PSi ultrasound emitter for applications in functional speaker devices such as super tweeters and parametric speakers. Tsubaki et al. (2004, 2006) believe that the use of dot electrode structures (in which the heat exchange is concentrated in small islands) should greatly improve the efficiency and output acoustic pressure of our devices. Moreover, such construction could prove useful in 2D array fabrication because it provides a large wiring space.

Regarding disadvantages of PSi-based emitters, one can say that they first of all include (1) the possible instability of parameters, and (2) small efficiency of the electrical power transformation to sound power (~10^{-9}) (Asamura et al. 2002). This means that the sound intensity is very low for acceptable input power (~0.1 Pa by 1 W), while prevailing transducers transmit 10 Pa ultrasound,

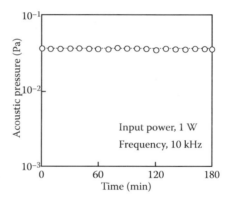

FIGURE 10.8 The time evolution of the output acoustic pressure amplitude. The input power and the frequency are 1 W and 10 kHz, respectively. (From Migita T. et al., *Jpn. J. Appl. Phys.* 41, 2588. 2002. Copyright 2002: The Japan Society of Applied Physics. With permission.)

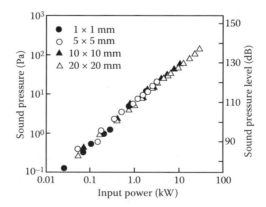

FIGURE 10.9 Device size dependence of sound pressure under impulse operation (16 ms pulse width). The microphone was located at a distance of 30 cm from the device surface. Generated sound pressure is independent of emitting area sizes. (From Tsubaki K. et al., *Jpn. J. Appl. Phys.* 44, 4436, 2005. Copyright 2005: The Japan Society of Applied Physics. With permission.)

and a large enough power is required to achieve the desired acoustic pressure. Koshida and co-workers (Shinoda et al. 1999; Koshida et al. 2001; Migita et al. 2002) stated that the stability of acoustic emission is sufficient for a long-time continuous wave operation because this device is free from the often-quoted instability of PSi. The PSi layer is protected by the aluminum electrode from the atmosphere, and is not exposed to an intense electric field. They reported that this device worked well after three months at room temperature, humidity, and pressure. However, additional studies for confirmation of this statement are required because real data are presented only for the operation within 3 h (see Figure 10.8). With regard to increasing the acoustic pressure, Koshida and co-workers (Asamura et al. 2002) proposed to use impulse mode of operation. They have shown that in this case the intensity of thermally induced ultrasound (TIU) can be heightened as strong as 10 kPa potentially and that a very small and strong sound beam emitter can be realized (Figure 10.9).

ACKNOWLEDGMENTS

This work was supported by the Ministry of Science, ICT and Future Planning (MSIP) of the Republic of Korea, and partly by the Moldova Government under grant 15.817.02.29F and ASM-STCU project #5937.

REFERENCES

Arnold, H.D. and Crandall, I.B. (1917). The thermophone as a precision source of sound. *Phys. Rev.* 10, 22–38.

Asamura, N., Saman Keerthi, U.K., Migita, T., Koshida, N., and Shinoda, H. (2002). Intensifying thermally induced ultrasound emission. In: *Proceedings of 19th Sensor Symposium on Micromachines and Applied Systems*, IEEJ, May 30–31, Kyoto, pp. 477–482.

Drinkwater, B.W. and Wilcox, P.D. (2006). Ultrasonic arrays for non-destructive evaluation: A review. *NDT&E Int.* 39(7), 525–541.

Gavrilchenko, I.V., Benilov, A.I., Shulimov, Yu.G., and Skryshe, V.A. (2009). Thermally induced acoustic waves in porous silicon. *Phys. Stat. Sol. (c)* 6(7), 1725–1728.

Gelloz, B., Sugawara, M., and Koshida, B. (2008). Acoustic wave manipulation by phased operation of two-dimensionally arrayed nanocrystalline silicon ultrasonic emitters. *Jpn. J. Appl. Phys.* 47(4), 3123–3126.

Haller, M.I. and Khuri-Yakub, B.T. (1998). A surface micromachined electrostatic ultrasonic air transducer. *IEEE Trans. Ultrasonics Ferroelectr Freq. Control* 43, 1–6.

Hirota, J., Shinoda, H., and Koshida, N. (2004). Generation of radiation oressure in thermally induced ultrasonic emitter based on nanocrystalline silicon. *Jpn. J. Appl. Phys.* 43, 2080–2082.

Holman, J.P. (1963). *Heat Transfer*. McGraw-Hill, New York.

Hu, H., Wang, Y., and Wang, Z. (2012). Wideband flat frequency response of thermo-acoustic emission. *J. Phys. D: Appl. Phys.* 45, 345–401.

Isozaki, A., Nakai, A., Matsumoto, K., and Shimoyama, I. (2007). Nanocrystalline porous silicon ultrasonic transmitter with patterned emission area. In: *Proceedings of the IEEE International Conference on Micro Electro Mechanical Systems (MEMS)*, January 21–25, Kobe, Japan, pp. 55–58.

Kihara, T., Harada, T., Hirota, J., and Koshida, N. (2004). Ultrasound emisson characteristics of a thermally induced sound emitter employing a nanocrystalline silicon layer. *Jpn. J. Appl. Phys.* 43, 2973–2975.

Koshida, N., Migita, T., Kishimoto, Y., Fuchigami, M., and Shinoda, H. (2001). Novel ultrasonic technology by nanocrystalline porous silicon. In: Schmuki, P., Lockwood, D.J., Ogata, Y.H., and Isaacs, H.S. (Eds.) *Pits and Pores II: Formation, Properties, and Significance for Advanced Materials*. ECS, Pennington, NJ, pp. 326–331.

Manthey, W., Kroemer, N., and Magori, V. (1992). Ultrasonic transducers and transducer arrays for applications in air. *Meas. Sci. Technol.* 3, 249–261.

McDonald, F.A. and Wetsel, G.C. Jr. (1978). Generalized theory of the photoacoustic effect. *J. Appl. Phys.* 49, 2313–2322.

Migita, T., Shinoda, H., and Koshida, N. (2002). Transient and stationary characteristics of thermally induced ultrasonic emission from nanocrystalline porous silicon. *Jpn. J. Appl. Phys.* 41, 2588–2590.

Mo, J.H., Fowlkes, J.B., Robinson, A.L., and Carson, P.L. (1992). Crosstalk reduction with a micromachined diaphragm structure for integrated ultrasonic transducer arrays. *IEEE Trans. Ultrasonics Ferroelectr. Freq. Control* 39, 48–53.

Shinoda, H., Nakajima, T., Ueno, K., and Koshida, N. (1999). Thermally induced ultrasonic emission from porous silicon. *Nature* 400, 853–855.

Tsubaki, K., Komoda, T., and Koshida, N. (2004). Improvement in the efficiency of thermally induced ultrasonic emission from porous silicon by nano-structural control. In: *Proceedings of 206th Meeting of the Electrochemical Society,* October 3–8, Honolulu, Hawaii, Abs. 1028.

Tsubaki, K., Yamanaka, H., Kitada, K., Komoda, T., and Koshida, N. (2005). Three-dimensional image sensing in air by thermally induced ultrasonic emitter based on nanocrystalline porous silicon. *Jpn. J. Appl. Phys.* 44, 4436–4439.

Tsubaki, K., Komoda, T., and Koshida, N. (2006). Acoustic emission characteristics of nanocrystalline porous silicon device driven as an ultrasonic speaker. *Jpn. J. Appl. Phys.* 45(4B), 3642–3644.

Wang, Z., Zhu, W., Tan, O.K., Chao, C., Zhu, H., and Miao, J. (2005). Ultrasound radiating performances of piezoelectric micromachined ultrasonic transmitter. *Appl. Phys. Lett.* 86(3), 033508.

Wang, Y.D., Hu, H.P., and Wang, D.D. (2014). Generalized theory of thermo-acoustic emission from nanocrystalline porous silicon. *Appl. Mech. Mater.* 472, 734–738.

Watabe, Y., Honda, Y. and Koshida, N. (2006a). Characteristics of thermally induced ultrasonic emission from nanocrystalline porous silicon device under impulse operation. *Jpn. J. Appl. Phys.* 45(4B), 3645–3647.

Watabe, Y., Honda, Y., and Koshida, N. (2006b). Tunable output directivity of thermally induced ultrasound generator based on nanocrystalline porous silicon. *Jpn. J. Appl. Phys.* 45(9A), 7240–7242.

Porous Silicon in Micromachining Hotplates Aimed for Sensor Applications

11

Ghenadii Korotcenkov and Beongki Cho

CONTENTS

11.1   Micromachining and Its Applications   208

11.2   Advantages of Porous Silicon in Micromachining Applications   210

11.3   Micro Hotplates with PSi Active Layer   217

Acknowledgments   220

References   220

11.1 MICROMACHINING AND ITS APPLICATIONS

Another promising direction in the use of porous silicon (PSi) in sensor design is micromachining applications (Lang et al. 1994; Steiner and Lang 1995; Lang 1996; Racine et al. 1997; Hedrich et al. 2000). Silicon-based micromachining is being used for fabrication of small-scale mechanical devices that are integrated with conventional microelectronics. In this way, it has become possible to produce miniaturized devices, which combine mechanical, electrical, and thermal functionalities within a single piece of silicon. Examples of micromachined devices include motors, movable mirrors, optical shutters, electromechanical filters, cantilevers, and a wide variety of micro hotplates for sensors (Lang 1996; Splinter et al. 2001; Goericke et al. 2008). Many of those structures have been fabricated on freestanding membranes, which can be relatively easily fabricated and used, and can interface with existing laboratory equipment and integrated microfluidic handling systems. For example, microcantilevers with integrated piezoresistive strain sensors are used mainly to replace optical deflection sensing (Tortonese et al. 1993), but they are also employed in various sensing applications such as gas flow sensing (Su et al. 2002), acceleration sensing (Roylance and Angell 1979), and microjet measurements (Lee et al. 2007). In addition, the portfolio of micromachining processes also includes an increasing variety of silicon-compatible functional materials to enable sensor and actuator functionalities that are impossible to realize using the silicon base material itself. Especially for use as bio/chemical sensors, piezoresistive microcantilevers or micro hotplates are often prepared with a selective coating that is sensitive to a specific analyte (Thundat et al. 1995). For example, analyte adsorption induces static deflection by creating a surface stress in microcantilevers; therefore, embedded piezoresistor microcantilevers can measure analyte adsorption (Boisen et al. 2000; Jensenius et al. 2000). It should be noted that the driving force in all these miniaturization attempts is arriving at small-size low-cost devices, either for use in handheld instruments or as miniaturized low-power consumption devices in distributed and bus-connected sensor networks. In all these applications, micromachining fabrication technologies are not only interesting because of their miniaturization potential but also because of their potential of providing high performance at low cost in mass production scenarios (Spannhake et al. 2009).

Thus, micromachining technology has a very wide area of application. However, in this chapter we will discuss micromachining technology only with reference to micro hotplates: a basic micro-electromechanical system (MEMS) thermal microstructuctures on a silicon substrate with miniature thermally isolated areas. Different types of micro hotplates on silicon substrate have been reported in the literature. Most common are micro hotplates based on (a) suspended membranes (Spannhake et al. 2009) and (b) cantilevers (Lee and King 2007). Detailed description of the features of such platforms' manufacturing can be found in Gardner et al. (2001), Simon et al. (2001), Laconte et al. (2006), and Vasiliev et al. (2009). Miniaturized hotplates are very important parts of pellistor (Furjes et al. 2004), resistor-type gas sensors (Maccagnani et al. 1999; Simon et al. 2001; Splinter et al. 2001), thermometric sensors (Kaltsas and Nassiopoulou 1999; Furjes et al. 2004; Nassiopoulou 2005), piezoresistive force sensors (Adam et al. 2008), capacitance gas and humidity sensors (Hille and Strack 1992), MOSFET gas sensors (Briand et al. 2001), flow sensors (Kim et al. 2003), and so on. (see Figure 11.1). In all indicated sensing technologies, thermal microstructures often play the dual role of a sensor and an actuator within the same device (Spannhake et al. 2009). Such a situation, for instance, arises in the case of thermal conductivity gas sensors (Nassiopoulou 2005; Spannhake et al. 2009). Such sensors consist of a heated membrane or cantilever, which is cooled by the ambient atmosphere. As an actuator, this microstructure produces the heat that is carried away by the molecules in the ambient atmosphere. As a sensor, this microstructure senses the heat that is being carried away and thus provides information concerning the molecular composition of the ambient atmosphere. This dual role can also be observed in catalytic gas sensors, that is, pellistors (Furjes et al. 2004). Such devices first produce the heat that is required to initiate chemical reactions of the analyte molecules at a catalytically active surface and second they detect the amounts of heat that are generated as the analyte molecules are being burnt at the sensor surface.

In all those applications, the thermal properties of the sensing elements determine the functional operation. In order to operate thermal sensors and actuators efficiently, it is necessary to produce maximum temperature changes from a minimum amount of input energy. It should be noted that minimizing power consumption is one of the main requirements for sensor technology, especially when the devices are intended for use in portable systems. For example,

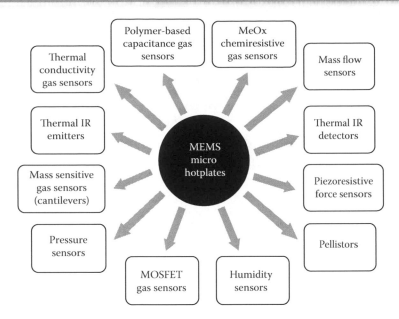

FIGURE 11.1   Possible fields of micro hotplate applications.

conventional combustive and semiconductive oxide gas sensors, which operate at elevated temperatures between 300 and 600°C, typically consume 0.2–1.0 W, while the single sensing element aimed for application in portable devices must have a power dissipation of less than 10–50 mW. The use of thermal microstructures having small heat capacitance with a high degree of thermal insulation gives the possibility to resolve this problem. Therefore, incorporation of thermally isolated micro hotplates in the construction of the indicated devices allows reducing considerably both the power consumption and the thermal transient time.

Experiment has shown that the best way of obtaining active device structures with a low heat capacitance and a high degree of thermal insulation is building thin-film stacks on silicon wafers with predeposited dielectric passivation layers. Thermal insulation of the active device structures is attained by removing the silicon substrate underneath the active devices (Simon et al. 2001); for example, the closed-type membrane, where the membrane overlaps the silicon substrate along its periphery and the suspended-type membrane, where the membrane element is supported to the Si substrate by means of supporting beams, its central part being suspended over a cavity etched in the substrate. In the latter case, the thermal losses to the substrate take place only through the supporting beams, and they are thus minimized compared to the closed-type membrane. This permits a local rise of the temperature of several hundreds of degrees, while neighboring areas remain at room temperature. Thus, in addition to minimization of power consumption, it is possible to separate micro hotplate and hot sensing layers from contact pads.

The materials chosen for the membrane of the micro hotplate should combine low thermal conductivity (i.e., small thickness) with high mechanical strength (i.e., large thickness). Therefore, standard materials for forming such suspended membrane structures are silicon dioxide (SiO_2) and silicon nitride (Si_3N_4)—either in the form of layer stacks or in the form of thin-film alloys—to arrive at a small level of tensile mechanical stress inside the suspended membrane (Rossi et al. 1998). Of course, silicon membrane devices also can be designed (Muller et al. 2003). In this case, the membrane structures and their mechanical suspensions consist of mono-crystalline silicon. An easy and efficient way of producing such devices is employing silicon-on-insulator (SOI) techniques. However, because the thermal conductivity of bulk silicon is roughly 100 times higher than that of SiO_2 or Si_3N_4, ultra-low-power-consumption devices cannot be made with this kind of SOI technology. This means that the power consumption of a silicon membrane is much more than that of SiO_2 membrane for the same maximum temperature. For SiO_2 and Si_3N_4 films forming a chemical vapor deposition (CVD) or other methods can be used. For achievement of maximal thermal insulation, the thickness of the suspended membrane should be minimal. For example, Laconte et al. (2006) designed micro hotplates with 1-μm thick $SiO_2/Si_3N_4/SiO_2$ (400 nm/300 nm/300 nm) membrane. Due to the indicated structure, these membranes showed very low residual stress (tensile stress of

80 MPa) and low strain gradient. Newer developments also include silicon carbide membranes, which are additionally often suspended from the substrate to reduce thermal capacity and losses even further. Despite very good power efficiency and low heating times, these thin membranes are mechanically rather fragile, which significantly limits their use.

Forming a Pt meander on top of such a membrane, the structure can be used as both a thermal sensor and an actuator device. This double function is enabled by the fact that Pt exhibits a positive and almost linear temperature coefficient of resistivity (TCR). Pt meanders, therefore, can serve both as active heating as well as passive temperature sensing elements. In order to avoid catalytic interactions with the heated Pt meander, the meander is coated with a thin, chemically inert layer of SiO_2. Very often, such passivation layers form the substrate for further functional layers that may be required to arrive at a particular sensor or actuator function (Spannhake et al. 2009). Other materials such as passivated polysilicon can also be used for design of internal heaters (Tsamis et al. 2003). However, using polysilicon has an important disadvantage. Polysilicon heater has slow drift of resistance (about 30% per year at 450–500°C). This drift is due to silicon oxidation, which is not blocked completely by protective layers of silicon oxide or nitride of ~1 µm typical of microelectronics. Another problem is the stability of contacts to poly-silicon.

It should be noted that during the last decades, different research groups were involved in the design and fabrication of various hotplate platforms for sensor applications. However, there are several problems not solved completely until now in the silicon micromachining. These problems are as follows (Vasiliev et al. 2009):

1. Stability of the membrane at high temperature. This problem is due to very strong stresses in silicon oxide and silicon nitride used for the fabrication of the sensor. These stresses have opposite signs for both materials, and the application of a multilayer structure permits decreasing overall stress in the membrane. However, fatigue of the membrane leads finally to its destruction even at constant temperature.
2. Stability of the heater material and its adhesion to membrane material at high temperature.
3. Stability of silicon nitride at high temperature in humid atmosphere.
4. Adhesion of sensing layer to the membrane material.
5. High production cost.

11.2 ADVANTAGES OF POROUS SILICON IN MICROMACHINING APPLICATIONS

It was shown that both silicon bulk and surface micromachining could be used for hotplate fabrication (Mlcak et al. 1994; Lang 1996; Gardner et al. 2001; Simon et al. 2001). In addition, there is epi-micromachining which is a variation on the surface micromachining. Bulk micromachining can be divided into two main groups: wet and dry. There are also other techniques such as laser drilling and sand blasting. The first to be developed was wet etching. In wet bulk micromachining, the silicon wafer is etched from the back in defined regions (Lang et al. 1994; Lang 1996; Gardner et al. 2001; Splinter et al. 2001). Therefore, a masking layer structured with photolithography on the back of a double-sided polished wafer is needed. The anisotropic chemical wet etch can be realized with potassium hydroxide (KOH), tetramethyl ammonium hydroxide (TMAH), hydrazine, or ethylenediamine pyrocatechol (EDP) into a single-crystalline silicon (see Table 11.1). All of these processes are relatively low temperature and therefore can be used as postprocessing after IC processing, although care should be taken to protect the front side of the wafer during etching. The etch process is defined by the crystal planes of the silicon, the etch rates of (111)-planes are less than the other directions (Bean 1978). Thereby, perfectly oriented silicon substrates and accurately justified masking layers are necessary. Because of the crystal planes and the etch slope of 54.7°, the needed substrate place is larger than the membrane dimensions. This limits the maximum packaging density of the membrane structures.

Dry deep reactive ion etching (DRIE) addressed some of the limitations of wet etching, although the process is more expensive. Description of this process can be found in *Porous Silicon: Formation and Properties* (Chapter 10). Two main processes of DRIE are cryogenic and

TABLE 11.1   Properties of Main Anisotropic Etchants

		Etch Rate				
Etchant	Mask	(100) μm/min	(100)/(111)	SiO₂ (nm/h)	SiNₓ (nm/h)	Comments
Hydrazine	SiO₂, SiNₓ, Metals	0.5–3	16:1	10	≪10	Toxic, potentially explosive
EDP	Au, Cr, Ag, Ta, SiO₂, SiNₓ	0.3–1.5		12	6	Toxic
KOH	SiNₓ, Au	0.5–2	Up to 200:1	170–360	<1	Not cleanroom compatible
TMAH + IPA	SiO₂, SiNₓ	0.2–1	Up to 35:1	<10	<1	Expensive

Source: Data extracted from French P.J. and Sarro P.M., In: *Micromachining Techniques for Fabrication of Micro and Nano Structures*, Kahrizi M. (ed.). InTech, Rijeka, Croatia, 253, 2012.

Note: IPA = isopropyl alcohol.

Bosch processes (Roozeboom et al. 2008). The cryogenic process works at about −100°C and uses oxygen to passivation of the sidewall during etching to maintain vertical etching. The Bosch process uses a switching between isotropic etching, passivation, and ion bombardment. This results in a rippled sidewall, although recent developments allow faster switching without losing etch rate, thus significantly reducing the ripples. The etching can be performed from both frontside and backside and can be combined with the electronics.

Surface micromachining is quite different from bulk micromachining in terms of both processing steps and dimensions. This involves the deposition of thin films and selective removal to yield freestanding structures (Gardner et al. 2001). The basic process is given in Figure 11.2, although this can be augmented with additional sacrificial and mechanical layers. An easily etchable sacrificial layer is deposited on to the substrate surface, followed by a second layer that will form the membrane. Thus, sacrificial layer is deposited only to support another layer on top of it during part of the process. Then it is removed and a cavity is left behind. The sacrificial layer is removed at the end of process with a liquid or gaseous etching medium. There are many possible combinations of sacrificial and mechanical layers and a few examples are given in Table 11.2.

The drawback of this method is the limited distance that can be obtained between the membrane and the substrate. The limiting factor is defined by the maximum thickness obtainable for the sacrificial layer, which is typically limited to several micrometers. Although this distance may be sufficient for such applications as micromotors, sensing often requires the thickness of the sacrificial layer to be several tens of micrometers, in order to reduce heat transfer to the substrate. Particulars of bulk and surface micromachining techniques are shown in Figure 11.3. Both techniques have advantages and disadvantages (Steiner and Lang 1995; Dusco et al. 1997; Hedrich et al. 2000; Simon et al. 2001; Splinter et al. 2001). For example, the anisotropic etching of bulk

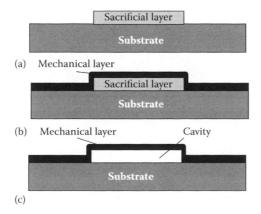

FIGURE 11.2   Basic surface micromachining process: (a) deposition and patterning of sacrificial layer, (b) deposition and patterning of mechanical layer, and (c) sacrificial etching.

TABLE 11.2   Examples of Combinations of Sacrificial and Mechanical Layers

Sacrificial Layer	Mechanical Layer	Sacrificial Etchant
Silicon dioxide	Polysilicon, silicon nitride, silicon carbide	HF
Silicon dioxide	Aluminum	Pad etch, 73% HF
Polysilicon	Silicon nitride, silicon carbide	KOH
Polysilicon	Silicon dioxide	TMAH
Resist, polymers	Aluminum, silicon carbide	Acetone, oxygen plasma

Source: Data extracted from French P.J. and Sarro P.M., In: *Micromachining Techniques for Fabrication of Micro and Nano Structures*, Kahrizi M. (ed.). InTech, Rijeka, Croatia, 253, 2012.

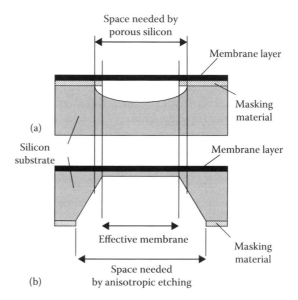

FIGURE 11.3   Comparison of micromachined membranes fabricated using (a) PSi as a sacrificial layer material and (b) bulk micromachining technology. (Adapted from Splinter A. et al., *Sens. Actuators B* 76, 354, 2001. Copyright 2001 Elsevier. With permission.)

micromachining offers the possibility of different membrane materials because the alkaline etching solution is selective to different oxide and nitride layers and the membrane can be constructed with bulk silicon as well, while the surface micromachining technique provides greater flexibility in lateral geometry design (i.e., size, density) and offers more possibilities for integrating sensors and read-out circuits in a large array (Adam et al. 2008).

Steiner and Lang (Lang et al. 1994; Steiner and Lang 1995; Lang 1996) and Splinter et al. (2001) believed that PSi technology is an ideal approach to combine the advantages of both technologies. According to their analyses (Lang et al. 1994; Steiner and Lang 1995; Lang 1996; Hedrich et al. 2000; Splinter et al. 2001), the following features are particular to using PSi layers:

■ The ability to produce very thick layers that extend over the whole wafer thickness. This property is employed in manufacturing thermal sensor membranes, which require a sacrificial layer with a thickness not less than 50 μm. Sacrificial layers with this thickness cannot be obtained with such commonly used layers as silicon dioxide or doped silicate glasses. Because of its greater thickness, PSi offers the possibility of creating a large air gap between the membrane and the substrate.
■ Fast chemical reactions caused by the large inner surface of PSi. This permits quick and easy removal of the layer at room temperature. This means that the sacrificial layer can be removed using low-concentration alkaline solutions at room temperature instead of aggressive HF.

- No reverse side lithography necessary. In this case, the applied technology (1) provides easier packaging of the final sensor; (2) reduces problems connected with the sensor mounting and the use of an adhesive; (3) produces higher yields; and (4) leads to better mechanical stability of the wafer.
- Lower heat conductivity than with bulk silicon (two or three orders of magnitude, depending on the preparation of the PSi layers) (Tsamis et al. 2003).
- No geometric restrictions caused by the crystallographic orientation of the substrate used. Unlike anisotropic etching, the geometry of the PSi layers is not limited to certain planes, and so they can be formed locally on a wafer with controlled undercutting. Geometry has been a major hindrance in the development of special sensors, for example, for optical applications that require circular designs.
- High selectivity of PSi formation in terms of type and doping level of the substrate. This property can be used to define an etching stop in silicon.
- Geometrically arbitrary structures can be designed using appropriate surface masking layers.
- Good possibilities for further thin-film deposition on the surface of PSi layers. For example, a high-quality epitaxial layer can be formed over PSi. These epitaxial layers can be used in silicon-on-insulator epitaxial (SOI-Epi) devices as well as in complementary-metal-oxide-semiconductor (CMOS) devices. The CMOS compatibility has been demonstrated by manufacturing electronic circuits on silicon islands isolated by oxidized PSi.

Thus, surface micromachining on PSi layers can lead to a considerable decrease in chip size, an increase in the number of chips per wafer, and consequently a decrease in the chip unit cost. Typical processes for membrane and cantilever fabrication using PSi sacrificial layers are presented in Figure 11.4 and Table 11.3.

As it was shown before, the PSi technique is extremely simple and can be applied as a postprocessing step and it is therefore fully compatible with the electronic circuitry. The only remaining problem is to protect the areas of electronics and metallization from the HF etchant. One disadvantage of this technique is the added process complexity introduced by the requirement of a backside electrical contact during etching. According to Lang (1996), other important technical problems in using PSi layers are the stress in the film structures and the possible sticking of the structures to the substrate because of the small gap. Since the structures are set free by lateral undercutting, freestanding planes cannot be realized without holes for the etchant.

According to the research (Steiner and Lang 1995; Dusco et al. 1997; Splinter et al. 2001), for application of PSi as a sacrificial layer material, it is necessary to form pores with sizes in the range

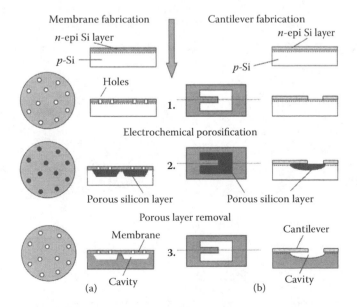

FIGURE 11.4 Process sequence for the fabrication of (a) membrane and (b) cantilever using PSi as a sacrificial layer. 1, photolithography; 2, Si porosification; 3, PSi etching.

TABLE 11.3   Typical Technological Parameters of Micro Hotplates Fabrication

Silicon Substrate	Mask	Electrolyte	Cavity Thickness	Membrane	PSi Removal	References
p-Si(100)	Si_3N_4	HF:ethanol = 7:3	~72 μm	Si-100 × 100 μm 5 μm thick Si membrane	2 wt.% KOH	Dusco et al. 1997
p-Si(100)	Si_3N_4	HF:ethanol (25%) $J = 15$–120 mA/cm²	5–30 μm	PSi-2.2 × 2.2 mm² 50 μm thick		Maccagnani et al. 1998
p-Si(100)	Si_3N_4	HF:ethanol (60%) $J = 80$ mA/cm²	~60 μm	PSi-100 × 100 μm² ~4 μm thick	SF_6 plasma	Tsamis et al. 2003
p-Si(100)	SiO_2/polySi	HF:ethanol = 4:1 $J = 70$ mA/cm²	>100 μm	Si_3N_4 2 μm	0.5 wt.% KOH	Splinter et al. 2001

between 4 and 50 nm (mesoporous silicon). These pore sizes guarantee a physically stable layer. If the dimensions are too small, the porous structure becomes unstable. Removal of the PSi sacrificial layer can be accomplished with highly dilute alkaline solutions, especially KOH or NaOH at room temperature in a concentration of 0.5–5 wt%. High-density plasma etching can also be used for this purpose (Tsamis et al. 2003).

Before the sacrificial layer is removed, however, the wafer should be diced. Splinter et al. (2001) have shown that for freestanding structures, dicing is a critical step in the micromachining process. The rising particles and the cleaning water destroy the membrane. The best solution for this problem may be a fabrication process in which removing the PSi is the last stage in the fabrication of devices, after all necessary components have been applied on the membrane's surface. If the sacrificial layer stays under the membrane during dicing, the device is protected from mechanical strain. At the same time, PSi allows removing of the sacrificial layer after all the other layers are formed because KOH in low concentration at room temperature does not attack the other standard layers. The dicing tape holds the single chips, but removal can be done on the wafer scale.

FIGURE 11.5 (a) Suspended PSi micro hotplate with a Pt heater. The thickness of the membrane is 4 μm. The depth of the cavity under the membrane is more that 60 μm. (Reprinted from *Sens. Actuators B* 95, Tsamis C. et al., 78, Copyright 2003, with permission from Elsevier.) (b) SEM view of thermally decoupled membrane array. (Reprinted from *Sens. Actuators B* 76, Splinter A. et al., 354, Copyright 2001, with permission from Elsevier.) (c) SEM view of a freestanding sensor platform fabricated using a sacrificial layer of PSi. (Reprinted from *Superlattice Microstruct.* 35, Furjes P. et al., 455, Copyright 2004, with permission from Elsevier.) (d) SEM image of a micro hotplate on the top of its active area. SnO_2:Pd is deposited by micro-dropping. (Reprinted from *Microelectron. Eng.* 85, Triantafyllopoulou R. et al., 1116, Copyright 2008, with permission from Elsevier.)

TABLE 11.4   Thermal Characterization of Designed Micro Hotplates

		Parameters			
References	Type	Size of Sensing Element	Max. Temp. (°C)	Diss. Power (°C/mW)	t_T(ms)
Dusco et al. 1997	Membrane	$100 \times 100\ \mu m^2$	350	13	
Tsamis et al. 2003	Membrane	$100 \times 100\ \mu m^2$	600	20	
Furjes et al. 2004	Membrane	$100 \times 100\ \mu m^2$	550	27	4.2
Adam et al. 2008	Membrane	$300 \times 300\ \mu m^2$			
Triantafyllopoulou et al. 2006	Cantilever	$100 \times 100\ \mu m^2$	900	180–340	

Note: t_T = time to maximum temperature.

Typical configurations of freestanding gas sensor platforms fabricated using PSi are shown in Figure 11.5.

Some achievements in the field of micro hotplate fabrication are presented in Table 11.4. For example, using the hotplate shown in Figure 11.6b, temperatures above 800°C are achieved with power consumption below 35 mW (Han et al. 2001). In platforms designed by Furjes et al. (2002, 2004), the required measuring temperature range of 100–800°C was achieved with 3–30 mW. Hotplates with such low power consumption are an ideal platform for portable sensors. At that, Csíkvári et al. (2009) showed that to achieve these parameters, a microfilament heater structure should be encapsulated in the appropriate material. For example, the use of diamond layers is not optimal because of their high thermal conductivity (see Figure 11.6a).

In addition, freestanding membranes have very small thermal capacity, which allows rapid variation of the operation temperature. For example, the temperature response time of the hotplate designed by Furjes et al. (2004) was less than 5 ms. Splinter et al. (2001) believed that this fast temperature modulation was the starting point for a completely new measurement concept, with a multitude of new and advanced possibilities to optimize gas sensor selectivity in the presence of multicomponent gas mixtures.

Contacts in micro hotplates can be formed using various materials. However, Furjes et al. (2002) believe that Pt/Ti, polysilicon, and $TiSi_2$ are the most suitable materials for the contacts forming. The main issue in the selection of contact material is the high temperature of the contacts during subsequent processing and functional operation. There are several alternatives; however, materials of ideal properties cannot be found when considering their high thermal and low electrical conductivity, unfavorable TCR dependence versus temperature, high residual stress, their chemical reactivity with Si at high temperature and non-compatibility with silicon processing. Most of the materials are soluble in HF used in PSi processing (in the case of Pt the adhesive

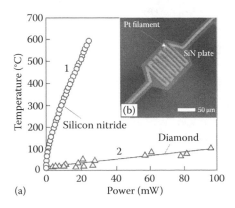

FIGURE 11.6   (a) Temperature versus power characteristics of the analyzed silicon-nitride and diamond-coated heater filaments; (b) SEM image of hotplate with Pt filament. ([a] Reprinted from *Microelectron. J.* 40, Csíkvári P. et al., 1393, Copyright 2009, with permission from Elsevier. [b] From Furjes P. et al., *J. Micromech. Microeng.* 12, 425, 2002; Copyright 2002: IOP. With permission.)

Ti layer); therefore, the wires have to be encapsulated by another SiN$_{1.05}$ layer. Unfortunately, the high deposition temperature of the second SiN$_{1.05}$ layer makes the application of Al impossible. NiFe is far from being a CMOS-compatible material at all.

Experiment has shown that PSi-based micro hotplates with low power consumption are also a good platform for designing gas sensor arrays (see Figures 11.7 and 11.8) aimed for application in electronic noise (Triantafyllopoulou et al. 2008; Barsony et al. 2009). The fact that the sensor element on the top of the membrane is thermally isolated from the bulk silicon allows the generation of a sensor array with thermally decoupled sensors on one chip. Each sensor of this array can operate at a different temperature or can use different gas sensing material. In comparison to a hybrid integrated sensor array, the advantages of this single-chip monolithic integration concept are the very small dimensions of the array, low production costs, and low power consumption of the complete array.

However, we need to note that the hotplate is one of the simplest technical elements whose fabrication is based on micromachining technology. How complicated the structure of devices fabricated using PSi-based micromachining technology can be, one can see in the results presented by Lammel et al. (2002).

Recently, it has been suggested that for some applications, the sacrificial PSi layer may not be removed (Mondal et al. 2011; Lucklum et al. 2013, 2014), as was demonstrated over almost

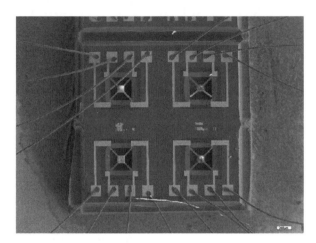

FIGURE 11.7   SEM image of sensors array of various sensitive materials (SnO$_2$:Pd and WO$_3$:Cr) designed on the base of PSi micro hotplates. The deposition of two different sensitive materials took place, using the micro-dropping technique. (Reprinted from *Microelectron. Eng.* 85, Triantafyllopoulou R. et al., 1116, Copyright 2008, with permission from Elsevier.)

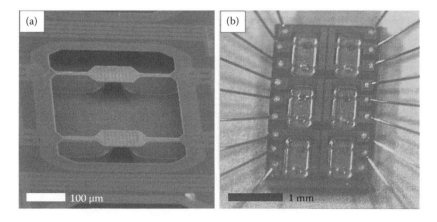

FIGURE 11.8   (a) The porous silicon bulk micromachined, suspended, and mechanically supported hotplates forming a micropellistor pair, and (b) the assembled, integrated microfilament array with six hotplate pairs described in detail in Furjes et al. (2002). (From Barsony I. et al., *Meas. Sci. Technol.* 20, 124009, 2009. Copyright 2009: IOP. With permission.)

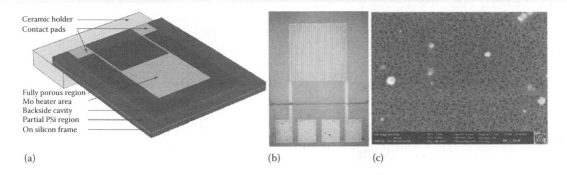

FIGURE 11.9 Design of microheater measurement setup (a), micrograph of fabricated heater meander on fully porous region (b), and SEM image of porous surface with around 15-nm pore sizes and sponge-like morphology (c). (Reprinted from *Sens. Actuators A* 213, Lucklum F. et al., 35, Copyright 2014, with permission from Elsevier.)

two decades ago for flow sensors (Tabata 1986). The low thermal conductivity of *p*-porous silicon means that the PSi may provide sufficient thermal isolation from the substrate. This makes removing the PSi to form the air gap unnecessary. As a result, such an approach improves the mechanical strength of the device while providing almost identical thermal isolation. However, it should be noted that such platforms by their operating parameters such as power consumption and the temperature response time are inferior to platforms based on the membranes (Mondal et al. 2011; Lucklum et al. 2013, 2014). According to Lucklum et al. (2014), the power efficiency of the 10-mm² molybdenum heaters was about 0.40°C/mW. A micro heater designed by Lucklum et al. (2014) is shown in Figure 11.9. Micro hotplates with Fe/Ni/Co heater designed by Mondal et al. (2011) had the power efficiency equal to 1.6°C/mW. The maximum temperature of 150°C was achieved at a power consumption of 95 mW. It should be noted that in such micro hotplates in contrast to membrane-based hotplates, the power efficiency increases with increasing thickness of the PSi layer (Mondal et al. 2011). For membrane-based hotplates for achievement of maximum power efficiency, the membrane thickness should be minimal.

11.3 MICRO HOTPLATES WITH PSi ACTIVE LAYER

In the previous section we have shown that PSi can be successfully used in fabrication of silicon microstructures (Garel et al. 2007). However, experiments showed that PSi can also be used as a sensing element incorporated in micro hotplate of bio and chemical sensors. As it was demonstrated, the PSi exhibits extremely high chemical reactivity due to a well-developed system of microscopic pores. Different variants of this approach in the development of sensors have been described by Domanski et al. (2002, 2003) and Hwang et al. (2010, 2012). Usually, the main element of such sensors is the nanometer scale resonator made as a cantilever (Domanski et al. 2002, 2003; Meltzman et al. 2014; Lee et al. 2015) or a torsional beam resonator (Hwang et al. 2010, 2012). These elements are shown in Figure 11.10.

The basic principle of operation of such sensors is binding events on the resonator surface altering the mechanical stress and total mass on the sensor, allowing for measurement of the mass via a shift in resonant frequency. Miniaturization of resonant-based sensors is one of the most popular strategies for improvement of sensitivity for chemical detection. Because of the relatively small mass of the microstructures, a very high mass sensitivity can be achieved. For example, Domanski et al. (2003) proposed a piezoresistive cantilever beam structure with PSi adsorbing spot as a high sensitive gas sensor. In this structure piezoresistors, necessary for resonance measurement, are placed at the support of the cantilever (Tortonese et al. 1991), while PSi spot is formed on the free end of the beam to obtain maximum sensitivity. Decrease of the resonance frequency due to gas adsorption on the free end of the cantilever beam can be estimated using the formula:

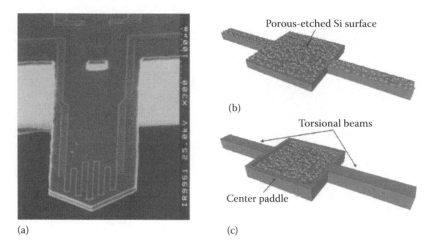

FIGURE 11.10 (a) Cantilever beam integrated with piezoresistors and microheater, (b, c) schematic diagrams of (b) the fully porous silicon resonator, and (c) the partially porous silicon resonator. While both center paddle and torsional beams of the fully porous structures consist of porous-etched silicon, for the partially porous resonator, only the center paddle is porous-etched, and the torsional beams and edges of the center paddle remain nonporous-etched. ([a] Reprinted from *Microelectron. Eng.* 57–58, Zaborowski M. et al., 787, Copyright 2001, with permission from Elsevier. [b] Idea from Hwang Y. et al., *J. Microelectromech. Syst.* 21(1), 235, 2012.)

$$\Delta f_0 = \frac{1}{2\pi}\left(\sqrt{\frac{k}{M+\Delta m}} - \sqrt{\frac{k}{M}} \right), \tag{11.1}$$

where M is cantilever weight, Δm is the weight of the adsorbed substance, and k is the beam elasticity coefficient.

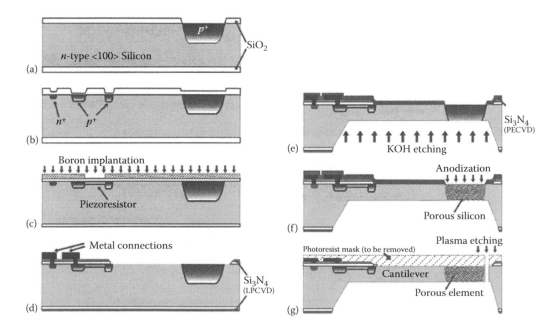

FIGURE 11.11 Sequence of fabrication of a piezoresistive cantilever with a porous element: (a) deep boron diffusion, (b) diffusion of connection paths, (c) implantation for piezoresistors forming, (d) metal connection fabrication, (e) anisotropic deep etching of silicon, (f) formation of porous silicon, and (g) cantilever formation. (Adapted from Domanski K. et al., *J. Vac. Sci. Technol. B* 21(1), 48, 2003. Copyright 2003: American Vacuum Society. With permission.)

Fabrication of the piezoresistive cantilever-based devices designed by Domanski et al. (2003) was based on double-side silicon bulk/surface micromachining combined with standard CMOS processing. The fabrication process is schematically illustrated in Figure 11.11. PSi (porosity ~65%, pore size ~20 nm, surface area ~200 m²/cm³) was obtained by anodization of monocrystalline silicon in hydrofluoric acid solutions.

The crucial problem of the process and its novelty consists in integration of the standard microprobe fabrication sequence with formation of the PSi element. The fragile structure of the well-developed PSi structure cannot withstand the high temperature required for manufacturing of the microelectronic structure of the device. Therefore, the PSi has to be formed at the last steps of the technological sequence. A micro heater made of resistive metal meander placed over the PSi area is integrated with the sensor to outgas the porous area after each measurement, thus enabling a permanent monitoring of rapid changes of the atmosphere. It should be pointed out that different expansion coefficients of two layers (silicon beam coated with oxide) cause the cantilever deflection induced by the adsorption of infrared radiation or electrical power supplied to the micro heater. Operating characteristics of humidity sensors designed using the above-mentioned approach are shown in Figure 11.12a. The same figure (Figure 11.12b) presents results related to sensor response of gas sensors designed on the basis of torsional beam resonator. Hwang et al. (2012) have demonstrated that response of sensors with fully porous resonators and partially porous resonators increased up to 261% and 165% as compared to nonporous silicon resonators.

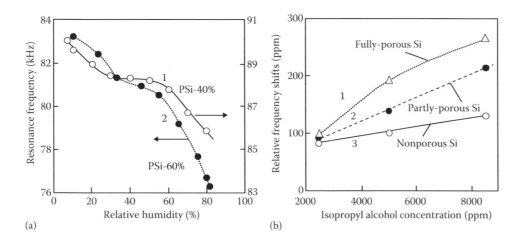

(a)   (b)

FIGURE 11.12 (a) Variation of resonance frequency as a function of relative humidity for the T-shaped cantilevers with different porosity of porous element. (From Domanski K. et al., *J. Vac. Sci. Technol. B* 21(1), 48, 2003. Copyright 2003: American Vacuum Society. With permission.) (b) Comparison of relative frequency shifts ($\Delta f/f_o$) of torsional beam resonant gas sensor versus concentration of isopropyl alcohol (IPA) vapor. (Data extracted from Hwang Y. et al., *J. Microelectromech. Syst.* 21(1) 235, 2012.)

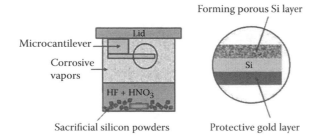

FIGURE 11.13 Vapor phase stain-etching fabrication of the porous silicon-over-silicon (PSOS) layer on the lower unmasked side of the microcantilevers. To start the process, sacrificial silicon was added to the solution, initiating the formation of brownish NO_x fumes. (Meltzman S. et al.: *J. Polymer Sci. B: Popymer Phys.* 2014. 52. 141. Copyright Wiley-VCH Verlag GmbH & Co. KGaA. Reproduced with permission.)

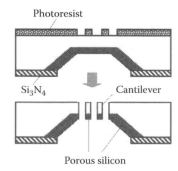

FIGURE 11.14 Cantilever process fabrication. (From Garel O. et al., *Micromech. Microeng.* 17, S164, 2007. Copyright 2007: IOP Publishing. With permission.)

We need to note that for the formation of PSilayers on the surface of mass sensitive micromachining sensors in addition to the anodization process (Domanski et al 2003; Lee et al. 2015), other methods of silicon porosification such as the electroless metal-assisted etching (Hwang et al. 2012) and the vapor phase stain-etching (Meltzman et al. 2014) can be used (read Chapter 10). An example of vapor phase stain-etching used for fabrication of the PSi on the surface of Si cantilever is shown in Figure 11.13.

Fundamentally different approaches to manufacturing microcantilever with incorporated layers of PSi was proposed by Garel et al. (2007). Cantilever beams were created by DRIE anisotropic etching of n-type porous membranes with a resist masking layer. The fabrication process is schematically illustrated in Figure 11.14. The PSi has to be protected by c-Si during photolithography because the AZ resist remover quickly dissolves the PSi. This process is delicate to implement because the cantilevers can be destroyed during all wet steps. PSi layers on the surface of n-Si(100) were grown in an electrolytic cell with a 1:1 (98 wt% ethanol:48 wt% HF) ethanoic-HF acid electrolyte using a DC current source (J = 80 mA/cm^2 and 160 mA/cm^2), at room temperature and in the dark.

ACKNOWLEDGMENTS

This work was supported by the Ministry of Science, ICT and Future Planning (MSIP) of the Republic of Korea, and partly by the National Research Foundation (NRF) grants funded by the Korean government (No. 2011-0028736 and No. 2013-K000315).

REFERENCES

Adam, M., Mohacsy, T., Jonas, P., Ducso, C., Vazsonyi, E., and Barsony, I. (2008). CMOS integrated tactile sensor array by porous Si bulk micromachining. *Sens. Actuators A* 142, 192–195.

Barsony, I., Adam, M., Furjes, P. et al. (2009). Efficient catalytic combustion in integrated micropellistors. *Meas. Sci. Technol.* 20, 124009.

Bean, K.E. (1978). Anisotropic etching of silicon. *IEEE Trans. Electron. Dev. ED*-25, 1185–1193.

Boisen, A., Thaysen, J., Jensenius, H., and Hansen, O. (2000). Environmental sensors based on micromachined cantilevers with integrated read-out. *Ultramicroscopy* 82, 11–16.

Briand, D., Sundgren, H., van der Schoot, B., Lundstrom, I., and de Rooij, N.F. (2001). Thermally isolated MOSFET for gas sending application. *IEEE Electron Device Letters* 2001(22), 11–13.

Csíkvári, P., Furjes, P., Ducso, Cs., and Barsony, I. (2009). Micro-hotplates for thermal characterisation of structural materials of MEMS. *Microelectron. J.* 40, 1393–1397.

Domanski, K., Tomaszewski, D., Grabiec, P. et al. (2002). Silicon piezoresistive cantilever beam with porous silicon element. In: *Proceedings of International Conference on Physics of Semiconductor Devices*, December 11-15, 2001, Delhi, India. Proc. of SPIE, Vol. 47461, 523–526.

Domanski, K., Grabiec, P., Marczewski, J. et al. (2003). Fabrication and properties of piezoresistive cantilever beam with porous silicon element. *J. Vac. Sci. Technol. B* 21(1), 48–52.

Dusco, Cs., Vazzonyi, E., Adam, M. et al. (1997). Porous silicon bulk micromachining for thermally isolated membrane formation. *Sens. Actuators A* 60, 235–239.

French, P.J. and Sarro, P.M. (2012). Integrated MEMS: Opportunities and challenges. In: Kahrizi M. (ed.) *Micromachining Techniques for Fabrication of Micro and Nano Stru*ctures. InTech, Rijeka, Croatia, pp. 253–276.

Furjes, P., Vizvary, Zs., Adam, M., Barsony, I., Morrissey, A., and Ducso, Cs. (2002). Materials and processing for realization of micro-hotplates operated at elevated temperature. *J. Micromech. Microeng.* 12, 425–429.

Furjes, P., Ducso, Cs., Adam, M., Zettner, J., and Barsony, I. (2004). Thermal characterisation of micro-hotplates used in sensor structures. *Superlattice Microstruct.* 35, 455–464.

Gardner, J.W., Varadan, V.K., and Awadelkarim, O.O. (2001). *Microsensors, MEMS, and Smart Devices.* John Wiley & Sons, New York.

Garel, O., Breluzeau, C., Dufour-Gergam, E. et al. (2007). Fabrication of free-standing porous silicon microstructures. *J. Micromech. Microeng.* 17, S164–S167.

Goericke, F., Lee, J., and King, W.P. (2008). Microcantilever hotplates with temperature-compensated piezoresistive strain sensors. *Sens. Actuators A* 143, 181–190.

Han, P.G., Wong, H., and Poon, M.C. (2001). Sensitivity and stability of porous polycrystalline silicon gas sensor. *Colloids Surf. A: Physicochem. Eng. Asp.* 179, 171–175.

Hedrich, F., Billat, S., and Lang, W. (2000). Structuring of membrane sensors using sacrificial porous silicon. *Sens. Actuators A* 84, 315–323.

Hille, P. and Strack, H. (1992). A heated membrane for a capacitive gas sensor. *Sens. Actuators A* 32, 321–325.

Hwang, Y.-H., Gao, F., and Candler, R.N. (2010). Porous silicon resonators for sensitive vapor detection. In: *Proc. Hilton Head Workshop on a Solid-State Sensors, Actuators, and Mycrosystems,* June 6–10, Hilton Head Island, SC, pp. 150–153.

Hwang, Y., Gao, F., Hong, A.J., and Candler, R.N. (2012). Porous silicon resonators for improved vapor detection. *J. Microelectromech. Syst.* 21(1), 235–242.

Jensenius, H., Thaysen, J., Rasmussen, A.A., Veje, L.H., Hansen, O., and Boisen, A. (2000). A microcantilever-based alcohol vapor sensor-application and response model. *Appl. Phys. Lett.* 76, 2615–2617.

Kaltsas, G. and Nassiopoulou, A.G. (1999). Novel C-MOS compatible monolithic silicon gas flow sensor with porous silicon thermal isolation. *Sens. Actuators A* 76, 133–138.

Kim, Y.-M., Seo, C.-T., Eun, D.-S., Park, S.-G., Jo, C.-S., and Lee, J.-H. (2003). Characteristics of cantilever beam fabricated by silicon micromachining for flow sensor application. In: *Proceedings of IEEE Sensors 2003 Conference,* Oct. 22–24, *Proc. IEEE,* 1, 642–646.

Laconte, J., Flandre, D., and Raskin, J.-P. (2006). *Micromachined Thin-Film Sensors for SOI-CMOS Co-Integration.* Springer, Dordrecht, the Netherlands.

Lammel, G., Schweizer, S., Schiesser, S., and Renaud, P. (2002). Tunable optical filter of porous silicon as key component for a MEMS spectrometer. *J. Microelectromech. Syst.* 11(6), 815–827.

Lang, W. (1996) Silicon microstructuring technology. *Mater. Sci. Eng. R* 17, 1–55.

Lang, W., Steiner, P., Richter, A., Marusczyk, K., Weimann, G., and Sandmaier, H. (1994). Applications of porous silicon as a sacrificial layer. *Sens. Actuators A* 43, 239–242.

Lee, J. and King, W.P. (2007). Microcantilever hotplates: Design, fabrication, and characterization. *Sens. Actuators A* 136, 291–298.

Lee, J., Naeli, K., Hunter, H., Berg, J., Wright, T., Courcimault, C., Naik, N., Allen, M., Brand, O., Glezer, A., and King, W.P. (2007). Characterization of liquid and gaseous micro- and nanojets using microcantilever sensors. *Sens. Actuators A* 134, 128–139.

Lee, D., Zandieh, O., Kim, S., Jeon, S., and Thundat, T. (2015). Sensitive and selective detection of hydrocarbon/water vapor mixtures with a nanoporous silicon microcantilever. *Sens. Actuators B* 206, 84–89.

Lucklum, F., Schwaiger, A., and Jakoby, B. (2013). High temperature micro-hotplates on porous silicon substrates. In: *Proceedings of Transducers 2013,* June 16–20, Barcelona, Spain, pp. 1907–1910.

Lucklum, F., Schwaiger, A., and Jakoby, B. (2014). Highly insulating, fully porous silicon substrates for high temperature micro-hotplates. *Sens. Actuators A* 213, 35–42.

Maccagnani, P., Angelucci, R., Pozzi, P. et al. (1998). Thick oxidised porous silicon layer as a thermo-insulating membrane for high-temperature operating thin- and thick-film gas sensors. *Sens. Actuators B* 49, 22–29.

Maccagnani, P., Dori, L., and Negrini, P. (1999). Thick porous silicon thermo-insulating membranes for gas sensor applications. *Sens. Mater.* 11, 131–147.

Meltzman, S., Shemesh, A., Stolyarova, S., Nemirovsky, Y., and Eichen, Y. (2014). Microcantilevers as gas-phase sensing platforms: Simplification and optimization of the production of polymer coated porous-silicon-over-silicon microcantilevers. *J. Polymer Sci. B: Popymer Phys.* 52, 141–146.

Mlcak, R., Tuller, H.L., Greiff, P., Sohn, J., and Niles, L. (1994). Photoassisted electrochemical micromachining of silicon in HF electrolytes. *Sens. Actuators A* 40, 49–55.

Mondal, B., Mahanta, M.M., Phukan, P., Roychoudhury, C., and Saha, H. (2011). Oxidized macro porous silicon based thermal isolation in the design of microheater for MEMS based gas sensors. In: *Proceedings of International Symposium on Devices MEMS, Intelligent Systems & Communication (ISDMISC) 2011.* IJCA, pp. 7–9.

Muller, G., Friedberger, A., Kreisl, P., Ahlers, S., Schulz, O., and Becker, T. (2003). A MEMS toolkit for metal-oxide-based gas-sensing systems. *Thin Solid Films* 436, 34–45.

Nassiopoulou, A. (2005). Porous silicon for sensor applications. In: Veseashta A., Dimova-Malinovska D., and Marshall J.M. (eds.) *Nanostructured and Advanced Materials*, NATO Science Series, Vol. 204. Springer, Berlin, pp. 189–204.

Racine, G.A., Genolet, G., Clerc, P.A. Despont, M., Vettiger, P., and De Rooij, N.F. (1997). Porous silicon sacrificial layer technique for the fabrication of free standing membrane resonators and cantilever arrays. In: *Proceeding of the 11th European Conference on Solid State Transducers, Eurosensors XI*, Warsaw, September 21–24, 1997, 1, 285–288.

Roozeboom, F., Blauw, M.A., Lamy, Y. et al. (2008). Deep reactive ion etching of through-silicon vias. In: Garrou P., Bower C., and Ramm P. (eds.) *Handbook of 3-D Integration: Technology and Applications of 3D Integrated Circuits*. Wiley-VCH Verlag, Weinheim, pp. 47–91.

Rossi, C., Temple-Boyer, P., and Estve, D. (1998). Realization and performance of thin SiO_2-SiN_x membrane for microheater applications. *Sens. Actuators A* 64, 241–245.

Roylance, L.M. and Angell, J.B. (1979). A batch fabricated silicon accelerometer. *IEEE Trans. Electron. Dev.* 26(12), 1911–1917.

Simon, I., Barsan, N., Bauer, M., and Weimar, U. (2001). Micromachined metal oxide gas sensors: Opportunities to improve sensor performance. *Sens. Actuators B* 73, 1–26.

Spannhake, J., Helwig, A., Schulz, O., and Muller, G. (2009). Micro-fabrication of gas sensors. In: Comini E., Faglia G., and Sverbeglieri G. (eds.) *Solid State Gas Sensing*. Springer, New York, pp. 1–46.

Splinter, A., Bartels, O., and Benecke, W. (2001). Thick porous silicon formation using implanted mask technology. *Sens. Actuators B* 76, 354–360.

Steiner, P. and Lang, W. (1995). Micromachining applications of porous silicon. *Thin Solid Films* 255, 52–58.

Su, Y., Evans, A.G.R., Brunnschweiler, A., and Ensell, G. (2002). Characterization of a highly sensitive ultra-thin piezoresistive silicon cantilever probe and its application in gas flow velocity sensing. *J. Micromech. Microeng.* 12, 780–785.

Tabata, O. (1986). Fast-response silicon flow sensor with an on-chop fluid temperature sensing element. *IEEE Trans. Electron. Dev.* 33(3), 361–365.

Thundat, T., Wachter, E.A., Sharp, S.L., and Warmack, R.J. (1995). Detection of mercury vapor using resonating microcantilevers. *Appl. Phys. Lett.* 66, 1695–1697.

Tortonese, M., Yamada, H., Barrett, R.C., and Quate, F.C. (1991). Atomic force microscopy using a piezoresistive cantilever. *In: Proceedings of International Conference on Solid-State Sensors and Actuators, TRANSDUCERS '91*, June 24–27, San Francisco, IEEE, pp. 448–451.

Tortonese, M., Barrett, R., and Quate, C. (1993). Atomic resolution with an atomic force microscope using piezoresistive detection. *Appl. Phys. Lett.* 62, 834–836.

Triantafyllopoulou, R., Chatzandroulis, S., Tsamis, C., and Tserepi, A. (2006). Alternative micro-hotplate design for low power sensor arrays. *Microelectron. Eng.* 83, 1189–1191.

Triantafyllopoulou, R., Illa, X., Casals, O. et al. (2008). Nanostructured oxides on porous silicon micro-hotplates for NH_3 sensing. *Microelectron. Eng.* 85, 1116–1119.

Tsamis, C., Nassiopoulou, A.G., and Tserepi, A. (2003). Thermal properties of suspended porous silicon micro-hotplates for sensor applications. *Sens. Actuators B* 95, 78–82.

Vasiliev, A., Pavelko, R., Gogish-Klushin, S. et al. (2009). Sensors based on technology "Nano-on-micro" for wireless instruments preventing ecological and industrial catastrophes. In: Baraton M.-I. (ed.) *Sensors for Environment, Health and Security*. NATO Science for Peace and Security Series C: Environmental Security. Springer, New York, pp. 205–227.

Zaborowski, M., Grabiec, P., Gotszalk, T., Romanowska, E., and Rangelow, I. (2001). A temperature microsensor for biological investigation. *Microelectron. Eng.* 57–58, 787–792.

PSi-Based Preconcentrators, Filters, and Gas Sources

12

Ghenadii Korotcenkov, Vladimir Brinzari, and Beongki Cho

CONTENTS

12.1 Preconcentrators and Filters 224

    12.1.1 Preconcentration: General View 224

    12.1.2 Micro Preconcentrator Design 225

12.2 Gas Sources for μGC and Gas Sensor Calibration 232

    12.2.1 Diffusion Systems 232

    12.2.2 Permeation Systems 232

    12.2.3 PSi-Based Controllable Vapor Generators 233

Acknowledgments 235

References 235

12.1 PRECONCENTRATORS AND FILTERS

12.1.1 PRECONCENTRATION: GENERAL VIEW

Preconditioning systems (preconcentrators) today are highly required in trace detection, either for conventional analytical methods such as gas chromatography or for gas sensors (Littlewood 1970; Van Es 1992; Filho et al. 2006; Chiriac 2007; Pijolat et al. 2007; Camara et al. 2009, 2010, 2011; Seo 2012). It was established that the preconcentration is a useful way to advance the selectivity and detection limits of these devices versus certain analytes of interest. Preconcentrators are also needed for the miniaturization and the higher mobility of such detection devices. The preconcentrator is conventionally placed at the front end of the analytical system based not only on the gas sensors (Moseley and Tofield 1987; Gopel et al. 1991) but also on the ion-mobility spectrometer and gas chromatograph (GC) (Janata 1989; Madou and Morrison 1989; Moseley et al. 1991). It concentrates and purifies the analyte at the sorption material. As a rule, physisorption is the phenomenon governing the preconcentration process, where weak intermolecular interactions between analyte molecules and an adsorbent surface are caused by means of Van der Waals force. Physisorption processes can easily be reversed by heating that raises the internal energy and breaks the weak bonds. Therefore, the method based on physisorption is preferred for preconcentration due to its simplicity and reversibility. Ideally, the sorptive material must absorb selectively one or more chemical species of interest over a time period necessary to concentrate the chemical compound in the absorptive material (Figure 12.1a) (Filho et al. 2006). Then, the sorptive layer must be heated with a pulse of temperature for providing narrow desorption peaks with relatively high concentration (Figure 12.1b) to the connecting detectors (e.g., gas sensors, electronic noise, or conventional analytical detectors such as gas chromatograph/mass spectrometers, etc). This process must allow the analytes present in a large air volume to be purified and concentrated, so increasing the efficiency of detection.

However, it is necessary to take into account that the preconcentrator obviously needs time to collect the analyte. This limits the application to cases when the analyte detection is carried out over a rather long period of time. Typical sample collection times from the inlet gas stream are 30 to 60 sec. The preconcentrators are also required to operate at high flow rates, to be quickly heated up, and to have a selective and high capacitive sampling of the gases of interest. In addition, overall, regardless of preconcentrator type or group, all the existing μ-preconcentrators require active pumping to draw the vapor sample into the adsorbents. The power required for pumping in a μGC or gas sensor system becomes more significant along with the increase of the required sample volume and analysis time depending on needs.

A potential problem encountered in these techniques is reactions occurring during the sampling between the concentrated organic gasses of interest and other reactive air constituents such as ozone, halogens, the hydroxyl radical, nitrogen oxides, water, or hydrogen peroxide. Reactions of this kind may alter the quantities of the trace gases of interest and may contribute to the formation of artifact products, which may mistakenly be interpreted as atmospheric constituents (Helmig 1997). Therefore, the investigation of interferences from reactions during sampling is an important part of method development and improvement strategies for atmospheric monitoring.

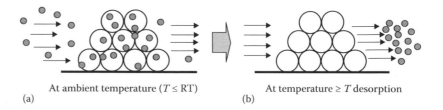

At ambient temperature ($T \le$ RT)   At temperature $\ge T$ desorption
(a)   (b)

FIGURE 12.1 Operating principle of a gas preconcentrator based on thermal programmed adsorption/desorption process: (a) adsorption stage, (b) desorption stage. (Idea from Lahlou H., PhD Thesis, Universitat Rovira I Virgili, Tarragona, Spain, 2011.)

12.1.2 MICRO PRECONCENTRATOR DESIGN

Different approaches can be used for preconcentrator fabrication. For example, the conventional preconcentrators, so called microtraps, comprise a stainless steel tube or glass-capillary tube packed with one or more granular absorbent materials (Gopel et al. 1991; Moseley et al. 1991; Janata et al. 1998; Bakker and Telting-Diaz 2002; Hierlemann and Baltes 2002; Pierce et al. 2003; Hu et al. 2003). To heat the unit and to desorb the gas, a current is passed through the stainless-steel tube or a metal spiral heater coiled around the glass-capillary tube. These units are characterized by large dead volumes and limited heating efficiency due to their larger thermal mass, which subsequently contributes to rather long time at the analyte pulse. However, design of the preconcentrators based on silicon micromachining technology and hotplate platforms is the most promising (Voiculescu et al. 2006; Manginell et al. 2008, 2011; Lahlou 2011). The microfabricated preconcentrator devices based on the micro hotplates, discussed in previous chapters, overcome disadvantages of conventional preconcentrators through a significant reduction of dead volume and thermal mass. The suspended membrane structure of the preconcentrator has an extremely low heat capacity, which allows for very rapid heating. As a result, the application of a current pulse to the heater causes the film layer to heat rapidly and uniformly; this thermally desorbs the collected analytes in a narrow concentrated chemical pulse, which can be delivered for analysis (Simon et al. 2001). As a rule, this desorbed pulse occurs over approximately up to 0.2–2 seconds. In this case, using a sample collection time of 30 to 60 sec causes a 10- to 100-fold concentration enhancement in the desorbed pulse over the inlet stream. This effect is illustrated in Figure 12.2. In addition, microfabricated technology has allowed the fabrication of miniaturized preconcentrating devices with extremely low power consumption suitable for use with micro gas chromatographs and MEMS-based gas sensors (Voiculescu 2005; Lahlou 2011). In particular, the microfabrication has allowed the incorporation of the microconcentrator with the detection analytical systems or MEMS-based gas sensors on the same chip. The preconcentrator can perform sample extraction and injection into a single device for on-site analysis without human intervention in the transfer of sample from extraction media to the detector inlet. Several authors have developed such systems and have implemented integrated prototypes based on the coupling of a concentrator-focuser in front of a microcolumn and either a microsensor or a microsensor array (Bianchi et al. 2003; Zampolli et al. 2009). These systems have been devoted to the selective analysis of gas phase mixtures at the ppb level. One such fully microfabricated GC system was recently developed in Sandia National Laboratories (see Figure 12.3). In this system, all components were micromachined. This system incorporated a microfabricated preconcentrator with a micromachined-Si separation column and an integrated acoustic-wave sensor array into a hybrid microanalyzer (Lu et al. 2005; Lewis et al. 2006; Kim et al. 2011; Manginell et al. 2008, 2011). One such system is shown in Figure 12.4. More recently, Zampolli et al. (2009), a research group at the Institute of Microelectronics and Microsystems of the Italian National Research Council

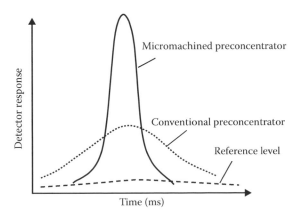

FIGURE 12.2   Kinetics of delivery of analytes from gas preconcentrators of two kinds. (Idea from Voiculescu I. et al., *Trends Anal. Chem.* 27, 327, 2008.)

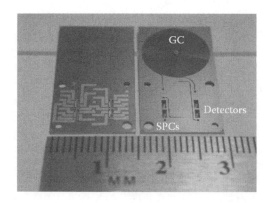

FIGURE 12.3   Two monolithically integrated μGCs chips side-by-side. The chip on the left-hand side has the metallic traces used to electrically connect the PC and sensor shown face up. The chip on the right-hand side is metal-side down to show the fluidic channels and deep reactive ion etched (DRIE) μGC channel. (From Manginell R.R. et al., *Sensors* 11, 6517, 2011. Published by MDPI as open access.)

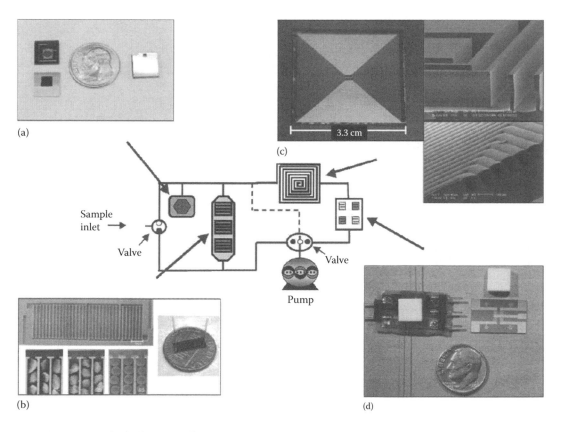

FIGURE 12.4   Block diagram of the MEMS μGC prototype analytical system: (a) calibration-vapor source before (left) and after (right) assembly; diffusion channel and headspace aperture can be seen in the top section and macro-PSi reservoir can be seen in the bottom section; (b) 3-stage adsorbent μPCF prior to loading and sealing (top left), with close-up SEM images of each section loaded with adsorbents (lower left) and assembled structure with capillary interconnects on a U.S. penny; (c) 3-m separation-column chip (left) with close up views of the channel cross-sections prior to (top right) and after (lower right) sealing; (d) detector assembly with 4-chemiresistor array chip (right), Macor lid (white square structure), and sealed detector with connecting capillaries mounted on a custom mounting fixture (left). The dashed line is a flow-splitter. (From Lu C.-J. et al., *Lab on a Chip*. 5, 1123, 2005. Reproduced by permission of The Royal Society of Chemistry.)

(CNR-IMM), developed a complete µGC system comprising a preconcentration column, a GC separation column, and a metal oxide semiconductor (MOX) solid-state gas sensor.

At present, the main microfabricated preconcentrators, which were reported in literature, have planar structure or complex 3D structures depending on the type of vapors to be detected (Guihen and Glennon 2003; Voiculescu et al. 2008; Seo 2012). Planar hotplate preconcentrators have simpler structures. These systems are based on a planar hotplate covered with an adsorbent coating (see Figure 12.5). The layer of the sorption material is deposited on the active area of membrane adjacent to the heating element. Planar microconcentrators were generally used for the trapping of a single analyte. In this case, the hotplate is covered with a selective coating to concentrate this specific analyte (Frye-Mason et al. 1999; Manginell et al. 2000; Lewis et al. 2006). Such preconcentrators can have very low consumable power (~100 mW) and very short desorption time (up to 4 ms) (Lewis et al. 2006). The disadvantage of this type of preconcentrator is the limited area of the hotplate that restricts the collection capability of the device.

According to Alfeeli et al. (2008), this problem can be resolved by

- Designing a preconcentrator with high heated area in order to host a sufficient amount of adsorbent, including big area planar micropreconcentrators.
- Using multiple preconcentration stages.
- Choosing an adsorbent with high adsorption capacity and high surface area.
- Increasing the gas adsorbed volume by optimizing the adsorption pressure or using high adsorption durations. During the desorption process, the dead volumes must be as low as possible to not dilute the concentrated analyte.

3D structures give the possibility to increase the active surface area while maintaining good heating efficiency and low dead volume (Lahlou 2011). In this case, the structure is based on thick hotplates (>500 µm) etched in silicon with deep trenches that can hold a large quantity of material and large heating area. Microcavities in silicon wafers are usually etched using deep reactive-ion etching (DRIE). The microheaters are surrounded by air-gaps in order to promote the thermal insulation of the microstructure. The adsorbent packing is achieved by filling the microconcentrator manually with adsorbent slurry using a syringe or with the help of pumps at low to moderate pressure. However, the microstructures, fittings, and device packaging usually cannot tolerate the harsh conditions of granular adsorbent packing such as high pressures and the application of ultrasonication, which is crucial to achieve good packing density (Jung et al. 2009).

It is important to note that 3D preconcentrators with adsorbing materials can also be used as chemical filters for gas sensors. As it is known that the use of additional various filters incorporated into the measuring unit or directly into the sensor construction is one of the most effective methods to improve both sensor selectivity and sensor resistivity to poisoning (Feng et al. 1994; Pijolat et al. 1999, 2005; Sauvan and Pijolat 1999; Fleischer et al. 2000; Kitsukawa et al. 2000; Mandayo et al. 2002; Prasad et al. 2010). In particular, chemical filters can eliminate interfering gases via a chemical reaction (Fleischer et al. 2000) or preferentially transform the target gas into a more active species (Hubalek et al. 2004). Additional filters could also be employed to protect the sensor surface from contamination, although they not only influence metal oxide surface properties, but could also provoke an electrical short circuit (Billi et al. 2002). However, the wide

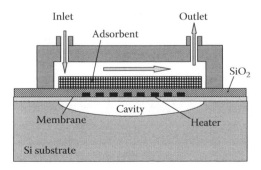

FIGURE 12.5   Schematic view of a planar preconcentrator.

application of conventional filters is limited by its short lifetime. Therefore, in the absence of control for their conditions, changes in their filtering capacities could be a reason for the observed drift in sensor characteristics. In addition, the adsorption capacity of filters is often decreased with temperature (Schweizer-Berberich et al. 2000). Thus, the filter sorption capacity becomes dependent on the surrounding temperature. This means that the use of conventional filters requires both the control for their parameters and frequent change of them. This greatly complicates the operation of gas sensors. Using preconcentrators as filters can resolve this problem because preconcentrators contain the heater that gives ability to make periodically the annealing of adsorbents and restore its initial properties.

3D structures used in preconcentrators can have different configurations (see Figure 12.6). For example, Camara et al. (2009) proposed "Neutral," "Straight," and "Zigzag" designs (Figure 12.6a,b). The pattern was etched using DRIE to form 240-μm deep 3D structures. Camara et al. (2009) believe that the "Zigzag" design contributes to improve the carbon deposition by stopping it when injected into the microchannel, while the "Straight" design leads to an increase of the available area for deposition since the surface of the walls constitutes a binding area for adsorbent particles. Microconcentrators can also have spiral shape. For example, Figure 12.6c shows such a micropreconcentrator proposed by Gracia et al. (2008). At that, micropreconcentrators can be classified into packed and wall-coated structures. Packed structures have the advantage of large surface area, but they suffer from limitations such as large pressure drop, high dead volume, gas residuals, and poor heat transfer. Wall-coated

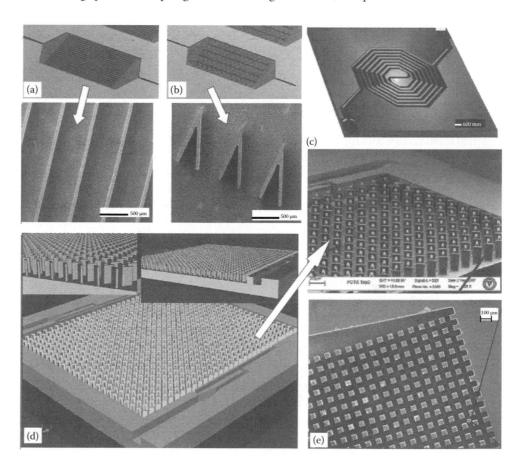

FIGURE 12.6 SEM images of preconcentrators designed in various groups: (a,b) "Straight" and "Zigzag" preconcentrators designed by Camara et al. (2010). Images obtained at different magnifications, (c) preconcentrator proposed by Gracia et al. (2008), (d) 3D structures design for microthermal preconcentrators by Alfeeli et al. (2008), and (e) preconcentrator fabricated by Akbar et al. (2013). (Reprinted from *Sens. Actuators B* 148, Camara E.H.M. et al., 610, Copyright 2010, with permission from Elsevier; Reprinted from *Sens. Actuators B* 132, Gracia I. et al., 149, Copyright 2008, with permission from Elsevier; Reprinted from *Sens. Actuators B* 133, Alfeeli B. et al., 24, Copyright 2008, with permission from Elsevier; Reprinted from *J. Chromatorg. A* 1322, Akbar M. et al., 1, Copyright 2013, with permission from Elsevier.)

structures overcome these disadvantages but their surface area is limited. Alfeeli et al. (2008) believe that the best way to overcome the current challenge of increased surface area without obstruction of the flow in a µTPC is to fabricate micropillars within the preconcentrator structure as shown in Figure 12.6d. The same approach was also used by Akbar et al. (2013) (see Figure 12.6e).

One should note that, in general, the 3D preconcentrator allows holding a higher amount of material compared to the planar structure, which allows slowing the breakthrough of the preconcentrator. This structure is useful for the trapping of low volatile compounds that are put in closer contact with the adsorbent, by contrary to the case of planar configuration. However, the fabrication technology of the planar structure is simpler, the thermal isolation is better, and the power consumption is generally much lower (Lewis et al. 2006; Lahlou 2011). In addition, it is necessary to take into account that due to the large adsorbent mass in 3D preconcentrators, thermal desorption with a thin-film micro hotplate design is not appropriate. Therefore, usually special micro heaters with large surface area for efficient thermal conduction of heat from a Joule heater to the adsorbents are utilized in 3D preconcentrators (Tian and Pang 2002, 2003; Tian et al. 2003, 2005; Camara et al. 2011).

Different materials can be used as adsorbant in preconcentrators. According to Yang (1997), Dettmer and Engewald (2002), Knaebel (2004), and Park and Yang (2005), for application as adsorbent these materials must have the following:

- Small pore diameters, which results in higher exposed surface area and hence high surface capacity for adsorption towards the analyte of interest
- High abrasion resistance and mechanical stability
- Low affinity to water, to reduce the effects of humidity
- Low desorption temperature to avoid the thermal degradation of the analytes but also to ensure a low power consumption of the device
- Complete desorption with fast kinetics for further rapid analysis
- Good regenerability
- Good thermal stability and multiple usability to avoid drift with time
- Tuneable physico-chemical adsorption properties for adjusting its selectivity toward the analyte of interest
- The adsorbents must also have a distinct pore structure, which enables fast transport of the gaseous vapors

Currently, a large number of materials has been tested. It was established that numerous inorganic and organic materials, such as $CaCl_2$, CaO, MgO, ZnO, $MgSiO_3$, activated carbon, cellulose, collagen, wool, zeolite, silica gel, and so on, can be used as adsorbents (Knaebel 2004). However, until now the main adsorbents are polymeric materials (PDMS, etc.) and carbon-based particles, in particular activated carbon, having extremely large specific surface area (Dettmer and Engewald 2002; Krol et al. 2010a,b). Adsorbents most widely used are shown in Table 12.1.

Activated carbon is a highly porous, amorphous solid consisting of microcrystallites with a graphite lattice. The base materials that comprise activated carbons include wood, coal, peat, coconut shells, saran, recycled tires, and others. The final adsorbents all look the same, that is, black granules or pellets. Effective surface area and pore size are strongly dependent on the activation conditions. The contents of inorganic components in activated carbon varied between 2 and 25%, but the average is about 7%. Effective surface area generally ranges from 300 to 1500 m²/g, depending on the base material, activation method, density, and so on. Some typical applications are removing volatile organic compounds or other specific contaminants in gas that impart odor or are hazardous, upgrading methane from substandard natural gas wells, and so on. It is the most widely used adsorbent since most of its chemical (e.g., surface groups) and physical properties (e.g., pore size distribution and surface area) can be tuned according to what is needed. Pretreatment for gas-phase applications requires heating to about 200°C. At present, there is a new type of product called a "carbon molecular sieve." It is analogous to the zeolite molecular sieve. While micropores in zeolites tend to have rounded apertures, the carbon-based counterparts are more slit-like, as in the space between layers of graphite. This material is used for separation of nitrogen from air. This pressure swing process exploits the difference between sizes of oxygen (0.343 nm) and nitrogen (0.368 nm), and can achieve 99.9% nitrogen purity. One

TABLE 12.1   Sorbents Usually Used for Atmospheric Air Control

Type of Sorbent	Specific Area (m²/g)	T_{max} (°C)	VOC Analytes Retained	Preferred Analyte-Liberation Technique
Porous organic polymers				
Tenax TA and GR	35	<350	Aromatic compounds, non-polar VOC	Thermal desorption
Chromosorb 102 (*styrene-divinylbenzene copolymer*)	750	20–250	Alcohols, ketones, aliphatic hydrocarbons	Thermal desorption
Chromosorb 106 (*polystyrene*)	350	20–250	C_5-C_{12} hydro-carbons, oxygenated VOCs	Thermal desorption
Porapak N (*polyvinylpyrrolidone*)	300	20–190	Acetylene, aliphatic hydrocarbons	Thermal desorption
Porapak Q (ethylvinyl-benzene-divinylbenzene copolymer)	550	20–190	Acetylene, aliphatic hydrocarbons, and oxygenated VOCs	Solvent extraction
Graphitized carbon blacks (specific surface depends on the degree of graphitization)				
Carbotrap	12	>400	Hydrocarbons less than C_{20},	Thermal desorption
Carbograph	100	>400	alkylbenzenes	
Carbopack	12	>400	Ketones, alcohols and aldehydes	
Molecular sieves (formed during the pyrolysis of organic polymers, e.g., polyvinyl chloride)				
Spherocarb	1200	>400	Very volatile organic	Thermal desorption
Unicarb	>1000	>400	compounds, methanol,	
Carbosieve III	800	400	acetone	
Carboxen	>1200	400	1,3-butadiene	
Molecular Sieve 13X, 5A		350–400		
Activated carbon (formed by the low temperature oxidation of vegetable carbon)				
Activated carbon	>1000	400	Aliphatic and aromatic hydrocarbons, e.g., C_2–C_4,	Solvent extraction or thermal desorption

Source: Reprinted from *TrAC Trends Anal. Chem.* 29(9), Krol S. et al., 1092, Copyright 2010, with permission from Elsevier.

of its main drawbacks of activated carbon is that it reacts with oxygen at moderate temperatures (over 300°C).

Instead of being limited to styrene/divinylbenzine, polymer adsorbents are also made from polymethacrylate, divinylbenzine/ethylvinylbenzene, or vivylpyridine and are sometimes sulfonated or chloromethylated, much as are ion exchange resins. As a result, some are sufficiently hydrophilic to be used as desiccant, while others are quite hydrophobic. The effective surface area is usually smaller than for activated carbon, for example, 5 to 800 m²/g. Pore diameters range from about 0.3 to 200 nm. The range of application is somewhat restricted due to higher price in comparison with other adsorbents. The current main application includes removing VOCs from off-gas.

The presence of water vapor in ambient media is a problem in gas sensing because it can cause a premature saturation of polymer-based adsorbent, leading to a lower preconcentration of the target gas. Zheng et al. (2006) have found that the use of other adsorbents like carbon nanotubes is promising in this case because of their hydrophobic properties, large specific surface area, and low desorption temperature.

One should note that usually preconcentrators' content is one type of adsorbent. However, in order to increase the range of detected vapors, multiple-stage preconcentrators were also proposed (Kim and Mitra 2003; Tian et al. 2005). The authors believe that such an approach provides larger adsorption capacity to a larger range of compounds with different volatilities. For example, Tian et al. (2005) used in designed preconcentrator three carbon adsorbents: Carbopack B, Carbopack X, and Carboxen 1000 (see Figure 12.4). An interesting work was reported by Markowitz et al. (2010). Markowitz et al. (2010) presented a cascade sorbent plate array preconcentrator consisting of a series of stacked thin membrane hotplates, each with a coating of a sorbent material. Effectively, the multiple stage hotplate based structure, when combined with a high adsorption capacity adsorbent, allows increasing significantly the concentration factor, and enables the detection of a wide range of complex vapors, compared to the single stage preconcentrator. The disadvantage of this type of device is its high power consumption and the complexity of its fabrication technique.

Pijolat et al. (Pijolat et al. 2007; Camara et al. 2010) have shown that PSi in addition to manufacturing of silicon membranes may be used for the formation of a micro-channel. The fabrication process is shown in Figure 12.7. Following a photolithographic step with a thick AZ4562 photoresist, the inlet, outlet, and preconcentration chamber of the micro-device are DRIE etched in the silicon wafer to a depth of 325 µm. PSi was formed on all the unprotected areas of the silicon micro-devices, including the sidewalls of the micro-channels. The micro-channels were sealed with a Borofloat glass wafer by anodic bonding and diced according to the dashed line shown in Figure 12.7d,e, revealing the inlet and outlet for the fluidic connectors. It is necessary to note that during preconcentrator designing and fabrication, the ability of hermetic sealing, mechanical reliability, and low dead volume in interconnections between the cover plate, preconcentrator housing, and capillary tubes should be considered. The benefits of PSi use for the formation of a micro-channel was to ease the fixing of the carbon absorbent in the micro-channels and to modify the gas desorption kinetics. Besides, the high specific area offered by PSi can be used as an interesting support to increase the quantity of adsorbent materials fixed in the micro-channels when made of this material. The device was tested with a gas sensor for benzene preconcentration. PSi presented in itself a high adsorption capacity of molecules, which has been already applied directly to the detection of gases. An adsorption time of 5 min provided a "practical" preconcentration factor of 55.

It was shown that the technology of Si porosification can be used for fabrication of a top layer of preconcentrators as well (Seo 2012; Seo et al. 2012). Figure 12.8a shows the cross-sectional view of the micro-preconcentrator/injector (µPPI) structure and Figure 12.8b provides an image of the top layer of a preconcentrator. The top layer contains a grid of square diffusion channels through which vapors pass into the device. This layer also has a through-hole used for fluidic interconnection to downstream components on thermal desorption/injection of capture VOC samples. The fabrication process of the top layer is presented in Figure 12.9. The DRIE technique was used for deep etching.

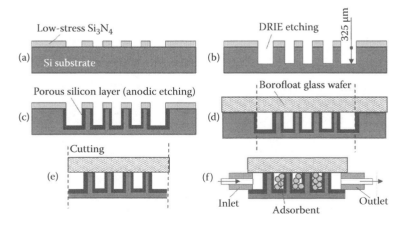

FIGURE 12.7 Schematic diagrams from (a) to (f), which illustrate one of the possible processes of the preconcentrator fabrication. (Reprinted from *Sens. Actuators B* 148, Camara E.H.M. et al., 610, Copyright 2010, with permission from Elsevier.)

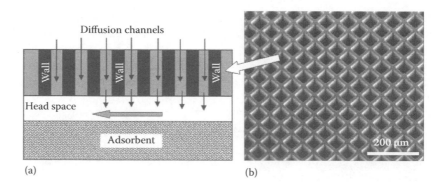

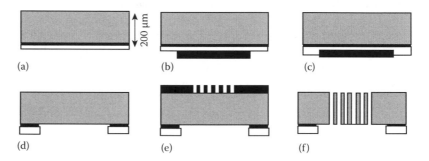

FIGURE 12.8   (a) Schematic cross-sectional view of the μ-PPI structure consisting of diffusion channel array, head space, and absorbent layer. (b) Images of top layer. (From Seo J.H. et al., *Lab on a Chip* 12, 717, 2012. Reproduced by permission of The Royal Society of Chemistry.)

FIGURE 12.9   Top layer fabrication processes: (a) deposition of Cr/Au on backside, (b) pattern for plating, (c) gold electro-plating on backside, (d) etch seed layer, (e) pattern for Si DRIE, and (f) etch grid and hole and remove photoresist. (Idea from Seo J.H. et al., *Lab on a Chip* 12, 717, 2012.)

12.2 GAS SOURCES FOR μGC AND GAS SENSOR CALIBRATION

The sources of calibration gas mixtures are an essential part of any gas chromatographs and systems used for gas sensors testing. There are many methods of preparing such gas mixes, but the most common and convenient methods are based on using diffusion systems and permeation systems.

12.2.1 DIFFUSION SYSTEMS

Diffusion systems have been used for a long time in analytical chemistry for preparing calibration gas mixtures (Nakamoto et al. 1991; ISO 2005; Zhou and Gai 2006; Pratzler et al. 2010). The component is filled into a diffusion vessel, which is provided with a diffusion tube and is surrounded by the carrier gas flow; see Figure 12.10a. Molecules from the liquid phase evaporate into the gaseous phase until the saturated vapor pressure is reached. Then, molecules in the gas phase rise into the diffusion tube and, after diffusion through the diffusion tube, mix with the carrier gas in one step. The quantity of liquid which diffusion uses into the carrier gas depends on the temperature, the geometry of the diffusion tube, and the vapor pressure of the component. The diffusion rate can be determined by weighing the diffusion vessel and is thus directly traceable to the mass. Zhou and Gai (2006) have shown that the relative uncertainty of the composition of the calibration gas mixture prepared by the use of diffusion tubes is about 5%.

12.2.2 PERMEATION SYSTEMS

The operation of permeation systems is very similar to that of diffusion systems (Staude 1992). However, their application is much handier. A *permeation device* is a sealed small container filled

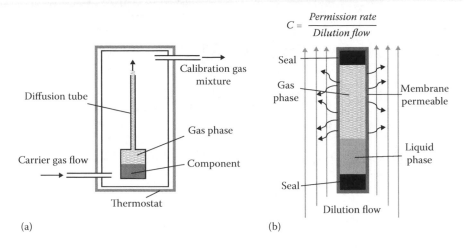

FIGURE 12.10   (a) Schematic diagram of a diffusion cell. (b) Diagram illustrating construction and operation of permeation tube.

with a pure chemical compound in a two-phase equilibrium between its gas phase and liquid phase (see Figure 12.10b). The gas molecules are permeated either through a permeable container wall or through the end cap and mixed with a carrier gas to obtain a known gas mixture used as a reference in gas testing equipment. This phenomenon leads to a very versatile method of producing ultralow concentrations of calibration gas mixtures. In preparing a permeation device, the liquid form of the gas is usually sealed inside a polymeric container, typically, polytetrafluoroethylene (Teflon). The rate at which the gas molecules permeate depends on the permeability of the material and the temperature. Therefore, permeation tubes are held in a carrier gas stream of dry air or nitrogen at a constant temperature. At a known rate of permeation at a given temperature, a constant flow rate of air mixed with the permeated chemicals forms a constant stream of calibration gas. The rate of permeation is constant over long periods of time if the temperature, concentration gradient, and tube geometry remain constants. Thus, a calibrator with constant temperature and flow regulation is needed for this method. The substances that work best with this method are those that have a critical temperature above 20–25°C. Permeation devices are ideal for field use, and high precision and excellent accuracy may be obtained for many gases (Mitchell 1987). In addition, after they are depleted, they can be disposed of easily. However, permeation tubes continuously emit chemicals at a constant rate, thus creating a storage and safety problem (Chou 2000). In addition, the rate of permeation for a given gas of interest can be too high or too low for a given application. For example, high-vapor-pressure gases permeate too quickly, while very low-vapor-pressure chemicals have a permeation rate that is too slow to be of any use.

12.2.3 PSi-BASED CONTROLLABLE VAPOR GENERATORS

Oborny et al. (2003) and Wallner et al. (2007) have shown that gas sources can be designed on the basis of PSi as well. The first variant of controllable vapor generation has been achieved by a three-layer diffusion device consisting of a PSi reservoir, a spacer, and a cap with a micromachined diffusion channel and an outlet port (Oborny et al. 2003). A diagram illustrating this device is shown in Figure 12.11. The calibrant liquid is immobilized in the PSi reservoir. The PSi reservoir (3.5×3.5 mm^2) was made from Si(100) substrates via a standard anodization procedure (HF [49%]:ethanol:water [1:1:1] and $J = 85$ mA/cm^2). Mesoporous PSi layers were of 180-μm deepness with a broad pore size distribution ranging from 5 to 200 nm in diameter. Vapors saturate the headspace and pass down the narrow-bore diffusion channel into the sample inlet channel of the μGC. Three diffusion channel lengths were chosen to provide inlet concentrations ranging from 45 to 122 ppb of n-decane at 25°C. A diffusion channel cross-section of ~0.04 mm^2 and lengths of 7.5, 13.0, and 20.3 mm have been utilized. Oborny et al. (2003) have shown that this source

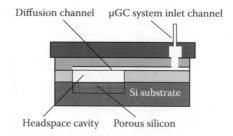

FIGURE 12.11   Diagram of integrated diffusion calibration source. (From Oborny M.C. et al., In: *Proceedings of 7th International Conference on Miniaturized Chemical and Biochemical Analysts Systems*, October 5–9, Squaw Valley, CA, 1243, 2003.)

design is suitable for extended μGC field deployment and provided an average generation rate at 25°C within 5% of theoretical predictions.

However, that design has two drawbacks (Wallner et al. 2007). Because the PSi reservoir has to be filled before the device is put together, room temperature polymer-based bonding is the only choice for device assembly. Unfortunately, those polymer-based bonds not only introduce contaminative vapors, but also suffer deterioration in the presence of the calibrant liquid, such as *n*-decane. Second, since the PSi reservoir cannot be refilled after assembly, the vapor source's lifetime is limited by the depth of the PSi layer and the storage time before utilization. Devices proposed by Wallner et al. (2007) are devoid of these disadvantages. The new design of the integrated calibration source is shown in Figure 12.12. It consists of three layers. The top layer has a deep recess and a through hole, which serves as the reservoir and the filling port of the calibration source. The middle layer is a macroporous silicon membrane, which serves as a wick for the reservoir. The bottom layer has a shallow recess of the same size of the PSi wick, a groove, and a small hole in it. When the device is assembled, the shallow recess will form a headspace over the PSi wick; the groove will form a diffusion channel; and the small hole becomes an outlet port for the diffusion channel. In operation, the reservoir will be filled with calibrant liquid through the filling port. The PSi wick holds the calibrant liquid by capillary forces while allowing the calibrant vapor to enter into the headspace. The calibrant vapor then transports through the attached diffusion channel and the outlet port into the sample inlet of the μGC, where the calibrant vapor will mix with the carrier gas stream.

Both the reservoir layer and the diffusion channel layer are made from #7740 Pyrex glass using micromilling. The reservoir has a capacity of 6.7 μL. Diffusion channels of three different lengths (7.5, 13, and 20.5 mm) and a cross-section of 200×200 μm^2 have been fabricated. The PSi wick is made from *p*-type (100) substrates ($\rho = 17$–23 Ω·cm) using an aqueous electrolyte with 10^{-3} M cetyltrimethylammonium chloride (CTAC) and HF (49%), ethanol, and H$_2$O at volume ratio 1:2:3 ($J = 27$ mA/cm^2). The PSi region has a diameter of 3.5 mm, defined by an SiO$_2$/Si double

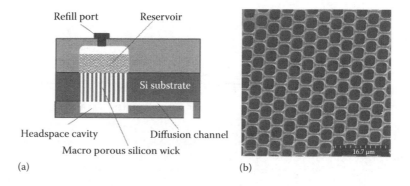

FIGURE 12.12   (a) Schematic of the integrated calibration source with a macroporous silicon wick. (b) Top view of the macroporous silicon wick. (Wallner J.Z. et al.: *Phys. Stat. Sol. (a)*. 2007. 204 (5). 1449. Copyright Wiley-VCH Verlag GmbH & Co. KGaA. Reproduced with permission.)

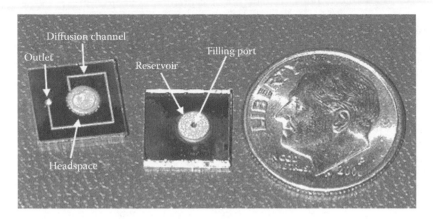

FIGURE 12.13   Picture of two assembled calibration sources, showing the reservoir, the filling port, the headspace, the diffusion channel, and the outlet port. (Wallner J.Z. et al.: *Phys. Stat. Sol. (a)*. 2007. 204(5). 1449. Copyright Wiley-VCH Verlag GmbH & Co. KGaA. Reproduced with permission.)

layer mask fabricated by RF sputter deposition. Standard photolithography and subsequent KOH etching were used to form arrays of inverted pyramid pits as initial pores. The pores in the entire thickness of the wafer (>500 μm long) were obtained by simply stacking a sacrificial sample at the backside of the macroporous sample during the anodization (Zheng et al. 2005). The surface of the macroporous membrane with a pores open at both sides and the view of final devices are presented in Figures 12.12b and 12.13.

Simulations have shown that for better stability, the silicon wick should be macroporous. It was established that when the pore diameter is less than 0.5 μm, a small variation in pore size would result in a significant variation in the vapor pressure. As the pore size increases, the impact of pore size on the vapor pressure reduces. Therefore, Wallner et al. (2007) proposed to use the macroporous silicon wick with a pore size of 3.5 μm, which can provide sufficient capillary force without having a huge impact on the vapor pressure. In addition, the pore size in this case is uniform along the full thickness of the PSi wick. Therefore, a constant vapor pressure can be obtained in the headspace, in spite of the liquid-gas interface moving in the wick during the operation of the calibration source. Experiments carried out by Wallner et al. (2007) confirmed their assumptions: vapor generation was very stable and reproducible. Only small variations of 0.1% over a 9-h period and 0.5% over a week have been observed.

ACKNOWLEDGMENTS

This work was supported by the Ministry of Science, ICT and Future Planning (MSIP) of the Republic of Korea, by the National Research Foundation (NRF) grants funded by the Korean government (No. 2011-0028736 and No. 2013-K000315), and partly by the Moldova Government under grant 15.817.02.29F and ASM-STCU project #5937.

REFERENCES

Akbar, M., Wang, D., Goodman, R. et al. (2013). Improved performance of micro-fabricated preconcentrators usingsilica nanoparticles as a surface template. *J. Chromatorg. A* 1322, 1–7.

Alfeeli, B., Cho, D., Ashraf-Khorassani, M., Taylor, L.T., and Agah, M. (2008). MEMS-based multi-inlet/outlet preconcentrator coated by inkjet printing of polymer adsorbents. *Sens. Actuators B* 133, 24–32.

Bakker, E. and Telting-Diaz, M. (2002). Electrochemical sensors. *Anal. Chem.* 74, 2781–2800.

Bianchi, F., Pinalli, R., Ugozzoli, F., Spera, S., Careri, M., and Dalcanale, E. (2003). Cavitands as superior sorbents for BTX detection at trace level. *New J. Chem.* 27, 502–509.

Billi, E., Viricelle, J.-P., Montanaro, L., and Pijolat, C. (2002). Development of a protected gas sensor for exhaust automotive applications. *IEEE Sensor J.* 2(4), 342–348.

Camara, E.H.M., Breuil, P., Briand, D. et al. (2009). Influence of the adsorbent material in the performances of a micro gas preconcentrator. In: *Proceedings of the 13th International Symposium on Olfaction and Electronic Nose*, Brescia, Italy. *AIP Conf. Proc.* 1137(1), 323–326.

Camara, E.H.M., Breuil, P., Briand, D., Guillot, L., Pijolat, C., and de Rooij, N.F. (2010). Micro gas preconcentrator in porous silicon filled with a carbon absorbent. *Sens. Actuators B* 148, 610–619.

Camara, E.H.M., Breuil, P., Briand, D., de Rooij, N.F., and Pijolat, C. (2011). A micro gas preconcentrator with improved performance for pollution monitoring and explosives detection. *Anal. Chim. Acta* 688, 175–182.

Chiriac, R.E. (2007). Development of a pre-concentrator-thermo-desorber/micro-gas chromatograph/mass spectrometer coupling for on-site analyses of emissions of volatile organic compounds from landfills. *Int. J. Environ. Anal. Chem.* 87(1), 43–55.

Chou, J. (2000). *Hazardous Gas Monitors: A Practical Guide to Selection, Operation and Application*. McGraw-Hill, New York.

Dettmer, K. and Engewald, W. (2002). Adsorbent materials commonly used in air analysis for adsorptive enrichment and thermal desorption of volatile organic compounds. *Anal. Bioanal. Chem.* 373, 490–500.

Feng, C.D., Shimizu, Y., and Egashira, M. (1994). Effect of gas diffusion process on sensing properties of SnO_2 thin film sensors in a SiO_2/SnO_2 layer-built structure fabricated by sol-gel process. *J. Electrochem. Soc.* 141, 220–225.

Filho, A.P.N., de Carvalho, A.T., da Silva, M.L.P., and Demarquette, N.R. (2006). Preconcentration in gas or liquid phases using adsorbent thin films. *Mater. Res.* 9(1), 33–40.

Fleischer, M., Kornely, S., Weh, T., Frank, J., and Meixner, H. (2000). Selective gas detection with high-temperature operated metal oxides using catalytic filters. *Sens. Actuators B* 69, 205–210.

Frye-Mason, G.C., Heller, E.J., Hietala, V.M. et al. (1999). Microfabricated gas phase chemical analysis systems. In: *Proceedings of Microprocesses and Nanotechnology Conference*, Yokohama, Japan, pp. 60–61.

Gopel, W., Hesse, J., and Zemel, J.N. (eds.) (1991). *Sensors: A Comprehensive Survey*, Vols. 1–3. VCH, Weinheim, Germany.

Gracia, I., Ivanov, P., Blanco, F. et al. (2008). Sub-ppm gas sensor detection via spiral μ-preconcentrator. *Sens. Actuators B* 132, 149–154.

Guihen, E. and Glennon, J.D. (2003). Nanoparticles in separation science—Recent developments. *Anal. Lett.* 36, 3309–3336.

Helmig, D. (1997). Ozone removal techniques in the sampling of atmospheric volatile organic trace gases. *Atmos. Environ.* 31, 3635–3651.

Hierlemann, A. and Baltes, H. (2002). CMOS-based chemical microsensors. *Analyst* 128, 15–28.

Hu, J., Zhu, F., Zhang, J., and Gong, H. (2003). A room temperature indium tin oxide/quartz crystal microbalance gas sensor for nitric oxide. *Sens. Actuators B* 93, 175–180.

Hubalek, J., Malysz, K., Prasek, J. et al. (2004). Pt-loaded Al_2O_3 catalytic filters for screen-printed WO_3 sensors highly selective to benzene. *Sens. Actuators B* 101, 277–283.

ISO 6145-8 (2005). Gas Analysis—Preparation of Calibration Gas Mixtures Using Dynamic Volumetric Methods: Part 8. Diffusion Method. http://www.iso.org.

Janata, J. (1989). *Principle of Chemical Sensors*. Plenum Press, New York.

Janata, J., Josowicz, M., Vanysek, P., and DeVaney, D.M. (1998). Chemical sensors. *Anal. Chem.* 70, 179R–208R.

Jung, S., Ehlert, S., Mora, J.-A. et al. (2009). Packing density, permeability, and separation efficiency of packed microchips at different particle-aspect ratios. *J. Chromatogr. A* 1216, 264–273.

Kim, M. and Mitra, S. (2003). A microfabricated microconcentrator for sensors and gas chromatography. *J. Chromatogr. A* 996, 1–11.

Kim, S.K., Chang, H., and Zellers, E.T. (2011). Microfabricated gas chromatograph for the selective determination of trichloroethlyene vapor at sub-parts-per-billion concentrations in complex mixtures. *Anal. Chem.* 83, 7198–7206.

Kitsukawa, S., Nakagawa, H., Fukuda, K., Asakura, S., Takahashi, S., and Shigemori, T. (2000). The interference elimination for gas sensor by catalyst filters. *Sens. Actuators B* 65(1), 120–121.

Knaebel, K.S. (2004). Adsorbent selection. Adsorption Research Inc. (http://www.adsorption.com/publications/AdsorbentSel1B.pdf)

Krol, S., Zabiegata, B., and Namiesnik, J. (2010a). Monitoring VOCs in atmospheric air. I. On-line gas analyzers. *TrAC Trends Anal. Chem.* 29(9), 1092–1100.

Krol, S., Zabiegata, B., and Namiesnik, J. (2010b). Monitoring VOCs in atmospheric air II. Sample collection and preparation. *TrAC Trends Anal. Chem.* 29(9), 1101–1112.

Lahlou, H. (2011). Design, Fabrication and Characterization of Gas Preconcentrator based on Thermal Programmed Adsorption/Desorption for Gas Phase Microdetection Systems. PhD Thesis, Universitat Rovira I Virgili, Tarragona, Spain.

Lewis, P.R., Manginell, R.P., Adkins, D.R. et al. (2006). Recent advancements in the gas-phase microChemlab. *IEEE Sensors J.* 6, 784–795.

Littlewood, A.B. (1970). *Gas Chromatography: Principles, Techniques, and Applications*. Academic, New York.

Lu, C.-J., Steinecker, W.H., Tian, W.-C. et al. (2005). First-generation hybrid MEMS gas chromatograph. *Lab on a Chip*. 5, 1123–1131.

Madou, M.J. and Morrison, S.R. (1989). *Chemical Sensing with Solid State Devices*. Academic Press, Boston.

Mandayo, G.C., Castano, E., Gracia, F.J., Cirera, A., Cornet, A., and Morante, J.R. (2002). Built-in active filter for an improved response to carbon monoxide combining thin and thick-film technologies. *Sens. Actuators B* 87, 88–94.

Manginell, R.P., Frye-Mason, G.C., Kottenstette, R.J., Lewis, P.R., and Wong, C.C. (2000). Microfabricated planar preconcentrator. In: *Technical Digest of the 2000 Solid-State Sensor and Actuator Workshop*, Cleveland, OH, pp. 179–182.

Manginell, R.P., Adkins, D.R., Moorman, M.W. et al. (2008). Mass-sensitive microfabricated chemical preconcentrator. *J. Microelectromech. Syst.* 17, 1396–1407.

Manginell, R.P., Bauer, J.M., Moorman, M.W. et al. (2011). A monolithically-integrated μGC chemical sensor system. *Sensors* 11, 6517–6532.

Markowitz, M.A., Zeinali, M., McGill, R.A., Kusterbeck, A.W., and Stepnowski, J.L. (2010). Hybrid preconcentrator for detection of materials, US Patent No 2010/00837 36 A1.

Mitchell, G.D. (1987). Trace gas calibration systems using permeation devices. In: Taylor, J.K. (ed.), *Sampling and Calibration for Atmospheric Measurements*, ASTM STP 957. American Society for Testing and Materials, Philadelphia, pp. 110–120.

Moseley, P.T. and Tofield, B. C. (eds.) (1987). *Solid State Gas Sensors*. Adam Hilger, Bristol, UK.

Moseley, P.T., Norris, J.O.W., and Williams, D.E. (1991). *Techniques and Mechanisms in Gas Sensing*. Adam Hilger, Bristol, UK.

Nakamoto, T., Fukuda, T., and Moriizumi, T. (1991). Gas identification system using plural sensors with characteristics of plasticity. *Sens. Actuators B* 3, 1–6.

Oborny, M.C., Zheng, J., Nichols, J.M. et al. (2003). Passive callibrator-vapor source for a micro gas chromatograph. In: *Proceedings of 7th Internatonal Conference on Miniaturized Chemical and Biochemical Analysts Systems*, October 5–9, Squaw Valley, CA, pp. 1243–1246.

Park, J.-H. and Yang, R.T. (2005). Simple criterion for adsorbent selection for gas purification by pressure swing adsorption processes. *Ind. Eng. Chem. Res.* 44, 1914–1921.

Pierce, T.C., Schiffma, S.S., Nagle, H.T., and Gardner, J.W. (eds.) (2003). *Handbook of Machine Olfaction: Electronic Nose Technology*. Wiley-VCH, Weinheim, Germany.

Pijolat, C., Pupier, C., Sauvan, M., Tournier, G., and Lalauze, R. (1999). Gas detection for automotive pollution control. *Sens. Actuators B* 59, 195–202.

Pijolat, C., Viricelle, J.P., Tournier, G., and Montment, P. (2005). Application of membranes and filtering films for gas sensors improvements. *Thin Solid Films* 490, 7–16.

Pijolat, C., Camara, M., Courbat, J., Viricelle, J.-P., Briand, D., and de Rooij, N.F. (2007). Application of carbon nano-powders for a gas micro-preconcentrator. *Sens. Actuators B* 127, 179–185.

Prasad, R.M., Gurlo, A., Riedel, R., Hübner, M., Barsan, N., and Weimar, U. (2010). Microporous ceramic coated SnO_2 sensors for hydrogen and carbon monoxide sensing in harsh reducing conditions. *Sens. Actuators B* 149, 105–109.

Pratzler, S., Knopf, D., Ulbig, P., and Scholl, S. (2010). Preparation of calibration gas mixtures for the measurement of breath alcohol concentration. *J. Breath Res.* 4, 036004.

Sauvan, M. and Pijolat, C. (1999). Selectivity improvement of SnO_2 films by superficial metallic films. *Sens. Actuators B* 58, 295–301.

Schweizer-Berberich, M., Strathmann, S., Gopel, W., Sharma, R., and Peyre-Lavigne, A. (2000). Filters for tin dioxide CO gas sensors to pass the UL2034 standard. *Sens. Actuators B* 66, 34–36.

Seo, J.H. (2012). Microfabricted Passive Preconcentrator/Injector Designed for Microscale Gas Chromatography. PhD Thesis, the University of Michigan.

Seo, J.H., Kim, S.K., Zellers, E.T., and Kurabayashi, K. (2012). Microfabricated passive vapor preconcentrator/injector designed for microscale gas chromatography. *Lab on a Chip* 12, 717–724.

Simon, I., Bârsan, N., Bauer, M., and Weimar, U. (2001). Micromachined metal oxide gas sensors: Opportunities to improve sensor performance. *Sens. Actuators B* 73, 1–26.

Staude, E. (1992). *Membranen und Membranprozesse*. Verlag Chemie,Weinheim, Germany.

Tian, W.-C. and Pang, S.W. (2002). Freestanding microheaters in Si with high aspect ratio microstructures. *J. Vac. Sci. Technol. B* 20, 1008–1012.

Tian, W.-C. and Pang, S.W. (2003). Thick and thermally isolated Si microheaters for microfabricated preconcentrators. *J. Vac. Sci. Technol. B* 21, 274–279.

Tian, W.-C., Pang, S.W., Lu, C.-J., and Zellers, E.T. (2003). Microfabricated preconcentrator-focuser for a micro scale gas chromatograph. *J. Microelectromech. Syst.* 12, 264–272.

Tian, W.-C., Chan, H.K.L., Lu, C.-J., Pang, S.W. (2005). Multiple-stage microfabricated preconcentrator-focuser for micro gas chromatography system. *IEEE J. Microelectromech. Syst.* 14, 498–507.

Van Es, A. (1992). *High-Speed Narrow Bore Capillary Gas Chromatography*. Huthig Buch Verlag Heidelberg, Germany.

Voiculescu, I. (2005). Design and Development of MEMS Devices for Trace Detection of Hazardous Materials. PhD Thesis, George Washington University, Washington, DC.

Voiculescu, I., McGrill, R.A., and Zaghoul, M.E. (2006). Micropreconcentrator for enhanced trace detection of explosives and chemical agents. *IEEE Sensor J.* 6(5), 1094–1104.

Voiculescu, I., Zaghloul, M., and Narasimhan, N. (2008). Microfabricated chemical preconcentrators for gas-phase microanalytical detection systems. *Trends Anal. Chem.* 27, 327–342.

Wallner, J.Z., Kunt, K.S., Obanionwu, H., Oborny, M.C., Bergstrom, P.L., and Zellers, E.T. (2007). An integrated vapor source with a porous silicon wick. *Phys. Stat. Sol. (a)* 204(5), 1449–1453.

Yang, R.T. (1997). *Gas Separation by Adsorption Processes*. Imperial College Press, London.

Zampolli, S., Elmi, I., Mancarella, F. et al. (2009). Realtime monitoring of sub-ppb concentrations of aromatic volatiles with a MEMS enabled miniaturized gas chromatograph. *Sens. Actuators B* 141, 322–328.

Zheng, J., Christophersen, M., and Bergstrom, P.L. (2005). Thick macroporous membranes made of *p*-type silicon. *Phys. Stat. Sol. (a)* 202(8), 1402–1406.

Zheng, F., Baldwin, D.L., Fiffield, L.S. et al. (2006). Single-walled carbon nanotube paper as a sorbent for organic vapour preconcentration. *Anal. Chem.* 78, 2442–2446.

Zhou, Z. and Gai, L. (2006). Preparation of calibration gas mixtures of water and nitrogen by using diffusion tubes. *Accred. Qual. Assur.* 11, 205–207.

Silicon Nanostructures for Laser Desorption/Ionization Mass Spectrometry

Sergei Alekseev

CONTENTS

13.1 Soft Ionization Methods in Mass Spectrometry   240

13.2 The Morphology of Silicon Nanostructures, Used as LDI-MS Substrates   242

13.3 Mechanisms of the Desorption-Ionization   247

13.4 Chemical Functionalization of Silicon LDI-MS Substrates   251

13.5 Practical Applications of nSi-LDI-MS   256

13.6 Conclusions and Future Trends   260

Acknowledgments   260

References   260

13.1 SOFT IONIZATION METHODS IN MASS SPECTROMETRY

Since the late 1980s, so-called soft ionization techniques, such as laser desorption ionization (LDI) (Karas and Hillenkamp 1988; Tanaka et al. 1988) and electrospray ionization (ESI) (Fenn et al. 1989) were introduced in mass spectrometric analysis. Differently from classical ionization techniques, such as electronic impact and chemical ionization, soft ionization methods allow producing low fragmented ions from polymers and large biomolecules (with molecular masses exceeding 100,000 Da) and other nonvolatile and labile molecules (Siuzdak 2006). Due to their importance for chemical analysis, the development of soft ionization techniques was rewarded the Nobel Prize in Chemistry in 2002.

One of the most frequently used soft ionization approaches is matrix-assisted laser desorption/ionization mass spectrometry (MALDI) (Karas and Hillenkamp 1988), involving cocrystallizing of the analyte with a matrix, which is generally a low molecular weight compound that can absorb the energy of a laser light pulse. Absorption of the laser energy resulted in evaporation of the matrix (along with the analyte) followed by analyte ionization within the plume by intermolecular interactions, usually proton transfer. However, the use of organic matrix led to the appearance of background ions in the low mass range, also the interaction between matrix and analyte could produce combined ions, which interferes with the analysis. That is why MALDI is rarely used for analysis of low molecular weight compounds (<1500 m/z). On the other hand, the MALDI method appears very sensitive to the presence of salts and other contaminants in the probe: even low concentrations of salts resulted in the formation of complexes with molecular ions and also in peak broadening and resolution decrease. That is why preliminary preconcentration and purification of the probe is required for MALDI MS analysis. In its turn the ESI, which is readily applicable for small molecules analysis, is also intolerant of contaminants and mixtures without prior separation such as with liquid chromatography (LC–ESI–MS).

Because of above-mentioned limitations of MALDI and ESI, there has been a significant amount of work to develop LDI techniques that can be performed without an organic matrix. A number of matrix-free LDI approaches based on application of different inorganic nanostructures (metals, carbon materials, oxides, and semiconductors) as ionization substrates were proposed in the past two decades. In the mass-spectrometric literature, they are known on a general abbreviation of surface-assisted laser desorption/ionization (SALDI) and they are comprehensively reviewed and compared in Peterson (2007), Arakawa and Kawasaki (2010), Kuzema (2011), Law and Larkin (2011), and Chiang et al. (2011).

Silicon-based nanomaterials, in particular the porous silicon (PSi), silicon nanowires (NWs), nanoparticles (NPs), and thin films seem to be the most popular among all inorganic LDI substrates due to their availability and advantageous characteristics (low thermal and high electrical conductivity, efficient light absorption, and tunable surface chemistry). The silicon acts as a well for the energy pulses from the laser, which in turn causes the desorption/ionization of the analyte without pyrolysis, fragmentation, or a background of matrix ions. Starting from the pioneering work of Wei et al. (1999), where the efficiency of the PSi in LDI MS was demonstrated, the desorption/ionization on PSi mass spectrometry is referenced under the abbreviation DIOS MS. For the application of any type of nanostructured silicon (no matter the PSi, Si NWs, NPs, or the thin layer) in LDI MS, the abbreviation nSi-LDI-MS will be used. The nanosilicon ionization targets based on the PSi and Si NWs were successfully commercialized by Waters (www.waters.com) under the trademark DIOS™ and by Bruker (Daniels et al. 2008) under the trademark NALDI™. To get an idea on different ionization techniques' applicability, their mass ranges and typical sensitivities are shown in the Table 13.1. In general, the nSi-LDI-MS surpasses MALDI MS by the sensitivity and overall performance in the low mass range, at the same time the efficiency of MALDI for high mass range is superior.

In fact, DIOS MS on the commercial targets allows reliable (signal-to-noise [S/N] ratio ~40) detection of 50 pg (~100 fmol) of model compound (verapamil) in 1 μL of solution, which corresponds to $1 \cdot 10^{-7}$ mol/L concentration (www.waters.com). To work with the commercial DIOS targets, no correction of existing MALDI MS setup is required, and differently from the classical MALDI, no low-molecular mass background appeared (Figure 13.1). The characterization of the efficiency of laboratory-made nSi-LDI-MS substrates usually requires only some minimal

TABLE 13.1   Typical Mass Ranges and Sensitivities Allowed by Different Ionization Techniques

Ionization Technique	Mass Range (DA)	Sensitivity (moles)
Electron ionization (EI)	1–1000	10^{-13}
Atmospheric pressure chemical ionization (APCI)	20–5000	10^{-13}
Electrospray ionization (ESI)	50–100000	10^{-13}
Nanospray ionization (nanoESI)	50–100000	10^{-15}
Matrix assisted laser desorption ionization (MALDI)	500–1000000	10^{-15}
Laser desorption ionization on Si nanostructures (nSi-LDI)	50–5000	10^{-14}–10^{-17}

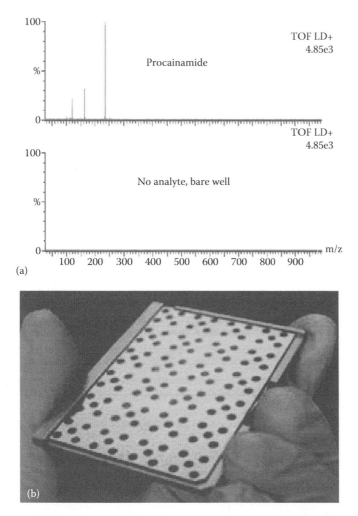

FIGURE 13.1   (a) The mass spectrum of 125 pg (460 fmol) of procainamide compared with the spectrum from a blank DIOS-target well and (b) commercial 120 wells DIOS-target™ plate from Waters secured in standard MALDI adapter. (Data extracted from http://www.waters.com, MassPREP DIOS-target Plates Care and Use Manual.)

modification of the commercial MALDI sample holders. Besides that, the irradiation of nitrogen laser (337 nm wavelength) of standard MALDI-TOF MS instrument efficiently adsorbs in Si activating the LDI processes. That is why the nSi-LDI-MS is one of the most intensively studied fields of the nano-silicon application and commercialization.

In this chapter the following aspects will be emphasized: (1) types of nano-Si materials used as the LDI substrates and the correlations between their structural parameters and the LDI efficiency; (2) the mechanisms of physicochemical processes taking place in the LDI from Si surface; (3) the influence of silicon surface chemical functionalization on the LDI performance as well as the possibility of specific analyte capture to combine the stages of sample preconcentration and consecutive LDI-MS analysis; and (4) examples of practical application of nSi-MS in chemical analysis, forensics, and biomedical research.

13.2 THE MORPHOLOGY OF SILICON NANOSTRUCTURES, USED AS LDI-MS SUBSTRATES

The influence of the PSi preparation conditions and morphology on the DIOS MS efficiency was extensively studied in early works from Scripps Research Institute carried out under supervision of Gary Siuzdak (Shen et al. 2001; Lewis et al. 2003). However, some important aspects are probably missing in these papers due to patenting (e.g., Siuzdak et al. 2001) and commercialization of DIOS. Up to 30 different analytes (peptides, carbohydrates, glycolipids, alkaloids, drugs, etc.) with molecular weights in a range from 150 to 20,000 Da were used for DIOS plates testing in these works.

According to the presented results, the morphology of the PSi layer is a key point to the LDI MS efficiency. Optimal sensitivity together with highest S/N ratio was demonstrated on the macroporous ($D_{por} = 10 - 1000$ nm) thin ($h = 0.1 - 1$ μm) PSi layers with predominantly vertical pores, similar to one demonstrated in Figure 13.2. This layer can be produced by electrochemical etching or by any other possible method. The doping level and type of Si wafer, its crystallographic orientation, and etching solution composition as well as the current density and time have an influence on the LDI performance only as the parameters determining the morphology of the layer. Commonly the n-type Si wafers are preferable compared to the p-type because they can be photopatterned to create the spots of the PSi on the plate surface by etching under illumination through a mask. For example, the PSi layer presented in Figure 13.2 was prepared by 1 min etching of low resistivity (0.005−0.02 Ω·cm) n-type Si wafer in a 25% HF/ethanol solution at extremely low (~5 mA/cm^2) anodic current under moderate (50 mV/cm^2) white light illumination followed by so-called postetching procedure including the PSi oxidation by O$_3$ or H$_2$O$_2$ and subsequent oxide removal by 5% HF.

Compared to standard photoluminescent PSi, the DIOS-active layers have much larger pores, smaller thickness, and lower overall porosity (~30−40% vol., from SEM data). In contrast, highly luminescent PSi layers are usually not very effective as LDI targets. Increase of the porous layer thickness (to tens of microns) as well as decrease of the pore size resulted in significant decrease of the LDI substrate sensitivity, possibly just due to the dissipation of the analyte in the thick

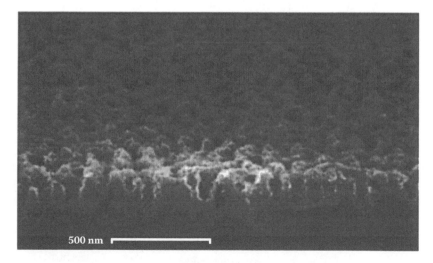

FIGURE 13.2 SEM image of the PSi sample, optimized for DIOS. The macropores (70–120 nm) are spaced ~100 nm apart. The thickness of the PSi layer is in the range of 200 nm. (Reprinted with permission from Shen Z. et al., *Anal. Chem.* 73(3), 612. Copyright 2001. American Chemical Society.)

porous layer and its confinement into the narrow pores. Besides the morphology, the surface chemistry of the PSi was shown to be crucial for its LDI performance. In particular, ambient air storage of the PSi without any chemical functionalization resulted in a fast drop of the LDI activity, probably due to the pore blockage by the oxide. The issue of chemical functionalization and its influence on the LDI activity will be discussed in a Section 13.4.

Further works on the influence of the PSi preparation conditions on the LDI activity confirmed the above conclusions on the influence of PSi morphology effects. According to Tuomikoski et al. (2002), large (50–200 nm) pores of the PSi are optimal due to the best filling of them with sample solution and following uniform evaporation of the droplet. Seino et al. (2005) studied the influence of electrochemical etching conditions for both p- and n-type Si wafers in order to obtain DIOS supports for accurate determination of molecular weight distribution of polystyrene (certified standard sample, degree of polymerization [n] ranging $n = 9 - 40$ with a maximum at $n = 21$). They found that a 460-nm layer of p-type and 2.5-µm layer of n-type PSi gives the polymer mass distribution coinciding with the certified value, otherwise (for thinner as well as for thicker layers) the distribution maximum is shifted to the lower mass range, indicating more efficient desorption of low molecular mass olygomers in comparison with higher ones. Besides that, the supports with "optimal" layer thicknesses exhibited highest values of S/N ratio. The threshold value of the laser beam intensity sufficient for the LDI of polystyrene from p-type PSi appeared three times higher than from the n-type PSi (15 µJ/cm^2 vs. 5 µJ/cm^2); however, this difference had no significant influence on the quality of the resulted mass spectra. Gomez et al. (2007) optimized the PSi substrates for LDI-MS detection of cyclic adenosine monophosphate. Doping type, resistivity, and etching conditions (current density, time, illumination) were varied. As an overall result, the best LDI performance was obtained for the PSi samples with average pore sizes from 300 to 1000 nm prepared from the n-type Si wafers under tungsten lamp illumination.

In the past years, a lot of efforts were made in development of the LDI-active substrates based on Si nanowires and other nanostructures of columnar morphology. The NWs usually can be prepared by the method called metal-assisted stain etching (MASE), which consists of the action of HF-containing chemical oxidant on silicon with noble metal (Ag or Au) nanoparticles predeposited on the surface or forming during the etch from the metal salt present in the solution (Piret et al. 2010; Chen et al. 2011; Cheng et al. 2012; Dupré et al. 2012a; Tsao et al. 2012a,b). The process, similar to the usual electrochemical corrosion, resulted in the NPs penetration inside the Si wafer and formation of vertically aligned arrays of Si NWs, which are typically nanometers to tens of nanometers thick and micrometers long (Figure 13.3). Optionally, the residual metal NPs can be removed from the pores.

Besides that, the technique of vapor-liquid-solid (VLS) growth, which is based on catalytic thermal decomposition of the silane gas (SiH$_4$) on the gold nanoparticles deposited on the support surface, is widely applied for the preparation of LDI-active Si NWs (Go et al. 2005; Dupré et al. 2012a). According to Law and Larkin (2011), the silicon NWs on NALDI targets from Bruker Daltonics Inc. are grown exactly by this method. In fact, an application of any method such as RIE (Sainiemi et al. 2007; Gulbakan et al. 2010; Wang et al. 2013) or laser ablation (Chen et al. 2007; Suni et al. 2010) giving a structured Si surface with tens to hundreds of nanometers characteristic feature size can be successfully used to get an LDI-active substrate (Figure 13.4).

Compared to "standard" PSi, the LDI substrates with Si NWs possess much higher sensitivity (0.5 fmol for des-Arg^9-bradykinin is possible to detect) and very low laser fluence (0.3 µJ) required for the ionization (Go et al. 2005). However, SEM studies on these surfaces revealed that the exposure of the thin (10 nm) nanowires to this extremely low laser radiation resulted in their total

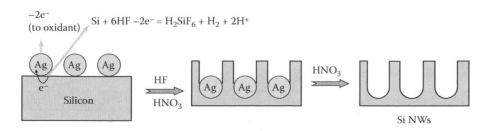

FIGURE 13.3   Schematic representation of the Si NWs growth by MASE.

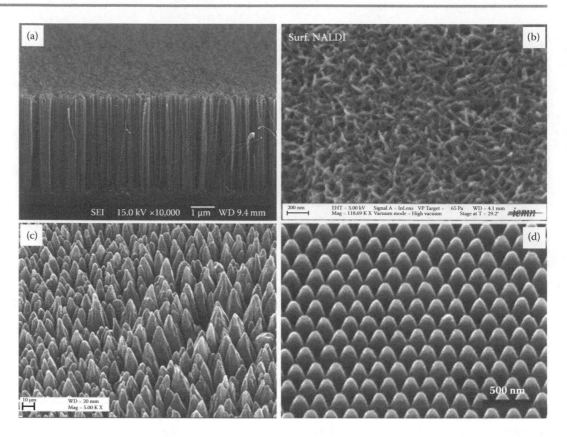

FIGURE 13.4   SEM images of different silicon nanostructures demonstrated high LDI activity: (a) Cross-section image of MASE-derived Si NWs. (From Cheng Y.-C. et al. *Analyst* 137(3), 654, 2012. Reproduced by permission of The Royal Society of Chemistry.) (b) The surface of NALDI plate from Bruker. (Reprinted with permission from Dupré M. et al., *Anal. Chem.* 84(24), 10637. Copyright 2012 American Chemical Society.) (c) Black silicon produced by laser irradiation of Si wafer, 10-µm scale bar (From Chen Y. et al., *J. Phys.: Conf. Ser.* 59, 548, 2007. Copyright 2007: IOP Publishing. With permission.) (d) Silicon nanocones produced by RIE. (With kind permission from Springer Science+Business Media: *J. Am. Soc. Mass Spectrom.* 24(1), 2013, 66. Wang Y. et al.)

destruction. Thus, the LDI-MS spectrum collection from a particular spot of such substrate damaged the surface and resulted in increasingly poor shot-to-shot reproducibility (Chen et al. 2007). The surface of the commercially available NALDI ionization targets (Figure 13.4b) is covered with Si nanowires (d = 20 nm, l = 100–500 nm and the surface density >100 wires·µm^{-2}) functionalized with perfluorophenyl groups. They demonstrate high sensitivity (femtomole range) to a wide number of small molecules (organic acids, amines, aromatic and heteroaromatic compounds, esters, peptides, carbohydrates, lipides, nucleozides, etc.) outperforming commercial DIOS targets and organic MALDI matrixes in several aspects, particularly in low molecular weight peptide analysis (Daniels et al. 2008; Guenin et al. 2009; Wyatt et al. 2010). However, NALDI mass spectra have characteristic background ions located at m/z 197, 235, and 243, and so on due to the formation of Au$^+$, [AuF$_2$]$^+$, [AuSi(H$_2$O)]$^+$, and other related cluster ions, and these obscure the analysis of small molecules (Guenin et al. 2009; Law and Larkin 2011).

Careful comparison of MASE-derived vertically aligned Si NWs arrays, "home-made" VLS nanowires (d = 50–100 nm, l = 12 µm), and commercial NALDI targets for the analysis of the peptide mixtures was performed by Dupré et al. (2012a). They found that the chips with vertical NWs demonstrate better LDI performance than do the ones with chaotically ordered VLS nanowires in terms of both the sensitivity and the quantity of different peptides in the mixture, which are possible to detect. An importance of wetting, analyte drop spreading, and sample concentration on drying, depending mainly not on the nanoscale morphological parameters (such the NWs thickness) but on the macroscopic surface roughness and surface chemistry (see Section 13.4) was particularly emphasized.

Any porous Si sample derived by an electrochemical etching as well as any Si NWs array produced by MASE or VLS has more or less wide distributions of the pore sizes and the Si feature size. That is why the precise study of morphology parameters influence on the LDI activity is possible only for uniformly structured samples, formed by the micromachining techniques. In particular, the nanocones presented in Figure 13.4d demonstrated highest S/N ratio in the LDI MS of polyethyleneglycol mixture for the height equal to 475 nm (at constant 580 nm distance between the nanocones). Both smaller (224 nm) and larger (632 nm) heights resulted in the appearance of low-molecular background of silicon cluster ions due to the decrease of the efficiency of energy transfer from Si to the analyte. For the different periods (with constant height 230 nm), the optimal S/N ratio was observed for a 220-nm period; however, smaller periods and higher aspect ratios were not tested due to the technological limitations of the nanocones preparations. The nanocone LDI substrates were shown to be efficient in the quantitative LDI MS analysis of glucose in human urine without any complicated probe pretreatment.

The ordered Si nanocavity arrays were prepared via e-beam lithography from 0.005–0.02 Ω/cm n-type Si wafer by Xiao et al. (2009) (Figure 13.5a). The parameters of the arrays were varied in the following ranges: porosity from 4 to 92% vol., pore diameter 130–500 nm, interpore spacing 16–1000 nm, and pore depth 200 and 459 nm (only these two values were tested). The authors found that the main parameter influencing the value of LDI fluence thresholds for studied compounds (so-called thermometric ions and model biological compounds, 1,2-dihexadecanoyl-sn-glycero-3-phosphocholine and angiotensin III) is the porosity of the array. Increase of the porosity and in some extent the pore depth resulted in decrease of laser fluence thresholds to yield appreciable ion intensity, the array with maximal (92%) porosity appears to be close to a conventional DIOS substrate. At the same time, the influence of pore size and interpore spacing takes place only when these parameters were influencing the porosity.

The silicon nanopost arrays with the dimensions (post diameter D = 50–600 nm, post height H = 200–1500 nm, and period P = 200–1200 nm) commensurate with the wavelength of mass spectrometer laser (337 nm) were produced by reactive ion etch (RIE) (Walker et al. 2010; Stolee and Vertes 2011) (Figure 13.5b). These substrates were shown to exhibit near-field effects and, as optical antennas, can couple laser radiation to the local environment. As a result, strong dependence on laser light polarization and significant enhancement of ion production was found for these novel substrates for p-polarized laser light (i.e., one, inducing the current oscillations along the post) compared with s-polarized light. At the same time, nearly no influence of Si resistivity

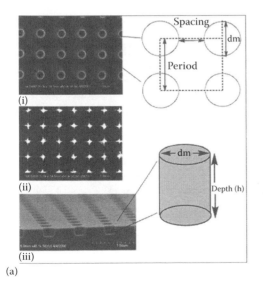

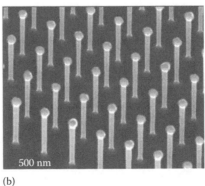

FIGURE 13.5   (a) Representative SEM images of ordered silicon nanocavity arrays, showing the top view of the ordered silicon nanocavity arrays with (i) low (9%) or (ii) high (91.6%) porosities and (iii) the cross-section of an ordered silicon nanocavity array with 10% porosity. (Reprinted with permission from Xiao Y. et al., *J. Phys. Chem. C* 13(8), 3076. Copyright 2009 American Chemical Society.) (b) SEM image of the silicon nanopost array. (From Stolee J.A. and Vertes A., *Phys. Chem. Chem. Phys.* 13(20), 9140, 2011. Copyright 2011: the PCCP Owner Society. With permission.)

(the range from $5 \cdot 10^{-3}$ to $100 \, \Omega \cdot \text{cm}$ was tested) on its LDI ion production was found, as the value of the DC resistivity has no effect on Si interaction with 337-nm laser irradiation. The post diameter increase resulted in significant growth of the threshold laser fluence required for the ionization; however, the character of the ion fragmentation dependencies from laser fluence remains practically the same for all diameters. The ion production (at optimal fluence) was found to increase slightly with P and it shows maximums for $H = 1000$ nm and $D = 200$ nm. Again, as in the case of PSi and Si NWs, the surface chemistry of Si support plays a crucial role on the LDI.

High LDI MS ionization activity was demonstrated for not only the "1D" and "3D" nanosized Si such as the nanowires and the PSi, respectively, but also for "0D" (the nanoparticles) and "2D" (thin films) nano-Si. The application of Si NPs for the LDI-MS combines the advantage of DIOS (absence of low-molecular ion background, high sensitivity, salt tolerance) and MALDI (the NPs can be used instead of the organic matrix without preparation of any special chip) (Wen et al. 2007; Dagan et al. 2009; Arakawa and Kawasaki 2010; Hua et al. 2010). The size of "optimal" Si NPs (30–50 nm) is in the range of the values found for the structural element (pore wall or the nanowires) thickness of the LDI-active PSi and NWs. Comparing to "classical" DIOS, the LDI from Si NPs requires lower laser fluences and gives less fragmented molecular ions. Homogeneous mixing of the analytes with the NPs allowed semi-quantitative analysis of the fatty acids in the milk (in negative ion mode) with minimal sample preparation (Hua et al. 2010).

The analysis of small molecules (drugs and peptides) deposited on top of the "2D" Si nanostructures such as silicon-on-insulator (SOI) wafers with 50 nm Si thickness (Kim et al. 2013, 2014) can be performed by two different types of the MS ionization (LDI and secondary ions mass-spectrometry [SIMS]) giving opportunity to combine advantages of both ionization techniques. According to Kim et al. (2014), the key cause of the LDI activity of SOI wafers is their low thermal conductivity because the highly thermally conductive bulk Si wafers were found LDI-inactive. A key advantage of the flat SOI substrates is an efficient ionization of relatively large molecules (such as insulin with 5729.6 Da mass) strongly adsorbing and thus poorly ionizing from other Si nanostructures with high surface area.

Comprehensive study of different silicon-based LDI substrates has been done by Alimpiev et al. (2008). Polished Si wafers, silicon wafers, sanded with low-disperse diamond powder (~1-μm grain size) and reactivated in HF prior to use, films (~0.5 μm thick) of amorphous silicon (α-Si) and hydrogenated amorphous silicon (α-SiH) prepared via RF sputtering of crystalline Si correspondingly in absence and presence of H_2, PSi produced by traditional HF etching under illumination (Shen et al. 2001), and iodine-modified PSi (PSi-I_2) prepared by anodization in the HF in the presence of I_2 were tested. No significant SALDI activity was detected for polished wafers and α-SiH films, the activity of sanded Si was rather modest, and other studied substrates showed excellent LDI activity. One important difference was found between porous and nonporous substrates. From any one spot on a nonporous substrate, that is, sanded or amorphous Si, the LDI signal is obtained only for the first few laser shots. In contrast, porous substrates gave ion current for some tens of the shots. The fact of activity of α-Si film, which is nearly flat in the nano- to micrometer range, but does show structural inhomogeneity with nanoscale "grainsize," and at the same time inactivity of (α-SiH) allows Alimpiev et al. (2008) to make a conclusion that comes in contrary to what has been stated repeatedly. They state that the porosity and roughness of the substrate are not important for the LDI activity, and the pores act only like reservoirs of analytes or proton donor molecules, and the key issue of the activity is the presence of so-called deep gap states in the material (see Section 13.3). Additionally, an outstanding stability to atmospheric conditions (up to a year without any significant decrease of ionization efficiency) of the PSi-I_2 in contrast to "standard" porous PSi was demonstrated.

Taking into account all of the aforementioned, it can be stated that efficient n-Si-LDI-MS substrates can usually be presented as relatively thin ($h = 0.1$–1 μm) and macroporous ($D_{por} = 10$–1000 nm) layers with predominantly columnar structure and rather high porosity. Too thick porous layers resulted in the decrease of the efficiency probably just due to the dissipation of the analyte inside the layer; too thin layers are also not very effective due to high thermal conductivity and weak light absorbance. The microporous substrates are not efficient due to strong adsorption of the analyte and its confinement inside the pores.

The size of Si features in the active layer should be in the range of 10–1000 nm. An evaporation of Si under laser light giving the LDI suppression as well as high intensity background ions is observed for small Si features; otherwise, a significant drop of the LDI activity takes place for the big ones. Probably the above issue is caused by the thermal conductivity of the substrate. One of the most important characteristics, which determines the LDI activity of the substrate, is the distribution of an analyte solution on its surface, governed by both the morphology and chemistry of the surface. At the same time, no appreciable influence of the electrical properties (doping type and level, resistivity) of Si was noted. Of course, due to the basic principles of the LDI-MS, the conductivity of supporting Si wafer should be high enough (we do not mean p- or n-type) in order to allow the charge drainage.

13.3 MECHANISMS OF THE DESORPTION-IONIZATION

The main purpose of any "soft" ionization mass-spectrometry technique is achievement of a maximal signal of molecular ion with its minimal fragmentation and background signals. This cannot be done without knowledge of the processes, happening with analyte under ionization. That is why great attention is paid to the investigation of desorption/ionization (DI) mechanisms in nSi-LDI-MS. However, until the present time, there is no generally accepted standpoint on DI mechanisms. Most likely they are rather ambiguous.

To analyze possible DI mechanisms, the processes happening under interaction of laser irradiation with nanostructured silicon should be considered. Remember that the PSi efficiently absorbs light with photon energy exceeding the band gap of Si, that is, 1.1 eV ($\alpha \approx 10^5$–10^6 cm^{-1} for common 337 nm laser (Li and Lipson 2013), it has low thermal (Perichon et al. 2000; Li and Lipson 2013) and relatively high electrical conductivity, its surface is covered with silane groups $Si_{4-x}SiH_x$ (x = 1–3) for as-prepared PSi and with O_3Si–H and silanol $\equiv Si$–OH groups for partially oxidized PSi (Alekseev et al. 2007). Due to the high surface area, the adsorbed molecules such as water and solvents are likely to be present on the surface of any Si nanostructure. In general, the absorption of laser light by the nano-Si can result in the surface heating, in the formation of electron-hole pairs or in the emission of electrons and in generation of local electric fields. Each of the above-mentioned phenomena may induce further physico-chemical transformations of the Si surface and analyte, finally giving the desorption/ionization event. The last in turn can be considered two stages (desorption and ionization), which may take place simultaneously or consequently.

Low thermal conductivity of the nano-Si resulted in significant (up to 600–1200 K according to Alimpiev et al. [2001] and Walker et al. [2010]) increase of local surface temperature during the laser shot. It should be mentioned that high surface area of the PSi decreases its melting point to 900°C (compared to 1410°C for bulk Si) due to the reduction of the surface energy (Herino et al. 1984). Consequently, the PSi surface restructurization and melting occurs even on low laser energies (Northen et al. 2007a) (Figure 13.6).

Threshold laser energy required for the LDI (10 mJ/cm^2 for electrochemically derived PSi) (Northen et al. 2007a), 0.3 mJ/cm^2 for thin Si nanowires (Go et al. 2005; Chen et al. 2007) correlates well with the beginning of surface restructurization process. Increase of laser fluence resulted in the formation of background silicon oxide (OSiH$^+$) and silicon cluster $\left(Si_x^+\right)$ ions in the mass-spectra (Northen et al. 2007a; Law 2010a). Very high laser fluences resulted in complete destruction of the porous surface layer, overscaled signals of Si clusters, and complete suppression of the analyte signal in the mass-spectra. The work under such conditions is strictly nonrecommendable due to the fast contamination and damaging of the LDI-MS ion source by Si ablation products.

Except restructurization and ablation of silicon, the surface heating resulted in desorption of hydrogen and other compounds (water, solvents) evolving due to thermal destruction of the surface layer. This gives rise to a plume that propagates against the background gas pressure in the pores and generates a shock wave, which can push the analyte molecules out of the pores (so-called explosive vaporization mechanism). If residual liquid solvent is present inside the pores, the pressure of the plume is relatively high and a shock wave is strong (so-called "wet" desorption), if the opposite is true, the plume pressure is low and the shock wave is weak ("dry" desorption) (Luo et al. 2005). The event of analyte ionization may happen on the surface of silicon due to electron transfer before the desorption or due to chemical or photochemical processes inside the plume.

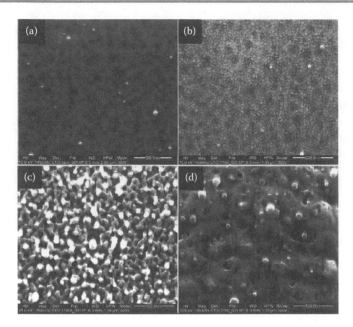

FIGURE 13.6 SEM images (scale bar 500 nm) of DIOS surface (0.01–0.02 Ω·cm n-type silicon [100] etched in 25% ethanolic HF at 5 mA for 3 min under illumination), exposed to single laser pulses at 337 nm (pulse width τ = 4 ns). (a) surface without irradiation; (b) laser energy 15 mJ/cm²; (c) 110 mJ/cm²; and (d) 970 mJ/cm². (With kind permission from Springer Science+Business Media: *J. Am. Soc. Mass. Spectrom.* 18(11), 2007, 1945, Northen T.R. et al.)

One of the most important objectives of DIOS (as any soft ionization technique) is the minimization of an analyte fragmentation. A concept of internal energy (IE) transfer, that is, the excessive energy of the desorbed ions can be used for quantitative comparison of the ionization "softness." The IE distributions can be calculated using a series of so-called thermometer ions. Typically, benzyl substituted benzyl pyridinium (BP) ions in the form of chloride salts are used (Figure 13.7).

The thermometer ions in series have similar and rather simple fragmentation pathways (loss of pyridine and formation of substituted benzyl cation for the BP ions for example) but different activation energy of the fragmentation. Doing the LDI MS (or other ionization experiment) at the same conditions (concentration, laser fluence, pulse length, etc.) for all the thermometer ion series and calculating their survival yields (SY), which are the ratios of the intensities of molecular ion currents to the sum of molecular and fragment ion currents, the IE distribution can be found (Luo et al. 2005). Since the thermometer ions already exist on the PSi surface as cations, no ionization stage is required to see them in the MS, so the experimental IE values relate to the energy transfer only in the desorption process.

The influence of laser wavelength (337 and 532 nm), fluence, and pulse length on the IE distribution of BP ions was studied for α-cyano-4-hydroxycynnamic acid (CHCA) MALDI matrix and DIOS targets with surface chemical functionalization of different polarity (Luo et al. 2005). It was found that the threshold value of laser fluence required for the LDI of 3MO-BP ions amount to 5 mJ/cm² for the CHCA matrix, to 21 mJ/cm² for the most hydrophobic perfluorophenyl PSi, and to 42 mJ/cm² for the hydrophilic PSi with surface amino groups. The survival yields of the BP ions show a sharp decline with increasing laser fluence for the CHCA matrix (from 90 to 50% for 3MO-BP), but they are nearly constant or only slowly declining (from 60 to 40% for 3MO-BP) for all studied PSi substrates. The above observations indicate that DIOS is a "less soft" ionization technique compared to MALDI and that their LDI mechanisms are fundamentally different.

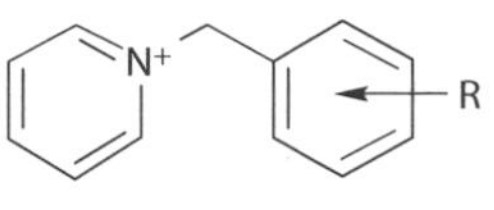

FIGURE 13.7 General structure of different BP ions (R = 4-chloro- (4C), 4-fluoro- (4F), 4-methoxy- (4MO), 3-methoxy- (3MO), 4-methyl- (4M), 3-methyl- (3M) and 2-methyl- (2M)).

According to Luo et al. (2005), the weak fluence dependence of the SY is an indicator of "dry" desorption, as the energy transfer between the hot surface of Si and the analyte species is limited due to extremely short interaction times. The large effect of derivatization on the threshold fluence shows that the activation energy of desorption for the thermometer ions indeed increases with the increase in the polarity of the surface. Changing the activation energy of desorption, however, does not result in appreciable changes in survival yields. This fact indicates that the energy portion, gotten by the analyte in the desorption stage, dispense into the activation of desorption and internal energy of the ion; further plume processes do not affect the IE transfer.

More recent work (Xiao et al. 2009) on the BP ions LDI MS from DIOS targets and ordered Si nanocavity arrays demonstrates sharp decline of the SYs with the fluence indicating a "wet desorption" mechanism, which is in contrast with the previously reported results of Luo et al. (2005). Careful examination of the experimental parameters allowed Xiao et al. (2009) to reveal that the residual pressure in their sample chamber was more than 2 orders of magnitude higher than the one used by Luo et al. (2005). Therefore, "wet desorption" takes place due to higher H_2O content in the sample chamber and on the PSi surface. The SYs of the BP ions in LDI from the Si NPs and NWs are usually significantly higher (100–60%) than for "classical" DIOS (Luo et al. 2006; Dagan et al. 2009). For the NWs, both "dry" and "wet" LDI can take place in dependence on the NWs morphology, surface chemistry, and pressure in the mass-spectrometer; and an energy transfer to the analyte depends not only on temperature but also on effects such as near electric field enhancement due to the light polarization (Luo et al. 2006; Stolee and Vertes 2011). For the SOI wafers, the IE of the BP ions is in the range of one for MALDI CHCA matrix and no other factors influencing the fragmentation except the substrate temperature were found (Kim et al. 2012).

If the analyte molecule is not already ionized (as the BP ions, for example), the stage of its ionization appeared equally important as the desorption stage. It is usually believed that in the nSi-LDI-MS, similarly to MALDI, the protonation of the analyte molecule is the main mechanism of the ionization and it is comprehensively studied (Ostman et al. 2006; Alimpiev et al. 2008; Chen et al. 2008; Liu and He 2008). Ostman et al. (2006) made a correlation between the proton affinity (PA) of the compound derived from quantum chemical calculations and atmospheric pressure DIOS-MS for a number of aromatic small molecules. The compounds with PA higher than 920–950 kJ/mol were easily detected in the form of $[M+H]^+$ ions and the signal intensities increased with the PA; otherwise, the compounds with lower PA usually were not seen in the MS or appeared as adducts with alkali metals $[M+Na]^+$ or as fragmentary ions. The radical-cations $[M]^{+\cdot}$ were observed only for the analytes with relatively low electron affinity, such as caffeine (Li and Lipson 2013).

Chen et al. (2008) demonstrate nearly linear dependence of the amino acids $[M+H]^+$ ion yield from their PA for porous Si (Figure 13.8) as well as for the graphite ionization substrates; at the same time, no correlation with the solution basisity constant (pK_b) was found.

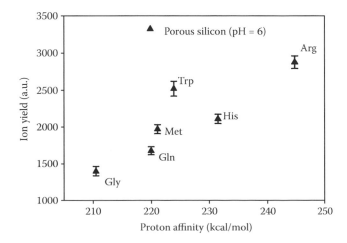

FIGURE 13.8   Comparison of amino acid proton affinities with their SALDI yields at an initial solution pH of 6 on porous Si. (Reprinted with permission from Chen Y.-F. et al., *J. Phys. Chem. C.* 112(17), 6953. Copyright 2008 American Chemical Society.)

As the measurements were done at ultrahigh vacuum (~10^{-9} Torr), the surface hydroxyls and physisorbed water were postulated as the proton source. The assumption of the LDI limitation by the protonation stage was confirmed by total quenching of the LDI activity of the PSi after thermal elimination of all surface Si-OH groups (and physisorbed H_2O as a result), by approximately 30-fold enhancement of the LDI signal for the amino acids, protonated by the HCl before the deposition on the LDI substrate and by very close values of ion currents found for the PSi and graphite. According to Chen et al. (2008), the role of porosity and high surface area of LDI substrate consisted of producing a contact between the analyte, coadsorbed water, and substrate surface groups, thus increasing the probability of proton transfer.

A detailed study by Liu and He (2008), performed with application of deuterated PSi and deuterated solvents, demonstrates that both the PSi surface groups and the physisorbed H_2O may supply a proton to analyte. As a result, slight oxidation of the PSi surface to transform surface Si–H groups to Si–OH resulted in significant (up to 17-fold) enhancement of ionization efficiency. However, the formation of an overoxidized silicon oxide layer resulted in the LDI reduce.

Comprehensive study of ionization mechanisms for the LDI MS on different nano-Si substrates was performed by a group of Alimpiev (Alimpiev et al. 2001, 2008; Zhabin et al. 2011; Grechnikov et al. 2013). They observed the increase of the LDI signal with the PA (similarly as discussed above) and also so-called activation (i.e., an increase of the LDI signal with repeated laser pulse irradiation at the same spot) of freshly etched PSi substrate covered mainly by the SiH groups. According to Alimpiev, the LDI activity of any Si-based substrate requires its low thermal conductivity, the presence of surface silanol groups, and the high concentration of defects (i.e., dangling bonds [DBs]) inside the crystallites. All of these features may be initially present in the substrate or may be generated by the laser irradiation (in case of activation) (Law 2010a). The porosity and roughness of the substrate determine not the LDI activity but the analyte and/or proton donor molecules distribution on the surface.

Based on the aforementioned, following a generalized ionization mechanism was proposed by Alimpiev et al. (2008) (Figure 13.9). Neutral analyte molecules adsorb on the PSi by formation of hydrogen bonds with the silanol groups. No protonation occurs, as the acidity of silanols is very weak. Absorbance of the laser irradiation resulted in the formation of free electron/hole pairs and substrate heating due to their subsequent electronic relaxation. The DBs inside the Si crystallites (deep gap states) efficiently trap the photogenerated electrons. Otherwise, the holes mostly occupy the near-surface sites, as the presence of Si-O surface species results in the weakening of adjacent Si–Si bonds. A net positive charge of the surface resulted in strong increase of Bronsted acidity of Si–OH groups and consequently in the proton transfer to adsorbed analyte molecules. The ejection of [M+H]$^+$ ions from the Si surface requires thermal activation or can be achieved due to local electric fields induced by laser irradiation.

If the positively charged analyte ion has an affinity to electrons, its LDI from the silicon surface is commonly accompanied by a reduction/protonation (R/P) process (Okuno et al. 2004; Shmygol et al. 2009). The distribution of M$^+$ and [M+H]$^{+\cdot}$ ions in LDI MS depends on the availability of reducing agents (protons and electrons) in the reaction space and on the reduction potential of the analyte (Okuno et al. 2004). Hence, the reduction/protonation via the electron transfer (Equation 13.1) is more reliable than via direct addition of an H$^\cdot$ atom.

(13.1) $$M^+ + H^+ + \bar{e} = [M+H]^{+\cdot}$$

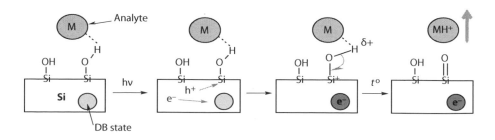

FIGURE 13.9   Schematic representation of the analyte protonation under the LDI from Si substrate.

The LDI MS of model compounds, methylene blue (MB^+Cl^-) and methyl orange (Na^+MO^-) from the PSi substrates with different surface terminations: as-prepared (PSi-H), oxidized (PSi-OX), and functionalized by cationic ($PSi-SO_3H$) and anionic ($PSi-ODMA^+Cl^-$) organic groups was studied by Shmygol et al. (2009). The methylene blue desorption/ionization occurs via both the R/P of the MB^+ to $[MB+H]^{+\bullet}$ and $[MB+2H]^+$ ions and by direct desorption of the MB^+ cation. The R/P, which predominates for PSi-H and PSi-OX substrates, requires lower laser fluences that is, it is more energetically favorable than direct desorption of the MB^+. The resulting $[MB+H]^{+\bullet}$ ion underwent significant fragmentation, which is higher for the PSi-OX than for the PSi-H. The detailed mechanism of the R/P is supposedly similar to the protonation mechanism shown in Figure 13.9 with the only difference of photoelectron location not in the DP state but on the MB^+. Higher $[MB+H]^{+\bullet}$ fragmentation on the PS-OX was explained by filtering-off the low energy photoelectrons by the oxide layer through the tunnel effect.

For the $PSi-SO_3H$, the R/P is strongly suppressed possibly due to the remoteness of the analyte from the Si surface and thus blocking of the MB^+ interaction with emitted electrons. Despite higher fluences required for the desorption, the fragmentation of the MB^+ from $PSi-SO_3H$ appeared to be significantly lower than the fragmentation of $[MB+H]^{+\bullet}$ from PSi-H and PSi-OX substrates. This fact indicates that the energy transfer in probably thermally induced "direct" desorption of the MB^+ is significantly lower than in the R/P, which requires the photoelectrons. Furthermore, the detection of the methyl orange (in negative MS mode) was possible only on the $PSi-ODMA^+Cl^-$ substrate, where this ion was efficiently protected from the protonation, causing its neutralization.

Except thermal and protonation processes, described above, the desorption/ionization from the surface of silicon may takes place due to the processes similar to the field ionization and field desorption. The electric field on the order of 10^7 V/cm, which is required for a field desorption event, is significantly higher than the operating field in standard MALDI mass spectrometer (8000 V/cm). However, the nano-sized tips of the PSi surface can act as sharp antennas focusing the laser energy onto a small area. Local enhancement of the electric fields up to 10^5 V/cm, which is sufficient for the field desorption, may take place as a result (Alimpiev et al. 2001; Royer et al. 2004). That is why the electric field effects should be considered in LDI mechanisms together with local heating and surface restructuring.

13.4 CHEMICAL FUNCTIONALIZATION OF SILICON LDI-MS SUBSTRATES

Despite its advantageous characteristics, the nanostructured Si with the "native" surface, usually covered by SiH_x groups, has very limited practical application as the LDI substrate due to its relatively fast oxidation and contamination in ambient conditions, resulting in irreproducibility of its LDI properties. That is why a stage of chemical functionalization resulted in stabilization of the Si surface to ambient environment is essential in the production of nSi-LDI-MS substrates. The surface chemistry of the PSi and the reactions used for its modification are comprehensively reviewed by Buriak (2002). In brief, two main approaches, that is, the reaction of the SiH_x groups with unsaturated compounds (silanization) and the oxidation of Si surface followed by the reaction of surface hydroxyls with anchor groups of the silane molecules (silanization), result in the organic grafted layers sufficiently stabilizing the surface of Si (Figure 13.10). Of course, further transformations of grafted groups can be performed.

Besides the LDI substrates stabilization, following main aims of chemical functionalization can be emphasized: (1) tuning of the wettability and overall analyte distribution in the porous layer; (2) creation of the "affinity chips," which are able to selectivly capture specific analytes and

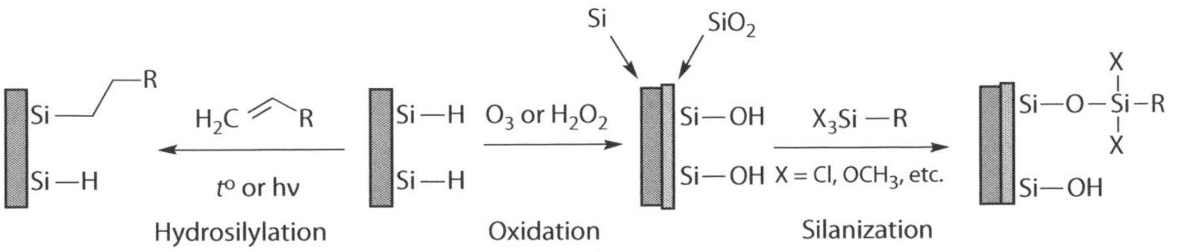

FIGURE 13.10   Chemical functionalizations of silicon surface.

thus allow the combination of sample preconcentration and purification with the MS analysis; and (3) improvement of overall mass-spectrometric sensitivity and suppression of the analyte fragmentation.

As it was mentioned above, wetting of the LDI substrate with the analyte solution (usually H_2O/CH_3OH or H_2O/CH_3CN 1/1 v/v) and its overall distribution in the porous active layer after droplet drying are crucial for overall LDI performance of the chip. The hydrophilic surface of the chip resulted in the analyte droplet spreading and LDI MS signal loss as a result. For very hydrophobic surfaces, droplets tend to roll off the surface making sample deposition difficult; also, the solution under evaporation is concentrated on a restricted area resulting in "hot spots" of deposited sample with strong LDI MS signal and very tiny ion intensities in the surroundings. That is why moderately hydrophobic surfaces with a droplet contact angle in a range of 100–130° are commonly used, and industry-made DIOS and NALDI targets satisfy these values (Law 2010b; Dupré et al. 2012a). As a result, the analyte molecules deposit uniformly in a local area of the chip and in the top part of the substrate porous layer after the droplet drying. According to Tuomikoski et al. (2002), uniform evaporation of the droplet on the PSi chip, derivatized with 10-undecylenic acid, requires pore sizes greater than 50 nm, allowing the solution to come inside.

Another important aspect of the chemical functionalization is tuning of an interaction between the surface and analyte molecules. If the analyte has no affinity to the surface, it may form separate crystals in different regions of the substrate, which result in poor reproducibility and sensitivity of the mass-spectra. As an example, slight oxidation of the PSi has no significant influence on the detection of hydrophilic amino acids but significantly degrades the signals of hydrophobic ones due to their low affinity to the hydrophilic PSi surface (Vaidyanathan et al. 2007). The carbohydrates (which are very hydrophilic) are selectively ionized on the hydrophilic amine-functionalized PSi, which demonstrates rather modest LDI activity to the hydrophobic analytes (Trauger et al. 2004). On the other hand, according to Luo et al. (2005), the increase of the surface hydrophylicity results in growth of an energy required for the analyte (benzylpyridinium [BP] salts) desorption, that is, the threshold laser fluence. No effect of the fluence on the fragmentation of the BP ions was found in this work; however, generally the increase of the fluence is undesirable due to the formation of Si clusters and other background ions and consequently the increase of the support affinity to the analyte resulting in overall suppression of its signal. As it was found by Dagan et al. (2009) for alkyl-modified Si nanoparticles, the increase of the alkyl chain length resulted in "softer" ionization due to the dissipation of energy, which is transferred from Si to analyte, in the organic grafted layer.

The presence of the surface hydroxyl groups as well as water and solvents inside the pores usually increases the LDI activity due to the enhancement of the analyte protonation (Alimpiev et al. 2008). That is why in order to remove possible contaminants it is recommended to keep the DIOS plates in ethanol for a night prior to application (www.waters.com). The water, which is usually present in EtOH in significant amounts, is able to oxidize the Si surface giving the silanol (SiOH) groups (Barabash et al. 2006). At the same time, the thick oxide layer on the Si surface resulted in significant quenching of the LDI activity (Lewis et al. 2003).

According to the literature analyzed above, optimal surface functionalization of nSi-LDI-MS substrates designed for the analysis of organic molecules of diverse nature (drugs, metabolites, peptides, etc.) is their coverage by long-chain alkyl ($-C_{10}H_{21}$, $-C_{18}H_{37}$, etc.) or perfluoroorganic groups ($-(CH_2)_3C_6F_5$, $-(CH_2)_2C_4F_9$, $-(CH_2)_2C_8F_{17}$), usually by means of silanization (Figure 13.10) of oxidized Si surfaces in the conditions excluding the formation of polysiloxane layers, which block the pores (Trauger et al. 2004). For the Si NWs with C18 groups, a short postmodification ozone oxidation was performed to adjust the hydrophobicity (Dupré et al. 2012a). Resulting surfaces demonstrate optimal (i.e., sufficient for uniform analyte deposition but not interfering its LDI) affinity to organic substances and low (to improve the salts tolerance compared with MALDI) affinity to inorganic salts. In particular, as it was shown independently from manufacturers (Law 2010b; Law and Larkin 2011), the surface of commercial DIOS and NALDI targets is covered by perfluorinated organic groups. Analysis of extremely polar compounds (in particular, carbohydrates) may require more hydrophilic Si surfaces such as lightly oxidized or functionalized with polar organic groups ($-(CH_2)_3NH_2$).

Additional advantages of perfluorinated organic functionalities is the combination of their hydrophobicity and oleophobicity. The preference of highly fluorinated molecules to reside in their perfluorinated environment rather than in organic or aqueous phase makes a name "fluorous" for such a phase. An application of the "fluorous" PSi-C_6F_5 chip allowed the LDI MS detection of des-Arg^9-bradykinin (relatively small peptide with m/z = 904, commonly used by instrument manufacturers as a test compound) with an extremely high sensitivity (800 yoctomoles, i.e., only 480 molecules) (Figure 13.11). Extreme hydrophobicity of the fluorous surface as well as small volume (0.2 µL) of analyte droplet resulted in the analyte concentration in the small area of a very thin top layer of the PSi, from where it was desorbed/ionized under the first laser shot.

The hydrophobic (alkyl as well as fluorous) nature of the nano-Si surface layer allowed the sample clean-up due to differential adsorption. Deposition of a droplet that contains organic molecules of interest (peptides, etc.) and hydrophilic contaminants (salts) and its subsequent (after a few seconds) removal with a pipettor allowed selective extraction of analyte on the Si surface. Several droplets can be applied consequently in order to concentrate the analyte. The LDI MS of the BSA tryptic digest in aqueous 8 mol/L urea obtained from the PSi-C_6F_5 chip using (a) differential adsorption sample cleanup and (b) dried drop deposition are shown in Figure 13.12. In the

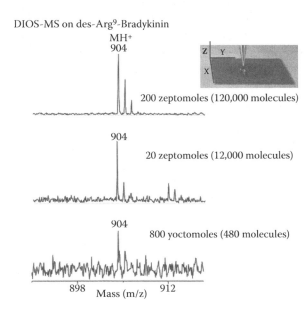

FIGURE 13.11 DIOS mass spectra obtained for different quantities of des-Arg^9-bradykinin using the PSi-C_6F_5 chip. (Reprinted with permission from Trauger S. A. et al., *Anal. Chem.* 76(15), 4484. Copyright 2004 American Chemical Society.)

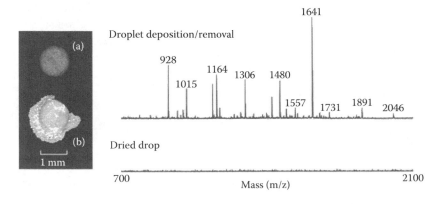

FIGURE 13.12 Optical images and DIOS mass-spectra of 500 amol of a BSA tryptic digest in 8 M urea using (a) differential adsorption sample cleanup on the PSi-C_6F_5 surface, and (b) dried drop deposition. (Reprinted with permission from Trauger S.A. et al., *Anal. Chem.* 76(15), 4484. Copyright 2004 American Chemical Society.)

first example, the mass spectrum contains peaks from multiple peptides that readily allowed for protein identification, while the second approach, conventional dried drop deposition, generated virtually no signal as urea crystallized out and covered the chip (see optical photos) (Trauger et al. 2004).

An important issue of the chemical functionalization is the preparation of "affine" surfaces capable of selective capture of specific analytes. Thus, the above-mentioned perfluorinated surfaces are affine for the molecules tagged with fluorous tails (Go et al. 2007). Washing of perfluorinated PSi wells with 50% MeOH in H_2O after sample deposition removes nontagged analytes (peptides and carbohydrates) and leaves tagged ones. Application of different tagging procedures allows introduction of fluorous tails into carboxylic acid functionalities or cysteinyl residues of the protein. Following tryptic digestion, tagged peptides capture and LDI MS analysis allowed selective identification of the protein chain fragments bearing $-CO_2H$ or $-SH$ functionalities into the complicated digest mixture.

The PSi (Pei et al. 2011) and Si NWs (Coffinier et al. 2012) with immobilized nickel complex of nitrilotriacetic acid (NTA) demonstrate good affinity to histidine-tagged peptides due to their coordination to Ni^{2+} ions (Figure 13.13). The NWs allow direct LDI-MS detection of captured peptides; otherwise, to get the MS of proteins complexed on the microporous photoluminescent PSi used by Pei et al. (2011) an addition of organic MALDI matrix was necessary. The same trick to get the MS signal was used by Tan et al. (2012) for the insulin (MW = 5.8 kDa), preconcentrated on the carboxylated PSi microparticles.

Among different types of chemical interactions the immunocapture, that is, the formation of complexes between the antibodies, which are large (4–14 nm in size) proteins and corresponding molecules (or their parts) named antigens seems the most selective one. The complicated task of immunocapture PSi LDI MS substrate design was successfully solved by Lowe et al. (2010). Highly selective targets for the MS detection of illicit drugs (benzodiazepines) were developed by this principle; however, their sensitivity appeared rather modest (5 pmol).

Capture of the ionic (or ionized in the solution) compounds such as histamine, arginine, methylene blue, sodium taurocholate, bilirubin, and so on by the PSi with ion-exchanging groups of opposite charge (sulfonic acid or quaternary ammonium salt) and their LDI MS detection was studied by our research group (Shmygol et al. 2007, 2009; Severinovskaya et al. 2010). The monocharged analytes were successfully preconcentrated and analyzed by LDI MS using this approach; furthermore, the fragmentation of their molecular ions was significantly diminished due to the suppression of reduction/protonation process (see Section 13.3). However, for the multiply charged compounds (such as anticancer drug Imatinib with five basic nitrogen atoms), which can be easily ionized from "usual" hydrophobic PSi, the oppositely charged ion-exchanging PSi appeared LDI inactive, probably due to very strong multipoint interaction between the substrate and analyte.

Except ionic and molecular interactions, the analyte capture on Si surface for LDI MS analysis can be performed by formation of photocleavable covalent bonds. In particular, the Diels-Alder adducts of grafted maleimide with active dienes undergone retro-[4+2] fragmentation under the laser illumination (Meng et al. 2004). A photocleavage of another type of linker, 2-nitro-benzyl alcoholcarbamate can be used for the LDI-MS analysis of amine-tagged compounds, in particular for tumor-associated antigen N3 minor octasaccharide (Lee et al. 2006). However, the multistage

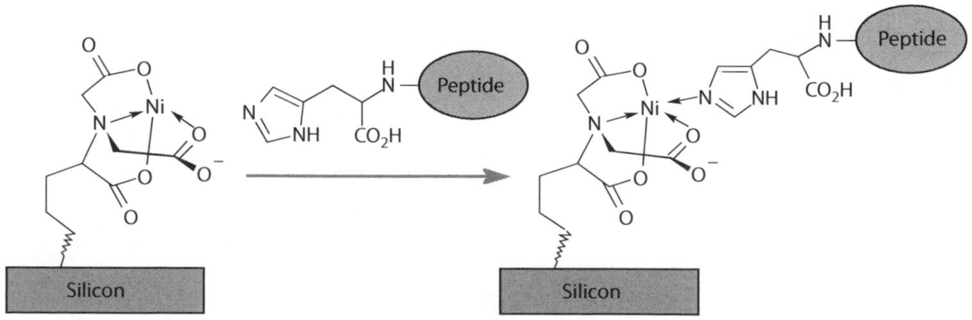

FIGURE 13.13   Complexation of His-tagged peptides with the NTA Ni^{2+} complex on Si surface.

chemical transformations giving desirable fragments on the surface of Si as well as corresponding linkers in the molecules of interest are rather complicated and the formation of by-products affecting the LDI MS is possible. That is why an approach of covalent tagging of the analytes with fluorous tails followed by their adsorption and LDI MS found much wider application in practice.

Usually the biological probes constitute very complex multicomponent mixtures of compounds of different and frequently unknown origin. That is why preliminary chromatographic separation is commonly required before their MS analysis, making it more complicated and time consuming. High surface area and porosity of the nanostructured silicon used as the LDI MS substrate make it similar to irreversible phases used in chromatography. Go et al. (2005) used the chip with Si NWs as a thin layer chromatography (TLC) plate and analyzed the compounds (model drugs, metabolites, and peptides) after separation directly by the LDI MS from the chip. More recently, the technique of electrowetting on dielectric, which is common for a Lab-on-Chip approach, was shown to be useful for the analyte displacement and rinsing prior to the LDI MS analysis on a chip patterned with hydrophobic and hydrophilic Si NWs (Lapierre et al. 2011).

Except analyte capture or separation, many efforts were made to improve overall mass-spectrometric performance of nSi-LDI MS substrates. One of the proposed approaches includes the combination of advantageous characteristics of silicon and nanostructured noble metals (chemical stability, optical properties, simplicity of surface modification via thiol chemistry) (Yan et al. 2009; Tsao et al. 2012b). According to Castellana et al. (2010), the metal-based nanostructures demonstrate enhanced LDI efficiency due to so-called plasmonic effect, that is, the excitation of collective electron motion under laser irradiation resulting in enhanced photon absorption and huge concentration of the optical near-field in a small volume. The same effect allowed detection of even a single molecule by means of surface enhanced Raman spectroscopy (SERS). However, according to Silina et al. (2014), the LDI MS performance of the Pd NPs on the PSi appeared significantly worse compared with the same NPs on the stainless steel due to the issues such as substrate memory, influence of adsorbed water, formation of the Pd/Si eutectic, and other complex effects.

Significant enhancement of the LDI MS signal compared with "classical" DIOS was achieved by Northen et al. (2007b) using a novel Si-based ionization approach called nanostructure initiated mass spectrometry (NIMS). The porous Si with liquid (or easily melted solid) initiator trapped inside the pores is used as ionization support in this method. Both laser irradiation and ion beam can be used for the ionization, that is, NIMS substrates are compatible with commercial MALDI and TOF SIMS instruments. The analyte after the deposition adsorbs on top of the NIMS surface. Heating of the substrate by laser irradiation (or ion beam) results in rapid expansion of the initiator, followed by its vaporization together with solvent trapped inside the PSi pores (Figure 13.14).

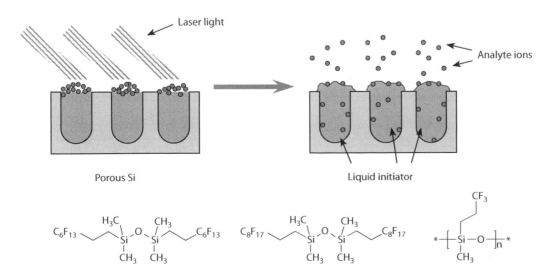

FIGURE 13.14 Principle of NIMS ionization and formulas of the initiators.

The vaporization and expansion of the initiator triggered desorption/ionization of the analyte adsorbed on top of the NIMS surface. Trapped initiator can migrate reversibly into and out of the NIMS surface with slight heating or cooling. This process lifts the analyte entrapped with the initiator inside the PSi pores, giving an analyte MS signal in consequent laser shots.

As a result of rigorous optimization work on the substrate morphology and surface chemistry, the relatively thick (approximately 100 µm) layers of the PSi with 10-nm pores and perfluorinated surface were found to be the most efficient NIMS substrates. They differ significantly from any nano-Si substrate optimized for "dry" nSi-LDI-MS, as the latter commonly have much larger pores and lower thicknesses (see Section 13.1). The perfluorinated siloxanes shown in Figure 13.14 are found to be the best NIMS initiators. Differently from organic MALDI matrixes, these initiators are completely inactive in LDI by itself without silicon.

Compared to DIOS, NIMS have better reproducibility and sensitivity (typically attomoles or femtomoles of the analyte can be detected), it covers much wider mass range (up to 30 kDa) and variety of the analytes (Northen et al. 2007b). Comparing to classical MALDI (with organic matrix), NIMS has the same advantages as DIOS: no background in small mass range and lower sensitivity to salts and other contaminants in the probe. For example, application of NIMS allowed registration of informative MS of fresh blood and urine without any sample preparation (such as desalting, fractionation, extraction, or derivatization), which is required if biological fluids are analyzed by any other MS techniques. However, as the MALDI method is better suited for the detection of large proteins than NIMS, these two methods are complementary in analysis of biological probes.

Similarly to other Si nanostructures, NIMS chips were used for selective capture of the analytes (Li et al. 2013). An addition of maleimide-functionalized fluorous tag into the peptide mixture allowed selective detection of the peptides containing cysteine residues.

13.5 PRACTICAL APPLICATIONS OF nSi-LDI-MS

As it was already discussed in Section 13.1, the main field of the nSi-LDI-MS application is an analysis of compounds with not very high MW (usually in 100–5000 Da range), which are commonly nonvolatile and labile (otherwise classical mass spectrometry can successfully be used). That is why low MW polymers and oligomers, drugs, metabolites, peptides, carbohydrates, steroids, and other similar small molecules attract the greatest attention as nSi-LDI-MS analytes (Compton and Siuzdak 2003).

Analysis of polymers by the LDI-MS is used to determine their chemical nature, MW, and polydispersity. In both MALDI and nSi-LDI-MS, the polymers usually desorb as silver $[M+Ag]^+$ or sodium $[M+Na]^+$ adducts due to the addition of cationizing salts or due to the natural abundance of Na^+ ions in the analyzing mixtures. The nSi-LDI-MS was found to be an efficient and precise tool to study the MW distributions of polyesters (Arakawa et al. 2004) and polystyrene (Seino et al. 2005) as well as for the quantitative analysis of diol and triol structure poly(propyleneglycol) mixtures (DPPG and TPPG), which are commonly applied as cryoprotectors (Okuno et al. 2005) (Figure 13.15). In comparison with MALDI, the DIOS MS demonstrates better reproducibility and lower standard deviation in PPG quantification.

One important case of low MW polymer analysis, where DIOS MS is successfully applied, is the analysis of ethoxylate polymers and polyethyleneglycols in biological media (Lewis et al. 2003; Shen et al. 2004a). These compounds are used as spermicides (nonoxynol-9, octoxynol-9, etc.) and lubricants in contraceptives. That is why their presence in body fluids can associate the victim and suspect in a sexual assault case, if no DNA sources are available. Even from the most contaminated forensic samples, DIOS MS allowed identification of these polymers directly from biological sources and produce data that are legally defensible.

Another forensic application of the nSi-LDI-MS is the analysis of illicit drugs in body fluids. Cheng et al. (2012) perform quantitative analysis of amphetamine, methamphetamine, codeine, morphine, and ketamine in urine samples taken from drug-abusers by the LDI-MS on Si NWs (see Figure 13.4a). The quantification was performed using the deuterated internal standards (e.g., MOR-d_3 for morphine) in order to diminish problems with nonuniform distribution of analyte on the support and other issues resulted in poor reproducibility of the MS peaks intensity.

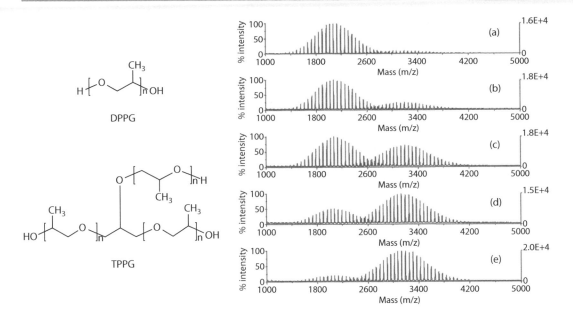

FIGURE 13.15 DIOS-MS of diol- and triol-type PPG mixture (1 mg/mL in THF/C$_2$H$_5$OH). (a) TPPG 10%wt, (b) 25%wt, (c) 50%wt, (d) 75%wt, and (e) 90%wt. (Reprinted from *Int. J. Mass Spectrom.* 241(1), Okuno S., Wada Y. and Arakawa R. 45, Copyright 2005, with permission from Elsevier.)

The limits of detection (LODs) were in the range of 10^{-7} M (10^{-10} M for ketamine!), which is significantly lower than the cut-off values (10^{-6} M) and high validity and precision of analysis results were achieved. In comparision with standard GS-MS and LC-MS procedures, the nSi-LDI-MS approach requires minimal analysis time (15 min), very simple probe pretreatment (only 1 min centrifugation), and small sample volume (10 μL). Guinan et al. (2012, 2014) analyzed Δ^9-tetrahydrocannabinol, methamphetamine, cocaine, and 3,4-methylenedioxy-methamphetamine in the samples of saliva using perfluorinated porous Si chips. Their results on the LODs and quantification of the drugs were highly coherent with the results of Cheng et al. (2012). An application of modern MALDI mass spectrometer with the MS/MS option significantly improves the reliability of the drug's identification.

In fact, the field where the nSi-LDI-MS found the widest application is biomedical research. An example of application of atmospheric pressure DIOS coupled with ion trap MS for pharmacokinetic studies was shown by Steenwyk et al. (2006). They perform quantification of a sedative drug midazolam in rat plasma with the verapamil as an internal standard (Figure 13.16). The molecular ions were chosen for fragmentation and the ratio of intensities of product ions with m/z = 291 and 303 was used as an analytical signal, giving 0.01–5 mg/L linear dynamic range for midazolam. In addition, the midazolam 1'-hydroxylation kinetics in pooled human liver microsomes and its inhibition by the cytochrome P450 3A4 model inhibitor ketoconazole were studied. However, according to Steenwyk et al. (2006), an application

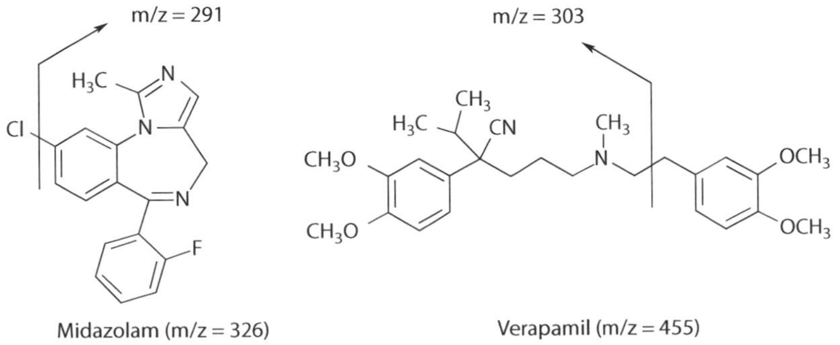

FIGURE 13.16 Formulas of the analyte (midazolam) and internal standard (verapamil) and structures of the product ions in the secondary MS, used for the quantification.

of this technique to routine analysis in the pharmaceutical environment requires significant improvement in quantitative precision and accuracy, which is a general bottleneck of LDI-MS-based methods.

One of the most relevant tasks of biochemical investigations and drug screening which is solved by nSi-LDI-MS is the monitoring of enzyme activity and inhibition. The fluorescence (PL) plate readers, typically used in biochemistry for this purpose, allow high throughput screening (10,000 inhibitors per day); however, some part of the false results may appear due to the PL properties of the inhibitors and other factors. The analysis of enzyme activity by DIOS MS includes the preparation of solutions containing the enzyme, substrate, possible inhibitor and buffers, their incubation, deposition onto the DIOS chip, drying, and LDI MS analysis (Shen et al. 2004b). Thousands of inhibitors of acetylcholine esterase, phenylalanine hydrolase, and sialyltransferase were successfully characterized because both the substrates and products of these enzymatic reactions are easily detectable in DIOS MS, and the enzyme macromolecules have no interference on the mass spectra. The procedures of the solutions preparation, deposition (done by the electrospray to enhance the reproducibility and accuracy), and MS screening were performed in fully automated mode, which allowed testing of 4000 inhibitors in 5 h with a standard commercial MALDI instrument.

Similarly, as it is shown above for DIOS, the NIMS approach was successfully applied to create enzyme activity assays, named Nimzyme (Northen et al. 2008). In this case, the enzymatic reaction was performed directly on the NIMS chip surface. The reaction substrate (lactose) was immobilized onto the NIMS chips via a fluorous affinity tag, which has high affinity to perfluorinated siloxane NIMS initiator. Additionally, an arginine unit was incorporated to the substrate to facilitate the ionization. Resulted Nimzyme assays were suitable for characterization of both galactosidase and sialyltransferase activity of biological complex mixtures. For example, temperature dependencies of galactosidase activity were tested for standard human galactosidase as well as for the lysate of thermophilic microbial community from Yellowstone hot springs. As it was expected, human galactosidase demonstrated maximal activity at 37°C, whereas the microbial community lysate was optimally active near the hot spring temperature (79°C) and retained activity even at 100°C.

Numerous examples of DIOS and NIMS application for the screening of enzyme activity and their comparison with other SALDI substrates and traditional MALDI MS are presented in a comprehensive review of de Rond et al. (2015) and references therein.

Besides the characterization of enzymes activity, the task of modern bioanalytical chemistry where the small molecules analysis is required, that is, the fields of potential application of nSi-LDI-MS, is an analysis of proteins and metabolites. A general approach of chemical identification of high MW proteins includes their site-specific digestion with enzymes followed by analysis of relatively low MW peptides in the resulted digest by any available method (usually different types of chromatography and mass spectrometry or their combinations are used) and the amino acids sequence reconstruction or the database search. An application of nSi-LDI-MS for the peptides mixture analysis was started from the pioneering work of Wei et al. (1999) and the peptide mixtures are the common probes to test any forthcoming LDI-MS substrates. Obviously, an application of the MS/MS approach allowed more reliable identification of the protein digests. An article of Dupré et al. (2012b) demonstrates an example of a good-quality work on the peptide analysis by LDI MS. The cytochrome C, β-casein, BSA and fibrinogen were identified in the femtomolar range and the complementariness of LDI-MS on the Si NWs (which is better for low MW proteins) and MALDI (which is more efficient for high MW proteins) for the peptide mass fingerprinting was demonstrated.

The term metabolomics means a field of bioanalytical chemistry consisting of identification and quantification of metabolites, tens or hundreds of small molecules appearing in biological fluids due to metabolism. The aims of metabolomics are diagnosis of diseases, following the response to drugs or even the evolution of patients. High throughput, sensitivity, and a wide range of the metabolits possible to detect make the nSi-LDI-MS extensively studied for metabolomics. Vaidyanathan et al. (2007) represent DIOS MS investigation of two metabolite cocktails, one consisting of 30 and the other of 20 metabolites, including aminoacids, carboxylic acids, and some other metabolites typically at 0.05 mmol/L concentration of each. Basic metabolites were detected mainly in positive mass spectra in the forms of protonated ions [M+H]⁺, sodium

[M+Na]⁺, or potassium [M+K]⁺ adducts while the acidic metabolites (as deprotonated ions [M−H]⁻) dominate in negative mass spectra. It was found that DIOS MS is suitable for detection of quantitative changes for individual analytes in mixtures; however, ion suppression effects are operational when one analyte is allowed to dominate the rest. Metabolomic application of different ionization methods such as ESI, AP chemical ionization, MALDI, and DIOS was analyzed and their complementariness was proved by Nordstrom et al. (2008). The potential of DIOS MS for metabolit analysis in human blood and urine with minimal sample pretreatment was demonstrated by Law (2010b).

The analysis of the distribution of the compounds of interest directly in the cells, tissues, or organs is a much more challenging task of mass-spectrometry (named MS-imaging) than the analysis of the same substances in the solution. As for any imaging technique, the lateral resolution is a crucial parameter for the MS imaging. The resolution of classical MALDI imaging is rather low due to the large size of matrix crystals. In the case of nSi-LDI-MS, the resolution is determined by the size of the laser beam (in the range of 10 μm) or the ion beam (about 150 nm) for the ion-NIMS (Northen et al. 2007b). That is why the nSi-LDI-MS imaging attracts a lot of attention. The recent literature on its application for the metabolites analysis is overviewed by Liu et al. (2014).

In the case of NIMS ionization substrate, the procedure of imaging experiment consisted of two steps. At the first one, a very thin (<12 μm) tissue slice is placed onto the NIMS surface and treated by a high power laser (0.4 J/cm² per pulse) or ion beam to etch the tissue out. Afterward the desorption/ionization of metabolites accumulated in the NIMS initiator liquid is performed by low-energy laser irradiation (0.01 J/cm² per pulse) or ion beam. Comparing the intensity of a particular analyte signal in different points of the surface, the MS image can be plotted.

The MS image of the sucrose distribution in the flower stem (Figure 13.17) was obtained by Patti et al. (2010a) using sodium-enhanced NIMS (nSi-LDI-MS detection of carbohydrates and steroids commonly required application of cationizing agents [NaCl for carbohydrates and AgNO₃ for steroids] due to low protonation ability [see Section 13.3] of these analytes).

The same research group (Patti et al. 2010b; Lee et al. 2012) described the use of NIMS for the detection and imaging of cholesterol, lipids, and phospholipids in mouse brain tissues, and the difference between the glial and neuronal cell enriched regions of the brain was clearly detected. However, the direct tissue analysis approach, which is described above, possesses one significant drawback. It requires very thin sections (<12 μm) of tissues to facilitate transmission of the laser pulse through the tissue to the active surfaces or to be etched out before analysis as described by Northen et al. (2007b). This means a complicated sample preparation and impossibility of analysis for some tissue types, which are not suitable for preparing such thin slices.

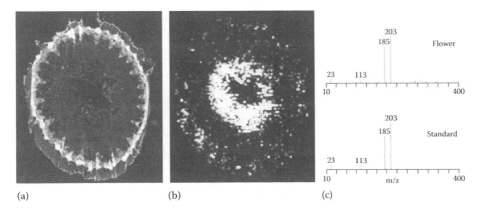

(a)   (b)   (c)

FIGURE 13.17 NIMS imaging of a *Gerbera jamesonii* flower stem. (a) Optical image of the stem mounted on the NIMS surface. (b) Corresponding sucrose [MNa⁺ m/z 365] distribution as obtained from sodium-enhanced NIMS. (c) MS/MS data from the flower stem and a commercial standard to confirm the identification of sucrose. (Reprinted with permission from Patti G.J. et al., *Anal. Chem.* 82(1), 121. Copyright 2010 American Chemical Society.)

Ronci et al. (2012) apply the LDI MS imaging on the PSi and on NALDI targets to study the distribution of bioactive compounds in marine mollusk tissues. To avoid the problem of thin slices described above, they used stamping of the tissue sections directly on the substrate surface. The low molecular compounds were adsorbed on the PSi surface and the imaging was performed after removing of the tissue slice.

Of course, in this brief section of the review, it is not possible to cover all the fields of nSi-LDI-MS practical applications. Undoubtedly, some of the references illustrating efficient and elegant solutions of different analytical tasks are missed here. However I hope it will give the readers an idea on the scope and limitations of nSi-LDI-MS and allow them to apply this ionization technique for new analytical tasks.

13.6 CONCLUSIONS AND FUTURE TRENDS

Despite a great attention made to DIOS and related ionization techniques since the first publication (Wei et al. 1999), the potential of nSi-LDI-MS is not fully revealed and this methodology extensively grows. Obviously, the research on further improvement of the LDI-MS performance of Si substrates as well as other SALDI-efficient inorganic nanostructures, which were beyond the scope of the present review, will proceed and new possibilities of DIOS-related ionization methods will be developed in the near future.

First, this can be related to the development of "on-chip" devices for the LDI-MS analysis with preliminary preconcentration of the specific classes of analytes by their selective adsorption or their separation on the principles of thin-layer chromatography or capillary electrophoresis. Likely, the attention to the Si thin films and Si nanoparticles as nSi-LDI-MS substrates will grow. The films allow ionization of high MW analytes, as the NPs can be used on ordinary MALDI targets instead of organic matrixes or they can be added to the analyte's solutions to achieve very high degree of the analyte concentration.

However, it is worthy to say that now nSi-LDI-MS (as any SALDI ionization technique) is not widely applied by practical chemists and biochemists. Possible reasons of this are poor awareness of practical specialists in the recent development of MS ionization methods and some inertness of their mind (traditional methods of the analysis such as LC MS or fluorescence are preferable due to their familiarity and reliability).

Taking into account all aforementioned, wider implementation of DIOS, NIMS, NALDI, and other SALDI ionization methods in analytical and bioanalytical practice should be expected during the following decade. That is why in the present time the LDI-MS can be considered one of the most successful fields of practical technological applications of the PSi and other silicon nanostructures.

ACKNOWLEDGMENTS

This work was partially supported by the Swedish Institute grant 00814/2011 "Surface-Assisted Laser Desorption Ionization of Biomolecules on Modified Porous Silicon/Silicon Carbide for Their Better Analysis."

REFERENCES

Alekseev, S.A., Lysenko, V., Zaitsev, V.N., and Barbier, D. (2007). Application of infra-red interferometry for quantitative analysis of chemical groups grafted onto internal surface of porous silicon nanostructures. *J. Phys. Chem. C.* **111**, 15217–15222.

Alimpiev, S., Nikiforov, S., Karavanskii, M. Minton, T., and Sunner, J. (2001). On the mechanism of laser-induced desorption/ionization of organic compounds from etched silicon and carbon surface. *J. Chem. Phys.* **115**(4), 1891–1901.

Alimpiev, S., Grechnikov, A., Sunner, J. et al. (2008). On the role of defects and surface chemistry for surface-assisted laser desorption ionization from silicon. *J. Chem. Phys.* **128**(1), 014711–014730.

Arakawa, R. and Kawasaki, H. (2010). Functionalized nanoparticles and nanostructured surfaces for surface-assisted laser desorption/ionization mass spectrometry. *Anal. Sci.* **26**, 1229–1240.

Arakawa, R., Shimonae, Y., Morikawa, H., Ohkara, K., and Okuno, S. (2004). Mass spectrometric analysis of low molecular mass polyesters by laser desorption/ionization on porous silicon. *J. Mass. Spectrom.* **39**, 961–965.

Barabash, R., Alekseev, S., Zaitsev, V., and Barbier, D. (2006). Oxidation stability and vinylsilanes modification of porous silicon. *Ukr. Khim. Zh.* **72**(10), 78–84 (in Russian).

Buriak, J.M. (2002). Organometallic chemistry on silicon and germanium surfaces. *Chem. Rev.* **102**(5), 1272–1306.

Castellana, E.T., Gamez, R.C., Gomez, M.E., and Russell, D.H. (2010). Longitudinal surface plasmon resonance based gold nanorod biosensors for mass spectrometry. *Langmuir* **26**(8), 6066–6070.

Chen, Y., Luo, G., Diao, J. et al. (2007). Laser desorption/ionization from nanostructured surfaces: Nanowires, nanoparticle films and silicon microcolumn arrays. *J. Phys.: Conf. Ser.* **59**, 548–554.

Chen, Y.-F., Chen, H., Aleksandrov, A., and Orlando, T.M. (2008). Roles of water, acidity, and surface morphology in surface-assisted laser desorption/ionization of amino acids. *J. Phys. Chem. C.* **112**(17), 6953–6960.

Chen, W.Y., Huang, J.T., Cheng, Y.C., Chien, C., and Tsao, C.W. (2011). Fabrication of nanostructured silicon by metal-assisted etching and its effects on matrix-free laser desorption/ionization mass spectrometry. *Anal. Chim. Acta* **687**, 97–104.

Cheng, Y.-C., Chen, K.-H., Wang, J.-S. et al. (2012). Rapid analysis of abused drugs using nanostructured silicon surface assisted laser desorption/ionization mass spectrometry. *Analyst* **137**(3), 654–661.

Chiang, C.-K., Chen, W.-T., and Chang, H.-T. (2011). Nanoparticle-based mass spectrometry for the analysis of biomolecules. *Chem. Soc. Rev.* **40**, 1269–1281.

Coffinier, Y., Nguyen, N., Drobecq, H. et al. (2012). Affinity surface-assisted laser desorption/ionization mass spectrometry for peptide enrichment. *Analyst* **137**(23), 5527–5532.

Compton, B.J. and Siuzdak, G. (2003). Mass spectrometry in nucleic acid, carbohydrate and steroid analisis. *Spectroscopy* **17**, 699–713.

Dagan, S., Hua, Y., Boday, D.J. et al. (2009). Internal energy deposition with silicon nanoparticle-assisted laser desorption/ionization (SPALDI) mass spectrometry. *Int. J. Mass Spectrom.* **283**(1–3), 200–205.

Daniels, R.H., Dikler, S., Li, E., and Stacey, C. (2008). Break free of the matrix: Sensitive and rapid analysis of small molecules using nanostructured surfaces and ldi-tof mass spectrometry. *JALA* **13**(6), 314–321.

de Rond, T., Danielewicz, M., and Northen, T. (2015). High throughput screening of enzyme activity with mass spectrometry imaging. *Curr. Opin. Biotechnol.* **31**, 1–9.

Dupré, M., Enjalbal, C., Cantel, S. et al. (2012a). Investigation of silicon-based nanostructure morphology and chemical termination on laser desorption ionization mass spectrometry performance. *Anal. Chem.* **84**(24), 10637–10644.

Dupré, M., Coffinier, Y., Boukherroub, R. et al. (2012b). Laser desorption ionization mass spectrometry of protein tryptic digests on nanostructured silicon plates. *J. Proteomics* **75**(7), 1973–1990.

Fenn, J.B., Mann, M., Meng, C.K., Wong, S.F., and Whitehouse, C.M. (1989). Electrospray ionization for mass spectrometry of large biomolecules. *Science* **246**, 64–71.

Go, E.P., Apon, J.V., Luo, G. et al. (2005). Desorption/ionization on silicon nanowires. *Anal. Chem.* **77**, 1641–1646.

Go, E.P., Uritboonthai, W., Apon, J.V. et al. (2007). Selective metabolite and peptide capture/mass detection using fluorous affinity tags. *J. Proteom. Res.* **6**, 1492–1499.

Gomez, D., Fernandez, J.A., Astigarraga, E., Marcaide, A., and Azcarate, S. (2007). Porous silicon surfaces for metabonomics: Detection and identification of nucleotides without matrix interference. *Phys. Stat. Sol. (c)* **4**(6), 2185–2189.

Grechnikov, A.A., Borodkov, A.S., Alimpiev, S.S., Nikiforov, S.M., and Simanovsky, Y.O. (2013). Gas-phase basicity: Parameter determining the efficiency of laser desorption/ionization from silicon surfaces. *J. Anal. Chem.* **68**(1), 19–26.

Guenin, E., Lecouvey, M., and Hardouin, J. (2009). Could a nano-assisted laser desorption/ionization target improve the study of small organic molecules by laser desorption/ionization time-of-flight mass spectrometry? *Rapid Commun. Mass Spectrom.* **23**(9), 1395–1400.

Guinan, T., Ronci, M., Kobus, H. et al. (2012). Rapid detection of illicit drugs in neat saliva using desorption/ionization on porous silicon. *Talanta* **99**, 791–798.

Guinan, T., Kirkbride, P., Pigou, P.E., and Voelcker, N.H. (2014). Surface-assisted laser desorption ionization mass spectrometry techniques for application in forensics. *Mass Spectrometry Rev.* doi: 10.1002/mas .21431.

Gulbakan, B., Park, D., Kang, M. et al. (2010). Laser desorption ionization mass spectrometry on silicon nanowell arrays. *Anal. Chem.* **82**, 7566–7575.

Herino, R., Perio, A., and Bomchil, G. (1984). Microstructure of porous silicon and its evolution with temperature. *Mater. Lett.* **2**, 519–523.

Hua, Y., Dagan, S., Wickramasekara, S., Boday, D.J., and Wysocki, V.H. (2010). Analysis of deprotonated acids with silicon nanoparticle-assisted laser desorption/ionization mass spectrometry. *J. Mass Spectrom.* **45**(12), 1394–1401.

Karas, M. and Hillenkamp, F. (1988). Laser desorption ionization of proteins with molecular masses exceeding 10,000 daltons. *Anal. Chem.* **60**, 2299–2301.

Kim, S.H., Lee, A., Song, J.Y., and Han, S.Y. (2012). Laser-induced thermal desorption facilitates postsource decay of peptide ions. *J. Am. Soc. Mass Spectrom.* **23**(5), 935–941.

Kim, S.H., Kim, J., Moon, D.W., and Han, S.Y. (2013). Commercial silicon-on-insulator (SOI) wafers as a versatile substrate for laser desorption/ionization mass spectrometry. *J. Am. Soc. Mass Spectrom.* **24**(1), 167–170.

Kim, S.H., Shon, H.K., Lee, T.G., and Han, S.Y. (2014). Thin film surfaces suitable to multimodal ionization for TOF-SIMS and LDI mass spectrometry of biomolecules. *Surf. Interface Anal.* **46**(1), 35–38.

Kuzema, P.A. (2011). Small-molecule analysis by surface-assisted laser desorption/ionization mass spectrometry. *J. Anal. Chem.* **66**(13), 28–43.

Lapierre, F., Piret, G., Drobecq, H. et al. (2011). High sensitive matrix-free mass spectrometry analysis of peptides using silicon nanowires-based digital microfluidic device. *Lab. Chip* **11**(9), 1620–1628.

Law, K.P. (2010a). Surface-assisted laser desorption/ionization mass spectrometry on nanostructured silicon substrates prepared by iodine-assisted etching. *Int. J. Mass. Spectrom.* **290**(1), 47–59.

Law, K.P. (2010b). Laser desorption/ionization mass spectrometry on nanostructured semiconductor substrates: DIOS™ and QuickMass™. *Int. J. Mass Spectrom.* **290**(1), 72–84.

Law, K.P. and Larkin, J.R. (2011). Recent advances in SALDI-MS techniques and their chemical and bioanalytical applications. *Anal. Bioanal. Chem.* **399**(8), 2597–2622.

Lee, J.-C., Wu, C.Y., Apon, J.V., Siuzdak, G., and Wong, C.-H. (2006). Reactivity-based one-pot synthesis of the tumor-associated antigen N3 minor octasaccharide for the development of a photocleavable DIOS-MS sugar array. *Angew. Chem. Int. Ed.* **45**(17), 2753–2757.

Lee, D.Y., Platt, V., Bowen, B. et al. (2012). Resolving brain regions using nanostructure initiator mass spectrometry imaging of phospholipids. *Integr. Biol.* **4**(6), 693–699.

Lewis, W.G., Shen, Z., Finn, M.G., and Siuzdak, G. (2003). Desorption/ionization on silicon (DIOS) mass spectrometry: Background and application. *Int. J. Mass Spectrom.* **226**, 107–116.

Li, J. and Lipson, R.H. (2013). Insights into desorption ionization on silicon (DIOS). *J. Phys. Chem. C* **117**(51), 27114–27119.

Li, J., Hu, X.K., and Lipson, R.H. (2013). On-chip enrichment and analysis of peptide subsets using a maleimide-functionalized fluorous affinity biochip and nanostructure initiator mass spectrometry. *Anal. Chem.* **85**(11), 5499–5505.

Liu, Q. and He, L. (2008). Quantitative study of solvent and surface effects on analyte ionization in desorption ionization on silicon (DIOS) mass spectrometry. *J. Am. Soc. Mass Spectrom.* **19**, 8–13.

Liu, Q., Brown, V.L., and He, L. (2014). Desorption/ionization on porous silicon (DIOS) for metabolite imaging. In: Santos H. (ed.) *Porous Silicon for Biomedical Applications.* Woodhead Publishing, Cambridge, pp. 270–285.

Lowe, R.D., Szili, E.J., Kirkbride, P. et al. (2010). Combined immunocapture and laser desorption/ionization mass spectrometry on porous silicon. *Anal. Chem.* **82**(10), 4201–4208.

Luo, G., Chen, Y., Siuzdak, G., and Vertes, A. (2005). Surface modification and laser pulse length on internal energy transfer in DIOS. *J. Phys. Chem. B* **109**(51), 24450–24456.

Luo, G., Chen, Y., Daniels, H., Dubrow, R., and Vertes, A. (2006). Internal energy transfer in laser desorption/ionization from silicon nanowires. *J. Phys. Chem. B* **110**(27), 13381–13386.

Meng, J.-C., Averbuj, C., Lewis, W.G., Siuzdak, G., and Finn, M.G. (2004). Cleavable linkers for porous silicon-based mass spectrometry. *Angew. Chem. Int. Ed.* **43**, 1255–1260.

Nordstrom, A., Want, E., Northen, T., Lehtio, J., and Siuzdak, G. (2008). Multiple ionization mass spectrometry strategy used to reveal the complexity of metabolomics. *Anal. Chem.* **80**(2), 421–429.

Northen, T.R., Woo, H.K., Northen, M.T. et al. (2007a). High surface area of porous silicon drives desorption of intact molecules. *J. Am. Soc. Mass. Spectrom.* **18**(11), 1945–1949.

Northen, T.R., Yanes, O., Northen, M.T. et al. (2007b). Clathrate nanostructure for mass spectrometry. *Nature* **449**, 1033–1036.

Northen, T.R., Lee, J.-C., Hoang, L. et al. (2008). A nanostructure-initiator mass spectrometry-based enzyme activity assay. *Proc. Natl. Acad. Sci. U.S.A.* **105**(10), 3678–3683.

Okuno, S., Nakano, M., Matsubayashi, G.E., Arakawa, R., and Wada, Y. (2004). Reduction of organic dyes in matrix-assisted laser desorption/ionization and desorption/ionization on porous silicon. *Rapid Commun. Mass Spectrom.* **18**(23), 2811–2817.

Okuno, S., Wada, Y., and Arakawa, R. (2005). Quantitative analysis of polypropyleneglycol mixtures by desorption/ionization on silicon mass spectrometry. *Int. J. Mass Spectrom.* **241**(1), 45–48.

Ostman, P., Pakarinen, J.M.H., Vainiotalo, P. et al. (2006). Minimum proton affinity for efficient ionization with atmospheric pressure desorption/ionization on silicon mass spectrometry. *Rapid Commun. Mass Spectrom.* **20**(24), 3669–3673.

Patti, G.J., Woo, H.-K., Yanes, O. et al. (2010a). Detection of carbohydrates and steroids by cation-enhanced nanostructure-initiator mass spectrometry (NIMS) for biofluid analysis and tissue imaging. *Anal. Chem.* **82**(1), 121–128.

Patti, G.J., Shriver, L.P., Wassif, C.A. et al. (2010b). Nanostructure-initiator mass spectrometry (NIMS) imaging of brain cholesterol metabolites in smith-lemli-opitz syndrome. *Neuroscience* **170**(3), 858–64.

Pei, J., Tang, Y., Xu, N. et al. (2011). Covalently derivatized NTA microarrays on porous silicon for multi-mode detection of His-tagged proteins. *Science China Chemistry* **54**(3), 526–535.

Perichon, S., Lysenko, V., Roussel, Ph. et al. (2000). Technology and micro-Raman characterization of thick meso-porous Silicon layers for thermal effect microsystems. *Sens. Actuators* A **85**(1–3), 335–339.

Peterson, D.S. (2007). Matrix-free methods for laser desorption/ionization mass spectrometry. *Mass Spectrom. Rev.* **26**(1), 19–34.

Piret, G., Drobecq, H., Coffinier, Y., Melnyk, O., and Boukherroub, R. (2010). Matrix-free laser desorption/ionization mass spectrometry on silicon nanowire arrays prepared by chemical etching of crystalline silicon. *Langmuir* **26**(2), 1354–1361.

Ronci, M., Rudd, D., Guinan, T., Benkendorff, K., and Voelcker, N.H. (2012). Mass spectrometry imaging on porous silicon: Investigating the distribution of bioactives in marine mollusk tissues. *Anal. Chem.* **84**(21), 8996–9001.

Royer, P., Barchiesi, D., Lerondel, G., and Bachelot, R. (2004) Near-field optical patterning and structuring based on local-field enhancement at the extremity of a metal tip. *Philos. Trans. R. Soc. London, Ser. A-Math. Phys. Eng. Sci.* **362**(1817), 821–842.

Sainiemi, L., Keskinen, H., Aromaa, M. et al. (2007). Rapid fabrication of high aspect ratio silicon nanopillars for chemical analysis. *Nanotechnology* **18**, 505303–505310.

Seino, T., Sato, H., Torimura, M. et al. (2005). Laser desorption/ionization on porous silicon mass spectrometry for accurately determining the molecular weight distribution of polymers evaluated using a certified polystyrene standard. *Anal. Sci.* **21**(5), 485–490.

Severinovskaya, O.V., Alekseev, S.A., Gurskaya, S.V., Lavrinenko, O.Yu., and Zaitsev, V.N. (2010). Mass-spectrometric determination of the bile components by DIOS method using modified porous silicon. *Rep. Nat. Acad. Sci. Ukraine* **6**, 164–168 (In Russian).

Shen, Z., Thomas, J., Averbuj, C. et al. (2001). Porous silicon as a versatile platform for laser desorption/ionization mass spectrometry. *Anal. Chem.* **73**(3), 612–619.

Shen, Z., Thomas, J.J., Siuzdak, G. et al. (2004a). A case study on forensic polymer analysis by DIOS-MS: The suspect who gave us the SLIP. *J. Forensic Sci.* **49**(5), 1–8.

Shen, Z., Go, E.P., Gamez, A., and Blackledge, R.D. (2004b). A mass spectrometry plate reader: Monitoring enzyme activity and inhibition with a desorption/ionization on silicon (DIOS) platform. *ChemBioChem* **5**(7), 921–927.

Shmygol, I.V., Severinovskaya, O.V., Vasylieva, N.S., Alekseev, S.A., and Pokrovskiy, V.A. (2007). Mass-spectrometric study on adsorption of histamine and arginine on various surface types of porous silicon by means of laser desorption-ionization. *Chem. Phys. Technol. Surf. (Transl. of A. A. Chuiko Instit. Surf. Chem.)* **13**, 341–348 (In Russian).

Shmygol, I.V., Alekseev, S.A., Lavrinenko, O.Yu. et al. (2009). Chemically modified porous silicon for laser desorption/ionization mass spectrometry of ionic dyes. *J. Mass Spectrom.* **44**(8), 1234–1240.

Silina, Y.E., Koch, M., and Volmer, D.A. (2014). The role of physical and chemical properties of Pd nano-structured materials immobilized on inorganic carriers on ion formation in atmospheric pressure laser desorption/ionization mass spectrometry. *J. Mass Spectrom.* **49**(6), 468–480.

Siuzdak, G. (2006). *The Expanding Role of Mass Spectrometry in Biotechnology*, 2nd ed. MCC Press, San Diego, CA.

Siuzdak, G., Buriak, J., and Wei, J. (2001). Desorption/ionization of analytes from porous light-adsorbing semiconductor. US patent 6,288,390 B1, Sept. 11, 2001.

Steenwyk, R.C., Hutzler, J.M., Sams, J., Shen, Z., and Suizdak, G. (2006). Atmospheric pressure desorption/ionization on silicon ion trap mass spectrometry applied to the quantification of midazolam in rat plasma and determination of midazolam 1'-hydroxylation kinetics in human liver microsomes. *Rapid Commun. Mass Spectrom.* **20**(24), 3717–3727.

Stolee, J.A. and Vertes, A. (2011). Polarization dependent fragmentation of ions produced by laser desorption from nanopost arrays. *Phys. Chem. Chem. Phys.* **13**(20), 9140–9146.

Suni, N.M., Haapala, M., Färm, E. et al. (2010). Fabrication of nanocluster silicon surface with electric discharge and the application in desorption/ionization on silicon-mass spectrometry, *Lab. Chip* **10**(13), 1689–1695.

Tan, J., Zhao, W.-J., Yu, J.-K. et al. (2012). Capture, enrichment, and mass spectrometric detection of low-molecular-weight biomarkers with nanoporous silicon microparticles. *Adv. Healthcare Mater.* **1**(6), 742–750.

Tanaka, K., Waki, H., Ido, Y. et al. (1988). Protein and polymer analysis up to m/z 100,000 by laser ionization time-of-flight mass spectrometry. *Rapid Commun. Mass Spectrom.* **2**(8), 151–153.

Trauger, S.A., Go, E.P., Shen, Z. et al. (2004). High sensitivity and analyte capture with desorption/ionization mass spectrometry on silylated porous silicon. *Anal. Chem.* **76**(15), 4484–4489.

Tsao, C.-W., Yang, Z.-J., and Chung, C.-W. (2012a). Preparation of nanostructured silicon surface for mass spectrometry analysis by an all-wet fabrication process using electroless metal deposition and metal assisted etching International. *J. Mass Spectrom.* **321–322**, 8–13.

Tsao, C.W., Lin, C.H., Cheng, Y.C. et al. (2012b). Nanostructured silicon surface modifications for as a selective matrix-free laser desorption/ionization mass spectrometry. *Analyst* **137**(11), 2643–2650.

Tuomikoski, S., Huikko, K., Grigoras, K. et al. (2002). Preparation of porous *n*-type silicon sample plates for desorption/ionization on silicon mass spectrometry (DIOS-MS). *Lab. Chip* **2**(4), 247–253.

Vaidyanathan, S., Jones, D., Ellis, J. et al. (2007). Laser desorption/ionization mass spectrometry on porous silicon for metabolome analyses: Influence of surface oxidation. *Rapid Commun. Mass Spectrom.* **21**(13), 2157–2166.

Walker, B.N., Stolee, J.A., Pickel, D.L., Retterer, S.T., and Vertes, A. (2010). Tailored silicon nanopost arrays for resonant nanophotonic ion production. *J. Phys. Chem. C* **114**(11), 4835–4840.

Wang, Y., Zeng, Z., Li, J. et al. (2013). Biomimetic antireflective silicon nanocones array for small molecules analysis. *J. Am. Soc. Mass Spectrom.* **24**(1), 66–73.

Wei, J., Buriak, J.M., and Siuzdak, G. (1999). Desorption-ionization mass spectrometry on porous silicon. *Nature* **399**, 243–246.

Wen, X., Dagan, S., and Wysocki, V.H. (2007). Small-molecule analysis with silicon-nanoparticle-assisted laser desorption/ionization mass spectrometry. *Anal. Chem.* **79**(2), 434–444.

www.waters.com. MassPREP DIOS-target Plates Care and Use Manual.

Wyatt, M.F., Ding, S., Stein, B.K., Brenton, A.G., and Daniels, R.H. (2010). Analysis of various organic and organometallic compounds using nanostructure assisted laser desorption/ionization time-of-flight mass spectrometry (NALDI-TOFMS). *J. Am. Soc. Mass Spectrom.* **21**(7), 1256–1259.

Xiao, Y., Retterer, S.T., Thomas, D.K., Tao, J.Y., and He, L. (2009). Impacts of surface morphology on ion desorption and ionization in desorption ionization on porous silicon (DIOS) mass spectrometry. *J. Phys. Chem. C* **113**(8), 3076–3083.

Yan, H., Xu, N., Huang, W.Y., Han, H.M., and Xiao, S.J. (2009). Electroless plating of silver nanoparticles on porous silicon for laser desorption/ionization mass spectrometry. *Int. J. Mass Spectrom.* **281**(1–2), 1–7.

Zhabin, S.N., Pento, A.V., Grechnikov, A.A. et al. (2011). On the role of laser irradiation in the processes of laser desorption/ionisation from silicon surfaces. *Quantum Electron.* **41**(9), 835–842.

PSi Substrates for Raman, Terahertz, and Atomic Absorption Spectroscopy

14

Ghenadii Korotcenkov and Songhee Han

CONTENTS

14.1	Porous Silicon in Raman Spectroscopy	266
14.2	Porous Silicon in Terahertz Spectroscopy	270
14.3	PSi Substrate as a Material for Pre-Enrichment of Trace Heavy Metal Ions	273
Acknowledgments		275
References		275

14.1 POROUS SILICON IN RAMAN SPECTROSCOPY

Raman spectroscopy is one of the most promising analytical methods for detection and identification of chemical and biological substances, due to the correspondence of the vibrational Raman frequencies to certain types of chemical bonds. However, due to the very low cross-section of the Raman scattering process, the detection of substances of low concentration is complicated without special enhancement. This is why techniques to enhance the Raman scattering, such as the surface-enhanced Raman scattering (SERS), have been attracting great attention in the last years (Moskovits 2005). Surface-enhanced Raman spectroscopy or surface-enhanced Raman scattering (SERS) is a surface-sensitive technique that enhances Raman scattering by molecules adsorbed on rough metal surfaces or by nanostructures such as plasmonic-magnetic silica nanotubes. The enhancement factor can be as much as 10^{10} to 10^{11} (Evan et al. 2007, 2009), which means this technique gives the possibility of single molecule detection (Kneipp et al. 2006; Le Ru et al. 2006). Development of efficient and reliable SERS substrates is especially important for biological analysis applications because Raman spectroscopy provides highly resolved vibrational information at room temperature and does not suffer from rapid photobleaching commonly observed in fluorescence spectroscopy (Chan et al. 2003). For example, today, methods for rapid quantitative medical analysis based on SERS are used already for the analysis of lactic acid in blood (Chiang and Hsu 2005), creatine (Stosch et al. 2005), and real-time and high-sensitivity determination of glucose (Stuart et al. 2005) and drugs (Hellsten et al. 2012).

However, the large-scale applications in chemistry, biology, and medicine require SERS-active substrates with optimal efficiency/price relation. In particular, the successful application of metal nanoparticles in SERS strongly depends on the metal characteristics, in terms of morphology (shape, size, and aggregation state) and the metal nature of the nanostructured metals employed as substrates (Cacamares et al. 2008). The shape and size of the metal nanoparticles strongly affect the strength of the enhancement because these factors influence the ratio of absorption and scattering events (Aroca 2006; Lu et al. 2011). One should note that for each experiment there is an ideal size for these particles, and an ideal surface thickness (Bao et al. 2003). Additionally, a key factor in SERS experiments is the metal–solution interface because it may affect the adherence of the adsorbate on the surface. This factor depends indeed on the preparation method employed to create metal nanoparticles.

The most popular methods of SERS substrates preparation in our days are deposition of Au or Ag nanoparticles on the Si single crystal substrate by laser ablation or from colloids (Cacamares et al. 2008). Evaporated silver and gold island films have also been found to provide controllable and quite reproducible SERS substrates, even though these structures are characterized by a certain size polydispersity (Tourrel and Corset 1996). In the past, several lithographic techniques (such as electron beam lithography, nanoimprinting, etc.) have been employed for the production of uniform and controllable SERS substrates with fully controlled shape and size of metal nanoparticles (Gunnarsson et al. 2001). But, among these substrates, colloidal nanoscale particles have been the most frequently used (Aroca et al. 2005; Cacamares et al. 2008). However, the low stability and difficulties involved in preparation of colloidal solutions with reproducible characteristics limit the application of such substrates. In addition, the above-mentioned methods do not allow obtaining stable substrates during any thermal treatments.

One of the promising candidates in this respect is formation of metallic nanocrystal structures on porous templates, for example, porous silicon (PSi) (Chan et al. 2003; Giorgis et al. 2008). Due to its large surface area and open porous structure, this semiconductor material allows obtaining highly sensitive SERS substrates (Chan et al. 2003). Introduction of metal, for example, silver, into the silicon nanoscale pores forms a nanocomposite material that uniquely combines the ability of metal surfaces to amplify Raman scattering signals with an enlarged surface area that allows achieving a magnitude of enhancement for the detection of various analyte, including biological molecules. It is also expected that such structures will have better stability (Lin et al. 2004). Morphology of PSi should prevent the agglomeration of nanoparticles. Chan et al. (2003) believe that the ability for a silicon substrate to detect low concentrations of target molecules opens the door to applications where it can be used as the detection tool for integrated, on-chip devices.

At present, PSi was used as a basis for SERS-active substrate preparation in several recent works (Sakka et al. 2000; Harraz et al. 2002; Chan et al. 2003; Lin et al. 2004; Giorgis et al. 2008; Ye et al. 2008; Panarin et al. 2009; Chursanova et al. 2010; Fukami et al. 2011; Virga et al. 2012; Sun et al. 2013). The majority of these works was devoted to the deposition of silver nanoparticles. Silver is probably the best choice of metal for this purpose, due to its broad plasmon resonance in visible-near IR spectral range, stability, and simple preparation procedure (Cacamares et al. 2008). However, studies relating formation of gold nanoparticles on silicon surfaces are also available (Jiao et al. 2010; Fukami et al. 2011; Sun et al. 2013). As a rule, SERS-active substrates were formed by gold and silver nanostructures synthesis on PSi surfaces (Lin et al. 2004; Panarin et al. 2009) or homogenous coatings of pore walls with metal layer (Chan et al. 2003). For these purposes, methods such as immersion plating (Harraz et al. 2002; Lin et al. 2004; Giorgis et al. 2008; Feng et al. 2009), spontaneous deposition (Zeiri et al. 2012), sputtering (Sun et al. 2013), electrochemical deposition (Fukami et al. 2011), metal infiltration, and thermal decomposition of Ag nitrate (Chan et al. 2003; Giorgis et al. 2008) were used. Of course, each method has its own features. For example, in the immersion plating synthesis, freshly etched PSi samples are used, while in the thermal decomposition synthesis, PSi samples are previously oxidized in order to improve hydrophilic properties of the spongiform matrix (Giorgis et al. 2008). In addition, in order to remove oxide complex and residual silver nitrate salt, the samples can be annealed at temperatures up to 500°C.

It is important to note that in each specific case, the deposition conditions should be selected very carefully because the parameters of the nanoparticles formed in the matrix of PSi not only depends on the morphology of the PSi, but also on the deposition conditions (Virga et al. 2012, 2013), including the salt concentration in the solution used, the dipping time, and the process temperature (see Figure 14.1). It is necessary also to take into account that PSi shows specific behavior during the deposition of the metal. For example, Harraz et al. (2002) found that PSi has an interesting chemical property: during immersion plating of metal, PSi acts as a modest reducing agent. Harraz et al. (2002) have also observed that metal plating occurs at a much higher rate on the porous surface compared to that on the Si wafer.

It should also be recognized that the process of PSi formation is very complex and still not completely understood and is influenced by numerous factors (read *Porous Silicon: Formation and Properties*). PSi used in SERS substrates usually is prepared by standard electrochemical etching. However, metal-assisted wet etching has been used as well (Sun et al. 2013). As it was shown in *Porous Silicon: Formation and Properties*, the morphology of PSi surfaces, independently on the

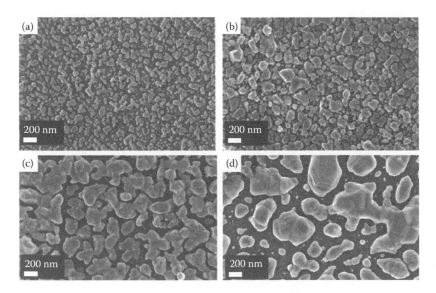

FIGURE 14.1   SEM images of Ag NPs obtained by immersion of *p*-Si in AgNO$_3$ solution (10^{-2} M) with different temperature/dipping time: (a) 60°C/30 s; (b) 30°C/60 s; (c) 50°C/60 s; (d) 50°C/90 s. (Reprinted with permission from Virga A. et al., *J. Phys. Chem. C* 117, 20139. Copyright 2013 American Chemical Society.)

method of preparing, is determined by crystal orientation, doping degree and type, current density and etching time, additional illumination, temperature, electrolyte composition, and so on, and, therefore, is quite difficult to trace the particular influence of each of them. This means that the morphology of the PSi substrate and the morphology of the final SERS-active film formed on this substrate as well as Raman spectra should be determined by silicon etching conditions, which are often very difficult to maintain. The examples of fabrication parameters influencing the Raman spectra are shown in Figure 14.2. Moreover, the morphology of the films depends not only on the conditions of silicon anodization but also on the conditions of the porous substrate storage (see Figure 14.3). In other words, improving the reproducibility of the results requires special attention.

As for the optimal conditions of electrochemical etching, in particular HF concentration in etchant, which can be used during preparing PSi for SERS applications, there are no specific recommendations. We have a situation due to the lack of such studies. We can only mention the opinion of Chursanova et al. (2010) that though the Raman intensity at lower HF concentration

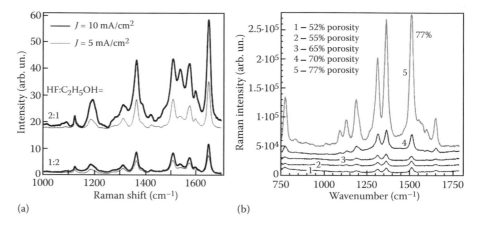

(a)   (b)

FIGURE 14.2   (a) Raman spectra of 10^{-5} M Rhodamine 6G (R6G) on samples PSi-Ag obtained by silver deposition during $t = 15$ min. Etching parameters for PSi formation: $J = 5$ mA/cm^2 (thin curves) and 10 mA/cm^2 (thick curves), etching time $t = 20$ min. Background level is shifted for visual clearness. Raman spectra were measured with the Renishaw Ramanscope 2000, using 514 nm line of Ar$^+$-ion laser for excitation. (Reprinted from *Appl. Surf. Sci.* 256, Chursanova M.V. et al., 3369, Copyright 2010, with permission from Elsevier.) (b) Surface-enhanced Raman spectra of 10^{-4} M Rhodamine 6G dye molecules on SCPSi substrates using 785 nm excitation. SERS spectra are obtained using LiCl as the chemical enhancer. SERS spectra were acquired using detector integration times that varied from 100 ms to 500 ms, depending on the Raman intensity of the target molecules. (Chan S. et al.: *Adv. Mater.* 2003, 15(19), 1595. Copyright Wiley-VCH Verlag GmbH & Co. KGaA. Reproduced with permission.)

FIGURE 14.3   Comparison of the morphology of the silver particles deposited onto fresh (a) and 2 years' stored PSi (b) substrates. The PSi was prepared using electrolyte HF:C$_2$H$_5$OH = 2:1 at etching time $t = 20$ min and current density $J = 10$ mA/cm^2. Silver deposition time $t = 10$ min in both cases. (Reprinted from *Appl. Surf. Sci.* 256, Chursanova M.V. et al., 3369, Copyright 2010, with permission from Elsevier.)

was somewhat smaller, those substrates are more mechanically stable and thus promising for multiple use. Earlier, Chan et al. (2003) reached the same conclusion. In their research, they used the PSi with a surface area of approximately 170 m^2/cm^3. Increasing the porosity (or pore size) of the PSi layer increases the surface area and Raman intensity (see Figure 14.2b); however, the mechanical stability of the layer is compromised. As for the other parameters of silicon porosification, it should be noted that the conclusions reported in many published papers relate to the specific conditions and applications, and therefore cannot have general character. For example, Zhang et al. (2012) found that optimal PSi substrate for Raman spectroscopy of Rhodamine 6G (R6G) can be prepared using n-Si (ρ = 1–10 Ω cm) at an anodization current density of 6 mA/cm^2 and an etching time of 8 min. They believe that the larger pore diameter of this Ag-PSi substrate (~1.2 μm) permits better biomolecule infiltration. The detection limit for R6G absorbed on such Ag-coated PSi was 10 nM.

It is known that the control in the shape and alignment of metal nanostructures is important for obtaining a high surface-enhancement. However, the results of published studies mostly do not answer the following question: What size of gold and silver nanoparticles is optimal for use in Raman spectroscopy? In particular, Giorgis et al. (2008) established that the highest Raman efficiency was obtained for sample Ag-PSi characterized by an average particle size of 50 nm, while for larger and smaller sizes, Raman signal intensity decreases or even disappears. SERS-active substrates were obtained synthesizing Ag nanoparticles through immersion of p-Si samples in AgNO$_3$ solutions and a successive thermal annealing. According to Giorgis et al. (2008), such behavior could be ascribed to the 514.5 nm excitation wavelength, which may fall within or outside the resonance plasmonic band of the nanostructured substrates. Optimal Ag-PSi nanostructures were checked for enhanced Raman activity on cyanine-based dyes and horseradish peroxidase. At the same time, Fukami et al. (2011) have found that the optimum length of the gold nanorods, formed in PSi-based matrix with ~100 nm pore size, was ~600 nm for surface-enhanced Raman spectroscopy using a He-Ne laser (see Figure 14.4). At that, the highest intensity was obtained when using the 600-nm nanorods regardless of the morphologies with and without PSi substrate, namely the gold nanocarpet and gold-embedded PSi. Fukami et al. (2011) also believe that the enhancement of the Raman intensity for such nanorods can be explained by the overlap of the plasmon resonance peak and the incident laser.

Another conclusion of Chan et al. (2003) is also important from our point of view. The amount of material introduced into the pores can vary largely, from complete filling of the pores to the adsorption of a few monolayers on the pore walls. However, Chan et al. (2003) believe that to take full advantage of the large surface area that PSi provides, complete pore filling is undesirable for SERS substrates because the exposed surface area is reduced and the target molecules will be merely sitting on top of a rough surface. Ideally, a reproducible and reliable nanoporous SERS substrate will have metal coating of the total internal surface area of the nanostructured silicon without pore blockage.

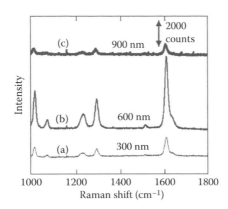

FIGURE 14.4   Raman spectra of 4,4′-bipyridine using the gold-embedded PSi substrates with different nanorod lengths (different depths of pores in PSi). The length of the gold nanorods in porous silicon is 300, 600, and 900 nm for (a), (b), and (c), respectively. (From Fukami K. et al., *Materials* 4, 791, 2011. Published by MDPI as open access.)

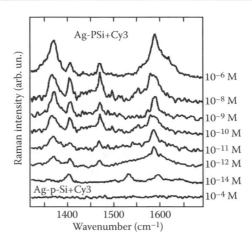

FIGURE 14.5   SERS spectra of Cy3 adsorbed on Ag-PSi substrate at different molar concentrations. Ag NPs were synthesized through the impregnation of mesoporous silicon in AgNO₃ solutions. As a reference, the bottom graph shows the detected Raman spectrum of 10⁻⁴ M dye molecules on bare *p*-Si. All of the spectra were obtained under 514.5 nm excitation. (Reprinted with permission from Virga A. et al., *J. Phys. Chem. C* 117, 20139. Copyright 2013 American Chemical Society.)

It should be noted that the concentrations detectable by SERS using PSi-based substrates in most of the works are estimated in the range of 10^{-6} to 10^{-8} M. However, the experiment showed that in optimal conditions, the sensitivity of SERS based on PSi substrates could be even higher. For example, Virga et al. (2013) reported that they were able to detect the cyanine dye (Cy3) adsorbed on silvered PSi samples at molar concentration 10^{-12} M (see Figure 14.5). This means that they had the possibility to detect single molecules.

14.2 POROUS SILICON IN TERAHERTZ SPECTROSCOPY

Terahertz (THz) frequencies are broadly defined as 0.1–30 THz (Dexheimer 2007; Lo 2010; Jepsen et al. 2011; McIntosh et al. 2012; Peiponen et al. 2013). The term thus refers to a relatively narrow part of the electromagnetic spectrum. Despite this narrowness, which it shares, for example, with visible light, terahertz radiation is of great importance in terms of fundamental research as well as in technology, environmental monitoring, chemistry research, and the life sciences because far infrared (IR) or terahertz spectroscopy is a diagnostic tool that can provide insight into the range of low-energy excitations in electronic materials, low-frequency vibrational modes of condensed phase media, and vibrational and rotational transitions in complex molecules (see Figure 14.6). This means that terahertz spectroscopy allows scientists both to study fundamental physical phenomena by probing low-energy electronic and molecular excitations, and to design devices for sensing, imaging, and future high-bandwidth communications. For example, in addition, detection of amplitude and phase as opposed to just intensity allows THz spectroscopy to easily measure important parameters such as refractive index and complex permittivity, enabling the investigation of molecular dynamics in liquids. In the solid state, THz spectroscopy has already shown itself to be a useful technique and is emerging as a powerful addition to other solid-state spectroscopies, X-ray diffraction (XRD), and differential scanning calorimetry (DSC). It has also shown that it can provide insights into the low energy dynamics of large biomolecules. This ability of THz spectroscopy to study low energy vibrational modes allows valuable insights into the little studied and understood area of intermolecular interactions. Increased understanding of such interactions and systems could have implications for a number of important topics in the chemical sciences, such as molecular crystallization, solvation, biochemistry, surface science, and supramolecular chemistry (McIntosh et al. 2012). Moreover, the short pulse durations achievable with terahertz techniques based on femtosecond laser sources allow time-resolved measurements on femtosecond timescales, which were previously inaccessible in this spectral range. It should be noted there is no doubt that the main reasons for the

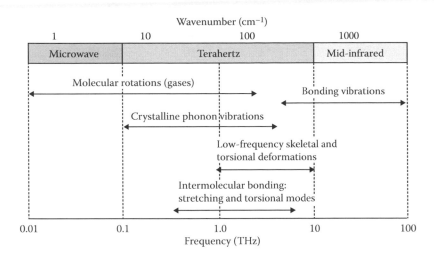

FIGURE 14.6 Molecular modes and activity in the terahertz region of the electromagnetic spectrum. (Adapted from McIntosh A.I. et al., *Chem. Soc. Rev.* 41, 2072, 2012. Reproduced by permission of The Royal Society of Chemistry.)

surge in interest in performing spectroscopy at terahertz frequencies were the development of such ultrafast lasers (Peiponen et al. 2013). Based on coherent and time-resolved detection of the electric field of ultrashort radiation bursts in the far IR, this technique has become known as terahertz time-domain spectroscopy (THz-TDS) (Jepsen et al. 2011). The brightness and coherence of THz sources allows spectroscopy to be undertaken with very good signal-to-noise and dynamic range compared to current IR techniques.

Generation and detection of THz pulses occur through nonlinear interaction of the driving optical pulse with a material with fast response. Therefore, one of the challenges with THz spectroscopy is finding a suitable substrate or host material that can support the substance to be analyzed without obscuring or impairing its absorption spectrum (Lo et al. 2013). Optical substrates that are commonly used in other spectral regimes are either opaque in the far IR, or exhibit their own characteristic absorption features that must be removed from the measurement. Even when a transparent terahertz substrate, like semi-insulating silicon, is employed, a thin surface layer of analyte precipitated or deposited on the flat substrate is often too thin to produce significant absorption at long wavelengths. Aqueous solutions are equally problematic because water has strong absorption throughout the far IR (Van Exter et al. 1989).

The most common method of preparing samples for far-IR or terahertz spectroscopy is to grind the analyte into a dry, fine powder, mix it with polyethylene (PE) powder or potassium bromide (KBr), and compress the resulting mixture into millimeter-thick pellets (Lo et al. 2013). PE and KBr are used because they exhibit reasonably low loss and few absorption bands across the mid- to far-IR spectral regions. This approach has been used to measure the far-IR or THz absorption spectra of a number of amino acids, crystalline peptides, proteins, nucleic acids, and other biomolecules (Kutteruf et al. 2003; Kumar et al. 2005; Heilweil and Plusquellic 2008). There is some evidence that the process of compression and consolidation can result in asymmetry, broadening and displacement of the absorption bands, compared to what is observed in bulk crystals (Meenatchi et al. 2012). Moreover, the pressed pellet method requires a large volume of analyte and is often a time-consuming process that is incompatible with rapid, highly sensitive spectral analysis and monitoring. Recently, waveguide-based spectroscopy methods have been demonstrated, in which a thin layer of analyte is precipitated onto one plate of a metallic waveguide or parallel plate transmission line (Harsha et al. 2012). This leads to a stronger absorption spectrum compared to what would be obtained by transmission through a single layer, and a sharpening of features in comparison to results obtained in a pressed pellet.

Murphy and co-workers (Lo 2010; Lo et al. 2013) have presented a novel method for terahertz spectroscopy that uses nanoporous silicon as a host substrate for capturing and collecting the analyte. PSi was formed through electrochemical etching of crystalline silicon, a process that readily forms a thick optical material with nanometer pore size. Such a nanoporous matrix with

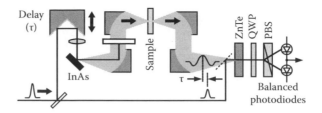

FIGURE 14.7 Experimental setup used to measure terahertz transmission spectrum of PSi samples. (Lo S.-Z.A. et al.: *Phys. Stat. Sol. (a)*. 2009. 206(6). 1273. Copyright Wiley-VCH Verlag GmbH & Co. KGaA. Reproduced with permission.)

extra large surface area allows a large quantity of surface-bound molecules to be distributed throughout an interaction volume that could be thicker than the wavelength—a tremendous advantage when conducting terahertz spectroscopy. Moreover, PSi is found to have several broad transparency windows in the mid- to far-IR spectral range, making it a suitable substrate for spectroscopy of a variety of substances.

Figure 14.7 depicts the terahertz time-domain system, which Lo et al. (2009) used to characterize the PSi samples. The terahertz signal was generated by illuminating an indium arsenide semiconductor surface at oblique incidence with 100 femtosecond optical pulses from a mode locked Ti:sapphire laser. The optical pulses are absorbed at the surface of the InAs, producing electrons and holes that diffuse at different rates into the semiconductor. Because the generated electromagnetic pulse has sub-picosecond duration, it contains frequency components extending into the terahertz regime. The terahertz signal was collimated and then focused at normal incidence to a size of approximately 1 mm onto the freestanding PSi membrane by a pair of off-axis gold-coated parabolic mirrors. A complementary pair of mirrors is then employed to collect the transmitted beam and direct it to the receiver. The terahertz receiver used a conventional electrooptic sampling technique, in which the terahertz pulse modulates the birefringence of a 1-mm thick, ⟨100⟩-oriented ZnTe crystal. This modulation is sensed by measuring the shift in polarization of a probe beam that is derived from the same Ti:sapphire laser used to generate the terahertz pulse. By scanning the delay τ between the pump pulse and the probe pulse, it is possible to directly measure the terahertz electric field temporal waveform. By computing the Fourier transform of this temporal waveform, one obtains the spectrum of the terahertz signal. By comparing the measured spectra with and without the sample in place, one can compute the complete terahertz transmission spectrum of the sample.

Murphy and co-workers (Lo et al. 2013) compared the absorption spectrum measured using a PSi substrate to conventional pressed-pellet and surface evaporation/crystallization methods for different molecules and found that in cases where the molecules can be efficiently dissolved and deposited inside of the pores, smaller quantities of analytes are used and the absorption spectrum can be both sharper and more intense than for the pressed-pellet or surface precipitation methods. Murphy and co-workers (Lo et al. 2013) established also that the PSi substrate used in IR absorption vibrational spectroscopy had indeed enhanced and in many cases sharpened the higher energy absorption modes for several crystallized organic and biological analytes. For low energy modes, the absorption features remained relatively broad and unamplified. Murphy and co-workers (Lo et al. 2013) attributed this discrepancy to a reduction of intermolecular interaction, which lead to the reduction of intermolecular and lattice modes usually dominant in the low wavenumber regime. This characteristic of porous substrate was also observed by Ueno and co-workers (Ueno et al. 2006; Ueno and Ajito 2007) for multiple analytes.

According to Lo et al. (2013), one potential setback for this method is that the absorption features of the substrate can obstruct or interfere with measuring absorption features from the analyte. It was found that the absorption band from 360–1250 cm^{-1}, which is caused by the oxidized and hydrogenated surface of the PSi, could be observed in the spectra. However, Lo et al. (2013) believe that absorption by the substrate could be significantly reduced using a thinner substrate. Such use would not significantly impact the experiment, as only micrograms of analyte are apparently required for terahertz absorption spectroscopy work. Lo et al. (2013) observed also that the residual solvent in the PSi substrate clearly affects the crystallization properties of analytes, especially when aqueous solutions are used. According to Lo et al. (2013), this drawback could potentially be alleviated by applying better environmental control of the FTIR sample chamber.

Knab et al. (2014) established that broadband optical properties also make nanoporous silicon a strong candidate for an all-optical THz modulator. Investigation of the broadband terahertz optical properties of nanoporous silicon samples with different porosities and the ultrafast carrier dynamics of photogenerated charge carriers in these materials has shown that nanoporous silicon had the ultrafast carrier recovery time (below ~10 ps), favorable modulation depth, and broadband optical properties. In particular, all samples exhibited initially low THz absorption, and the refractive index depended on porosity but showed little frequency dependence. It should be noted that the decrease of frequency-dependent refractive index at terahertz (THz) frequencies in PSi compared to bulk material was first reported by Labbé-Lavigne et al. (1998). In addition, PSi is relatively inexpensive and easy to fabricate, and the fabrication process can be easily up-scaled for large-scale production. Nanoporous silicon is also structurally simple because deposition of electrodes, apertures, or other structures is not necessary for a functioning modulator. Knab et al. (2014) believe that not many modulator materials share all of these characteristics. The low attenuation and relatively flat refractive index extends the useful frequency range for applications to ~7 THz, depending on how much loss is acceptable for a given application. These favorable ultra-broadband optical properties indicate that nanoporous silicon can be utilized in a number of THz optical and photonic components throughout the THz frequency range. In particular, the ability to engineer the effective dielectric constant by adjusting the porosity and to construct complex multilayer porous structures on a single silicon substrate provide unparalleled flexibility in the fabrication of THz components such as THz filters in this underexploited part of the electromagnetic spectrum (Lo and Murphy 2009). In addition, it has been shown that the attenuation of THz light in the 0.1–2.5 THz range is quite low (<4 cm^{-1}) (Lo et al. 2009), making nanoporous silicon an ideal material for use in applications where low insertion loss is important. Knab et al. (2014) established that absorption loss becomes more prominent at higher frequencies. This may limit the usefulness of nanoporous silicon in applications where low insertion loss is desired (e.g., THz filters), but could still allow use below ~10 THz.

14.3 PSi SUBSTRATE AS A MATERIAL FOR PRE-ENRICHMENT OF TRACE HEAVY METAL IONS

Determination of heavy metal ions in environmental samples is of great importance because the toxic elements can be accumulative and maintained in the environment and biological systems (Gruiz et al. 1998; Wilken 1992). For quantitative analysis of heavy metals in aquatic environment, there are a number of sensitive instrumental methods, for example, atomic absorption spectroscopy (Welz and Sperling 1999) is available. However, modern analytical methods often fail to determine heavy metal ions at trace levels. Therefore, it is necessary to enrich the metal ions prior to the determination, especially for atomic absorption spectroscopy. Therefore, a great deal of interest has been devoted to various materials (Pisareva et al. 2004; Wan Ngah et al. 2011) including mesoporous silica-based inorganic hybrid materials (Pu et al. 1998; Mahmoud et al. 2000; Zougagh et al. 2005) for enriching or detecting heavy metal ions from aqueous medium. The advances performed in silica surface chemistry allowed for the attachment of different functional groups, which can be extremely useful for studying biomolecules and cell immobilization, interfacial electron transfer, ion binding, coupled chemical reactions, and so on.

Sam et al. (2010) and Li et al. (2012) established that the PSi, which has extremely large surface area, also can be used for these purposes. In particular, Sam et al. (2010) have found that a glycine-modified PSi is sufficient to capture copper ions. Glycine amino acid was covalently incorporated into the PSi structure using multistep chemistry consisting of PSi formation, thermal hydrosilylation of undecylenic acid, activation of acid-terminated surface by formation of succinimidyl ester, glycine methyl ester anchoring by amidation reaction, and finally, deprotection reaction to form glycine-modified PSi. Amino acids were used because amino acids or peptides are well known as metal ion chelating agents (Sigel and Martin 1982). As is known, the application of modified surfaces for metal ion pre-enrichment requires the anchoring of probes that have affinities for different metal ions (Arrigan and Bihan 1999; Gooding et al. 2001). Moreover, an efficient immobilization method must lead to a high surface coverage of molecules attached through a stable and easily formed linkage (Borisov and Wolfbeis 2008).

Li et al. (2012) established that the effect of PSi enrichment by heavy metals also significantly increased after modifying the surface of PSi by cleavable groups. The cleavable groups (benzimidazoledithio) were grafted on the PSi surface by a stepwise covalent process. PSi was first obtained by anodization of bulk silicon wafers. For this purpose, the pretreatment samples were electrochemically etched in a 1/1 (v/v) mixture solution of 40% aqueous HF and absolute ethanol for 20 min at the current density of 10 mA/cm². After etching, the PSi was rinsed with absolute ethanol and dried under a stream of dry nitrogen. Then PSi was subsequently silanized by 3-mercaptopropyltriethoxysilane (MPTS) to synthesize MPTS-PSi, the MPTS-PSi was further converted into pyridyldithio-terminated PSi (PDT-PSi), and finally, the PDT-PSi reacted with 2-mercaptobenzimidazole to form the benzimidazoledithio-modified PSi (BDT-PSi). These processes are shown in Figure 14.8.

The efficiency of a PSi layer for pre-enrichment of trace heavy metal ions such as Cd, Hg, Pb, Cu, and Co was estimated using UV–vis absorption spectroscopy. These results are presented in Figure 14.9. The results show that the BDT-PSi possesses a similar preferential adsorption trend (Cd > Cu ≫ Hg ~ Pb ~ Co) at different pH (from 2.0 to 6.0). At pH 5.0, the best pre-enrichment

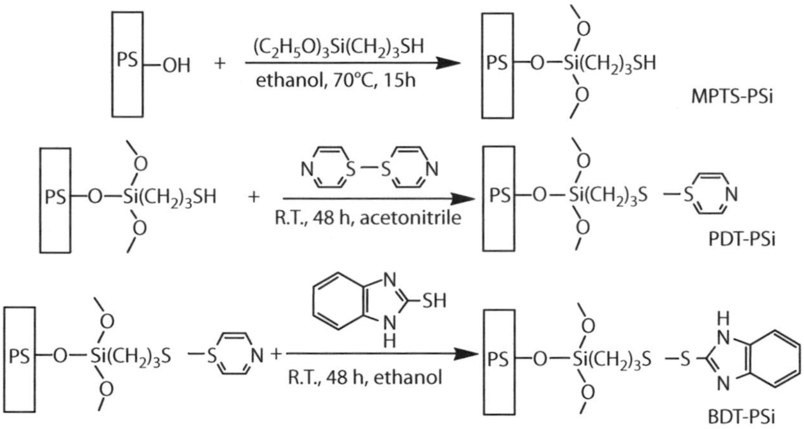

FIGURE 14.8   The scheme for the preparation of the BDT-PSi. (Reprinted from *Appl. Surf. Sci.* 258, Li S. et al., 5538, Copyright 2012, with permission from Elsevier.)

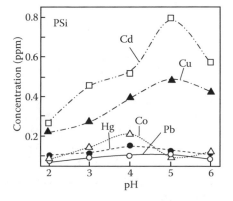

FIGURE 14.9   The effect of pH on the pre-enrichment for different metal ions. Initial metal ions concentration, 0.05 ppm; solution volume, 50 ml; enrichment time, 5 h; eluent, 3 ml GSH solution; desorption time, 12 h; temperature = 25°C. (Reprinted from *Appl. Surf. Sci.* 258, Li S. et al., 5538, Copyright 2012, with permission from Elsevier.)

efficiency for Cd ions is observed, the concentration of Cd is increased more than 10 times, and the recovery is found to be 95.4%.

Taking into account experience received during silica and silica gel using for the pre-enrichment of trace heavy metal ions, and the simplicity of PSi oxidation even in ambient air, that is, transformation of silicon in silicon dioxide, one can conclude that besides benzimidazoledithio and glycine-modification for PSi functionalizing other modifiers can be applied as well. For example, for preconcentration of heavy metals in environmental waters one can use various compounds containing S and N functional groups. These organic compounds are (3-aminopropyl)trimethoxysilane (APTMS) (Hassanien and Abou-El-Sherbini 2006), (3-mercaptopropyl)trimethoxysilane (MPTMS) (Ekinci and Köklü 2000), 2–aminophenol, 2-aminothiophenol, 2-aminobenzothiazole, 2-aminothiazole (Roldan et al. 2005), dithizone (Mahmoud et al. 2000), 2-mercaptobenzothiazole (Pu et al. 1998), 2-mercaptothiazoline (Evangelista et al. 2007), and so on. At that it was shown that rare earth elements (REEs) can be sorbed only by APTMS and MPTMS modified silica gel quantitatively in a broad pH range (pH >3). Silica gel phase impregnated with dithizone is selective for extraction and solid phase preconcentration of mercury(II). (3-Aminopropyl)triethoxysilane modified silica-gel is effective for the separation and preconcentration of vanadium, silver, manganese, and lead (Ekinci and Köklü 2000). o-Vanillin-immobilized silica gel can be used for the adsorption and estimation of copper, cobalt, iron, and zinc (Garg et al. 1999). Information about other functional groups, which can be used for PSi functionalizing, can be found in Pisareva et al. (2004) and Zougagh et al. (2005).

ACKNOWLEDGMENTS

This work was supported by the Ministry of Science, ICT and Future Planning (MSIP) of the Republic of Korea.

REFERENCES

Aroca, R. (2006). *Surface-Enhanced Vibrational Spectroscopy*. John Wiley & Sons, Chichester, England.

Aroca, R.F., Alvarez-Puebla, R.A., Pieczonka, N., Sanchez-Cortes, S., and Garcia-Ramos, J.V. (2005). Surface-enhanced Raman scattering on colloidal nanostructures. *Adv. Colloid Interface Sci.* 116, 45–61.

Arrigan, D.W.M. and Le Bihan, L. (1999). A study of L-cysteine adsorption on gold via electrochemical desorption and copper(II) ion complexation. *Analyst* 124, 1645–1649.

Bao, L.-L., Mahurin, S.M., Liang, C.-D., and Dai, S. (2003). Study of silver films over silica beads as a surface-enhanced Raman scattering (SERS) substrate for detection of benzoic acid. *J. Raman Spectrosc.* 34(5), 394–398.

Borisov, S.M. and Wolfbeis, O.S. (2008). Optical biosensors. *Chem. Rev.* 108, 423–461.

Cacamares, M.V., Garcia-Ramos, J.V., Sanchez-Cortes, S., Castillejo, M., and Oujja, M. (2008). Comparative SERS effectiveness of silver nanoparticles prepared by different methods: A study of the enhancement factor and the interfacial properties. *J. Colloid Interface Sci.* 326, 103–109.

Chan, S., Kwon, S., Koo, T.-W., Lee, L.P., and Berlin, A.A. (2003). Surface-enhanced Raman scattering of small molecules from silver-coated silicon nanopores. *Adv. Mater.* 15(19), 1595–1598.

Chiang, H.H.K. and Hsu, P.H. (2005). Surface-enhanced Raman scattering (SERS) spectroscopy technique for lactic acid in serum measurement. *Proc. SPIE* 5927, 395–402.

Chursanova, M.V., Germash, L.P., Yukhymchuk, V.O., Dzhagan, V.M., Khodasevich, I.A., and Cojoc, D. (2010). Optimization of porous silicon preparation technology for SERS applications. *Appl. Surf. Sci.* 256, 3369–3373.

Dexheimer, S.L. (ed.) (2007). *Terahertz Spectroscopy: Principles and Application*. CRC Press, Boca Raton, FL.

Ekinci, C. and Köklü, Ü. (2000). Determination of vanadium, manganese, silver and lead by graphite furnace atomic absorption spectrometry after preconcentration on silica-gel modified with (3-Aminopropyl) triethoxysilane. *Spectrochim. Acta B* 55, 1491–1495.

Evan, J., Le Ru, E.C., Meyer, M., and Etchegoin, P.G. (2007). Surface enhanced Raman scattering enhancement factors: A comprehensive study. *J. Phys. Chem. C* 111(37), 13794–13803.

Evan, J., Le Ru, E.C., and Etchegoin, P.G. (2009). Single-molecule surface-enhanced Raman spectroscopy of nonresonant molecules. *J. Am. Chem. Soc.* 131(40), 14466–14472.

Evangelista, S.M., DeOliveira, E., Castro, G.R., Zara, L.F., and Prado, A.G.S. (2007). Hexagonal mesoporous silica modified with 2-mercaptothiazoline for removing mercury from water solution. *Surf. Sci.* 601(10), 2194–2202.

Feng, F., Zhi, G., Jia, H.S., Cheng, L., Tian, Y.T., and Li, X.J. (2009). SERS detection of low-concentration adenine by a patterned silver structure immersion plated on a silicon nanoporous pillar array. *Nanotechnology* 20, 295501.

Fukami, K., Chourou, M.L., Miyagawa, R. et al. (2011). Gold nanostructures for surface-enhanced Raman spectroscopy, prepared by electrodeposition in porous silicon. *Materials* 4, 791–800.

Garg, B.S., Sharma, R.K., Bist, J.S., Bhojak, N., and Mittal, S. (1999). Separation and preconcentration of metal ions and their estimation in vitamin, steel and milk samples using *o*-vanillin-immobilized silica gel. *Talanta* 48, 49–55.

Giorgis, F., Descrovi, E., Chiodoni, A. et al. (2008). Porous silicon as efficient surface-enhanced Raman scattering (SERS) substrate. *Appl. Surf. Sci.* 254, 7494–7497.

Gooding, J.J., Hibbert, D.B., and Yang, W. (2001). Electrochemical metal ion sensors. Exploiting amino acids and peptides as recognition elements. *Sensors* 1, 75–90.

Gruiz, K., Murányi, A., Molnár, M., and Horváth, B. (1998). Risk assessment of heavy metal contamination in Danube sediments. *Wat. Sci. Tech.* 37, 273–3281.

Gunnarsson, L., Bjerneld, E.J., Xu, H., Petronis, S., Kasemo, B., and Käll, M. (2001). Interparticle coupling effects in nanofabricated substrates for surface-enhanced Raman scattering *Appl. Phys. Lett.* 2001, 78, 802–804.

Harraz, F.A., Tsuboi, T., Sasano, J., Sakka, T., and Ogata, Y.H. (2002). Metal deposition onto porous silicon layer by immersion plating from aqueous and nonaqueous solutions. *J. Electrochem. Soc.* 149(9), C456–C463.

Harsha, S.S., Melinger, J.S., Qadri, S.B., and Grischkowsky, D. (2012). Substrate independence of THz vibrational modes of polycrystalline thin films of molecular solids in waveguide THz-TDS. *J. Appl. Phys.* 111, 023105.

Hassanien, M.M. and Abou-El-Sherbini, K.S. (2006). Synthesis and characterisation of morin-functionalised silica gel for the enrichment of some precious metal ions. *Talanta* 68, 1550–1559.

Heilweil, E.J. and Plusquellic, D.F. (2008). Terahertz spectroscopy of biomolecules. In: Dexheimer S. (ed.) *Terahertz Spectroscopy: Principles and Applications.* CRC Press, Boca Raton, pp. 269–298.

Hellsten, S., Qu, H., Heikkila, T., Kohonen, J., Reinikainen, S.-P., and Louhi-Kultanen, M. (2012). Raman spectroscopic imaging of indomethacin loaded in porous silica. *CrystEngComm.* 14, 1582–1587.

Jepsen, P.U., Cooke, D.G., and Koc, M. (2011). Terahertz spectroscopy and imaging—Modern techniques and applications. *Laser Photon. Rev.* 5(1), 124–166.

Jiao, Y., Koktysh, D.S., Phambu, N., and Weiss, S.M. (2010). Dual-mode sensing platform based on colloidal gold functionalized porous silicon. *Appl. Phys. Lett.* 97, 153125.

Knab, J.R., Lu, X., Vallejo, F.A., Kumar, G., Murphy, T.E., and Hayden, L.M. (2014). Ultrafast carrier dynamics and optical properties of nanoporous silicon at terahertz frequencies. *Opt. Mater. Express* 4(2), 300–307.

Kneipp, K., Moskovits, M., and Kneipp, H. (eds.) (2006). *Surface-Enhanced Raman Scattering: Physics and Applications.* Springer, Berlin.

Kumar, S., K. Rai, A.K., Singh, V., and Rai, S. (2005). Vibrational spectrum of glycine molecule. *Spectrochim. Acta* 61, 2741–2746.

Kutteruf, M., Brown, C., Iwaki, I., Campbell, M., Korter, T., and Heilweil, E. (2003). Terahertz spectroscopy of short-chain polypeptides. *Chem. Phys. Lett.* 375, 337–343.

Labbé-Lavigne, S., Barret, S., Garet, F., Duvillaret, L., and Coutaz, J.L. (1998). Far-infrared dielectric constant of porous silicon layers measured by terahertz time-domain spectroscopy. *J. Appl. Phys.* 83(11), 6007–6010.

Le Ru, E.C., Meyer, M., and Etchegoin, P.G. (2006). Proof of single-molecule sensitivity in surface enhanced Raman scattering (SERS) by means of a two-analyte technique. *J. Phys. Chem. B* 110(4), 1944–1948.

Li, S., Ma, W., Zhou, Y., Wang, Y., Li W., and Chen, X. (2012). Cleavable porous silicon based hybrid material for pre-enrichment of trace heavy metal ions. *Appl. Surf. Sci.* 258, 5538–5542.

Lin, H., Mock, J., Smith, D., Gao, T., and Sailor, M.J. (2004). Surface-enhanced Raman scattering from silver-plated porous silicon. *J. Phys. Chem. B* 108(31), 11654–11659.

Lo, S.-Z.A. (2010). Application of Porous Silicon in Terahertz Technology, PhD Thesis, University of Maryland.

Lo, S.-Z.A. and Murphy, T.E. (2009). Nanoporous silicon multilayers for terahertz filtering. *Opt. Lett.* 34(19), 2921–2923.

Lo, S.-Z.A., Rossi, A.M., and Murphy, T.E. (2009). Terahertz transmission through p+ porous silicon membranes. *Phys. Stat. Sol. (a)* 206(6), 1273–1277.

Lo, S.-Z.A., Kumar, G., Murphy, T.E., and Heilweil, E.J. (2013). Application of nanoporous silicon substrates for terahertz spectroscopy. *Opt. Mater. Express* 3(1), 114–125

Lu, H., Zhang, H., Yu, X., Zeng, S., Yong, K.-T., and Ho, H.-P. (2011). Seed-mediated Plasmon-driven regrowth of silver nanodecahedrons (NDs). *Plasmonics* 7(1), 167–173.

Mahmoud, M.E., Osman, M.M., Mohamed, E., and Amer, M.E. (2000). Selective preconcentration and solid phase extraction of mercury(II) from natural water by silica gel-loaded dithizone phases. *Anal. Chim. Acta.* 415, 33–40.

McIntosh, A.I., Yang, B., Goldup, S.M., Watkinson, M., and Donnan, R.S. (2012). Terahertz spectroscopy: A powerful new tool for the chemical sciences? *Chem. Soc. Rev.* 41, 2072–2082.

Meenatchi, V., Muthu, K., Rajasekar, M., Meenakshisundaram, S., and Mojumdar, S. (2012). Crystal growth, structure and characterization of o-hydroxybenzoic acid single crystals. *J. Therm. Anal. Calorim.* 108, 895–900.

Moskovits, M. (2005). Surface-enhanced Raman spectroscopy: A brief retrospective. *J. Raman Spectrosc.* 36, 485–496.

Panarin, A.Yu., Chirvony, V.S., Kholostov, K.I., Turpin, P.-Y., and Terekhov, S.N. (2009). Formation of SERS-active silver structures on the surface of mesoporous silicon. *J. Appl. Spectrosc.* 76(2), 280–287.

Peiponen, K.-E., Zeitler, A., and Kuwata-Gonokami, M. (eds.) (2013). *Terahertz Spectroscopy and Imaging* (Springer Series in Optical Sciences, Vol. 171). Springer, Berlin.

Pisareva, V.P., Tsizin, G.I., and Zolotov, Yu.A. (2004). Filters for the preconcentration of elements from solutions. *J. Anal. Chem.* 59(10), 912–929.

Pu, Q., Sun, Q., Hu, Z., and Su, Z. (1998). Application of 2-mercaptobenzothiazole-modified silica gel to on-line preconcentration and separation of silver for its atomic absorption spectrometric determination. *Analyst* 123, 239–243.

Roldan, P.S., Ilton, L., Alcăntara, I.L., Padilha, C.C.F., and Padilha, P.M. (2005). Determination of copper, iron, nickel and zinc in gasoline by FAAS after sorption and preconcentration on silica modified with 2-aminotiazole groups. *Fuel* 84, 305–309.

Sakka, T., Tsuboi, T., Ogata, Y.H., and Mabuchi, M. (2000). Raman scattering from metal-deposited porous silicon. *J. Porous Mater.* 7, 397–400.

Sam, S., Chazalviel, J.-N., Gouget-Laemmel, A.C. et al. (2010). Covalent immobilization of amino acids on the porous silicon surface. *Surf. Interface Anal.* 42, 515–518.

Sigel, H. and Martin, R.B. (1982). Coordinating properties of the amide bond. Stability and structure of metal ion complexes of peptides and related ligands. *Chem. Rev.* 82, 385–426.

Stosch, R., Henrion, A., Schiel, D., and Guttler, B. (2005). Surface-enhanced Raman scattering based approach for quantitative determination of creatinine in human serum. *Anal. Chem.* 77, 7386–7392.

Stuart, D.A., Yonzon, C.R., Zhang, X.Y. et al. (2005). Glucose sensing using near-infrared surface-enhanced Raman spectroscopy: Gold surfaces, 10-day stability, and improved accuracy. *Anal. Chem.* 77, 4013–4019.

Sun, X., Wang, N., and Li, H. (2013). Deep etched porous Si decorated with Au nanoparticles for surface-enhanced Raman spectroscopy (SERS). *Appl. Surf. Sci.* 284, 549–555.

Tourrel, G. and Corset, J. (1996). *Raman Microscopy: Developments and Applications.* Elsevier Academic Press, San Diego, CA.

Ueno, Y. and Ajito, K. (2007). Terahertz time-domain spectra of aromatic carboxylic acids incorporated in nano-sized pores of mesoporous silicate. *Anal. Sci.* 23, 803–807.

Ueno, Y., Rungsawang, R., Tomita, I., and Ajito, K. (2006). Terahertz time-domain spectra of inter- and intra-molecular hydrogen bonds of fumaric and maleic acids. *Chem. Lett.* 35, 1128–1129.

Van Exter, N., Fattinger, C., and Grischkowsky, D. (1989). Terahertz time-domain spectroscopy of water vapor. *Opt. Lett.* 14, 1128–1130.

Virga, A., Rivolo, P., Descrovi, E. et al. (2012). SERS active Ag nanoparticles in mesoporous silicon: Detection of organic molecules and peptide–antibody assays. *J. Raman Spectrosc.* 43, 730–736.

Virga, A., Rivolo, P., Frascella, F. et al. (2013). Silver nanoparticles on porous silicon: Approaching single molecule detection in resonant SERS regime. *J. Phys. Chem. C* 117, 20139–20145.

Wan Ngah, W.S., Teong, L.C., and Hanafiah, M.A.K.M. (2011). Adsorption of dyes and heavy metal ions by chitosan composites: A review. *Carbohydrate Polymers* 83, 1446–1456.

Welz, B. and Sperling, M. (1999). *Atomic Absorption Spectrometry*, 3rd ed. Wiley-VCH, Weinheim, Germany.

Wilken, R.D. (1992). Mercury analysis: A special example of species analysis. *Fresenius J. Anal. Chem.* 342, 795–801.

Ye, W., Shen, C., Tian, J., Wang, C., Bao, L., and Gao, H. (2008). Self-assembled synthesis of SERS active silver dendrites and photoluminescence properties of a thin porous silicon layer. *Electrochem. Commun.* 10, 625–629.

Zeiri, L., Rechav, K., Porat, Z., and Zeiri, Y. (2012). Silver nanoparticles deposited on porous silicon as a surface-enhanced Raman scattering (SERS) active substrate. *Appl. Spectrosc.* 66(3), 294–299.

Zhang, H., Liu, X., Liu, C., and Jia, Z. (2012). n-Type porous silicon as an efficient surface enhancement Raman scattering substrate. *Opt. Eng.* 51(9), 099003.

Zougagh, M., Cano Pavon, J.M., and Garcia de Torres, A. (2005). Chelating sorbents based on silica gel and their application in atomic spectrometry. *Anal. Bioanal. Chem.* 381, 1103–1113.

PSi-Based Diffusion Membranes

15

Andras Kovacs and Ulrich Mescheder

CONTENTS

15.1 Introduction 280

15.2 Membrane Fabrication 280

15.3 Membrane Properties 286

15.4 Transport in Porous Membranes 287

15.5 Applications 293

15.6 Conclusion 294

References 295

15.1 INTRODUCTION

Due to the intensive research of electrochemical etching and development of stable and well-controllable silicon anodization processes, research activities since the 1990s show a growing interest in fabrication and application of porous silicon (PSi)-based membranes. Application of PSi membranes offers many advantages compared to conventional membrane materials: low cost fabrication process (in comparison to other microtechnology processes), precise control of pore size and pore size distribution depending on the fabrication conditions (e.g., doping, electrolyte concentration, anodization current density, light intensity), high chemical and temperature stability, applicability at high transmembrane pressure, high permeability, high selectivity, sterilizability, easy system and process integration in silicon-based MEMS/NEMS, smart sensors, Lab-on-a-Chips (LOC) or micro total analysis systems (μTAS). PSi is perfectly suited for formation of channels with different properties in nanometer as well as in micrometer scale. Flexible electrochemical and micromachining processes allow the control of membrane properties and realization of complex membrane architectures, including pore size and pore form variations. PSi membranes exhibit high surface-to-volume ratio resulting in a large inner surface, which can be activated in a postprocessing step for catalytic, immobilization, or functionalization purposes. However, the membrane properties are considerably changing due to the anodization and postprocessing process. Investigation of PSi membrane properties is a key issue for proper membrane design and for applications. Low cost processes, access to both membrane sides and compact system integrations open new applications of PSi-based membranes, especially in filtration and separation technology, fuel cells, fluidic systems, smart sensors, optical sensors, biosensors, and miniaturized LOC and μTAS systems.

In this chapter, the fabrication and properties of PSi-based membranes will be summarized. In addition, transport processes in PSi membranes and different membrane applications will be presented.

15.2 MEMBRANE FABRICATION

PSi membranes can be fabricated by a combination of anodization and bulk or surface micromachining processes. The formation of porous silicon is a typical top-down electrochemical process. To provide the anodization current flow through the silicon layer in the anodization setup, high p-doped silicon or low doped p-type silicon with additional highly doped backside contact is needed. N-type doped silicon can be anodized with additional backside illumination and pre-patterning to achieve ordered pore structure. Typically, low concentrated hydrofluoric acid (HF) with surfactant is used as an electrolyte.

PSi membranes can be fabricated with different configurations: without, with full or partial support, in-plane or out-of-plane membrane configuration, vertical or horizontal pore orientation, periodic pore arrangement (ordered pores) or self-organized (statistically distributed) pore architecture, straight or tortuous channels, different porosity levels (30–85%), homogeneous or inhomogeneous porosity distributions, constant or modulated pore sizes, subsequent postprocessing steps (surface stabilization, additional coating), macro- or mesoporous pore size, and different membrane thicknesses (typically 1–500 μm). A crucial process step in the fabrication of macro- or mesoporous silicon membranes is the release of the membrane from the substrate, including the opening of the backside (substrate side) pores. Figure 15.1 shows some basic concepts for formation of macro- and mesoporous silicon membranes using bulk or surface micromachining.

Two different schemes for realizing macro- or mesoporous silicon membranes are possible:

- Bulk micromachining and anodization; definition of the membrane thickness, that is, wafer thinning can be carried out before or after the anodization process. Thick porous membrane can be fabricated without wafer thinning.
- Surface micromachining by application of lift-off technique or two-step electrochemical process (PSi formation and electropolishing).

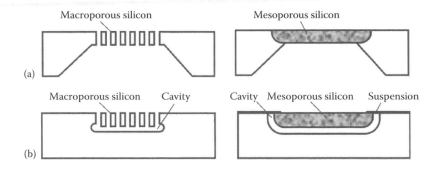

FIGURE 15.1 Schematic pictures of basic concepts for formation of macro- or mesoporous silicon membranes using (a) bulk or (b) surface micromachining.

The challenge of the successful membrane fabrication is the compensation of the length inhomogeneity in the pore formation and complete opening of all pores from the backside. Alternatively, ultrathin porous membranes can be fabricated by annealing of deposited films.

Thick macroporous membranes (>500 μm) have been fabricated by Zheng et al. (2005a) using p-type materials, electrolytes with HF (49%):ethanol:H_2O at volume ratio 1:1:1 and 1:2:3 with ammonium chloride and anodization current densities from 7 to 230 mA/cm². The nonuniform electrochemical etching was compensated by a sacrificial wafer, which was placed into direct contact with the backside of the membrane wafer and averages out all process inhomogeneities. Bias-assisted KOH etching has been developed by Mathwig et al. (2011a) for fabrication of ordered macroporous membranes. Macroporous silicon (n-type silicon with prepatterned 3-μm pore distance was anodized with 5% HF and backside illumination at 10°C using 2–2.8 V) was oxidized and backside etched with KOH. The SiO_2 layer serves as protection of the formed macroporous structure against the backside micromachining process. However, in conventional backside pore opening by KOH due to the inhomogeneity in the KOH etching and the limited masking ability of the SiO_2 in KOH only a short timeframe is available for process stop before the membrane is seriously attacked. With monitoring the current-time curve between the biased KOH and Si, the pore opening process can be stopped (current is reduced when etch front reaches the pores) to realize fully opened homogeneous pores. Alternatively, RIE etching can be used for opening process of the backside pores as shown by Grigoras et al. (2005) at fabrication of macroporous membranes using low-doped p-type silicon. RIE backside pore opening has been also used by Chu et al. (2006) for fabrication of n-type as well as p-type silicon PSi membranes. Large arrays of ratched-shaped macroporous silicon membranes with variation of pore diameter from 2.0 to 3.8 μm with 25 perfectly identical periods with a total depth of 200 μm and 4.2-μm pitch from n-type silicon, anodized in 6 wt% HF at 10°C by modulated photo-electrochemical etching has been presented by Müller et al. (2000). The variation of the pore diameter was achieved by controlling the light source intensity in a sawtooth-like profile. The total current is controlled by illumination intensity resulting in the modulation of pore size. Naturally ordered macroporous silicon (4–8 μm pore size) membrane has been fabricated by anodization (HF:ethanol = 1:1, 0.25 mA/cm² current density) with pre-etched low p-doped (3000 Ω·cm) silicon membranes by Splinter et al. (2002). The nanoporous layer was removed by 1 wt% KOH. The backside opening of the permeable membrane was realized by an advanced silicon etching (ASE) process. Figure 15.2 shows a close-up and cross-section of the fabricated porous membrane.

For better control of the membrane thickness (here 3 μm) silicon on insulator (SOI) n-type wafer with backside KOH etching has been applied by Tantawi et al. (2013) for fabrication of macroporous silicon membranes. KOH backside etching was performed with SiO_2 etch stop layer to form the thin membrane. The anodization process was performed after the free-etching of the thin silicon layer in 30:15:55 = 40% HF:95% Eethanol:DI water with current density of 10–15 mA/cm² and application of backside illumination. However, after the porosification process, porous membrane thicknesses between 1 and 3 μm were measured because partially anodization was working under electropolishing conditions, thus removing part of the material. Macroporous silicon membranes with different morphology have been realized by Zheng et al. (2005b) using crystal

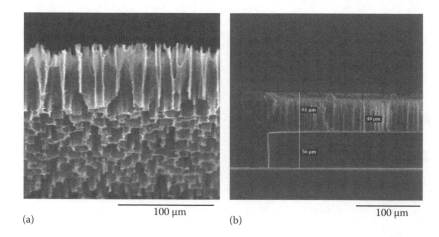

(a)   (b)

FIGURE 15.2   (a) Close-up of PSi membrane for micro membrane reactor application; and (b) cross-section of PSi membrane. (Reprinted from *Sens. Actuators B* 83, Splinter A. et al., 169, Copyright 2002, with permission from Elsevier.)

oriented and electric current dependent growth of macropores. *P*-type {111} Si substrates were used as substrate for porosification in HF (49%) and dimethylsulf-oxide (DMSO) (2:25 and 6:26 ration) and HF (49%):ethanol:H_2O = 1:2:3 with cetyltrimethyl-ammonium chloride (CTAC). A macroporous membrane with periodical variation of a stack from crystal orientation dependent and current direction dependent pore growth was manufactured for particle filter applications. Freestanding macroporous silicon membranes can be fabricated by two-step lift-off anodization process (PSi formation and electropolishing) as described by Thakur et al. (2012). In the first step, a macroporous silicon layer was delivered from low-doped (14–22 Ω·cm) *p*-type Si-wafer in dimethylformamide (DMF) and 49% HF (30 mL:4 mL) electrolyte at constant current density of 2 mA/cm² for 1 h. In the second step, the anodization current was changed in incremental stages of 1 mA/cm² up to 15 mA/cm². Each increment step was held for 10 min. With this process, porous membrane with pore diameters between 500 nm and 2 µm and thicknesses between 10 and 50 µm were realized by adjusting the anodization conditions. Due to the reduced availability of fluoride ions at the pore tips under these conditions, isotropic etching can be observed resulting in pore wall weakening and finally in separation process. Freestanding macroporous silicon membranes have been reported by Pagonis and Nassiopoulou (2006) using a combination of macro- and nanoporous silicon layer. Both nonordered and ordered macroporous 50-µm thick layers were formed in a mixture of HF:DMF (1:9 volume ratio) at current density of 10 mA/cm². The 30-µm thick nanoprous layer is etched in ethanoic HF solution at current density of 80 mA/cm² directly underneath the macroporous layer. The buried nanoporous layer acted as a sacrificial layer and was removed in subsequent oxidation and oxide etching process without etching the macroporous material to build the cavity and to release the macroporous membrane from the substrate. With this technique, a freestanding macroporous membrane side length in the millimeter scale was realized (Figure 15.3).

Mesoporous silicon membranes even completely through the thickness of a 500-µm wafer can be formed by long anodization processes (Searson 1991). Both *n*-type (10 and 0.1 Ω·cm, As-doped) and *p*-type (0.003 Ω·cm, B doped) wafers of {100} orientation were electrochemically etched in HF (49 wt%)-C_2H_5OH solution with current density of 50 mA/cm² and 25 mA/cm², respectively. The complete anodization of the whole substrate thickness was identified by the sharp increase in the electrode potential as well as by the presence of electrolyte leaking through the wafer. The inhomogeneity of the morphology between the surface and bulk layer was analyzed. The influences of substrate doping, metal catalyst, and etching solution temperature have been investigated by Cruz et al. (2005) for diffusion membrane applications. Both *n*- and *p*-doped substrates with doping level from 10^{14} to 10^{19} cm^{-3} have been anodized in a mixture of 49 wt% HF, 39 wt% H_2O_2, and 90 wt% ethanol with different etching times and temperatures (1 to 4 h, at 20°C and 40°C) using thin (1 to 8 nm) catalyst materials of Au and Pt. Deep and fine intercolumnar porous layers were achieved by application of Au catalyst metal in contrast to Pt assisted anodization, which delivers random structures with low etch rate. Porous membranes have been formed by

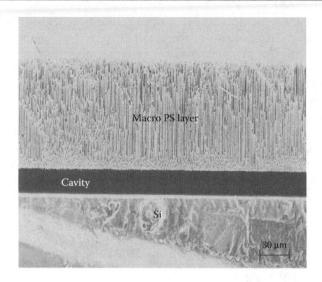

FIGURE 15.3   Cross-sectional SEM image of freestanding macro PS membrane over the cavity. The macroporous membrane is released from the substrate by selective dissolution in the same anodization process flow. (Reprinted from *Microelectron. Eng.* 83, Pagonis D.N. and Nassiopoulou A.G., 1421, Copyright 2006, with permission from Elsevier.)

backside etching and substrate thinning down to 50 μm by deep reactive ion etching in SF_6 with Ar and O_2 at 800 W. Reinforced mesoporous silicon membranes have been developed by Kovács et al. (2007). Porous membranes were realized by KOH backside micromachining and application of a special silicon nitride masking layer to form during the anodization process single crystalline support structure. Due to the applied masking structure, the anodization current is mainly directed in a vertical direction and inverted Y-shaped single crystalline columns were formed below the masking structures. The porous layer is formed in a thin KOH etched p-type silicon membrane from the front side by electrochemical etching in an electrolyte of HF/ethanol/H_2O (1:2:1), using a current density of 10 mA/cm² to achieve pore size in the lower mesoporous range. RIE etching process was applied as postprocessing to open the backside pores. The edge of the PSi membrane is placed outside of the KOH window to reduce the edge effect during PSi formation and to achieve a homogeneous PSi etch front formation (i.e., uniform porous membrane thickness). Due to the integrated crystalline support layer, bursting pressures of porous membranes were measured up to 1 bar. Figure 15.4 shows the schematic cross-section and SEM picture of the reinforced PSi membrane.

Gauthier-Manuel and Pichonat (2005) fabricated PSi membrane with Cr-Au masking layer and subsequent double-side KOH etching of n^+-type silicon substrate to form a 50-μm thick single crystalline membrane in the middle of the wafer. Anodization was carried out in the dark in 50% HF(48%):50% ethanol electrolyte with current density from 18 to 36 mA/cm² to transfer the single crystalline material in porous layer. After the anodization, only a few channels were totally opened because of the inhomogeneous etching front development and the automatically stopped electrochemical etching process. The backside pore opening process was realized by a short (150 s) RIE postprocessing (Pichonat and Gauthier-Manuel 2006) to achieve a completely opened porous membrane.

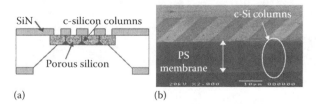

FIGURE 15.4   (a) Schematic cross-section and (b) close-up of reinforced PSi membrane. (Reprinted from *Sens. Actuators B* 127, Kovács A. et al., 120, Copyright 2007, with permission from Elsevier.)

An interesting approach to increase the specific surface area of PSi membranes was presented by Cavalaglio et al. (2011). They used a combined masking process of Cr and Au layer and double-side KOH micromachining for preparation of the surface to form double V-grooves. With this solution, a 1.7 times larger specific area is obtained compared to a flat (in-plane) membrane. The porosification of the n^+-doped 3D-accordion structured silicon membrane was done in 1:1 = HF:ethanol solution in the dark with current densities from 50 to 250 mA/cm² resulting in a pore diameter from 6 to 30 nm and porosity from 20 to 50%. The anodization was carried out up to a sharp voltage drop between the electrodes indicating the opening of backside pores. Figure 15.5 shows SEM views of the accordion-shaped porous membrane.

The flexibility and the integration possibilities of the electrochemical process have been shown by Ghulinyan et al. (2003) in the realization of freestanding porous membranes for optical applications. In this solution, a self-supporting porous membrane is manufactured by a two-step anodization process. In the first process sequence, the porous membrane is defined. After that, in the second process sequence, anodization current is switched to electropolishing mode, which serves as a postprocessing step for free-etching of back side pores and for release of the membrane from substrate. By this approach, ambient atmosphere has access to both sides of the porous membrane thus allowing effective diffusion processes. By an appropriate choice of anodization current density, the first process sequence is divided into single process steps to fabricate an inhomogeneous porosity distribution by step-like variation of the anodization current for optical sensor applications. In the process, p^+-type substrate was anodized in 30% volumetric fraction of aqueous HF (48%) using periodical variation of the current density (50 mA/cm² for 5.9 s and 7 mA/cm² for 21.5 s) with a final electropolishing process (0.4 A/cm² for 2 s). Application of electropolishing for lift-off process (Turner 1958) or for release of porous layer from the substrate (Sagnes et al. 1993) is a widely used separation technique. However, a complete release from substrate by electropolishing will result in a totally separated very sensitive component. This membrane type needs a careful and special mechanical handling. Fully released PSi films can be manufactured even by a one-step anodization process as described by Solanki et al. (2004): Lift-off (separation) of thin PSi layers (from a few to tens of micrometers) was carried out in an HF-based solution under specific settings of current density and HF concentration. The *in situ* change of fluoride ion concentration leads to formation of high porosity layer beneath the low porosity layer. The morphological mismatches at the layer boundary of low and high porosity layers lead to a lift-off (separation) process and allow the fabrication of self-standing thin porous membranes. However, additional suspensions were preferred to hold the porous membrane and to build a compact micromachined sensor unit. Lammel and Renaud (2000, 2001) developed a freestanding PSi microstructure by application of an electropolishing process to release the porous membrane while still keeping a suspension to the substrate (p^+-type silicon is anodized in HF (50%):ethanol = 1:1). Lammel and Renaud (2000, 2001) also discussed the influence of the masking layers on the release process. When metal mask is used, the electropolishing process releases the porous layer from the substrate but not from the boarders, so keeping suspensions to substrate. Using insulating masks instead of metal masks, the anodization current is forced to squeeze through the mask window resulting in lateral current

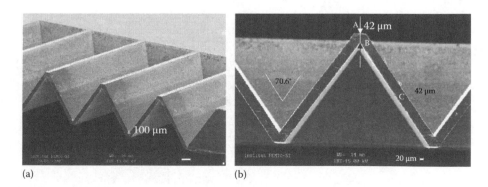

(a)   (b)

FIGURE 15.5   (a) SEM view of the cut of an accordion-shaped PSi membrane suspended on a silicon frame. (b) Close-up of the structured membrane. (Reprinted from *J. Membr. Sci.* 383, Cavalaglio S. et al., 250, Copyright 2011, with permission from Elsevier.)

flow and an isotropic lateral underetching. With this concept, the PSi membrane is held only by the suspension layer, which is fully separated from the substrate. However, the influence of the edge effect on the membrane thickness evolution during the PSi formation has to be considered (Krüger et al. 1996; Ivanov et al. 2011). The influence of the masking layer on the underetching characteristics at the formation of thick macroporous membranes was also discussed by Bassu et al. (2012). Using an HF-resistant metal layer (Au/Ti) both overetching and layer adherence can be solved. With one- or double-side support and application of Cr/Au/Ni/Si$_3$N$_4$ or Au/Si$_3$N$_4$ actuator out-of-plane configurations (Lammel and Renaud 2000, 2001; Lammel et al. 2002; Song et al. 2008) with suspended PSi membranes can be realized as shown in Figure 15.6.

Multi-walled micro channels (MWµC) have been presented by Tjerkstra et al. (2000) consisting of two or more coaxially constructed micro channels separated by a suspended PSi membrane. In the buried construction solution, the porous membrane is suspended halfway through an etched cavity surrounded by silicon nitride walls. The micro channels and the suspended PSi membrane were formed from p^+-type silicon substrate in aqueous 5% HF solution by switching of the current density (180 mA/cm^2 for 10 min, 40 mA/cm^2 for 15 min, and 180 mA/cm^2 for 15 min) at different potential versus Ag/AgCl reference electrode (3 V, 0 V, 3 V). By periodically switching the potential between high and low, that is, channel and membrane formation phase, MWµC and rounded PSi membranes were realized. By trench etching, flat PSi membranes were produced with similar anodizaton conditions.

Mesoporous silicon membrane with horizontal pore orientation has been reported by Dubosc et al. (2012). P^+-type silicon wafer was etched by deep reactive ion etching to form a silicon stamp. Anode is locally created by evaporation and lift-off of a Cr/Au layer (50 nm/150 nm) on the side of the silicon stamp. A silicon nitride masking layer was used for protection and definition of the anodization window on the other side of the silicon stamp. Lateral porosification of the silicon stamp was carried out by HF (48%):ethanol = 1:1 electrolyte at a current density of 50 mA/cm^2 for 5 min to build a 10-µm long and 3-µm thick horizontally oriented nanoporous membrane with pore size distribution in the range of 10–30 nm. Horizontally oriented membranes enable the monolithic integration into planar microfluidic channels.

A novel fabrication technique using porous nanocrystalline silicon (pnc-Si) has been presented by Striemer et al. (2007). Porous nanocrystalline silicon membranes exhibit extraordinary permeability and transparency due to an extreme thinness (in the molecular dimension), typically in the range of 10–50 nm. The key process step in the fabrication is the deposition of a 3-layer film stack consisting of silicon dioxide (20 nm)/amorphous silicon (15 nm)/silicon dioxide (20 nm) and a subsequent anneal process at 1000°C for 60 s to transform the amorphous silicon to a nanocrystalline state. The backside anisotropic micromachining is done by etchant ethylenediamine pyrocatechol (EDP) and the membrane is finally released in buffered oxide etchant (BOE) (Striemer et al. 2007; Agrawal et al. 2010).

The development of well-controlled and stable anodization processes or novel techniques allows the fabrication of PSi membranes even as commercial product in a wide range of configurations,

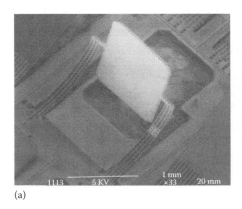
(a)

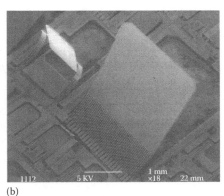
(b)

FIGURE 15.6   SEM pictures of tunable interference filters (a) with two actuator arms and (b) with ribbon layer for actuation and suspension. (Reprinted from *Sens. Actuators* 92, Lammel G. and Renaud Ph., 52, Copyright 2001, with permission from Elsevier.)

thicknesses (from ultrathin 10 nm up to thick 500 μm), porosities (30–85%), pore sizes (macro- or mesoporous), pore arrangements, and postprocessing steps.

15.3 MEMBRANE PROPERTIES

Investigation of PSi membrane properties is a key issue for successful membrane applications. The corresponding properties have to be considered for proper membrane design and fabrication process.

The basic geometrical parameter of the fabricated membrane, such as membrane size, pore distance, and pit width at macroporous layers are controlled by design while the membrane thickness is defined by KOH preprocessing or by PSi formation rate and anodization time. Anodization causes a material transformation from bulk single crystalline silicon to PSi and results in significant change of the membrane's material properties. The porous layer properties, such as porosity, pore size, pore size distribution, specific surface area, and morphology can be controlled by bulk material and anodization process conditions, such as substrate doping, electrolyte concentration, anodization current, and backside illumination. The structural characterization of the porous membranes can be done using scanning electron microscopy (SEM), transmission electron microscopy (TEM), or atomic force microscopy (AFM). The porosity measurement can be carried out by gravimetric technique. Pore size, pore size distribution, and specific surface area (SSA) can be evaluated from measured adsorption/desorption isotherms using BET and BJH theory (Brunauer et al. 1938; Barrett et al. 1951). Pore size and pore size distribution, which play an essential role in the membrane technology, are controlled mainly by substrate doping level (Lehmann and Grüning 1997), anodization current density (Canham 1997; Kovacs et al. 2011), electrolyte properties (Canham 1997; Föll et al. 2002), and intensity of backside illumination (Müller et al. 2000). The possibility of fine-tuning pore size and pore size distribution by adjusting current density is shown in Kovacs et al. (2011). The surface wettability, that is, hydrophilic/hydrophobic surface properties of the porous layer, was characterized by (macroscopic) water contact angle measurements (De Stefano et al. 2008; Kovacs et al. 2009; Velleman et al. 2010).

Optimization of the mechanical properties of PSi membranes as well as stress control is a main issue for successful membrane applications in filtration and separation processes.

A crucial aspect in the successful application of porous membranes—especially in filtration and separation systems with high transmembrane pressure—is the investigation and optimization of the mechanical properties: stability, that is, bursting pressure, elastic modulus (E), and membrane deflection. Mechanical investigations of porous reinforced membranes were presented by Kovács et al. (2007) and Kovács and Kovacs (2014). Porous membranes were modeled as a non-perforated membrane with adjusted Young's modulus (Van Rijn 2004; Kovács et al. 2008; Kovács and Kovacs 2014) (Equation 15.1):

$$(15.1) \qquad E_p = (1 - \varepsilon)E$$

where ε is the porosity and E is Young's modulus.

Fracture strength (σ_f) is calculated by the maximum normal stress theory commonly applied for brittle materials (Equation 15.2):

$$(15.2) \qquad \sigma_f = |\sigma_l|_{max}$$

where $\sigma_{l\,max}$ is the maximum principal stress in the plate. The ultimate strength of reinforced PSi membranes were determined from the bursting (critical) pressure, which is the principal quantity of load bearing capacity (Kovács and Kovacs 2014). Micro-Raman spectroscopy was used for stress characterization of supported and suspended PSi layers under various thermal treatments by Papadimitriou et al. (2004). Stress-analysis reveals that membranes treated in an inert ambient and at moderate temperatures are less strained and do not break when they are released from the substrate. Mechanical properties of PSi are discussed in detail in other chapters of this book.

Surface postprocessing steps of the PSi layer, such as stabilizations, activations, functionalization, or coatings have strong influence on the selectivity and permeability of PSi membranes.

Furthermore, postprocessing steps have a strong effect on the pore size and porosity; however, long-term stability of the porous membrane can be improved considerably. Additionally, surface processing has influence on the transport processes and surface interactions. Native porous layers are hydride terminated and thus show aging effects in air and aqueous solutions. Typically, a simple oxidation process is commonly used for surface stabilization of the porous layer. Oxidation causes material conversion from Si to SiO_2 and a corresponding pore size reduction (Kovacs et al. 2011). This pore size reduction should be taken into account in membrane design and diffusion processes. On the other hand, oxidation involves an essential reduction of porosity (Charrier et al. 2007), which leads to reduction of the permeability.

Self-supporting PSi membranes have been functionalized by Velleman et al. 2010 using silanes (heptadecafluoro-1,1,2,2-tetrahydrodecyl)dimethylchlorosilane (PFDS), and N-(triethoxysilyl-propyl)-o-polyethylene oxide urethane (PEGS) to provide hydrophobic (by PFDS) and hydrophilic (by PEGS) membrane properties. A two-step etching process has been used for the fabrication of the PSi membranes with an etching area of 1.767 cm². The first step anodization was carried out at 100 mA for 30 min in a solution of 48% aqueous HF/ethanol (3:1) followed by a second etching phase at 450 mA for 60 s in HF/ethanol (1:1). After subsequent drying with ethanol and thermal anneal process at 400°C for 1 h, the porous membranes were functionalized with silane. The porous membranes were treated with the dimethylchlorosilane (PFDS) via a neat silanization method and with the hydrophilic PEGS via a solution phase deposition method.

Nafion-filled porous membranes were used for fuel cell applications to improve the photon conduction by Pichonat et al. (2004) and Pichonat and Gauthier-Manuel (2006). First, a hydrophilic surface was realized by immersion of the porous membrane in "Piranha" solution (80% pure sulfic acid and 20% aqueous solution of hydrogen peroxide) for 10 min followed by a simple capillary pore filling with Nafion (5% Nafion-117 solution) with complete pore filling. Deposition of a catalyzing agent (700-nm thick Palladium) into the porous membrane using wet-chemical processing to convert CO into CO_2 was presented by Splinter et al. (2002). A nickel layer was deposited in macroporous silicon membranes for high density capacitor devices by Vega et al. (2014a,b).

Surface chemistry and surface functionalizing of PSi are discussed in detail in other chapters of this book.

15.4 TRANSPORT IN POROUS MEMBRANES

Diffusion or mass transport resulting from the concentration gradient can be commonly described by Fick's first law under the assumption of steady state condition (Equation 15.3 in one-dimensional form). The flux J, in the direction of flow is proportional to the concentration gradient:

$$(15.3) \qquad J = -D\left(\frac{\partial c}{\partial x}\right)$$

where D is the diffusion coefficient, c is the concentration, and x is the coordinate along diffusion. Considering conservation of matter, Fick's second law describes the nonsteady state for transport process, which is given by the rate of change of the concentration at a position within the membrane (Equation 15.4):

$$(15.4) \qquad \frac{\partial c}{\partial t} = D\left(\frac{\partial c^2}{\partial x^2}\right)$$

where c is the concentration, t is the time. This is valid for an ideal case where the membrane is isotropic and the diffusion coefficient is independent of distance, time, and concentration.

Transport processes in micro- or mesoporous systems are considerably more complex than in macroporous systems. Understanding of transport in narrow pores and confined spaces dates back to the seminal works of Knudsen (1909) and Smoluchowski (1910). Nowadays, molecular dynamics (MD) or Monte Carlo (MC) simulations are useful tools to determine the diffusivity values in nanoporous media for different guest molecules (Lu et al. 2006; Aksimentiev et al. 2009; Krishna and van Baten 2009, 2011a,b; Cablares-Navarro et al. 2013; Ebro et al. 2013; Naserifar et al. 2015). Molecular simulation based investigations showed that the Knudsen model fails when significant surface adsorption is occurring (Bhatia and Nicholson 2003; Jepps et al. 2003; Bhatia et al. 2004; Krishna and van Baten 2009, 2011a,b). Experimental permeance data for several light gases in a mesoporous silica membrane were analyzed in detail by Ruthven et al. (2009). This study demonstrated that the simple Knudsen model still provides a good representation of the permeance data even under conditions of significant adsorption. The effect of pore surface roughness on Knudsen diffusion in nanoporous media was investigated by Malek and Coppens (2003) using dynamic Monte Carlo simulations and analytical calculations. The investigations indicate that a rougher surface corresponds to more molecular traps along the surface and leads to a decrease in self-diffusivity, whereas transport diffusion or transmission probability are unaffected by the presence of these traps. Diffusion in nanoporous materials is well studied and is still an important research topic today for different applications and materials (Webb 1998; Conner and Fraissard 2003; Malek and Coppens 2003; Karniadakis et al. 2005; Ruthven 2005, 2007; Ho and Webb 2006; Beerdsen and Smit 2007; Heinke 2007; Roque-Malherbe 2007; Ruthven et al. 2008; Higgins et al. 2009; Chmelik et al. 2010; Civan 2011; Frentrup et al. 2012; Kärger et al. 2012; Valiullin 2013; Krishna 2014).

In porous materials, different diffusion mechanisms have to be considered. These are molecular or gaseous flow, Knudsen flow, surface diffusion, multilayer diffusion, capillary condensation, and configurational diffusion (Roque-Malherbe 2007). The different diffusion processes in porous membranes are schematically illustrated in Figure 15.7.

Diffusion mechanisms depend on the relation between the characteristic length of the geometry (L) and mean free path of the molecules (λ). The Knudsen number is defined as a ratio of the two parameters (Equation 15.5):

(15.5)
$$K_n = \frac{\lambda}{L}$$

Viscous flow occurs when the characteristic length is considerably larger than the mean free path of the molecule ($K_n < 0.01$). In the case of flow through a porous material, the following behavior is found: With increasing mean free path and/or decreasing pore size, more and more collisions occur with the pore wall ($10 > K_n > 0.01$). Thus, transport is described as Knudsen diffusion. In this region of K_n further subclassification was introduced by Karniadakis et al. (2005) such as slip flow ($0.01 < K_n < 0.1$) and transition flow ($0.1 < K_n < 10$). However, when the mean free path is much larger than the pore diameter ($K_n > 10$), free molecular diffusion is the main contribution to the transport. Furthermore, surface diffusion has to be considered when diffusing molecules are effectively adsorbed on the pore surfaces. A monolayer adsorption is followed by a multilayer formation up to completely pore filling, then capillary condensation takes place

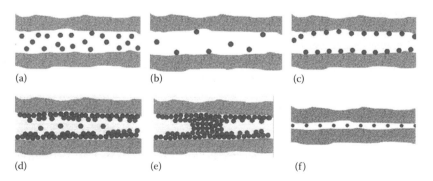

(a)　　　　　(b)　　　　　(c)

(d)　　　　　(e)　　　　　(f)

FIGURE 15.7 Diffusion processes in porous membranes: (a) gaseous or molecular flow, (b) Knudsen diffusion, (c) surface diffusion, (d) multilayer diffusion, (e) capillary condensation, and (f) configurational diffusion.

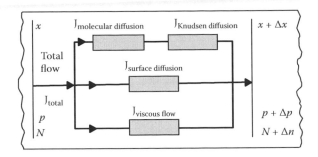

FIGURE 15.8 Schematic picture of electric analog circuit with contribution of different diffusion mechanisms, after Mason and Malinauskas (1983). (From Kovacs A. et al., *ECS Trans.* **50**(37), 207, 2013. Copyright 2013: The Electrochemical Society. With permission.)

in pores with radii less than the specific Kelvin radius. Finally, configurational diffusion might contribute to diffusion if the pore diameter turns out to be similar to the size of the molecules (Roque-Malherbe 2007). Additionally, transport processes are depending on the layer morphology (pore size, porosity, tortuosity), surface treatment, surface interactions between adsorbate and adsorbent, adsorbate properties, and vapor pressure of the investigated adsorbates.

Diffusion in nanostructured porous layers based on Mason and Malinauskas (1983) can be described using an electric analog circuit in which the various transport mechanisms are summarized (Figure 15.8). In this model, the Knudsen and molecular diffusions are considered series resistances, whereas viscous flow and surface diffusion are considered independent parallel resistances.

The permeance in the case of pure Knudsen flow (Π_{Kn}) is given by Equation 15.6 (Higgins et al. 2009; Ruthven et al. 2009):

$$(15.6) \qquad \Pi_{Kn} = \frac{\varepsilon}{\tau} \cdot \frac{D_{Kn}}{l} \cdot \frac{1}{RT}$$

where ε is the porosity, τ is the tortuosity, D_{Kn} is the Knudsen diffusivity, l is the membrane thickness, and R and T are the gas constant and absolute temperature, respectively.

The Knudsen diffusivity for a cylindrical pore is given by the classic Knudsen formula (Equation 15.7):

$$(15.7) \qquad D_{Kn} = \frac{d}{3} \sqrt{\frac{8RT}{\pi M}}$$

where d is the pore diameter, M is the molecular weight of the investigated species, and π is the Archimedes constant.

In general, the surface flux (J_s) is related to the surface diffusion coefficient and the local gradient of surface concentration and is given by Fick's law (Equation 15.8) (Roque-Malherbe 2007):

$$(15.8) \qquad J_s = -D_s \cdot \frac{dc_s}{dx}$$

where D_s and c_s are the surface diffusivity and surface concentration, respectively.

The variation of diffusivity in porous materials with respect to pore size (r) is schematically illustrated in Figure 15.9 with indication of the different diffusion regimes.

A real porous network mostly consists of a random network of interconnected pores with different pore arrangement, pore size, orientation, and finally, surface roughness and specific surface area. The influences of pore structure, randomness, and pore size on the effective mass diffusivity ($<D_m>$) were investigated by Mezedur et al. (2002). Two-dimensional pore network models were used to calculate the effective mass diffusivity in porous materials. The variations

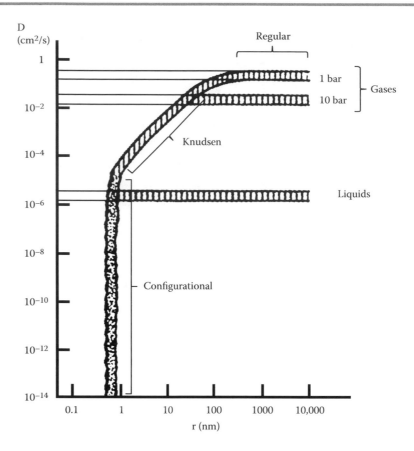

FIGURE 15.9   Schematic picture of the diffusivity in porous materials as a function of the pore radius with indication of the different diffusion regimes. (Kärger J. et al.: *Diffusion in Nanoporous Materials.* 2012. Copyright Wiley-VCH Verlag GmbH & Co. KGaA. Reproduced with permission.)

of the effective diffusivity with respect to porosity (ε) for tetragonal, hexagonal network models, ordered periodic unit cells, and spherical shells (model from Neale and Nader 1973) are shown in Figure 15.10.

Membranes can be operated in the so-called dead-end or in the cross-flow mode. In the dead-end mode, a feed stream is completely transported through the membrane. Here, components

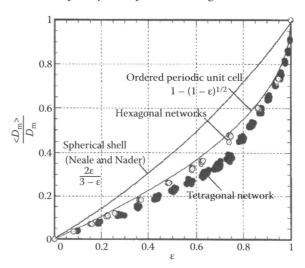

FIGURE 15.10   Effective diffusitivity as a function of the porosity for tetragonal, hexagonal network, ordered periodic unit cell, and spherical shell. (From Mezedur M.M. et al., *AIChE J.* **48**, 15, 2002. Copyright 2002: AICHE. With permission.)

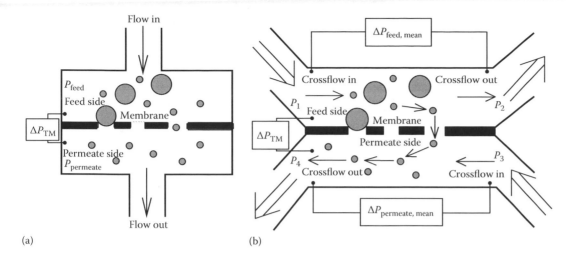

FIGURE 15.11 Measurement techniques of (a) dead-end and (b) cross-flow filtration. (Reprinted from *Sens. Actuators B* 175, Kovacs A. and Mescheder U., 179, Copyright 2012, with permission from Elsevier.)

rejected by the membrane will accumulate at the membrane surface leading to so-called cake formation and thus to pore blocking and permeability reduction. In the continuous cross-flow mode, both feed and permeate flow along the membrane. In cross-flow filtration, the cake builds up along the membrane, its thickness increases in flow direction unless the shear rate is high enough to drag this immobile layer. Figure 15.11 shows the schematic picture of dead-end and cross-flow filtration technique. The transmembrane pressure (ΔP_{TM}) is defined as the difference between the average feed/concentrate pressure and the permeate pressure (Equation 15.9); however, the influence of the cake layer should be considered depending on applied filtration technique:

$$(15.9) \qquad \Delta P_{TM} = \frac{(P_{feed} + P_{concentrate})}{2} - P_{permeate}$$

Gas diffusion in PSi membranes was investigated by different authors (Lysenko et al. 2004; Cruz et al. 2005; Kovacs and Mescheder 2012; Kovacs et al. 2013). Lysenko et al. (2004) studied the permeability of gas (air and hydrogen) in native and oxidized porous silicon nanostructures. The used transmembrane pressure gradient through the PSi layer was in the low-pressure range up to 2000 Pa. In this measurement range, an ideal linear behavior of flow rate as a function of the applied pressure has been observed. Permeability values of 10^{-16}–10^{-15} m^2 have been measured, corresponding to the 50–70% porosity range. Permeability values of hydrogen are much larger compared to air because of their viscosity difference (28.6 µPa for air and 9 µPa for hydrogen). Permeabilities in the range of 2600·10^9–100·10^9 m^3/m^{2}s Pa have been determined by Cruz et al. (2005) for N$_2$ flow in porous membrane thickness range of 50–200 µm, respectively. Permeability diminishes with increasing membrane thickness. Eight percent lower permeability values have been measured using CO$_2$ and only 2–10·10^9 m^3/m^{2}s Pa using an additional 3-µm plasma polymerized Teflon-like coating.

Transport in surface passivated PSi membranes was investigated by Kovacs et al. (2013). Here, permeability was measured for helium, nitrogen, and argon to explore the influence of molecular weight and to compare the corresponding permeance values for different surface conditions. Surface treatments (e.g., oxidation, functionalization, or other postprocessing surface covering processes) have a strong influence both on the adsorption processes and on surface interactions. Thereby, these treatments have a strong influence on the proportions of the different diffusion mechanisms concerning transport. Stronger adsorption and higher interaction energy in oxidized layers lead to increasing adsorption. In this case, the species are bound after collision to the wall rather than return to the bulk as investigations of molecular dynamics simulations have shown (Krishna and van Baten 2011a,b). Gas transport using surface passivated (oxidized) porous membranes is controlled by the Knudsen molecular and surface diffusion whereas in the investigated native porous membrane occurs primarily by molecular and Knudsen diffusion (Kovacs et al. 2013). The contribution of surface diffusion in surface passivated (oxidized) porous membranes is attributed to the decreasing pore size and strong adsorption of heavier species (N_2 and Ar). Surface interactions and adhesion forces are increased by thermal oxidation of PSi. Three times larger c-values from the BET theory and two times larger total adhesion forces have been observed for passivated (oxidized) PSi membranes (Kovacs et al. 2013).

During the electrochemical formation process of the porous membranes diffusion or mass transport, limited effects lead to inhomogeneities in the pore size and porosity distribution (Searson 1991; Canham 1997). This effect will influence the transport processes. A 20% higher etch rate was observed by Cruz et al. (2005) in the center of the wafer compared to its periphery, which directly reflected in the permeability measurements. Small differences between gas flow rates in direction of the pore formation and that in the opposite direction were observed by Kovacs et al. (2013). This effect is attributed to the inhomogeneity in the pore size and porosity distribution.

Water transport through nanoporous materials has been investigated by Shearer et al. (2010). For the measurements, PSi membranes were fabricated by a two-step process in ethanoic HF (100 mA/cm² for 30 min in 3:1 48% HF/ethanol and 450 mA/cm² for 60 sec in 1:1 HF/ethanol). The formed porous membranes were annealed (1 h, 400°C), then either investigated for water transport or surface-treated further with the hydrophobic silane PFDS ((heptadecafluoro-1,1,2,2-tetrahydrodecyl)dimethylchlorosilane, Gelest) via the neat silanization and annealed at 80°C for 15 min. Water permeability of 16,926 mm³ s⁻¹ atm⁻¹ cm⁻² was measured using oxidized porous membranes at constant pressure of 0.1 bar. This value is much higher compared to hydrophobic (PFDS surface treated) membranes, where a value of 408 mm³ s⁻¹ atm⁻¹ cm⁻² was derived. Hydrophilic surfaces of nanostructured membranes support the wetting and frictionless transport of water resulting in increased water transport. Fluid transport in PSi host systems with complex pore structure was studied by Dvoyashkin et al. (2009) to investigate the influence of inkbottle effect of alternating pore diameter.

Permeability of mesoporous silicon membranes for liquids was investigated by Kovacs and Mescheder (2012) using aqueous NaCl solution. Cross-flow filtration technique with prefiltration was used to reduce the so-called "cake formation" on the feed side and to obtain high throughput. The concentration of aqueous NaCl solution was determined from resistivity measurements of the solutions. The resistivity (i.e., the NaCl concentration) was continuously measured with a computer-controlled system both on the feed and the permeate side. The NaCl concentration on the permeate side was changing by about 6 orders of magnitude by wetting the feed side with the NaCl-solution (Figure 15.12), thus proving the large permeability for an NaCl solution. Transport rates of $v \leq 56$–78 ppm/sec for NaCl were derived using the investigated mesoporous silicon membranes with membrane thickness of 15 µm after activation of the transmembrane pressure (TMP) with $TMP_{on} = 500$ mbar between the feed and permeate side (Kovacs et al. 2013).

The influence of the surface functionalization on the selectivity and separation ability of PSi membranes was explored by Velleman et al. (2010). Self-supporting PSi membranes were fabricated by modification with the highly hydrophobic silane, PFDS, and the hydrophilic silane,

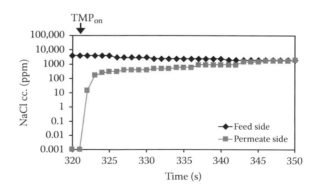

FIGURE 15.12   Change of the NaCl concentration in the $TMP_{on} = 500$ mbar phase through the mesoporous silicon membrane using cross-flow filtration. (From Kovacs A. et al., *ECS Trans.* **50**(37), 207, 2013. Copyright 2013: The Electrochemical Society. With permission.)

PEGS. To investigate the transport and selectivity properties of the modified membranes, the transport of dyes, tris(2,20-bipyridyl)dichlororuthenium (II) hexahydrate, Rubpy (hydrophobic) and Rose Bengal, RB (hydrophilic) through the functionalized membranes (hydrophobic and hydrophilic) was measured. The flux and permeability measurements showed that for a PFDS hydrophobic functionalization, the transport of hydrophobic molecules is facilitated. On the other hand, diffusion of hydrophilic molecules is hindered. For PEGS hydrophilic-modified membrane, the transport of hydrophilic species was facilitated while the transport of the hydrophobic species was reduced. The role of surface functionalization on the selectivity and separation ability of PSi membranes was effectively demonstrated and the measurement results can be used for membrane-based bioengineering sciences.

15.5 APPLICATIONS

PSi membranes are widely investigated and used for different applications. Due to the high permeability of gases, PSi-based membranes were preferably applied in gas, vapor, solvent or water systems such as micromembrane reactors (Splinter et al. 2002), gas sensors (Taliercio et al. 1995), loop heat pipe evaporators (Cytrynowicz et al. 2002), microvapor sources (Wallner et al. 2007), solvent detector with a macroporous silicon field-effect sensor (Clarkson et al. 2007), fiber optic sensors for the simultaneous detection of pressure and organic gases (Cho et al. 2013) or measurement of dissolved-gas concentration in water (Lee and Lee 2014). Until now, large effort has been put into the system integration where porous membranes could be an important system component especially in microfluidic, analysis systems, or miniaturized pumps (Müller et al. 2000; Tjerkstra et al. 2000; Yao et al. 2006; Saharil et al. 2012). Macroporous silicon membrane with periodic asymmetric variation in pore diameter can act as massively parallel and multiply stacked Brownian ratchets and can be used for large-scale particle separations. Brownian molecular motion can be induced by application of a periodic pressure profile through a porous membrane with asymmetric pore arrangement (Figure 15.13) (Matthias and Müller 2003). Particle transport in asymmetrically modulated pores was also investigated by Mathwig et al. (2011b).

Due to the precise control and wide variation range of pore size, PSi membranes are perfectly suited for particle separation (Wallner and Bergstrom 2007; Saharil et al. 2012) and contamination detection, for example, of volatile organic compounds by FTIR spectroscopy (Cavalaglio et al. 2011).

PSi membranes are very important in the development of nano-bio applications such as molecular-sensing, bio-sensing, separation and purification of biomolecules, immobilization, drug delivery, and tissue engineering. PSi membranes were used for selective bio-organism capture (Létant et al. 2003), for charge- and size-based separation of macromolecules using ultrathin membranes (Striemer et al. 2007; Snyder et al. 2011), for cell culture on highly permeable and molecularly thin substrates (Agrawal et al. 2010), for transmembrane protein biological

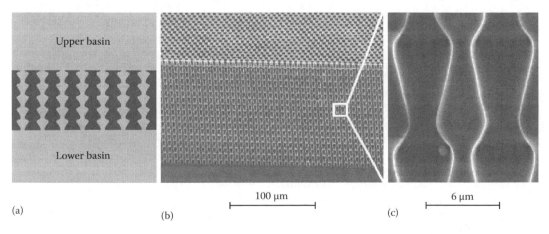

FIGURE 15.13   (a) Experimental setup and (b and c) SEM pictures of cleaved modulated macroporous silicon ratchet membrane for investigation of Brownian motion. (Reprinted by permission from Macmillan Publishers Ltd., *Nature*, Matthias S. and Müller F., **424**, 53, 2003, copyright 2003.)

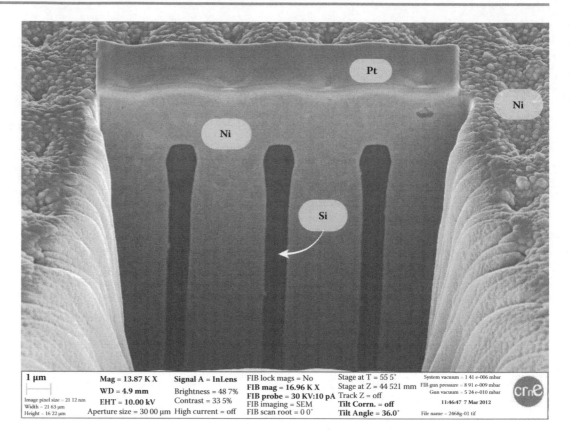

FIGURE 15.14   SEM micrograph of a macroporous silicon membrane filled with nickel for high-density capacitor devices. (With kind permission from Springer Science+Business Media: *Nanoscale Res. Lett.* 9, 2014, 473, Vega D. et al.)

membranes as support layer (Tantawi et al. 2013), or for cell culture and tissue engineering applications (McInnes and Voelcker 2014).

Development of miniaturized fuel cells is a topic of great interest where nanostructured membranes have been explored in detail (Hayase et al. 2004; Pichonat et al. 2004; Pichonat and Gauthier-Manuel 2005, 2006; Chu et al. 2006; Pichonat and Gauthier-Manuel 2006; Lee et al. 2008; Moghaddam et al. 2010; Gautier et al. 2012). Furthermore, macroporous silicon membranes were successfully applied for the fabrication of high-capacity energy storage devices (Thakur et al. 2012; Vega et al. 2014a,b) (Figure 15.14). Silicon-based technology allows an easy integration scheme for fuel cells or batteries, and thus provides an inexpensive fabrication process which opens new perspectives especially for the development of portable electronic devices.

Optical applications of suspended out-of-plane (see Figure 15.6) or freestanding porous silicon membranes with integrated tunable optical filters have been reported by Lammel and Renaud (2000, 2001), Lammel et al. (2002), Ghulinyan et al. (2003), and Song et al. (2008).

Further applications of PSi-based membranes are discussed in detail in other chapters of this book.

15.6 CONCLUSION

PSi membrane fabrication is established as a standard process for different purposes. Combinations of electrochemical process and bulk or surface micromachining allow the formation of both macro- and mesoporous silicon membranes over a wide range of thickness, porosity, and pore size with different merits and limitations. The huge internal surface of porous layer can be activated with additional postprocessing steps, such as stabilizations, activations, functionalization, or coatings for different applications. However, electrochemical etching of porous membranes and postprocessing steps lead to modified membrane properties which should be considered in the membrane design. Properties of porous materials can be described by effective medium approximation. Analysis of transport processes in nanostructured membranes is a key issue for proper function of the device. PSi-based membranes have great potential in different micro-nano-bio system applications, especially in filtration and separation technology, fuel cells, fluidic systems, smart sensors, optical sensors, biosensors, and miniaturized LOC and μTAS systems.

REFERENCES

Agrawal A.A., Nehilla B.J., Reisig K.V., Gaborski T.R., Fang D.Z., Striemer C.C., Fauchet P.M., and McGrath J.L. (2010). Porous nanocrystalline silicon membranes as highly permeable and molecularly thin substrates for cell culture. *Biomater.* **31**, 5408–5417.

Aksimentiev A., Brunner R.K., Cruz-Chu E., Comer J., and Schulten K. (2009). Modeling transport through synthetic nanopores. *IEEE Nanotechnol. Mag.* **3**, 20–28.

Barrett E.P., Joyner L.G., and Halenda P.H. (1951). The determination of pore volume and area distribution in porous substances. I. Computations from nitrogen isotherms. *J. Am. Chem. Soc.* **73**, 373–380.

Bassu M., Scheen G., and Francis L.A. (2012). Thick macroporous silicon membranes: Influence of the masking layer on the underetching characteristics. *Sens. Actuators A* **185**, 66–72.

Beerdsen E. and Smit B. (2007). Understanding diffusion in nanoporous materials. In: *From Zeolites to Porous MOF Materials—Proceeding of the 40th Anniversary of International Zeolite Conference*, August 12–17, 2007, Beijing, P. R. China, pp. 1646–1651.

Bhatia S.K. and Nicholson D. (2003). Molecular transport in nanopores. *J. Chem. Phys.* **119**, 1719–1730.

Bhatia S.K., Jepps O., and Nicholson D. (2004). Tractable molecular theory of transport of Lennard-Jones fluids in nanopores. *J. Chem. Phys.* **120**, 4472–4485.

Brunauer S., Emmett P.H., and Teller E. (1938). Adsorption of gases in multimolecular layers. *J. Am. Chem. Soc.* **60**, 309–319.

Cablares-Navarro F.A., Gómez-Ballesteros J.L., and Balbuena P.B. (2013). Molecular dynamics simulations of metal-organic frameworks as membranes for gas mixtures separation. *J. Membr. Sci.* **428**, 241–250.

Canham L. (ed.) (1997). *Properties of Porous Silicon.* INSPEC, London.

Cavalaglio S., Robert L., and Gauthier-Manuel B. (2011). Accordion structured porous silicon membranes: Application to contamination detection of organic volatile by FTIR spectroscopy. *J. Membr. Sci.* **383**, 250–253.

Charrier J., Alaiwan V., Pirasteh P., Najar A., and Gadonna M. (2007). Influence of experimental parameters on physical properties of porous silicon and oxidized porous silicon layers. *Appl. Surf. Sci.* **253**, 8632–8636.

Chmelik C., Heinke L., Valiullin, and Kärger J. (2010). A new view of diffusion in nanoporous materials. *Chem. Ing. Tech.* **82**(6), 779–804.

Cho S.-Y., Lee K.-W., Kim J.-W., and Kim D.-H. (2013). Rugate-structured free-standing porous silicon-based fiber-optic sensor for the simultaneous detection of pressure and organic gases. *Sens. Actuators B* **183**, 428–433.

Chu K.-L, Shannon M.A., and Masel R.I. (2006). Porous silicon fuel cells for micro power generation. In: Proceedings of the 6th International Workshop on Micro and Nanotechnology for Power Generation and Energy Conversion Applications, Nov. 20–Dec. 1, 2006, Berkeley, pp. 255–258.

Civan F. (2011). *Porous Media Transport Phenomena.* John Wiley & Sons, Hoboken, NJ.

Clarkson J.P., Fauchet P.M., Rajalingam V., and Hirschman K.D. (2007). Solvent detection and water monitoring with a macroporous silicon field-effect sensor. *IEEE Sen. J.* **7**(3), 329–335.

Conner Wm.C. and Fraissard J. (eds.) (2003). *Fluid Transport in Nanoporous Materials.* Springer, La Colle sur Loup.

Cruz S., Hönig-d′Orville A., and Müller J. (2005). Fabrication and optimization of porous silicon substrates for diffusion membrane applications. *J. Electrochem. Soc.* **152**(6), C418–C424.

Cytrynowicz D., Hamdan M., Medis P., Shuja A., Henderson H. T., Gerner F. M., and Golliher E. (2002). MEMS loop heat pipe based on coherent porous silicon technology. In: *AIP Conference Proceedings*, Feb. 3–6, 2002, Albuquerque, NM, pp. 220–232.

De Stefano L., Rea I., Giardina P., Armenante A., and Rendina I. (2008). Protein modified porous silicon nanostructures. *Adv. Mater.* **20**, 1529–1533.

Dubosc F., Bourrier D., and Leichlé T. (2012). Fabrication of lateral porous silicon membranes for planar microfluidic devices. *Procedia Eng.* **47**, 801–804.

Dvoyashkin M., Khokhlov A., Valiullin R., Kärger J., and Thommes M. (2009). Fluid behavior in porous silicon channels with complex pore structure. *Diffusion-Fundamentals* **11**, 1–2.

Ebro H., Kim Y.M., and Kim J.H. (2013). Molecular dynamics simulations in membrane-based water treatment processes: A systematic overview. *J. Membr. Sci.* **438**, 112–125.

Föll H., Christophersen M., Carstensen J., and Hasse G. (2002). Formation and application of porous silicon. *Mater. Sci. Eng. R* **280**, 1–49.

Frentrup H., Avendano C., Horsch M., Salin A., and Müller E.A. (2012). Transport diffusivities of fluids in nanopores by non-equilibrium molecular dynamics simulation. *Mol. Simul.* **38**(7), 540–553.

Gauthier-Manuel B. and Pichonat T. (2005). Nanostructured membranes: A new class of photonic conductor for miniature fuel cells. *J. Nanotechnol. Online*, Vol. 1 (July) doi: 10.2240/azojono0102.

Gautier G., Kouassi S., Desplobain S., and Ventura L. (2012). Macroporous silicon hydrogen diffusion layers for micro-fuel cells: From planar to 3D structures. *Microelelectron. Eng.* **90**, 79–82.

Ghulinyan M., Oton C. J., Bonetti G., Gaburro Z., and Pavesi L. (2003). Free-standing porous silicon single and multiple optical cavities. *J. Appl. Phys.* **93**, 9427–9429.

Grigoras K., Franssila S., Sikanen T., Kotiaho T., and Kostiainen R. (2005). Fabrication of porous membrane filter from *p*-type silicon. *Phys. Stat. Sol. (a)* **202**(8), 1624–1628.

Hayase M., Saito D., and Hatsuzawa T. (2004). Thin fuel cell with monolithically fabricated Si electrodes. In: Proceeding of the Fourth International Workshop on Micro and Nanotechnology for Power Generation and Energy Conversion Applications, Nov. 28–30, Kyoto, Japan, pp. 150–153.

Heinke L. (2007). Significance of concentration—Dependent intracrystalline diffusion and surface permeation for overall mass transfer. *Diffusion-Fundamentals* **4**, 12.1–12.11.

Higgins S., DeSisto W., and Ruthven D. (2009). Diffusive transport through mesoporous silica membranes. *Microporous Mesoporous Mater.* **117**, 268–277.

Ho C.K. and Webb S.W. (eds.) (2006). *Gas Transport in Porous Media.* Springer, Dordrecht.

Ivanov A., Kovacs A., and Mescheder U. (2011). High quality 3D shapes by silicon anodization. *Phys. Stat. Sol. (a)* **208**, 1383–1388.

Jepps O.G., Bhatia S.K., and Searles D.J. (2003). Wall mediated transport in confined spaces: Exact theory for low density. *Phys. Rev. Lett.* **91**(12), 126102.

Kärger J., Ruthven M., and Theodorou D.N. (2012). *Diffusion in Nanoporous Materials.* Wiley-VCH Verlag & Co. KGaA, Weinheim.

Karniadakis, G., Beskok A., and Aluru N. (2005). *Microflows and nanoflows: Fundamentals and Simulation.* Springer Science+Business Media, Inc., New York.

Knudsen M. (1909). Die Gesetze der molekularströmung und der inneren reibungsströmung der gase durch Röheren. *Ann. Phys. (Leipzig)* **28**, 75–130.

Kovacs A. and Mescheder U. (2012). Transport mechanisms in nanostructured porous silicon layers for sensor and filter applications. *Sens. Actuators B* **175**, 179–185.

Kovács Á. and Kovacs A. (2014). Increase of load bearing capacity of a square-form nanofilter. *Periodica Polytechnica Ser. Mech. Eng.* **58**(2), 83–86.

Kovács A., Kovács Á., Pogány M., and Mescheder U. (2007). Mechanical investigation of perforated and porous membranes for micro- and nanofilter applications. *Sens. Actuators B* **127**, 120–125.

Kovács Á., Kovács A., and Mescheder U. (2008). Estimation of elasticity modulus and fracture strength of thin perforated SiN membranes with finite element simulations. *Comput. Mater. Sci.* **43**, 59–64.

Kovacs A., Meister D., and Mescheder U. (2009). Investigation of humidity adsorption in porous silicon layers. *Phys. Status Solidi* **A 206**(6), 1343–1347.

Kovacs A., Jonnalagadda P., and Mescheder U. (2011). Optoelectrical detection system using porous silicon based optical multilayers. *IEEE Sens. J.* **11**(10), 2413–2420.

Kovacs A., Kronast W., Filbert A., and Mescheder U. (2013). Transport in surface passivated porous silicon membranes. *ECS Trans.* **50**(37), 207–216.

Krishna R. (2014). The Maxwell–Stefan description of mixture diffusion in nanoporous crystalline materials. *Microporous Mesoporous Mater.* **185**, 30–50.

Krishna R. and van Baten J.M. (2009). Unified Maxwell-Stefan description of binary mixture diffusion in micro- and meso-porous materials. *Chem. Eng. Sci.* **64**(13), 3159–3178.

Krishna R. and van Baten J.M. (2011a). Investigation the validity of the Knudsen prescription for diffusivities in a mesoporous covalent organic framework. *Ind. Eng. Chem. Res.* **50**, 7083–7087.

Krishna R. and van Baten J.M. (2011b). A molecular dynamics investigation of the unusual concentration dependencies of Fick diffusivities in silica mesopores. *Microporous Macroporous Mater.* **138**, 228–234.

Krüger M., Arens-Fischer R., Thönissen M. et al. (1996). Formation of porous silicon on patterned substrates. *Thin Solid Films* **276**, 257–260.

Lammel G. and Renaud Ph. (2000). Free-standing, mobile 3D porous silicon microstructures. *Sens. Actuators A* **85**, 356–360.

Lammel G. and Renaud Ph. (2001). Microspectrometer based on a tunable optical filter of porous silicon. *Sens. Actuators* **92**, 52–59.

Lammel G., Schweizer S., Schiesser S., and Renaud P. (2002). Tunable optical filter of porous silicon as key component for a MEMS spectrometer. *J. Microelectromech. Sys.* **11**, 815–827.

Langner A., Müller F., and Gösele U. (2011). Macroporous silicon. In: Hayden O. and Nielsch K. (eds.), *Molecular- and Nano-Tubes.* Springer, New York, pp. 431–460.

Lee Y. and Lee K. (2014). Sensing of dissolved-gas concentration in water using a rugate-structured porous silicon membrane. *Sens. Actuators B* **202**, 417–425.

Lee C.-Y., Lee S.-J., Dai C.-L., and Chuang C.-W. (2008). Application of porous silicon on the gas diffusion layer of micro fuel cells. *Key Eng. Mater.* **364–366**, 849–854.

Lehmann V., and Grüning U. (1997). The limits of macropore array fabrication. *Thin Solid Films* **297**, 13–17.

Létant S.E., Hart B.R., Van Buuren A.W., and Terminello L.J. (2003). Functionalized silicon membranes for selective bio-organism capture. *Nature Mater.* **2**, 391–395.

Lu D., Aksimentiev A., Shih A.Y., Cruz-Chu E., Freddolino P.L., Arkhipov A., and Schulten K. (2006). The role of molecular modeling in bionanotechnology. *Phys. Biol.* **3**, S40–S53.

Lysenko V., Vitiello J., Remaki B., and Barbier D. (2004). Gas permeability of porous silicon nanostructures. *Phys. Rev.* **70**, 017301.

Malek K. and Coppens M.-O. (2003). Knudsen self- and Fickian diffusion in rough nanoporous media. *J. Chem. Phys.* **119**(5), 2801–2811.

Mason E.A. and Malinauskas A.P. (1983). *Gas Transport in Porous Media: The Dusty Gas Model.* Elsevier, Amsterdam.

Mathwig K., Geilhufe M., Müller F., and Gösele U. (2011a). Bias-assisted KOH etching of macroporous silicon membranes. *J. Micromech. Microeng.* **21**, 035015.

Mathwig K., Müller F., and Gösele U. (2011b). Particle transport in asymmetrically modulated pores. *New J. Phys.* **13**, 033038.

Matthias S. and Müller F. (2003). Asymmetric pores in a silicon membrane acting as massively parallel Brownian ratchets. *Nature* **424**, 53–57.

McInnes S.J.P. and Voelcker N.H. (2014). Porous silicon–polymer composites for cell culture and tissue engineering applications. In: Santos H.A. (ed.), *Porous Silicon for Biomedical Applications.* Woodhead, Cambridge, UK, pp. 420–469.

Mezedur M. M., Kaviany M., and Moore W. (2002). Effect of pore structure, randomness and size on effective mass diffusivity. *AIChE J.* **48**(1), 15–24.

Moghaddam S., Pengwang E., Jiang Y.-B., Garcia A.R., Burnett D.J., Brinker C.J., Masel R.I., and Shannon M.A. (2010). An inorganic–organic proton exchange membrane for fuel cells with a controlled nanoscale pore structure. *Nature Nanotech.* **5**, 230–236.

Müller F., Birner A., Schilling J., Gösele U., Kettner Ch., and Hänggi P. (2000). Membranes for micropumps from macroporous silicon. *Phys. Stat. Sol. (a)* **182**, 585–590.

Naserifar S., Tsotsis T.T., Goddard W.A., and Sahimi M. (2015). Toward a process-based molecular model of SiC membranes: III. Prediction of transport and separation of binary gaseous mixtures based on the atomistic reactive force field. *J. Membr. Sci.* **437**, 85–93.

Neale G.H. and Nader W.K. (1973). Prediction of transport processes within porous media: Diffusive flow processes within homogeneous swarms of spherical particles. *AIChE J.* **19**, 112–119.

Pagonis D.N. and Nassiopoulou A.G. (2006). Free-standing macroporous silicon membranes over a large cavity for filtering and lab-on-chip applications. *Microelectron. Eng.* **83**, 1421–1425.

Papadimitriou D., Tsamis C., and Nassiopoulou (2004). The influence of thermal treatment on the stress characteristics of suspended porous silicon membranes on silicon. *Sens. Actuators B* **130**, 356–361.

Pichonat T. and Gauthier-Manuel B. (2005). Development of porous silicon-based miniature fuel cells. *J. Micromech. Microeng.* **15**, S179–S184.

Pichonat T. and Gauthier-Manuel B. (2006). A new process for the manufacturing of reproducible mesoporous silicon membranes. *J. Membr. Sci.* **280**, 494–500.

Pichonat T., Gauthier-Manuel B., and Hauden D. (2004). A new protonconducting porous silicon membrane for small fuel cells. *Chem. Eng. J.* **101**, 107–111.

Roque-Malherbe R.M.A. (2007). *Diffusion in Nanoporous Materials.* Taylor & Francis Group, Boca Raton, FL.

Ruthven D.M. (2005). The technological impact of diffusion in nanopores. *Diffusion-Fundamentals* **2**, 77.1–77.23.

Ruthven D.M. (2007). Adsorption and desorption kinetics for diffusion controlled systems with a strongly concentration dependent diffusivity. *Diffusion-Fundamentals* **6**, 51.1–51.11.

Ruthven D.M., Brandani S., and Eic M. (2008). Measurement of diffusion in microporous solids by macroscopic methods. *Adsorption and Diffusion*, Mol Sieves 7. Springer-Verlag, Berlin Heidelberg, pp. 45–84.

Ruthven D.M., DeSisto W.J., and Higgins S. (2009). Diffusion in a mesoporous silica membrane: Validity of the Knudsen diffusion model. *Chem. Eng. Sci.* **64**, 3201–3203.

Sagnes I., Halimaoui A., Vincent G., and Badoz P.A. (1993). Optical absorption evidence of a quantum size effect in porous silicon. *Appl. Phys. Lett.* **62**(10), 1155–1157.

Saharil F., Gylfason K.B., Liu Y., Haraldsson T., Bettotti P., Kumar N., and van der Wijngaart W. (2012). Dry transfer bonding of porous silicon membranes to oste (+) polymer microfluidic devices. In: *MEMS 2012*, 29 Jan.–2 Feb, 2012, Paris, France, pp. 232–234.

Searson P.C. (1991). Porous silicon membranes. *Appl. Phys. Lett.* **59**(7), 832–833.

Shearer C., Velleman L., Acosta F., Ellis A., Voelcker N., Mattia D., and Shapter J. (2010). Water transport through nanoporous materials: Porous silicon and single walled carbon nanotubes. In: *Proceedings of Nanoscience and Nanotechnology (ICONN)*, Feb. 22–26, 2010, Sydney, NSW, Australia, pp. 196–199.

Smoluchowski von M.V. (1910). Zur kinetischen theorie der transpiration und diffusion verdünnter gase. *Ann. Phys. (Leipzig)* **33**, 1559–1570.

Snyder J.L., Clark Jr. A., Fang D.Z., Gaborski T.R., Striemer C.C., Fauchet P.M., and McGrath J.L. (2011). An experimental and theoretical analysis of molecular separations by diffusion through ultrathin nanoporous membranes. *J. Membr. Sci.* **369**, 119–129.

Solanki C.S., Bilyalov R.R., Poortmans J., Celis J.-P., Nijs J., and Mertens R. (2004). Self-standing porous silicon films by one-step anodizing. *J. Electrochem. Soc.* **151**(5), C307–C314.

Song D., Tokranova N., Gracias A., and Castracane J. (2008). Integrating porous silicon with microsystems for chip-to-chip communications. *SPIE Newsroom* **01/2008**, 1–3.

Splinter A., Stürmann J., Bartels O., and Benecke W. (2002). Micro membrane reactor: A flow-through membrane for gas pre-combustion. *Sens. Actuators B* **83**, 169–174.

Striemer C.C., Gaborski T.R., McGrath J.L., and Fauchet P.M. (2007). Charge- and size-based separation of macromolecules using ultrathin silicon membranes. *Nature* **445**, 749–753.

Taliercio T., Dilhanb M., Massone E., Gu A.M., Fraisse B., and Foucaran A. (1995). Realization of porous silicon membranes for gas sensor applications. *Thin Solid Films* **255**, 310–312.

Tantawi K.H., Berdiev B., Cerro R., and Williams J.D. (2013). Porous silicon membrane for investigation of transmembrane proteins. *Superlattices Microstruct.* **58**, 72–80.

Thakur M., Pernites R.B., Nitta N., Isaacson M., Sinsabaugh S.L., Wong M.S., and Biswal S.L. (2012). Freestanding macroporous silicon and pyrolyzed polyacrylonitrile as a composite anode for lithium ion batteries. *Chem. Mater.* **24**, 2998–3003.

Tjerkstra R.W., Gardeniers J.G.E., Kelly J.J., and van der Berg A. (2000). Multi-walled microchannels: Freestanding porous silicon membranes for use in μTAS. *J. Microelectromech. Syst.* **9**, 495–501.

Turner D.R. (1958). Electropolishing silicon in hydrofluoric acid solutions. *J. Electrochem. Soc.* 105(7), 402–408.

Valiullin R. (2013). Diffusion in nanoporous host systems. *Ann. Rep. NMR Spectr.* **79**, 23–72.

Van Rijn C.J.M. (2004). *Nano and Micro Engineered Membrane Technology*. Elsevier, Amsterdam.

Vega D., Reina J., Martí F., Pavón R., and. Rodríguez Á. (2014a). Macroporous silicon for high-capacitance devices using metal electrodes. *Nanoscale Res. Lett.* **9**, 473.

Vega D., Reina J., Pavón R., and Rodríguez A. (2014b). High-density capacitor devices based on macroporous silicon and metal electroplating. *IEEE Trans. Electron. Dev.* **61**(1), 116–122.

Velleman L., Shearer C.J., Ellis A.V., Losic D., Voelcker N.H., and Shapter J.G. (2010). Fabrication of self-supporting porous silicon membranes and tuning transport properties by surface functionalization. *Nanoscale* **2**, 1756–1761.

Wallner J.Z. and Bergstrom P.L. (2007). A porous silicon based particle filter for microsystems. *Phys. Stat. Sol. (a)* **204**(5), 1469–1473.

Wallner J.Z., Kunt K.S., Obanionwu H., Oborny M.C., Bergstrom P.L., and Zellers E.T. (2007). An integrated vapor source with a porous silicon wick. *Phys. Stat. Sol. (a)* **204**(5), 1449–1453.

Webb S.W. (1998). Gas-phase diffusion in porous media: Comparison of models. In: *Proceedings of the TOUGH '98 Workshop*, May 4–6, 1998, Berkeley, CA, pp. 269–274.

Yao S., Myers A.M., Posner J.D., Rose K.A., and Santiago J.G. (2006). Electroosmotic pumps fabricated from porous silicon membranes. *J. Microelectromech. Syst.* **15**(3), 717–728.

Zheng J., Christopherensen M., and Bergstrom P.L. (2005a). Thick macroporous membranes made of p-type silicon. *Phys. Stat. Sol. (a)* **202**, 1402–1406.

Zheng J., Christopherensen M., and Bergstrom P.L. (2005b). Formation technique for macroporous morphology superlattice. *Phys. Stat. Sol. (a)* **202**, 1662–2667.

Liquid Microfluidic Devices

16

Ghenadii Korotcenkov and Beongki Cho

CONTENTS

16.1   Microfluidic Nozzles   300

16.2   Miniaturized Total Analysis Systems (µTAS): Liquid Chromatography on Chip   301

    16.2.1   Liquid Chromatography: General View   301

    16.2.2   Chromatographic Separation Columns   302

    16.2.3   Micropumps   307

16.3   Separation of Particles of Different Size   310

16.4   Filtering   312

References   314

16.1 MICROFLUIDIC NOZZLES

Microfluidics generally refers to the science and technology of manipulating minute amounts of fluids with volumes ranging from microliters (1 μL = 1 mm^3) to picoliters (1 pL = 10 μm^3). The first liquid microfluidic devices emerged in the 1970s. A microfluidic nozzle, used for ink jet printing, was one of the first fluidic applications that took advantage of silicon micromachining (Bassous et al. 1977). The first nozzles were simply through-wafer anisotropically wet etched holes. Since their initial release, ink jet nozzles have developed substantially and they still are probably the most widespread and most successfully commercialized miniaturized fluidic devices.

However, it should be noted that in recent years there have been great changes in the manufacture of microfluidic nozzles. New approaches to the design, manufacture, and use have been developed (Xie et al. 2005; Sainiemi et al. 2008). In particular, the technology designed for the formation of macroporous silicon become be more widely used, and microfluidic nozzles become used as electrospray ionization (ESI) chips for mass spectrometric analysis. For example, Sainiemi et al. (2008) designed a micropillar array electrospray ionization (μPESI) chip, where the liquid sample is transported through a 1-mm wide lidless flow channel by capillary forces facilitated by micropillars (see Figure 16.1).

Sainiemi et al. (2008) established that the micropillar array inside the channel has an essential role in the sample transport. Without the pillar array, wide lidless channels cannot be filled without external pumping. Therefore, in designed devices, no external pumping is required and the only high-voltage source needed is the one necessitated by electrospray ionization and mass spectroscopy. The whole chip was made of silicon, which allows the fabrication of high aspect ratio micropillars inside the channel. The μPESI chips were fabricated on 300-μm thick (100) silicon wafers that had resistivity of 1–50 Ω·cm. The depth of the channels was varied between 20 and 40 μm. Technology of fabrication was based on the use of deep reactive ion etching (DRIE). All

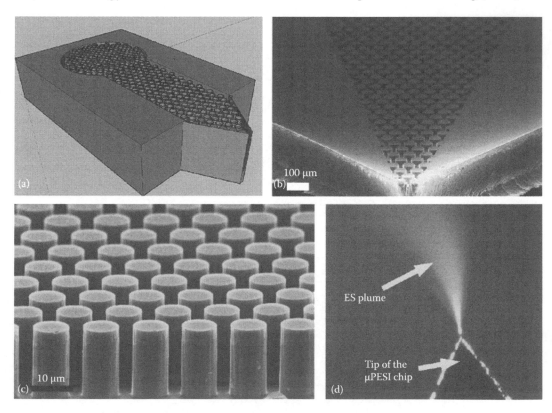

FIGURE 16.1   (a) Schematic view of the μPESI chip without the sharpening process. (b) Front side views of the ESI tip of the μPESI chip fabricated without sharpening process. (c) The cross-section of the flow channel of the μPESI chip (d). Taylor cone and electrospray plume formation. When the sample reaches the tip of the chip, the Taylor cone is formed and the sample is sprayed out. (Reprinted from *Sens. Actuators B* 132, Sainiemi L. et al., 380, Copyright 2008, with permission from Elsevier.)

silicon etchings were done in inductively coupled SF_6/O_2 plasma at cryogenic temperature. Al_2O_3 layers were used as a mask during the through wafer-etching process because of its exceptionally high selectivity in cryogenic DRIE (Sainiemi and Franssila 2007). Silicon micromachining made it possible to accurately define a truly three-dimensional ESI tip, which is in-plane. The channels were transformed to more hydrophilic using short oxygen plasma treatment or Piranha treatment (Suni et al. 2008).

16.2 MINIATURIZED TOTAL ANALYSIS SYSTEMS (μTAS): LIQUID CHROMATOGRAPHY ON CHIP

A new boost was given to the development of miniaturized fluidic systems capable of chemical analysis in the early 1990s, when Manz et al. (1990) introduced the concept of miniaturized total analysis systems (μTAS); the miniaturized microfluidic system that can automatically carry out all the necessary functions to transform chemical information into electronic information. The advantages of μTAS include significantly reduced size (portability), power consumption, sample and reagent consumption, and manufacture and operating costs (disposability). In addition, μTAS can achieve better performance in terms of speed, throughput, mass sensitivity, and automation. Since then, the goal of many research groups has been to integrate several functions such as sample concentration, separation, and analysis on a single fluidic chip (Erickson and Li 2004; Stachowiak et al. 2004; Xie et al. 2005). Liquid chromatography (LC) on chip is one such device where porous silicon (PSi) can be used. One should note that LC is one of the most powerful separation techniques, which has been widely used in research, environmental monitoring, and industry. Experiment has shown that the analytical method, using separation mechanisms based on the interaction of solutes with a stationary phase, is a powerful and robust technique applicable for analysis of a wide range of compounds. It was found that in liquid microchromatographs, the technology of PSi forming can be applied for fabrication of chromatographic separation columns (Clicq et al. 2004; De Malsche et al. 2007, 2008; Méry et al. 2009; Tiggelaar et al. 2009; Sukas et al. 2013), micropumps (Muller et al. 2000; Laser et al. 2003; Yao et al. 2006; Zheng 2006; Wallner et al. 2007; Vajandar et al. 2009; Vanga et al. 2011; Xu et al. 2011; Snyder et al. 2013), for separation of particles of different sizes (Kettner et al. 2000; Matthias and Muller 2003), filtering (Kuiper et al. 1998; Striemer et al. 2007; Campbell et al. 2008; Snyder et al. 2008; Eun et al. 2009; Dubosc et al. 2012), and so on.

16.2.1 LIQUID CHROMATOGRAPHY: GENERAL VIEW

Based on the definition given in Neue (1997), chromatography in general includes all separation techniques in which analytes partition between different phases that move relative to each other, or where analytes have different migration velocities. In most chromatographic techniques, one phase is stationary, while the other phase is mobile. When this mobile phase is gas, the chromatographic technique is called gas chromatography (GC). When this mobile phase is liquid, the chromatographic technique is called liquid chromatography (LC). The basic principle of the LC is to generate ions from either inorganic or organic compounds by any suitable method, to separate these ions, and to detect them qualitatively and quantitatively based on their respective properties such as m/z, color, fluorescence, absorbance, and so on. LC mainly utilizes a column that holds packing material (stationary phase) and a detector that shows the retention times of the molecules. In column chromatography a liquid solvent, the mobile phase, flows through the column continuously to carry the sample from the top to the bottom of the column. The LC stationary phase can be either solid or surface functional groups bonded on a solid. More sophisticated high-performance liquid chromatography instruments (HPLC) and LC on chip use a pump to force the mobile phase through the column, which provides faster analysis time. In short, during the elution process, the surface functional groups (stationary phase) on the particles interact with the sample and eluent (mobile phase). When the injected sample plug is carried through the column by the liquid eluent, different sample components interact with the stationary phase differently, partitioning their time in the stationary and mobile phases, and therefore migrate through the column at different speeds, and exit

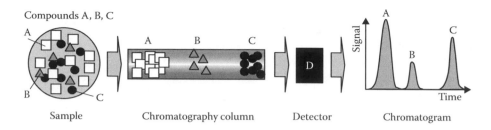

FIGURE 16.2   Illustration of the chromatographic separation process.

the column at different times. The partition coefficient of each component determines how much of it is in each phase at any time and how long it remains in that phase. After the sample traverses the column, a flow-through detector is employed to generate the chromatogram, which is the visual representation of the separation. Detectors can provide both quantitative and qualitative information about the separated components. This process is illustrated in Figure 16.2.

Depending on the separation mechanisms, LC is subdivided into normal phase chromatography, reversed phase chromatography, flash chromatography, size exclusion chromatography, partition chromatography, ion exchange chromatography, affinity chromatography, and chiral chromatography. For example, normal phase LC (NP-LC) separates analytes based on polarity. NP-LC uses a polar stationary phase and a nonpolar mobile phase. The polar analyte interacted with and is retained by the polar stationary phase. Adsorption strengths increase with increased analyte polarity, and the interaction between the polar analyte and the polar stationary phase increases the elution time. Almost any polar solid can be employed as a polar stationary phase. The choice of stationary phase is governed by the polarity of the feed components. If the feed components are adsorbed too strongly, they may be difficult to remove. Weakly polar mixtures should be separated on highly active absorbents, or little or no separation will occur. The choice of mobile phase is equally important. The polarity of the mobile phase should be chosen to compliment the choice of stationary phase. In general, good separation is achieved by using fairly polar stationary phases and low polarity mobile phases such as hexane. Water, it should be noted, is a very polar solvent. In reverse phase chromatography, the stationary phase is nonpolar (usually a hydrocarbon) and the mobile phase is polar such as water, methanol, acetonitrile, and so on. Here, most polar components appear first and increasing polarity increases the elution time. A detailed description of these chromatography types can be found in Snyder et al. (2009) and Lundanes et al. (2013).

Similarly to other miniaturized total analysis systems (μTAS), LC has great benefits from the miniaturization and design of chip-based devices (Stachowiak et al. 2004; Xie et al. 2005; Faure 2010; Kutter 2012). Integrated LC chips are valued because of their significant advantages over conventional systems. When downsizing conventional separation systems to the micron scale, many improvements are observed in terms of separation performances, instrument versatility, and effective costs. In most techniques, analysis time is reduced and efficiencies are improved. The integration of injectors minimizes the injected volumes to picoliters, which benefits both sample consumption and band broadening. The zero dead-volume interconnections to the detection systems contribute as well to this gain in efficiency. Moreover, microfabrication techniques allow the integration of multiple operations (sample preparation, microreactions) and the creation of multiple or parallel analytical devices, reducing the cost of the whole analytical chain. Mass production of these miniaturized devices and the extended use of thermoplastics participate in the reduction of costs and therefore encourage the development of disposable analytical microsystems, providing a suitable platform for samples that generate cross-contamination in conventional instruments. However, it is necessary to take into account that chip-based liquid chromatographs are very challenging to build due to the high complexity of LC systems and the need for high-level integration of many discrete microfluidic devices.

16.2.2 CHROMATOGRAPHIC SEPARATION COLUMNS

The main restriction of conventional open-channel LC is the limited specific surface in separation columns. For solving this problem, there are several approaches described (Neue 1997; He et al. 1998; Regnier 2000; Detobel et al. 2010; Faure 2010). As applied to PSi, these approaches are based on (1) the formation of PSi on the walls of the channel, and (2) the use of technologies peculiar for the formation of the macroporous silicon for creating pillars within the channel.

The first approach was designed by Clicq et al. (2004) and Méry et al. (2009). In particular, Clicq et al. (2004) reported on the possibility to strongly increase the mass loadability and retention capacity of shear-driven chromatography (SDC) channels by growing a thin PSi layer on the stationary wall part. As a rule, the fabrication (enabling) of PSi or silica inside the channels of microfluidic systems is mostly performed as a poststep after chip fabrication (a posttreatment to the inner walls of microchannels). For example, in (Clicq et al. 2004), a porous layer in the channels was formed using standard anodization process in a 5% (v/v) HF solution. The layers had a depth varying between 20 and 300 nm. In the final step, the PSi layers on the stationary channel plates were derivatized by putting them into a 30% (w/w) solution of dimethyloctylchlorosilane in toluene for 48 h at room temperature. During the treatment, the silicon plates were put in a PTFE holder and the coating solution was stirred to maximize the large-scale homogeneity of the coating. Afterward, the silicon plates were sequentially washed with toluene, methanol, and water. Clicq et al. (2004) have shown that using porous layers in combination with sub-micrometer thin flow-through channels, very large phase ratios, that is, much larger than what is commonly the case in open-tubular LC and the reversed-phase capillary electrochromatography (CEC) systems, could be obtained. At the same time, Clicq et al. (2004) have noted that a number of practical problems still need to be resolved (the presence of peak disturbing defects in the layers, poor mechanical resistance to cleaning) before it will be possible to use the PSi layer coated SDC channels on a routine basis.

As applied to reversed phase micro-column for conventional LC, the approach mentioned above was implemented by Méry et al. (2009). PSi samples were prepared by anodic etching of p^+ boron doped ($\rho = 10$ mΩ·cm) double-side polished silicon (100) wafers. Silicon samples were etched in a solution containing 1:1 volume mixture of HF (49%) and ethanol applying 150 mA/cm^2 anodic current density. Porous channels fabrication steps are described in Figure 16.3.

Méry et al. (2009) established that when using PSi, the analytical capabilities of reversed phase chromatography are improved. However, it was found that the use of PSi leads to peak broadening (see Figure 16.4). Méry et al. (2009) believe that the peak broadening is not predominantly

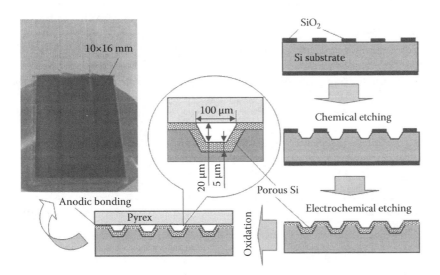

FIGURE 16.3 Porous channels fabrication steps: After photoresist strip, an anisotropic silicon etching in potassium hydroxide solution (KOH, 40% wt), defining the meander channel dimensions (depth = 20 μm, width = 100 μm, and length = 150 mm), was achieved. After the oxide removal, the wafer was electrochemically etched. A 5-μm thick PSi layer with a porosity of about 65% was obtained. The device was diced, and after a dry oxidation in oxygen at 300°C for 1 h, it was closed with Pyrex by anodic bonding (1000 V, 330°C, 15 min). Finally, the quartz capillaries were glued and silanisation was performed followed by successive toluene and methanol washing. (Méry E. et al.: *Phys. Stat. Sol.* (c) 6(7). 2009. 1777. Copyright Wiley-VCH Verlag GmbH & Co. KGaA. Reproduced with permission.)

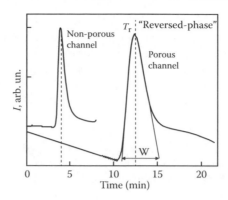

FIGURE 16.4   Elution profiles of dimethylphthalate, 20 nL, 5 × 10⁻³ mol/L, u_v = 60 nL/min. Mobile phase: pure methanol. (Méry E. et al.: *Phys. Stat. Sol.* (c) 6(7). 2009. 1777. Copyright Wiley-VCH Verlag GmbH & Co. KGaA. Reproduced with permission.)

due to the effect of PSi stationary phase. Broadening could be essentially explained by external dispersion (void volumes, capillary diameter, injected and detector cell volumes, etc.). In addition, the peak eluted from porous channels appeared much more tailed than from nonporous ones (Figure 16.4), and its asymmetry increased with decrease of the flow rate. According to Méry et al. (2009), there are two possible reasons of the peak tailing: (1) nonlinearity of dimethylphtalate adsorption isotherm in the concentration range used in experiments (overload phenomenon) and (2) penetration and retention of the analyte in the PSi deposited on top of the Si plate in interchannel space. This means that an optimization of LC chip and peripheral equipment is required for further increase of its efficiency.

Regarding the second approach, discussed in details in He et al. (1998), Regnier (2000), and De Pra et al. (2006), one can say that the microfabrication of monolith pillars is an elegant strategy, taking the advantages of microfabrication technology to combine small channel dimensions and large interacting surfaces. Microfabricated columns, which are extremely difficult to create on the macroscale, present many advantages from a chromatographic point of view. The microfabrication technology allows close-to-perfect, column-to-column reproducibility, large freedom in pillar shape, full control of pillar density, channel dimensions and geometry, and the ability to control the extent of mixing in the column. Variants of pillar shapes used in columns are shown in Figure 16.5. Analysis of these structures can be found in He et al. (1998). This means that microfabricated structures are inherently more regular than randomly packed particles. This difference has a direct influence on eddy diffusion. As a rule, such pillars are fabricated using DRIE. Theoretically, properly dimensioned pillars should hence result in higher efficiency (De Smet et al. 2004; Gzil et al. 2004; De Pra et al. 2006). The incorporation of pillars in columns leads to high efficiencies due to the reduction or even a complete elimination of the eddy-diffusion. In particular, De Smet et al. (2004, 2005) reported that the pressure drop for channels structured with pillars is strongly reduced in comparison with packed columns, when systems with the same porosity are considered. The possibility to decrease the flow resistance largely is very appealing for chromatography, as the pressure is often the limiting factor for the number of theoretical plates that can be obtained (Poppe 1997). Pressure becomes even more critical when chromatography is performed on a chip.

In addition, according to Peterson (2005), introducing structures into a microfluidic chip also has the advantages of allowing multiple functionalities to be incorporated into a single channel more easily. Introducing solid support structures into a chip makes incorporating multiple functionalities possible because each support can have a different surface chemistry and thus a different function. Fabricating multiple functionalities into a chip is important in producing more complex systems that incorporate multiple analytical operations into a single device. For example, these structures can be used as valves and mixers. Microfluidic streams are normally very laminar and species within fluids are not particularly mobile on these scales, which makes mixing in microfluidics a difficult problem. A microfabricated mixer is shown in Figure 16.6. The

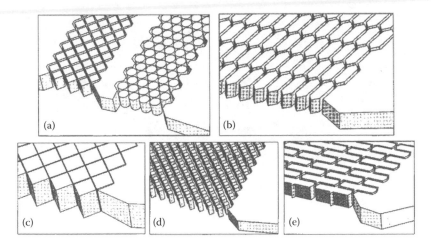

FIGURE 16.5   Illustrations of COMOSS structures that could both be micromachined *in situ* and would potentially be suitable for LC: (a) cylindrical and cubic monolith structures, (b) hexagonal monolith structure, (c) large "diamond"-shaped monolith structure, (d) small "diamond"-shaped monolith structure, and (e) horizontal "post"-monolith structure. Theory dictates that, in the ideal support, diffusive transport distances should be short, convective transport channels should be homogeneous, and there should be some degree of interchannel mixing. Any system that meets these criteria is acceptable. (Reprinted with permission from He B. et al., *Anal. Chem.* 70, 3790. Copyright 1998 American Chemical Society.)

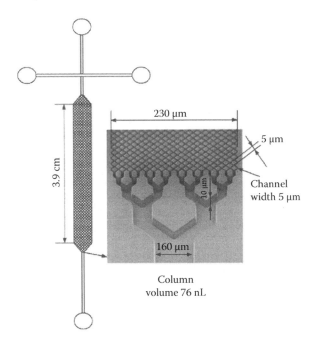

FIGURE 16.6   SEM images of microfabricated devices in chips, micro-LC column, and microfabricated mixer. (Reprinted from *J. Chromatogr. A* 948, Slentz B.E. et al., 225, Copyright 2002, with permission from Elsevier.)

geometry of these channels can be controlled to increase the mixing rate while at the same time reduce the sample consumed.

On the other hand, the use of monolithic pillars generates low phase ratio (in comparison with fully filled channels) and therefore low sample capacity, limiting the analysis of samples with a wide range of concentrations. This remains the main limitation of native material as stationary phase. Pushing pillar strategy to a further extent, De Malsche et al. (2007, 2008), Tiggelaar et al. (2009), and Sukas et al. (2013) recently proposed to form a porous layer on the surface of the

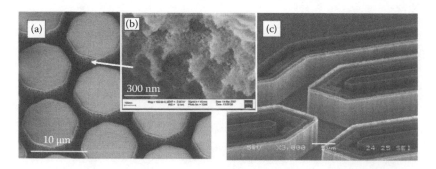

FIGURE 16.7   SEM images showing porous layers generated at the surface of various (a, c) PSi pillars. The core is non-porous, the outer rings are PSi. (b) The SEM image of the porous layer on the wall of Si pillar. ([a, b] Reprinted with permission from De Malsche W. et al., *Anal. Chem.* 80, 5391. Copyright 2007 American Chemical Society. [c] From Sukas S. et al., *Lab on a Chip* 13, 3061, 2013. Reproduced by permission of The Royal Society of Chemistry.)

silicon pillar (see Figure 16.7). They believed that via the use of porous pillars, the separation in columns could be optimized, for example, regarding the suppression of the sidewall effect. Being filled with pillars that are nonporous, only the outer surface is available for chromatographic exchange. Therefore, the phase ratio in these columns is orders of magnitude smaller than in the packed bed HPLC columns filled with porous silica beads that are currently used with great satisfaction in any analytical chemistry lab. The availability of a large exchange surface is an absolute prerequisite for the analysis of samples containing components that are present in a broad range of concentrations. In De Malsche et al. (2007), a 550-nm porous layer was generated by anodization with mesopores ranging between 5 and 15 nm, as shown in Figure 16.7. The process proposed for fabrication of microchannels with porous pillars is shown in Figure 16.8. As previously indicated, the value of the PSi layer thickness (550 nm) was used as a compromise between a sufficiently high retention capability on one hand and a limited additional band broadening due to the slow diffusion of analytes in this zone as compared to the fast diffusion outside the pillar on the other hand. Another reason to only anodize the outer pillar layer was that completely anodized pillars were much more fragile and displayed a high tendency to break during handling. De Malsche et al. (2007) established that this porous layer had no major influence on efficiency, but the authors calculated a 200-fold increase in the exchange surface, which could provide an

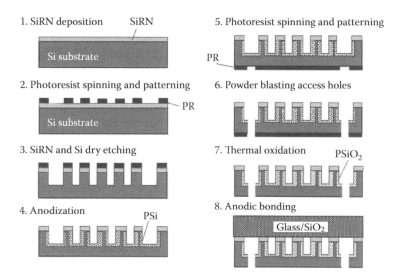

FIGURE 16.8   Fabrication process flow for microchip fabrication with integrated porous glass layers. Highly doped *p*-type Si(100) wafers (ρ = 0.010–0.025 Ω·cm) were used. 10-µm deep channels were etched with RIE or DRIE. PSi was obtained by anodic etching of silicon in hydrofluoric acid solution (HF:ethanol = 5:1) at room temperature. (Adapted from Sukas S. et al., *Lab on a Chip* 13, 3061, 2013. Reproduced by permission of The Royal Society of Chemistry.)

answer to limited sample capacity. Despite the results achieved, De Malsche et al. (2008) stated that a number of major hurdles still need to be crossed. Good solutions for the mechanical stability of the pillars, as well as a method to prevent that the layers are also formed on the bottom layer, need to be developed. In addition, the possibility to fabricate pure silica-based pillars needs to be investigated.

Of course there are alternative methods for porous pillars fabrication. For example, Kutter and co-workers (De Andrade Costa et al. 2005) published a method for the on-chip fabrication of porous glass (PG) by means of oxidizing Si needles (micrograss) obtained by a reactive ion etching method yielding so-called "black silicon," and used it for electrokinetic applications. Although it was possible to fabricate porous channels with this method, it did not provide a regular network of pores and it was not possible to apply this technique to structured microchannels.

16.2.3 MICROPUMPS

The pumping system is an essential instrumental part of any microfluidic system including the chip-LC (Faure 2010). Usually, it consists of micropumps that are connected to the chromatographic chip through a split connection. The description of micropumps acceptable for microfluidic systems is found in reviews that have been published previously (Laser and Santiago 2004; Woias 2005; Tsai and Sue 2007; Zhang et al. 2007; Nisar et al. 2008; Abhari et al. 2012). Their classification is shown in Figure 16.9.

In general, micropumps can be classified as either mechanical (reciprocating) or nonmechanical (dynamic) micropumps (Nisar et al. 2008; Abhari et al. 2012). The micropumps that have moving mechanical parts such as pumping diaphragm and check valves are referred to as mechanical micropumps whereas those involving no mechanical moving parts are referred to as nonmechanical micropumps. A mechanical type micropump needs a physical actuator or mechanism to perform a pumping function. From these pumps, pressure is generated by a periodic compression and expansion of a fluidic volume with the aid of moving surfaces that are interior to the pump. Nonmechanical types of micropumps have to transform certain available nonmechanical energy into kinetic momentum so that the fluid in microchannels can be driven. For applications that require highest pressures, reciprocating are usually preferred, in which the

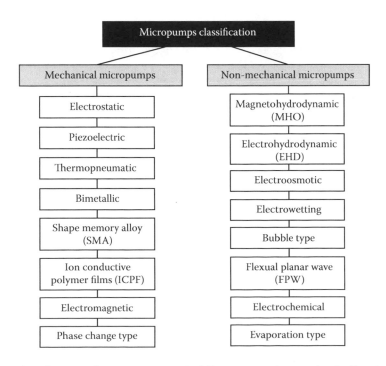

FIGURE 16.9   Classification of micropumps with different actuation methods. (Reprinted from *Sens. Actuators B* 130, Nisar A. et al., 917, Copyright 2008, with permission from Elsevier.)

fluid profile is described as a pulse-like motion. In contrary to reciprocating pumps, dynamic pumps produce pressures because of fluid that gains momentum as it moves through the pump. Advantages of using these pumps include that they produce constant pulseless flow and utilize fewer components and can be fabricated in a much smaller scale compared to reciprocating pumps. Continuous sampling of analytes without supervision is important in realizing efficient analysis and detection.

Experiment and simulation have shown that PSi (Laser et al. 2003; Zheng 2006; Wallner et al. 2007; Vanga 2012) along with other porous materials (Zeng et al. 2002; Yao et al. 2003; Wang et al. 2009) is promising as a pump media for electro-osmotic pumps (EOPs). However, with the increase in demand for miniaturized flow-through sensing systems, there is a high motivation for using silicon technology instead of other porous media due to its well-established knowledge base. EOPs work on the principle of creating a flow in a fluid by application of an electric field across a porous membrane and are highly desirable for miniature fluidic applications as they work without any mechanical parts. The flow rates in EOPs can be controlled accurately by the application of an electric field. This means that the flow rates achieved using EOPs are highly precise in the nano and micro range, which is a requirement in many biological and chemical applications. The operation of electro-osmotic micropumps is quite. Flow direction in electro-osmotic micropumps is controlled by switching the direction of the external electric field. Therefore, EOPs have evolved to play a significant role in applications such as micro-flow injection analysis (µ-FIA), micro-channel cooling system, volumetric nano-titrations, and micro-fluidic liquid chromatography systems (µ-LC) for fluid transport at the micro- and nano-scale (Pretorius et al. 1974; Liu and Dasgupta 1992; Harrison et al. 1993; Manz et al. 1994; Laser et al. 2003; Borowsky et al. 2008; Wang et al. 2009). The generation of fluid flow using electro-osmosis is perfectly suitable to miniaturization: (1) it delivers flow rates in the required range of a few hundred nanoliter/min, and (2) involving standard and cheap MEMS technology (no moving parts), EOP can be directly integrated on chip and can be discarded after use. For functioning, the EOP requires only sets of electrodes and it can operate off batteries.

Schematic diagrams illustrating functioning of EOPs are presented in Figure 16.10. Their construction usually uses fused silica or glass capillaries with electrodes that provide an electric field along the channel length. EOPs work on the phenomenon of fluid motion caused by the application of electric AC or DC fields on the ions that are bound to the sidewall of the capillary channels. The surface charge spontaneously develops when a liquid comes in contact with a solid (Hunter 1981; Adamson and Gast 1997). In the case of silica-based ceramics (e.g., glass), the working fluid needs to be a polar solvent having a pH value greater than 4, which is required for the formation of surface charge on the side walls of the pores (Adamson and Gast 1997; Chen et al. 2003). The thin layer of SiO_2 on the pore walls when in contact with a fluid of the above specifications achieves a negative surface charge due to the removal of H^+ ions of the acidic silanol groups and hence results in an electrical double layer (EDL) (Chen et al. 2003). The negatively charged surface attracts the positively charged ions of the solution. When an external electric field is applied along the length of the channel, the ions in the EDL will experience a coulombic

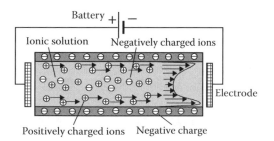

FIGURE 16.10 Schematic illustration of EOF in a channel. Chemical reactions at the interface leave the surface charged (shown as negative here). Counter ions in the liquid accumulate near the charged surface, forming the electric double layer. An externally applied electric field causes motion of counter ions that shield a negative wall charge. Ion drag forces the flow against a pressure gradient. (Idea from Wallner J.Z. et al., *Phys. Stat. Sol. (a)* 204, 1327, 2007.)

force. In this case, the thin layer of cation-rich fluid adjacent to the solid surfaces starts moving toward the cathode. This phenomenon is discussed in detail by Probstein (1994). The boundary layer like motion eventually sets the bulk liquid into motion through viscous interaction. This resultant flow is called electro-osmotic flow (EOF) (Chen et al. 2003; Wang et al. 2006). The EOF velocity of the electrolyte can be described by the Helmholtz–Smoluchowski equation and written as (Molho et al. 1998)

$$(16.1) \qquad\qquad U = -\frac{\varepsilon\varepsilon_0 \zeta E_{EOF}}{\mu},$$

where U is the EO flow velocity, E_{EOF} is the strength of the electric field, ζ is the zeta potential, μ is the dynamic viscosity of the electrolyte, ε_0 is the permittivity of vacuum, and ε is the relative dielectric permittivity of the medium (dielectric constant). The key parameters that dictate the performance of EOPs are (1) the magnitude of the applied electric field and applied voltage, (2) the cross-sectional dimensions of the structure in which flow is generated, (3) the surface charge density of the solid surface that is in contact with the working liquid, and (4) ion density and pH of the working fluid (Laser and Santiago 2004). Rice and Whitehead's analysis of EOF in a cylindrical capillary shows how these parameters relate to EOP performance. According to Rice and Whitehead (1965), the flow rate of the fluid in the capillary channels becomes significant as the dimensions of the EDL become closer to order of the capillary channel size. This means that EOF in channels is greatly affected by the dimensions of the capillary especially in the micro- and nanoscale, when the fluid interaction with the capillary walls becomes more pronounced and significant. Such a decrease in the diameter of the capillary channels is accompanied by an increase in the pressure generated by the pump, but a simultaneous reduction in the flow also takes place. Therefore, in general, for specific application, there could be a compromise between the flow rate and pressure generated by the EOPs. As a result, the flow rates produced by EOP based on individual capillaries or microchannels are typically very small ($Q_{max} < 1$ μL/min). Using PSi allows overcoming this problem because PSi membranes have ideal geometries for EO pumping: PSi membranes have hexagonally densely packed structures, uniform pores with near-unity tortuosity, and are well suited to maximize flow rate for a given applied voltage. In PSi membranes, each pore acts as a tortuous capillary for generating EO flow. These pumps can be modeled as a bundle of n capillaries (Zeng et al. 2001; Chen et al. 2003). This means that when using PSi, one can expect that this pumping media will produce reasonable flow rate and high pressure capacity. The last one should increase with decreasing pore diameter.

As for the main disadvantages of EO micropumps, the appearance of microchannel-blocking bubbles should be attributed to them. The large currents generated in the open channel may produce bubbles; electrolysis and reactions at the electrodes produce ions that can contaminate the sample and generate the bubbles. Other limitations of EO micropumps are high voltage required and electrically conductive solution.

Examples of PSi-based EOPs implementation can be found in the following references: Laser et al. 2003; Laser and Santiago 2004; Yao et al. 2004, 2006; Zheng 2006; Wallner et al. 2007; Vajandar et al. 2009; Snyder 2011; Vanga et al. 2011; Xu et al. 2011; Vanga 2012; Snyder et al. 2013. For example, Yao et al. (2006) designed an EOP that can generate high pressures of more than 340 atm and large flow rates per volt and unit area 0.13 mL/(min·cm²·V). They have shown that this pump can pump 40 mL/min of solution at 100 V in a pumping structure less than 1 cm³ in total volume. Snyder (2011) reported on PSi-based EOPs that demonstrated EO rates of 2.6×10^2 mL/(min·cm²·V), which are 2–3 orders of magnitude higher than other DC EOPs reported in the literature. A schematic diagram and possible views of designed EOPs are shown in Figure 16.11.

Mentioned above works have shown that the flow rates and pressure capacities obtained using PSi for EOP applications depends on the pore configuration that includes pore diameter, pore depths, and pore uniformity and can be optimized. It was also found that surface modifications could change the zeta potential of the material and influence the EOF rates (Vajandar et al. 2009; Wang et al. 2009; Snyder et al. 2013). For example, Snyder et al. (2013) established that oxidation increased flow rates by about two times over the untreated samples, whereas aminosilanization reduced the flow rates almost to zero.

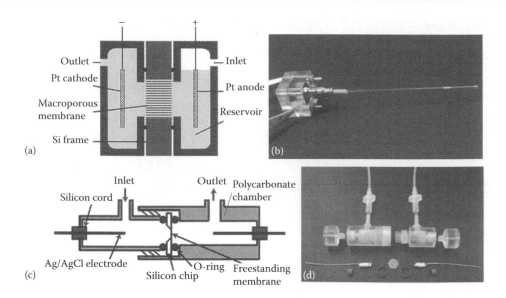

FIGURE 16.11 (a, c) Schematic of PSi EOPs assembly and (b, d) the view of the EOP pumps proto-types. ([a, b] Wallner J.Z. et al.: *Phys. Stat. Sol. (a)*. 2007. 204. 1327. Copyright Wiley-VCH Verlag GmbH & Co. KGaA. Reproduced with permission. [c, d] From Snyder J.L. et al., *PNAS* 110 (46), 18425, 2013. Copyright 2013: National Academy of Sciences. With permission.)

As a rule for EOP fabrication, the macroporous silicon with pore size of 1–15 μm is used (Yao et al. 2006; Wallner et al. 2007; Vajandar et al. 2009; Xu et al. 2011). Silicon porosification usually is conducted using electrochemical etching or DRIE. For PSi oxidation, either thermal or wet oxidation can be used (read *Porous Silicon: Formation and Properties*). For example, in their designs, Wallner et al. (2007) used macroporous silicon membranes with pore sizes of ~3 μm and depths of 500 μm. This EOP successfully demonstrated a maximum pressure capacity (5.2 kPa) and a high flow rate (11.9 μl/min·mm²) at 60 V. One should note that macroporous Si with pore sizes in the range of 3–5 μm is a material typically used for EOP fabrication. However, Vanga et al. (Vanga et al. 2011; Vanga 2012) believe that for specific applications, where high pressure is required, mesoporous Si with pore size of 50–200 nm will be more preferred. Using this approach, Snyder et al. (2013) designed EOP with extremely low operation voltage. They have shown that meso-porous ultrathin (15–30 nm) Si membranes with low active areas (0.36 mm²) can generate EOF rates of 10 μL/min at voltages of 20 V or lower. For example, EOPs developed using nanoporous Si membranes and Ag/AgCl electrodes have shown the possibility to pump fluids through capillary tubing at applied voltages as low as 250 mV. Characterization by electron microscopy determined that membranes used in this EOP had an average pore size of 19.5 nm and a porosity of 5.7%.

16.3 SEPARATION OF PARTICLES OF DIFFERENT SIZE

In microfluidics, materials are required with a well-defined surface, shape, and size. This means that separation of particles by size and shape is an important task. Müller and co-workers (Kettner et al. 2000; Muller et al. 2000; Matthias and Muller 2003; Langner et al. 2011) established that macroporous silicon is an excellent candidate for this aim. In particular, Muller et al. (2000), based on theoretical work (Kettner et al. 2000), assumed that the one-dimensional pores of a macroporous silicon membrane, etched to exhibit a periodic asymmetric variation in pore diam-eter, can act as massively parallel and multiply stacked Brownian ratchets that are potentially suitable for large-scale particle separations. The new concept of a drift ratchet uses the friction between the particles and a carrier liquid as the driving force. According to Kettner et al. (2000), the periodic modulation of the pore diameter leads to a modulation of the velocity and therefore of the driving force. Theoretical simulations indicate that for particles in the 1 μm range, a strong variation of the net drift velocity can be expected, depending on the particle size.

Experiments reported by Muller et al. (2000) and Matthias and Muller (2003) confirmed this assumption. It was shown that applying a periodic pressure profile with a mean value of zero to a basin separated by such a membrane induces a periodic flow of water and suspended particles through the pores resulting in a net motion of the particles from one side of the membrane to the other without moving the liquid itself. Moreover, it was found that the experimentally observed pressure dependence of the particle transport, including an inversion of the transport direction, agrees with calculations (Kettner et al. 2000) of the transport properties in the type of ratchet devices used here. In these experiments an upper and a lower basin were separated by a membrane with asymmetrically shaped pores (Figure 16.12a). A macroporous silicon sample with an asymmetric pore profile was grown and afterward the remaining silicon was removed from the backside with KOH. The technology of such macroporous silicon formation was discussed previously in *Porous Silicon: Formation and Properties*. The pore geometry can be seen in Figure 16.12c,d: It is a triangular pattern with 6 μm pore-to-pore distance, 150 μm in depth, and the maximum (minimum) diameter of the modulation is 4.8 μm (2.5 μm). The liquid with the dispersed microparticles was periodically pumped between upper and lower basins with an average net liquid flow of zero. The density of the spheres in the upper basin was measured via photoluminescence (PL). In Figure 16.12b, some experimental data are shown. The pressure oscillations were toggled on and off every 60 s to exclude long-term drifts. It is seen that during this period, the intensity in the upper basin—and therefore the concentration of particles—starts from a homogeneous particle distribution and increases during the pump-on times (curve U). Particles of 0.32 μm diameter could be transported to the upper basin with such a membrane when a periodic pressure oscillation was applied. The effect of the asymmetry of the pore shape can be seen

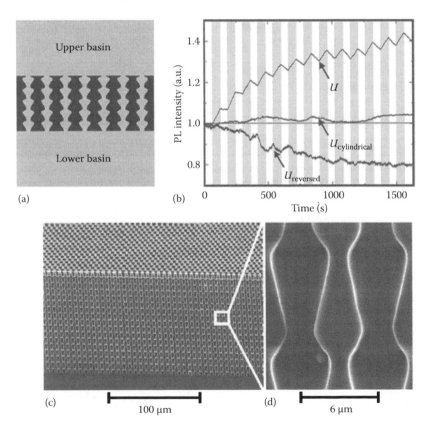

(a)   (b)

FIGURE 16.12   (a) Schematics of the horizontally mounted and asymmetric diameter-modulated membrane separating two basins. (b) Measured photoluminescence (PL) intensity in the upper basin (U), for the reverse mounted membrane ($U_{reversed}$), and a cylindrical pore shape for comparison ($U_{cylindrical}$). Membrane thickness 150 μm, basin thickness 100 μm. (c) SEM picture of cleaved macroporous silicon membrane used for the experiment. (d) Magnified version with a colloidal sphere of 1 μm in diameter inside the pore sticking to the pore wall. (Muller F. et al.: *Phys. Stat. Sol. (a)*. 2000. 182, 585. Copyright Wiley-VCH Verlag GmbH & Co. KGaA. Reproduced with permission.)

when turning the membrane upside down. Now the intensity decreases ($U_{reversed}$). It can be seen in the curve $U_{cylindrical}$ that this effect is clearly related to the asymmetric shape of the membrane: The measurement was performed with a sample of cylindrical pore shape (diameter 2.4 µm) but no significant particle transport could be observed.

Matthias and Muller (2003) believe that if problems associated with mixing and surface passivation will be resolved, the system could be well suited for the efficient and selective continuous separation of sensitive biological materials like viruses or cell fragments. As the external force is based on the flow resistance of the particles, it should be possible to separate elongated or flexible particles, provided their size be roughly in the 0.1–1 µm range. Of course, this assumption should be confirmed by further research.

16.4 FILTERING

Filtering the sample to remove unwanted components is one of the functions that can be incorporated in lab-on-chip devices. To this aim, the use of membranes with tunable pore sizes is of great interest. Standard PSi membranes with vertical micro-pores, discussed in Chapter 15, are a good candidate for such applications (Kuiper et al. 1998; Eun et al. 2008, 2009; Campbell et al. 2008). For example, Fauchet and co-workers (Striemer et al. 2007, 2010; Snyder et al. 2008) have shown that a PSi membrane that is a few nanometers thick (10–50 nm) can quickly filter liquids and separate molecules that are very close in size. In particular, the membrane could lead to efficient protein purification for use in research and drug discovery. Moreover, they established that air and water permeabilities through porous nanocrystalline silicon membranes are greater than both polycarbonate track etched membranes and carbon nanotube membranes. Membranes designed in Fauchet group had pore sizes in the range of ~5 to ~100 nm. However, the integration of such membranes with microfluidics is not straightforward and is usually done by creating a discrete device (Eun et al. 2009) or by laminating micromachined substrates separated by a porous layer (Dubosc et al. 2012). These hybrid processes raise bonding issues and are not suitable to complex designs. Whenever possible, it is thus preferable to work with planar fluidic systems (see Figure 16.13). Still, integration of porous membranes is thus far only done by manually assembling polymer pieces into microchannels, which is not a large-scale and highly reproducible method (Yang et al. 2011).

For resolving this problem, Dubosc et al. (2012) designed a fabrication process leading to the creation of lateral porous membranes compatible with silicon-based microfluidics (see

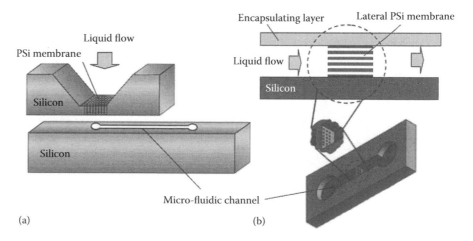

FIGURE 16.13   Schematics showing the differences between (a) transverse and (b) lateral porous membranes. ([a] Idea from Eun D.-S. et al., *Jpn. J. Appl. Phys.* 47 (6), 5236, 2008. [b] Reprinted from *Procedia Eng.* 47, Dubosc F. et al., 801, Copyright 2012, with permission from Elsevier.)

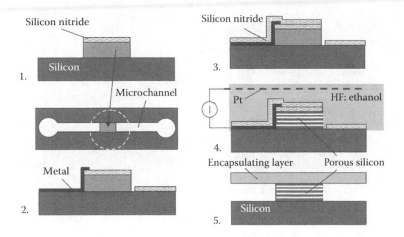

FIGURE 16.14   Fabrication process of lateral PSi membranes: (1) RIE of the silicon substrate to create microchannels; (2) conformal deposition of a metal layer for the working electrode; (3) patterning of a protective Si_3N_4 layer; (4) silicon anodization; (5) encapsulating.

Figure 16.14). This process relies on the local anodization of silicon, resulting in PSi membranes within planar channels. Lateral PSi membranes were fabricated in highly doped p-type Si(100) wafers. The fabrication process started by depositing an 80-nm LPCVD Si_3N_4 layer and by etching 3-μm thick channels in the silicon by deep reactive ion etching. Anodes were locally created by evaporation and liftoff of a 50 nm/150 nm Cr/Au layer in a conformal way to ensure the presence of metal on the sides of the channel parts to permeate. The metal was protected with a 1-μm PECVD Si_3N_4 layer, which was opened around the areas where pore formation was to be initiated. Finally, porosification of the membranes was carried out by anodic dissolution of silicon in an HF:ethanol = 1:1 solution at a current density of 50 mA/cm^2 and at room temperature during 5 min. As it was shown in *Porous Silicon: Formation and Properties*, pore formation by anodization of silicon occurs following the current density lines and propagates preferentially in the <100> direction. In the present configuration, the current flows parallel to the wafer surface, and the use of (100) silicon wafer offers <100> crystallographic directions for pore propagation in the bulk silicon. Thus, silicon anodization onto the etched walls using local electrodes results in the creation of lateral porous membranes within the microchannels. Dubosc et al. (2012) reported that using the above-mentioned technology, they had the possibility to fabricate lateral 10-μm long mesoporous membranes within 3-μm deep microchannels. Membranes had pores with diameters in the range of 10–30 nm.

One should note that these results are of interest. However, apparently this process requires optimization. For example, to avoid adverse effects during the manufacture of lateral PSi filters, it is desirable to use heavy-doped silicon layers formed on a surface of high-resistivity substrates.

Another interesting approach to integration of filtering element with lab-on-a-chip devices was proposed by Pagonis and Nassiopoulou (2006). The developed technique is based on a two-step electrochemical process of silicon dissolution, starting from thick macro-PSi membrane formation under the appropriate conditions and continuing with different anodization conditions that lead to formation of a nanoporous silicon layer underneath the macroporous membrane (see Figure 16.15). The obtained structure is a double-layer of a macroporous over a nanoporous layer. By selective dissolution of the nanoporous layer, a freestanding macroporous silicon membrane over a large cavity in silicon is obtained. Nanoporous silicon can be easily etched by wet etching with CMOS compatible chemicals (Kaltsas and Nassiopoulou 1998). The etching process involves chemical oxidation and oxide removal, which dissolves rapidly the silicon skeleton of interconnected nanowires. On the contrary, the macroporous material is practically not affected by the chemicals used for nanoporous Si dissolution because the pore walls are in the micrometer scale.

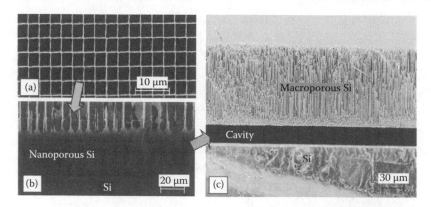

FIGURE 16.15   (a) SEM image of a typical ordered macro-PSi layer formed on a *p*-type silicon sub-strate. The anodization solution used was a mixture of HF:DMF (1:9 in volume) and the current density was 10 mA/cm². (b) Cross-sectional SEM images of structures consisting of macroporous silicon over nanoporous layers, formed on *p*-type silicon. The interfacial layer is relatively sharp. (c) Cross-sectional SEM images of typical freestanding macro-PSi membranes over cavity. The macroporous membranes are released from the substrate by selective dissolution of a nanoporous layer formed underneath the macroporous structure in the same anodization process flow. (Reprinted from *Microelectron. Eng.* 83, Pagonis D.N. and Nassiopoulou A.G., 1421, Copyright 2006, with permission from Elsevier.)

REFERENCES

Abhari, F., Jaafar, H., and Yunus, N.A.M. (2012). A comprehensive study of micropumps technologies. *Int. J. Electrochem. Sci.* 7, 9765–9780.

Adamson, A.W. and Gast, A.P. (1997). *Physical Chemistry of Surfaces.* Wiley, New York.

Bassous, E., Taub, H.H., and Kuhn, L. (1977). Ink jet printing nozzle arrays etched in silicon. *App. Phys. Lett.* 31, 135–137.

Borowsky, J.F., Giordano, B.C., Lu, Q., Terray, A., and Collins, G.E. (2008). Electroosmotic flow-based pump for liquid chromatography on a planar microchip. *Anal. Chem.* 80, 8287–8292.

Campbell, J., Corno, J.A., Larsen, N., and Gole, J.L. (2008). Development of porous-silicon-based active microfilters. *J. Electrochem. Soc.* 155(2), D128–D132.

Chen, L., Guan, Y., Ma, J., Shu, X., and Tan, F. (2003). Theory, controls parameter and application of the packed-bed electroosmotic pump. *Chinese Sci. Bull.* 48(23), 2572–2577.

Clicq, B., Tjerkstra, R.W., Gardeniers, J.G.E., van den Berg, A., Baron, G.V., and Desmet, G. (2004). Porous silicon as a stationary phase for shear-driven chromatography. *J. Chromatogr. A* 1032, 185–191.

De Andrade Costa, R., Mogensen, K., and Kutter, J. (2005). Microfabricated porous glass channels for electrokinetic separation devices. *Lab Chip* 5, 1310–1314.

De Malsche, W., Clicq, D., Verdoold, V., Gzil, P., Desmet, G., and Gardeniers, H. (2007). Integration of porous layers in ordered pillar arrays for liquid chromatography. *Lab Chip* 7, 1705–1711.

De Malsche, W., Gardeniers, H., and Desmet, G. (2008). Experimental study of porous silicon shell pillars under retentive conditions. *Anal. Chem.* 80, 5391–5400.

De Pra, M., Kok, W.T., Gardeniers, J.G.E. et al. (2006). Experimental study on band dispersion in channels structured with micropillars. *Anal. Chem.* 78, 6519–6525.

De Smet, J., Gzil, P., Vervoort, N., Verelst, H., Baron, G.V., and Desmet, G. (2004). Influence of the pillar shape on the band broadening and the separation impedance of perfectly ordered 2-D porous chromatographic media. *Anal. Chem.* 76, 3716–3726.

De Smet, J., Gzil, P., Vervoort, N., Verelst, H., Baron, G.V., and Desmet, G. (2005). On the optimisation of the bed porosity and the particle shape of ordered chromatographic separation media. *J. Chromatogr. A* 1073, 43–51.

Detobel, F., De Bruyne, S., Vangelooven, J. et al. (2010). Fabrication and chromatographic performance of porous-shell pillar-array columns. *Anal. Chem.* 82, 7208–7217.

Dubosc, F., Bourrier, D., and Leichlé, T. (2012). Fabrication of lateral porous silicon membranes for planar microfluidic devices. *Procedia Eng.* 47, 801–804.

Erickson, D. and Li, D. (2004). Integrated microfluidic devices. *Anal. Chim. Acta* 507, 11–26.

Eun, D.-S., Kong, D.-Y., Kong, S.H., Choi, P., Shin, J.-K., and Lee, J.-H. (2008). Fabrication of a based fluidic chip equipped with porous silicon filter and micro-channels. *Jpn. J. Appl. Phys.* 47(6), 5236–5241.

Eun, D.-S., Jeong, J.-H., Kong, D.-Y., Shin, J.-K., and Lee, J.-H. (2009). Fabrication of a micro-filter using porous layer on type-(110) silicon. *J. Korean Phys. Soc.* 55(3), 986–994.

Faure, K. (2010). Liquid chromatography on chip. *Electrophoresis* 31, 2499–2511.

Gzil, P., Vervoort, N., Baron, G.V., and Desmet, G. (2004). General rules for the optimal external porosity of LC supports. *Anal. Chem.* 76, 6707–6718.

Harrison, D.J., Fluri, K., Seiler, K., Fan, Z., Effenhauser, C.S., and Manz, A. (1993). Micromachining a miniaturized capillary electrophoresis-based chemical analysis system on a chip. *Science* 261, 895–897.

He, B., Tait, N., and Regnier, F. (1998). Fabrication of nanocolumns for liquid chromatography. *Anal. Chem.* 70, 3790–3797.

Hunter, R.J. (1981). *Zeta Potential in Colloid Science.* Academic, San Diego, CA.

Kaltsas, G. and Nassiopoulou, A.G. (1998). Fronside bulk siliocn micromachining using porous-silicon technology. *Sens. Actuators A* 65, 175–179.

Kettner, C., Reimann, P., Hanggi, P., and Muller, (2000). Drift ratchet. *Phys. Rev. E* 61(1), 312–323.

Kuiper, S., van Rijn C.J.M., Nijdam W., and Elwenspoek, M.C. (1998). Development and applications of very high flux microfiltration membranes. *J. Membrane Sci.* 150, 1–8.

Kutter, J.P. (2012). Liquid phase chromatography on microchips. *J. Chromatogr. A* 1221, 72– 82.

Langner, A., Müller, F., and Gösele, U. (2011). Macroporous silicon. In: O. Hayden, K. Nielsch (Eds.), *Molecular- and Nano-Tubes.* Springer Science+Business Media, New York, pp. 431–460.

Laser, D.J. and Santiago, J.D. (2004). A review of micropumps. *J. Micromech. Microeng.* 14, R35–R64.

Laser, D.J., Myers, A.M., and Yao, S. (2003). Silicon electroosmotic micropumps for integrated circuit thermal management. In: *Proceedings of 12th International Conference on Transducers, Solid-state Sensors, Actuators Microsystems,* June 8–12, Boston, Vol. 1, pp. 151–154.

Liu, S. and Dasgupta, P.K. (1992). Flow-injection analysis in the capillary format using electroosmotic pumping. *Anal. Chim. Acta* 268, 1–6.

Lundanes, E., Reubsaet, L., and Greibrokk, T. (2013). *Chromatography: Basic Principles, Sample Preparations and Related Methods.* Wiley, Weinheim.

Manz, A., Garber, N., and Widmer, H.M. (1990). Miniaturized total chemical analysis systems: A novel concept for chemical sensing. *Sens. Actuators B* 1, 244–248.

Manz, A., Effenhauser, C.S., Burggraf, N., Harrison, D.J., Seiker, K., and Fluri, K. (1994). Electroosmotic pumping and electrophoretic separations for miniaturized chemical-analysis systems. *J. Micromech. Microeng.* 4, 257–265.

Matthias, S., and Muller, F. (2003). Assymetric pores in silicon membrane acting as massively parallel brownian raychets. *Nature* 424(6944), 53–57.

Méry, E., Alekseev, S., Portet-Koltalo, F. et al. (2009). Porous silicon based microdevice for reversed phase liquid chromatography. *Phys. Stat. Sol. (c)* 6(7), 1777–1781.

Molho, J.I., Herr, A.E., Deshpande, M. et al. (1998). Fluid transport mechanisms in microfluidic devices. In: *Proceeding of ASME International Mechanical Engineering Congress and Exposition,* November 15–20, Anaheim, pp. 69–75.

Muller, F., Birner, A., Schilling, I., Gosele, U., Kettner, Ch., and Hanggi, P. (2000). Membranes for micropumps from macroporous silicon. *Phys. Stat. Sol. (a)* 182, 585–590.

Neue, U.D. (1997). *HPLC Columns: Theory, Technology, and Practice.* Wiley-VCH, New York.

Nisar, A., Afzulpurkar, N., Mahaisavariya, B., and Tuantranont, A. (2008). MEMS-based micropumps in drug delivery and biomedical applications. *Sens. Actuators B* 130, 917–942.

Pagonis, D.N. and Nassiopoulou, A.G. (2006). Free-standing macroporous silicon membranes over a large cavity for filtering and lab-on-chip applications. *Microelectron. Eng.* 83, 1421–1425.

Peterson, D.S. (2005). Solid supports for micro analytical systems. *Lab Chip* 5, 132–139.

Poppe, H. (1997). Some reflections on speed and efficiency of modern chromatographic methods. *J. Chromatogr. A* 778, 3–21.

Pretorius, V., Hopkins, B.J., and Schieke, J.D. (1974). Electro-osmosis: new concept for high-speed liquid-chromatography. *J. Chromatogr.* 99, 23–30.

Probstein, R.F. (1994). *Physicochemical Hydrodynamics.* Wiley, New York.

Regnier, F.E. (2000). Microfabricated monolith columns for liquid chromatography. *J. High Resol. Chromatogr.* 23(1), 19–26.

Rice, C.L. and Whitehead, R.J. (1965). Electrokinetic flow in a narrow cylindrical capillary *J. Phys. Chem.* 69, 4017–4024.

Sainiemi, L. and Franssila, S. (2007). Mask material effects in cryogenic deep reactive ion etching, *J. Vac. Sci. Technol. B* 25, 801–807.

Sainiemi, L., Nissilä, T., Jokinen, V. et al. (2008). Fabrication and fluidic characterization of silicon micropillar array electrospray ionization chip. *Sens. Actuators B* 132, 380–387.

Slentz, B.E., Penner, N.A., and Regnier, F.E. (2002). Capillary electrochromatography of peptides on microfabricated poly(dimethylsiloxane) chips modified by cerium(IV)-catalyzed polymerization. *J. Chromatogr. A* 948, 225–233.

Snyder, J.L (2011). *Porous Nanocrystalline Silicon Membranes as Sieves and Pumps.* Ph.D. Thesis, University of Rochester, NY.

Snyder, J.L., Kavalenka, M., Fang, D.Z., Streimer, C.C., Fauchet, P.M., McGrath, J.L. (2008). Permeability and protein separations: Functional studies of porous nanocrystalline silicon membranes. In: *Technical Proceedings of the 2008 NSTI Nanotechnology Conference and Trade Show,* Vol. 1, pp. 333–335.

Snyder, L.R., Kirkland, J.J., and Dolan, J.W. (2009). *Introduction to Modern Liquid Chromatography.* Wiley, Weinheim.

Snyder, J.L., Getpreecharsawas, J., Fang, D.Z. et al. (2013). High-performance, low-voltage electroosmotic pumps with molecularly thin silicon nanomembranes. *PNAS* 110(46), 18425–18430.

Stachowiak, T.B., Svec, F., and Fréchet, J.M.J. (2004). Chip electrochromatography. *J. Chromatogr. A* 1044, 97–111.

Striemer, C.C., Gaborski, T.R., McGrath, J.L., and Fauchet, P.M. (2007). Charge- and size-based separation of macromolecules using ultrathin silicon membranes. *Nature* 445, 749–753.

Striemer, C.C., Gaborski, T.R., Fang, D.Z., Snyder, J.L., McGrath, J.L., and Fauchet, P.M. (2010). Porous ultrathin silicon membranes for purification of nanoscale materials. *MRS Proc.* 1209, 95–100.

Sukas, S., Tiggelaar, R.M., Desmet, G., and Gardeniers, H.J.G.E. (2013). Fabrication of integrated porous glass for microfluidic applications. *Lab on a Chip* 13, 3061–3069.

Suni, N.M., Haapala, M., Mkinen, A. et al. (2008). Selective surface patterning with an electric discharge in the fabrication of microfluidic structures. *Angew. Chem. Int. Ed.* 47, 7442–7445.

Tiggelaar, R.M., Verdoold, V., Eghbali, H., Desmet, G., and Gardeniers, J.G.E. (2009). Characterization of porous silicon integrated in liquid chromatography chips. *Lab Chip* 9, 456–463.

Tsai, N.C. and Sue, C.Y. (2007). Review of MEMS based drug delivery and dosing systems. *Sens. Actuators A* 134, 555–564.

Vajandar, S.K., Xu, D., Sun, J., Markov, D.A., Hofmeister, W.H., and Li, D. (2009). Field-effect control of electroosmotic pumping using porous silicon–silicon nitride membranes. *J. Microelectromech. Syst.* 18(6), 1173–1183.

Vanga, K.L. (2012). Implementation of Porous Silicon Technology for a Fluidic Flow-through Optical Sensor for pH Measurements. PhD Thesis, Michigan Technological University, Houghton, MI.

Vanga, K.L., Hu, Q., Green, S.A., and Bergstrom, P.L. (2011). Implementation of porous silicon technology for flow-through sensing using electro-osmotic phenomenon. In: *Proceedings of 11th IEEE International Conference on Nanotechnology,* August 15–18, Portland, OR, pp. 1639–1643.

Wallner, J.Z., Nagar, N., Friedrich, C.R., and Bergstrom, P.L. (2007). Macro porous silicon as pump media for electro-osmotic pumps. *Phys. Stat. Sol. (a)* 204, 1327–1331.

Wang, P., Chen, Z., and Chang, H.C. (2006). A new electro-osmotic pump based on silica monoliths. *Sens. Actuators B* 113, 500–509.

Wang, X., Cheng, C., and Wang, S. (2009). Electroosmotic pumps and their applications in microfluidic systems. *Microfluid. Nanofluid.* 6, 145–162.

Woias, P. (2005) Micropumps-past, progress and future prospects. *Sens. Actuators B* 105, 28–38.

Xie, J., Miao, Y., Shih, J., Tai, Y.-C., and Lee, T.D. (2005). Microfluidic platform for liquid chromatography-tandem mass spectrometry analyses of complex peptide mixtures. *Anal. Chem.* 77, 6947–6953.

Xu, Z., Miao, J., Wang, N., Wen, W., and Sheng, P. (2011). Maximum efficiency of the electro-osmotic pump. *Phys. Rev. E* 83(6), 066303.

Yang, H., Shen, M., Sivagnanam, V., and Gijs, M.A.M. (2011). High-performance protein preconcentrator using microchannel-integrated Nafion strip. In: *Proceedings of the 16th International Conference on Solid-State Sensors, Actuators and Microsystems, Transducers'11,* June 5–9, Beijing, China, pp. 238–241.

Yao, S., Huber, D., and Mikkelsen, J.C. (2003). Porous glass electroosmotic pumps: design and experiments. *J. Coll. Interface Sci.* 268, 143–153.

Yao, S., Myers, A.M., Posner, J.D., and Santiago, J.G. (2004). Electroosmotic pumps fabricated from porous silicon membranes. In: *Proceedings of International Mechanical Engineering Congress and Exposition,* November 13–19, Anaheim, CA, pp. 185–190.

Yao, S., Myers, A.M., Posner, J.D., Rose, K.A., and Santiago, J.G. (2006). Electroosmotic pumps fabricated from porous silicon membranes. *J. Microelectromech. Syst.* 15(3), 717–728.

Zeng, S.L., Chen, C.H., Mikkelsen, J.C., and Santiago, J.G. (2001). Fabrication and characterization of electroosmotic micropumps. *Sens. Actuators B* 79, 107–114.

Zeng, S., Chen, C., and Santiago, J. (2002). Electroosmotic flow pumps with polymer frits. *Sens. Actuators B* 82, 209–212.

Zhang, C., Xing, D., and Li, Y. (2007). Micropumps, microvalves, and micromixers within PCR microfluidic chips: Advances and trends. *Biotechnol. Adv.* 25, 483–514.

Zheng, J. (2006). Porous Silicon Technology for Integrated Microsystems. PhD Thesis, Michigan Technological University, Houghton, MI.

Biomedical Applications

Biocompatibility and Bioactivity of Porous Silicon

Adel Dalilottojari, Wing Yin Tong, Steven J.P. McInnes, and Nicolas H. Voelcker

17

CONTENTS

17.1	Introduction	320
17.2	Assessing Biocompatibility	320
	17.2.1 *In Vitro* Assays and Assessment Methods	321
	17.2.2 *In Vivo* Assays and Assessment Methods	321
	17.2.3 Summary	322
17.3	Degradation of PSi in Aqueous Solution	322
17.4	Biological Responses to PSi	323
17.5	The Effect of Size, Shape, and Reactivity on the Biocompatibility of PSi	324
17.6	*In Vitro* Studies of PSi	326
17.7	*In Vivo* Studies of PSi	328
	17.7.1 *In Vivo* Administration Routes	330
	17.7.1.1 Intravenous Injection	330
	17.7.1.2 Oral Administration	330
	17.7.1.3 Summary	330
17.8	PSi in Clinical Trials	331
17.9	Conclusion	331
References		332

17.1 INTRODUCTION

One of the most important features that define a quality biomaterial is its ability to be in contact with biological tissue without generating an unacceptable degree of harm at the insertion/ implantation site (Williams 2008). Producing biocompatible implantable devices has been a major concern as they are required to remain for prolonged periods of time inside living tissues. Recent work in biomaterial development has focused on the generation of materials that can be fully resorbed or even tuned to degrade at a specified rate and be replaced by the surrounding tissue as it heals (Bitar and Zakhem 2014). These materials can even be tuned to promote the healing process by the addition of growth factors or other drugs (Chadwick et al. 2014). To address safety concerns, each individual biomaterial needs to be precisely evaluated both *in vitro* and *in vivo* before moving into clinical trials (Jones and Grainger 2009; Kohane and Langer 2010; Kunzmann et al. 2011; Jaganathan and Godin 2012).

The ability to assess a material's biocompatibility is reliant on the ability to detect the response elicited by the local environment both *in vitro* and *in vivo* (Liu et al. 2014). A material's overall biocompatibility will depend on a wide variety of properties including the intrinsic material properties, the local tissue environment, and the formulation and the administration route among others (Liu et al. 2014). The resulting interactions between a biomaterial and the host environment can be

Inert—having no effect on biological function
Advantageous—improving biological function
Detrimental—posing a toxic hazard to the local environment (Oberdorster 2010)

To ensure that all possible negative effects are considered, reliable and reproducible screening protocols are needed to assess the relationship between the material properties and the biological responses elicited (Nel et al. 2006; Bratlie et al. 2010). Material properties that affect biological response include size, shape, surface area, surface functionalization, crystallinity, and wettability, among others (Moghimi et al. 2005; Mitragotri and Lahann 2009).

Micro- and nanoparticle systems are now also beginning to be exploited as potential biomaterials as their nanoscale dimensions afford unique and interesting properties (Riehemann et al. 2009), which are already proving to be well suited toward medical applications in fields such as oncology (Hong et al. 2011) and ophthalmology (Cheng et al. 2008).

One promising new nanomaterial is porous silicon (PSi). PSi was discovered in 1956 by Uhlir in the context of silicon semiconductor research (Uhlir 1956). However, it was not until the subsequent pioneering work by Canham in the 1990s (Canham 1990) that PSi was revealed as a promising biomaterial. Over the last two decades, PSi has been extensively studied for biomedical applications due to its high biocompatibility and biodegradability. In addition, it is possible to generate a range of pore sizes, ranging from microporous (pore diameter < 2 nm) to macroporous (pore diameter > 50 nm up to 3 µm) (Gregg and Sing 1983; Canham 1997; Low et al. 2009). Moreover, the surface chemistry of PSi is easy to modify with a range of techniques including hydrosilylation, thermal carbonization, thermal hydrocarbonization, oxidation, and silanization (Buriak 2002; Salonen and Lehto 2008). These properties make PSi amenable to many different biological environments, once again increasing its versatility as a biomaterial for applications such as drug delivery (McInnes and Voelcker 2009), bioimaging (Park et al. 2009), and tissue engineering (McInnes and Voelcker 2014).

This chapter will summarize the research efforts currently being made to evaluate the results of exposure of PSi-based materials in both *in vitro* and *in vivo* biological environments.

17.2 ASSESSING BIOCOMPATIBILITY

An ideal biomaterial that can function at the intended site of action without causing any disturbance to the local environment will be nontoxic, nonimmunogenic, and fully biodegradable (Liu et al. 2014). This section will discuss the many different techniques available to assess the performance of nanomaterials both *in vitro* and *in vivo*.

17.2.1 *IN VITRO* ASSAYS AND ASSESSMENT METHODS

In vitro studies can be used to ascertain a general understanding of a biomaterial's cytotoxicity via the assessment of the ability of certain cell types to survive exposure to that material. Assessing cell behavior during interaction with a specific biomaterial is one of the main processes in order to get a biomaterial ready to be used in a clinical trial. While *in vitro* tests often fail to adequately recapitulate the complex responses *in vivo*, they enable faster and more reproducible screening of biomaterials than *in vivo* studies.

In vitro studies make an approximate evaluation of biomaterials biocompatibility in the presence of cultured cells (Lewinski et al. 2008). Cell behavior includes cell viability, proliferation, and phenotypic change. All of those can be important for the selection of a biomaterial (Masters 2000; Freshney 2005; Low and Voelcker 2014). In this regard, conventional cell viability assays have been commonly used to examine the cell's viability as well as the biomaterial toxicity, potentially indicating the suitability of a biomaterial for a specific application (Low and Voelcker 2014). Any toxic agent, which is released from the degradation process of a biomaterial within biological fluids, can also have a direct effect on cell behavior. The most severe cell responses to a cytotoxic component observed *in vitro* are apoptosis and necrosis (Freshney 2005; Liu et al. 2014; Low and Voelcker 2014).

Several cell types such as HeLa and HEK are commonly used, among a multitude of others, to assess this cytotoxicity (Lewinski et al. 2008). *In vitro* analysis often includes cell visualization techniques such as transmission electron microscopy (TEM), scanning electron microscopy (SEM), confocal microscopy, and flow cytometry (Liu et al. 2014). Current cell-based assays can be separated into three types of assessments including cytotoxicity, mechanistic, or cellular response (Liu et al. 2014). Assays that directly indicate cytotoxicity include viability, proliferation, and morphological assays (Liu et al. 2014). Assays that evaluate or indicate a mechanism of action include cell cycle, metabolistic, cytoskeletal, mitochondrial, and cell membrane integrity assays (Liu et al. 2014). Cellular responses to a material can be assessed by assays such as oxidative stress, DNA damage, and inflammatory and apoptosis/necrosis assays (Liu et al. 2014).

These *in vitro* methods are used as a first screening approach for the understanding of the biomaterials interactions with a biological environment. However, one must make sure that the assessment method used is compatible with the material under assessment. For example, it has been demonstrated that the 3-(4,5-dimethylthiazol-2-yl)-2,5-diphenyl tetrazolium bromide (MTT) assay is not compatible with PSi-based materials (Low et al. 2006; Laaksonen et al. 2007). MTT is reduced by mitochondrial dehydrogenases in metabolically active cells, but the PSi surface can also reduce MTT in the absence of cells. In addition, some PSi-based materials have been shown to produce reactive oxygen species (ROS) in cell culture media (Low et al. 2010). Maioli et al. (2009) suggests that assays like MTT but also other methods that work on the same principle, such as XTT, WST-8, and CCK-8, could also be interfered by ROS.

17.2.2 *IN VIVO* ASSAYS AND ASSESSMENT METHODS

In vivo assessment not only expands on the *in vitro* assessment, but also it is vital to validate the effect of a biomaterial on the animal body itself. This allows for the assessment of responses and reactions of biomaterials that may not have become apparent in *in vitro* cytotoxicity studies, but still have detrimental effects in animal models. To be determined as safe, the material must not only be nontoxic to the certain local cell types, but also it must be compatible with all tissues and organs with which it comes into contact throughout its entire lifetime in the body, including its resorption and excretion (Kunzmann et al. 2011). For example, *in vitro* assays cannot predict effects such as intravascular coagulation, embolic events, or chelation of ions among others (Liu et al. 2014). Most commonly, *in vivo* assessment on PSi-based particles is performed in mice, rats, or rabbits via routes including oral, subcutaneous, or intravenous administration (Liu et al. 2014). Essential aspects monitored postadministration include the biodistribution, accumulation/retention/excretion, clearance, body weight, gene expression, long-term effects, and immune response among others (Liu et al. 2014).

Opsonization *in vivo* is also an important factor that needs to be taken into consideration (Owens and Peppas 2006). The size, shape, and surface chemistry of particulate biomaterials can influence the opsonization in the bloodstream, and hence, affect the clearance and cellular uptake (Moghimi et al. 2005). Smaller particles appear to be less effective at activating the complement system and hence have prolonged lifetimes *in vivo* compared to particles above 200 nm in size (Moghimi et al. 2005; Liu et al. 2014).

The extent of biodegradation is also a vital parameter as nondegraded particles may accumulate and cause adverse effects *in vivo*. Hence, assessment of where these particles accumulate and how they are excreted or what potential issues they may cause if they are retained in the body is of vital importance.

17.2.3 SUMMARY

Most toxicity studies performed to date have been performed with immortalized cancer cell lines (Lewinski et al. 2008). It is more important now to begin to test these materials in primary cell cultures, which are more closely matched to *in vivo* conditions. These assessments must be performed for every surface modification used as this will affect the uptake, biological response, and eventual biodistribution.

17.3 DEGRADATION OF PSi IN AQUEOUS SOLUTION

Canham et al. (Canham 1995; Canham et al. 1996, 1999; Anderson et al. 2003) observed that PSi degrades into orthosilicic acid ($Si(OH)_4$) while being in direct contact with aqueous solutions such as blood plasma. Figure 17.1 shows the proposed mechanism of PSi degradation into $Si(OH)_4$ on reaction with water. Briefly, when the freshly etched PSi is immersed in aqueous solution, the Si-H bond is converted to an Si-OH bond and H_2 is evolved. This polarizes the surface and makes it susceptible to further hydrolytic attack, eventually resulting in the release of $Si(OH)_4$ (Allongue et al. 1993).

The degradation rate of PSi in aqueous solution can be influenced mainly by the percentage of porosity (void fraction) and other factors such as surface chemistry and pH of the aqueous environment (Canham et al. 1999, 2000; Anderson et al. 2003). It has been reported that PSi compositions with higher porosity commonly have a higher dissolution rate. In an investigation by Anderson et al. (2003), three different porosity percentages of PSi, medium 62%, high 83%, and very high 88%, have been compared in terms of their degradation rate in an aqueous buffer solution with a pH of 7. As shown in Figure 17.2a, the higher the porosity of PSi, the higher the rate of orthosilicic acid release. After 6 h, samples with the highest porosity (88%) showed 30% mass loss. It has also been shown that the degradation rate of PSi has a direct relationship with the pH value of the aqueous solvent. Dissolution rates increase with pH regardless of the porosity (Figure 17.2b) (Anderson et al. 2003) as the hydroxide ions will readily attack and dissolve the oxide layer (Sailor 2012).

Although orthosilicic acid is the major component released during PSi degradation, it is, however, not the only dissolution component, which might induce cytotoxicity. During the last

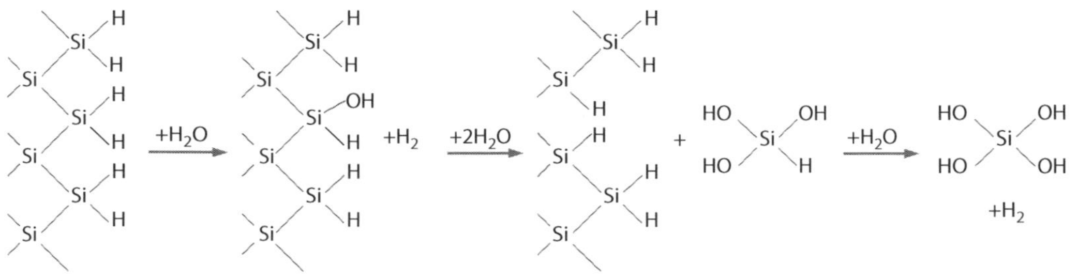

FIGURE 17.1   The proposed mechanism for PSi degradation in aqueous solution. (With kind permission from Springer Science+Business Media: In: Canham L.T. (Ed.), *Handbook of Porous Silicon*, 2014, Low S.P. and Voelcker N.H.)

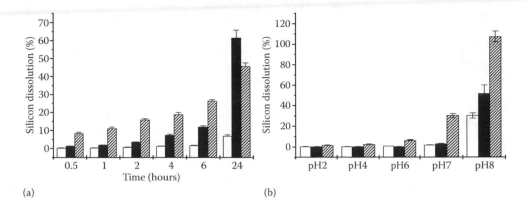

(a)   (b)

FIGURE 17.2   (a) Degradation of PSi with different porosities and (b) effect of increasing pH on PSi degradation. Medium (62%) porosity (white); high (83%) porosity (grey); and very high (88%) porosity (hatched). (Anderson S.H.C. et al.: *Phys. Stat. Sol. (a)*. 2003. 197(2). 331. Copyright Wiley-VCH Verlag GmbH & Co. KGaA. Reproduced with permission.)

decade, researchers have demonstrated that during the degradation process of PSi, ROS are also produced. This process is dependent on the surface chemistry of the degrading PSi (Kovalev et al. 2004; Belyakov et al. 2007). Recent studies have reported that freshly etched PSi can generate ROS at a concentration capable of causing cell death. However, simple oxidation of the PSi was successful at mitigating this effect (Low et al. 2010; Santos et al. 2010).

Surface modification of PSi can alter the rate of degradation, allowing for the crosslinking of targeting antibodies and controling the adsorption rate of biomolecules, including proteins. This may allow for the enhancement of cellular adhesion and proliferation (Dancil et al. 1999; Tinsley-Bown et al. 2000; Low et al. 2006; J. Yang et al. 2010). Utilizing different surface modification methods allows the dissolution time of PSi to be tuned from minutes to many months (Cheng et al. 2008). Being able to modify PSi with a wide range of surface modification methods allows for modulation of drug dissolution rate in a physiological media, highlighting the potential of PSi as a suitable material for drug delivery purposes (McInnes and Voelcker 2009). Salonen et al. (2005) have investigated the relation of two different PSi surface modifications including thermal carbonization (TCPSi) and oxidation on drug dissolution in a physiological relevant media. The authors observed that loading of TOPSi was much lower than TCPSi and that the influence of the PSi materials on drug solubility was most pronounced when using ibuprofen as a model drug at pH 5.5.

17.4 BIOLOGICAL RESPONSES TO PSi

The potential biological barriers, including adsorption of plasma protein (opsonization) and the mononuclear phagocyte system (MPS), that a biomaterial may encounter and need to overcome to perform optimally *in vivo* also need to be carefully considered. For example, the majority of drug vectors are eliminated quickly from blood circulation due to protein opsonization of the drug carrier surface allowing capture by the MPS in the liver (Gregoriadis and Ryman 1972). During opsonization, antigens were detected for immune responses promoting recognition by the MPS and enhancing the phagocytosis process (Patel 1991). It has been found that properties such as the overall surface charge and size of particles are significant factors in determining the rate and extent of phagocyte clearance (Liu et al. 2014). The biodistribution and chemical stability of PSi might be enhanced by coating with proteins, resulting in the generation of both negatively and positively charged PSi microparticle surfaces (Serda et al. 2011). The authors investigated the effect of serum proteins on cellular uptake using PSi microparticles and highlighted that the anionic-surface MPs accumulated equally in the liver and spleen while cationic-surface MPs were only observed in the spleen. Another study on nanoparticle biodistribution by Sarparanta et al. (2012a) certified that the distribution of THCPSi nanoparticles coated with HFBII favored the spleen and liver, compared to uncoated THCPSi nanoparticles. The authors attribute this to the liver and spleen containing certain apsonins and apolipoproteins (Figure 17.3).

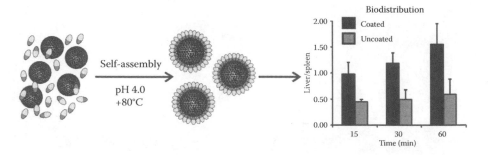

FIGURE 17.3 Biodistribution of HFBII coated on PSi nanoparticles. Uptake in the liver and spleen was higher than for uncoated PSi nanoparticles. (Reprinted with permission from Sarparanta M. et al., *Mol. Pharm.* 9(3), 654. Copyright 2012 American Chemical Society.)

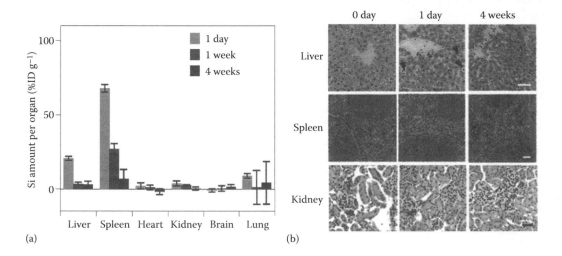

FIGURE 17.4 (a) Biodistribution and biodegradation of LPSiNPs for three different time points after intravenous injection into the mouse, (b) histology of three organs (liver, spleen, and kidney) after 0, 1, and 4 days of intravenous injection of LpSiNPs. The scale bar is 50 μm. (Reprinted by permission from Macmillan Publishers Ltd. *Nat. Mater.*, Park J.-H. et al., 8(4), 331. Copyright 2009.)

Different studies have reported that two parameters play an important role in the distribution of PSi particles within the living animal body including surface chemistry and surface electrical charge while the majority of intravenous injected PSi nanoparticles were rapidly removed from the intravascular system and accumulated inside the liver and spleen (Park et al. 2009; Bimbo et al. 2010; Tanaka et al. 2010a). In addition, the authors concluded that the degradation time of these particles was totally dependent on the organs in which the particles accumulated (Figure 17.4a). For example, Park et al. (2009) reported the biodegradation of luminescent PSi nanoparticles (LPSiNPs) within the liver and spleen within four weeks without any adverse effects on cell morphology (Figure 17.4b). These particles degraded into soluble orthosilicic acid followed by excretion in the urine. Another investigation by Tanaka et al. (2010a) showed that silanized PSi nanoparticles were totally cleared from the spleen within three weeks while it took much longer for the liver to clear the PSi nanoparticles.

17.5 THE EFFECT OF SIZE, SHAPE, AND REACTIVITY ON THE BIOCOMPATIBILITY OF PSi

Numerous studies have been performed to determine the importance of size and shape on chemical reactivity and biocompatibility of PSi. These factors are important as the surface area of PSi nanoparticles is much higher compared to microparticles; this enhances the chemical reactivity almost 1000-fold (Buzea et al. 2007). This effect makes it difficult to predict the interaction of the micro- and nanoparticles in biological environments.

Various studies have investigated the toxic effect of PSi microparticles (Laaksonen et al. 2007; Bimbo et al. 2010; Decuzzi et al. 2010; Low et al. 2010; Santos et al. 2010; Tanaka et al. 2010b) and PSi nanoparticles (Park et al. 2009; Bimbo et al. 2010, 2011). Bimbo et al. (2010, 2011) performed comprehensive investigations on different particle size of THCPSi and TOPSi (from nano to micro) on RAW 264.7 and Caco-2 macrophage cell lines to assess *in vitro* toxicity. Cell viability was evaluated using a luminescent method based on the quantitative analysis of ATP-content production (Santos et al. 2010), as shown in Figure 17.5a. The 1- to 10-µm THCPSi particles show the lowest percentage of cell viability for the Caco-2 cells compared to the other particle sizes, while the TOPSi particles in the range of 1–10 and 10–25 µm show the highest toxic effect in Caco-2 cells (Figure 17.5b). For the nano-sized particles, the THCPSi in the range of 97–188 nm represent more cytotoxicity compared to 142 nm in RAW 264.7 microphages (Figure 17.5c). TOPSi microparticles also displayed higher cytotoxicity in RAW 264.7 macrophage cell line than the nanoparticle formulations (Figure 17.5d). It might be concluded that PSi microparticles elicit more cytotoxicity responses than the nanoparticles in both cell lines and that, in general, the toxicity of all particle sizes increases with increasing particle concentrations (Bimbo et al. 2010, 2011).

In addition to size, it has been found that shape also has a strong influence on the fate of particles in a physiological environment. Typically, spherical particles are used for drug delivery applications. This is often due to the ease with which spheres are made but enhancements could be made with a basic understanding of shape influence on targeting efficiency (Shah et al. 2011). There are a limited number of investigations concluding that other particle shapes such as rod-like and cylindrical particles can be more effective in improving biological properties such as cell adhesion and specific organ accumulation (Shah et al. 2011; Wan et al. 2014). Other studies on PSi microparticles have highlighted the importance of shape in margination and internalization inside the living animal body (Gentile et al. 2008; Decuzzi et al. 2010). Based on their conclusions, discoidal particles showed larger margination compared to quasi-hemispherical and spherical particles, which nominated this particle shape in vascular carriers for optimal cellular uptake and biodistribution within the body.

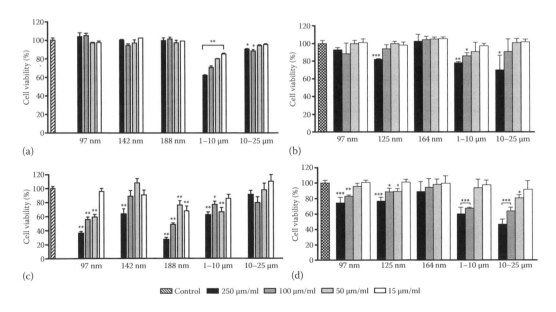

FIGURE 17.5 Cell viability after exposure to THC PSi and TO PSi micro- and nanoparticles in Caco-2 intestinal cells (a and b, respectively) and RAW 264.7 macrophage (c and d, respectively). Cell viability assessed based on luminescent assay after 24-h incubation with different particle concentrations (µg/ml). Error bars represent mean ± SEM (a and c) and s.d (b and d) (n ≥ 3). Statistical analysis by ANOVA represents the level of significance at probabilities of * $p < 0.05$, ** $p < 0.01$, and *** $p < 0.001$. (Reprinted with permission from Bimbo L.M. et al., *ACSNano* 4(6), 3023. Copyright 2011 American Chemical Society.)

17.6 *IN VITRO* STUDIES OF PSi

The first investigation on biocompatibility of PSi for cell delivery purposes was done in 1995. The experiment successfully showed the growth of hydroxyapatite on PSi (Canham 1995). Later on, Bayliss et al. (Bayliss et al. 1997, 1999; Sapelkin et al. 2006) investigated the viability of two different cell lines including neuronal cell line B50 and the Chinese Hamster Ovary (CHO) on freshly etched PSi surfaces. The cell viability was measured utilizing an MTT assay and by neutral red uptake. This was, of course, prior to the discovery that these assays are interfered with fresh PSi (Low et al. 2006). However, they found that cells preferred to grow on PSi compared to control glass surface (Bayliss et al. 1999). Various *in vitro* studies utilizing several cell types have now been performed with different modified PSi surfaces, which have resulted in varied cellular responses, demonstrating that cell growth and cell attachment on PSi materials can be manipulated by varying the surface chemistry. Table 17.1 summarizes a number of studies, which have assessed PSi-based materials in contact with different cell types.

Although orthosilicic acid is considered a nontoxic molecule, a study by Tanaka et al. (1994) revealed the potential cytotoxic effect of orthosilicic acid at concentrations of 2 mM on fibroblast and macrophages. The study suggests that the cytotoxic effect might be caused by the generation of a temporary silane gas (SiH_4) that is highly toxic. However, research by Mayne et al. (2000) showed the effect of different orthosilicic acid concentrations ranging from 10^{-4} to 10^2 mM on cell viability and no noticeable toxic effect was observed. Moreover, another study

TABLE 17.1   Interactions of Different Cell Types with PSi-Based Materials

Cell Type	PSi Surface Modification	Duration of Assay	Type of Viability Assay	Highlighted Conclusion	Ref.
Rat pheocheromocytoma cell	Functionalized with amine groups	1 day	Alamar blue and lysosomal incorporation assay	Cell attachment and proliferation were promoted. PSi degraded into nontoxic component.	Low et al. 2006
Chinese hamster ovary (CHO) cell	Oxidized silicon nanosponge	72 h	MTS and cell counting	Cell attachment and viability were much higher than control. Useful for complex tissue manipulating.	C.-Y. Yang et al. 2010
Human neuronal cell	Coated with aminosilane	5 h	Staining method	Cell attachment was enhanced.	Sweetman et al. 2011
Rat neural cell (B50)	Freshly etched	24 h	MTT assay	100% cell viability was observed. No toxic component was released.	Mayne et al. 2000
Rat neural tumor (B50) and CHO	Conjugated with trichlorosilane	4 days	MTT and neutral red (NR) assay	Cells remained viable in terms of respiratory and membrane viability. No toxic component was observed. Material assumed as an acceptable substrate.	Bayliss et al. 1999
Rat mesenchymal stem cell (MSC)	Functionalized with ethyl-6-bromohexanoate (EBH) to generate peptide density gradient	6 h	Staining by PICO green	Cell attachment on the peptide gradient surface has a direct relation with density.	Clements et al. 2011

(Continued)

TABLE 17.1 (CONTINUED)   Interactions of Different Cell Types with PSi-Based Materials

Cell Type	PSi Surface Modification	Duration of Assay	Type of Viability Assay	Highlighted Conclusion	Ref.
Human mesenchymal stem cell (hMSC)	4-µm thick sponge-like PSi layer	72 h	Actin and nuclear cytoskeletal staining and counting the cell nucleus	Cells preferentially attached to Si over PSi. Cell's nuclei were on PSi while cell's cytoskeleton were on Si stripes.	Noval et al. 2012
Human lens epithelial cells (HLE)	Functionalized with amine groups	7 days	Fluorescence and scanning electron microscopy	Cells grew out from the explants and formed a monolayer on the PSi membrane.	Low et al. 2009

examined the effects of orthosilicic acid in the human body; the results showed that orthosilicic acid is a weak acid with a pK_a of 9.5, which is likely to have an advantageous role inside the body (Canham 2007). It has been previously established that orthosilicic acid is readily absorbed within the gastrointestinal tract of humans and excreted via the urinary system without any accumulation inside the living environment (Popplewell et al. 1998).

A study by Bimbo et al. (2011) has investigated the effect of different concentration and size of thermally oxidized porous silicon (TOPSi) particles on intracellular ROS production in RAW 264.7 microphages cells after exposure to these particles. As shown in Figure 17.6, the samples with particle size between 1 and 10 µm and 164 nm had the highest and lowest percentage of intracellular ROS production in RAW 264.7 cells, respectively. Further literature evidence describing the increase of intracellular ROS after exposure to PSi micro- or nanoparticles has been provided by Mori et al. (2014). While changes in intracellular ROS could be a result of downstream signaling on many biochemical events, it may not implicate direct result of PSi particles exposure. However, despite evidence showing that extracellular ROS at certain levels induces apoptotic or necrotic cell death, less effort was devoted to investigating the extracellular ROS generated by PSi (Pal et al. 2012). Our research group has documented that freshly etched PSi generated extracellular ROS that induced significant cell death (Low et al. 2010). The lack of research in the role of extracellular ROS generated by PSi on cell behavior and the mechanism of ROS affecting cell behavior has thus far hindered our understanding of the cytocompatibility of PSi. For example, it has been confirmed that cellular tolerance to extracelluar ROS is highly dependent on protein concentration of the culture environment (Gülden et al. 2010). This renders the comparison of different *in vitro* studies on PSi reporting ROS generation but conducted using different cell density difficult.

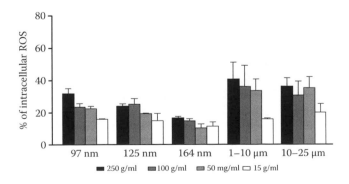

FIGURE 17.6   Percentage of intracellular ROS of RAW 264.7 microphage cells after 1 day incubation with different TOPSi concentrations measuring with a fluorescent 2′,7′-dichlorofluorescein diacetate assay. (Reprinted from *Biomaterials* 32(10), Bimbo L.M. et al., 2625, Copyright 2011, with permission from Elsevier.)

17.7 *IN VIVO* STUDIES OF PSi

The human immune system possesses a very complex set of mechanisms and barriers to defend against the entry of foreign objects, including degradation by enzymes, the mononuclear phagocyte system, interstitial pressure, cell membranes, and efflux pumps (Serda et al. 2009; Liu et al. 2014). Sustained inflammation may result in fibrotic capsules formation, while degradation products or systematically administered PSi particles may also induce a cascade of unpredictable events *in vivo*. Thus, it is difficult to predict the exact response that might be attained while examining the performance of new biomaterials based solely on *in vitro* studies (Liu et al. 2014). Therefore, there is always a high necessity of performing *in vivo* studies to confirm crucial information, such as the biocompatibility, biodistribution, chronic and acute toxicity, and resorption rate of an implantable/injectable biomaterial (Liu et al. 2014).

Rosengren et al. have investigated cellular reactions of soft tissue in contact with different types of materials as implants, including porous and planar silicon. They used titanium as a control and collected cell response data at 1, 6, and 12 weeks (Rosengren et al. 2000). Their results have shown that there were no differences in the density of immunologically activated macrophages as indicated by ED1 and ED2 positive macrophage populations. In addition, fibrous capsule formation has been observed around the cell-PSi interface, and the tissue response was similar for all types of implants. In most cases, the fibrous capsules might cause a complex situation in which they act as barriers around the implants resulting in poor tissue-biomaterial contact, and, even worse, implant loosening. Formation of these capsules is found to also affect the rate of drug release from PSi drug carriers (Ratner and Bryant 2004). Cheng et al. (2008) have compared the stability and cytotoxicity of fresh, hydrosilylated, and oxidized PSi microparticles injected into rabbit vitreous. After remaining for a period of four months inside the rabbit vitreous, no toxic effects were observed while the half life of the particles in the vitreous increased by 5 times for oxidized and 16 times for hydrosilylated PSi from its original 7-d half-life for unmodified PSi particles.

Low et al. (2009) oxidized and aminosilanized PSi membranes prior to examining their potential as an implantable biomaterial in the rat conjunctiva (Figure 17.7). The implant was observed for 8 weeks and showed a slow degradation rate. In addition, a thin fibrous capsule and little evidence of acute inflammatory cells were found at the interface of the tissue and the implanted PSi. In addition, no tissue erosion or neovascularization were observed.

In another study, the biocompatibility of PSi implants was assessed in a nerve regeneration setting (Johansson et al. 2009). A double-sided silicon chip with PSi on one side and smooth silicon

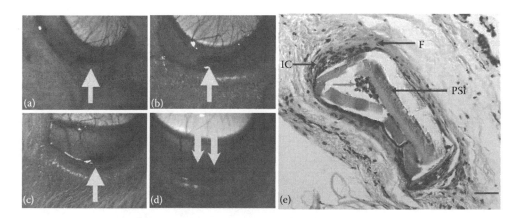

FIGURE 17.7   Modified surface of PSi membrane in rat conjunctiva after (a) implantation for (b) 3, (c) 6, and (d) 9 weeks. The arrows show the PSi implant. Degradation of the implant is completely observable during 9 weeks of implementation. (e) PSi remaining in hematoxylin and eosin-stained sections of eye after 9 weeks. IC, F, and PSi represent inflammatory cells, fibrous capsule, and PSi membrane, respectively (scale bar = 50 μm). (Reprinted from *Biomaterials* 30(15), Low S.P. et al., 2873, Copyright 2009, with permission from Elsevier.)

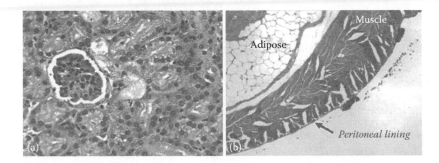

FIGURE 17.8   Histological images of necrosis in Balb/c mice kidney (a) apoptitic irregular cell and (b) cross-section of the peritoneal lining of a nu/nu mouse. (From Park J.S., PhD Dissertation, UCSD, 2010. With permission.)

on the other side was prepared and placed on a 5-mm rat sciatic nerve defect. After 6 to 12 weeks, new nerve tissue was formed on both sides of the Si chip surrounded by perineurium-like capsules. There was no difference in the number of capsules on both sides. However, the thickness of capsule formation on the porous side was significantly lower than on the smooth side. Moreover, lower inflammatory response and stronger adhesion of the tissue on the PSi side led them to conclude that PSi would be a suitable material for enhancement of nerve tissue regeneration. Nevertheless, there is still a lack of research data about the mechanism that allowed the porous side to induce less capsule formation. Future research should shine light into the molecular characteristics of cellular population and tissue integrity around PSi electrodes.

Park et al. (2010) investigated the highest tolerable dose of PSi microparticles intraperitoneally injected into Balb/c mice. Body mass, activity, and general healthiness were used as signs of vitality. They recommended that the maximum tolerated dose (MTD) for freshly etched PSi is 0.23 mg per gram of mouse body weight. In addition, as shown in Figure 17.8, they observed that injecting a high dose of PSi can cause noticeable toxic effects, produced by an accumulation of a great number of irregular cells with condensed nuclei and possibly necrotic cell bodies, due to the obstruction of kidney capillaries.

Table 17.2 summarizes the information on the most recent *in vivo* studies on implantable PSi-based biomaterials.

TABLE 17.2   Recent *In Vivo* Studies of Implantable PSi-Based Biomaterials

Type of PSi Modification	Administration Method	Animal Model	Highlighted Results	Ref.
Thermally hydrocarbonized (THC) porous silicon (THC PSi)	Intravenous	Rat	The majority of particles were in the liver and spleen showing rapid removal from circulation. Superior stability, low toxicity.	Bimbo et al. 2010
Functionalized with silane group	Tail vein	Mice	No infiltration of leukocyte was observed in liver, spleen. Suitable to use as drug delivery carries.	Tanaka et al. 2010a
Monodisperse oxidized PSi microparticles	Tail vein	Mice	No accumulation in kidney and heart were observed. Accumulation was observed in breast tumor, which enabled them to be used in drug delivery.	Godin et al. 2012
THC PSi nanoparticles	Tail vein	Rat	No necrosis was detected. Less toxicity was observed in kidney.	Shahbazi et al. 2013
THC PSi nanoparticles coated with chitosan	Intravenous injection	Mice	No evidence of inflammation was observed. No tissue necrosis or lesions were observed.	Kafshgari et al. 2015

17.7.1 *IN VIVO* ADMINISTRATION ROUTES

The aforementioned *in vivo* biocompatibility studies were conducted based on various administration routes to introduce PSi-derived materials into test animals. Each administration route will have differences in efficiency and bioavailability of the introduced PSi materials, and hence the same material could result in dramatically different biological responses. Intravenous injection (I.V.) is commonly used to administer PSi nanoparticles, although, small particles could be injected subcutaneously (S.C.) and absorbed into circulation. Bimbo et al. (2010) have demonstrated that PSi nanoparticles distributed differently when administered via these two routes. They observed via radiolabeling that PSi nanoparticles accumulated in spleen and liver when they are delivered I.V., while they subside mostly in bone and GI in the case of S.C. and oral administration. The same particles when administered orally were found in urine secretions. There are studies conducted using intraperitoneal injections (I.P.), especially in small animal models such as mice, given that absorption into circulation via peritoneum is proposed as efficient as I.V. for general drugs (Lukas et al. 1971) and even nanoparticles (Abdelhalim and Mady 2011).

17.7.1.1 INTRAVENOUS INJECTION

Intravenous administration of PSi drug carriers are receiving attention due to their ability for targeted and controlled drug delivery (Liu et al. 2014). In order to prevent the accumulation of these drug delivery nano-carriers within the targeted organ, the surface needs to be functionalized with biological recognition ligands. Therefore, there are some challenges in bioconjugation of the host molecules, biodistribution within the organs, and creation of fibrosis tissue. As an example, THCPSi and TOPSi micro- and nanoparticles were injected into the rat myocardium and no significant differences in creation of myocardial fibrosis were observed for both particle types (Tölli et al. 2014). At the end, the authors suggested that these two micro- and nanoparticle formulations were promising platforms for the future of cardiovascular disease treatment. In another study, PSi nanocarriers were decorated with Ly6C antibody and injected intravenously into the stroma of panceratic tumor mouse (Yokoi et al. 2013). They reported that these nanocarriers were accumulated within the tumor-associated endothelial cells within 15 min while it took 4 h to be accumulated within the pancreatic tumor. These results also suggested that Ly6C (or CD59) might serve as a novel dual target to deliver therapeutic agents to the stroma of pancreatic tumors.

17.7.1.2 ORAL ADMINISTRATION

This route is the most common for drug administration because it is most convenient and usually the safest and most inexpensive way. However, some limitations exist with this method due to the enzymes secreted by the digestive tract. In this method, the drug adsorption might be begun in the mouth, then in stomach, and finally in the small intestine. The gastrointestinal (GI) system also possesses wide fluctuations in pH and the efficiency of permeability across the intestinal wall can be low. PSi particles are more stable in acidic pHs compared to alkali pHs and are therefore somewhat resistant to the low pH of the stomach. Capping of PSi with enteric polymeric coatings can allow for the tunable release at the desired pH (Salonen et al. 2008; Santos et al. 2011; McInnes et al. 2012; Santos and Hirvonen 2012). Albrecht et al. (2007) reported that PSi functionalized with alkyl group had a superior durability within simulated gastric fluids compared to unfunctionalized PSi. Another investigation by Sarparanta et al. (2012b) recommended that THCPSi coated with class II hydrophobin (HFBII) nanoparticles remained in the stomach up to 3 h after oral administration as well as the mucoadhesive properties of HFBII coated layer were lost once the PSi particles entered the small intestine, resulting in the localized release to the small intestine.

17.7.1.3 SUMMARY

The accumulation and bioavailability of PSi particles are highly dependent on the route of administration and the size of the particles. There is a plethora of studies showing that various cell types have different susceptibilities to various types of stress-induced damage, including ROS (Morita et al. 2009; Burova et al. 2013). Lastly, it has also been proven that indirect DNA damage induced by the use of nanoparticles depends on the thickness of a cellular barrier, and is mediated through gap junction proteins followed by generation of mitochondrial free radicals (Sood et al. 2011). Therefore, the future progress of *in vivo* studies on biocompatibility of PSi particles has to consider the toxicity mediated by the preferential accumulation in different organs. To date, there are hardly any studies on the *in vivo* biocompatibility of PSi films and membranes. Since PSi is also a potential base material for applications such as *in vivo* biosensing and tissue engineering, future research should shine light into the related biocompatibility issues.

17.8 PSi IN CLINICAL TRIALS

Goh et al. (2007) performed the first in-man study of the safety of a PSi-based biomaterial, called BrachySil™, developed by PSiMedica Ltd. BrachySil is PSi with the ^{32}P radioactive isotope incorporated. This material is used for brachytherapy, the treatment of tumors via the emission of β-radiation from inside the tumor. BrachySil™ can be introduced to the tumor environment via intratumoral injection, under local anesthesia, and deliver the radiation dose through approximately 8 mm of tissue (Canham and Ferguson 2014). Zhang et al. (2005) demonstrated its ability to reduce tumors in nude mice models of pancreatic (2119) and hepatocellular carcinoma xenografts. The first in-man study was performed on unresectable hepatocellular carcinomas (Goh et al. 2007) and CT scans showed successful reduction in the tumor size at 12 and 24 weeks. Advantages of this system include 2-week half-life of the radioactive ^{32}P isotope and the degradation and excretion of the PSi by the body post-dose delivery. Changes observed during the trial were minimal and the targeted tumors had responded within 12 weeks. Three lesions showed 100% regression. It was concluded that percutaneous implantation of BrachySil is safe and well tolerated. In 2008, Ross et al. (2008) published the results of a second safety study of BrachySil (Phase IIa clinical trial) on inoperable pancreatic cancer. The BrachySil was delivered to tumors via endoscopy and CT assessments were performed at 8, 16, and 24 weeks. The treatment was well tolerated and the disease was controlled in 82% of patients; the median survival time was 309 days.

17.9 CONCLUSION

Since early investigations on the properties of PSi and its behavior in contact with biological environments, several studies have examined the biocompatibility of this material for different biomedical applications. Some early *in vitro* studies proposed that orthosilisic acid, the main degradation component of PSi, could have toxic effects on the biological environment. However, further and more extensive investigations proved orthosilicic acid to be safe and cleared in the urine. Hence, properly modified and designed PSi appears to be highly biocompatible. Evidence of enhancement in cell attachment and proliferation has also been observed. Subsequent *in vivo* experiments verified these findings. However, all of the *in vivo* studies were conducted using animal models. The need for further more extensive *in vivo* studies is clear as *in vitro* testing can only determine the response from single cell types at a time and is dependent on the size, shape, and chemistry of the PSi as well as what cell types they are interacting with. *In vivo* testing is also required to investigate the optimal administration route and subsequent biodistribution. PSi has been studied in clinical trials for brachytherapy, and proved effective. The properties of PSi including its biodegradability, high surface area, pore size tunability, and ease of surface modification have raised the demand for further *in vitro* and *in vivo* investigations on PSi-based injectable/implantable biomaterials. The successful design of these materials will lead to interesting applications in different fields such as oncology, ophthalmology, orthpedics, tissue engineering, and drug delivery.

REFERENCES

Abdelhalim, M.A.K. and Mady, M.M. (2011). Liver uptake of gold nanoparticles after intraperitoneal administration in vivo: A fluorescence study. *Lipids Health Dis.* 10, 195–204.

Albrecht, D.S., Lee, J.T., Molby, N. et al. (2007). Functionalized porous silicon in a simulated gastrointestinal tract: Modeling the biocompatibility of a monolayer protected nanostructured material. *MRS Proc.* 1063. http://dx.doi.org/10.1557/PROC-1063-OO06-01.

Allongue, P., Costa-Kieling, V., and Gerischer, H. (1993). Etching of silicon in NaOH solutions I. in situ scanning tunneling microscopic investigation of *n*-Si (111). *J. Electrochem. Soc.* 140(4), 1009–1018.

Anderson, S.H.C., Elliott, H., Wallis, D., Canham, L., and Powell, J. (2003). Dissolution of different forms of partially porous silicon wafers under simulated physiological conditions. *Phys. Stat. Sol. (a)* 197(2), 331–335.

Bayliss, S., Buckberry, L., Harris, P., and Rousseau, C. (1997). Nanostructured semiconductors: Compatibility with biomaterials. *Thin Solid Films* 297(1), 308–310.

Bayliss, S.C., Heald, R., Fletcher, D.I., and Buckberry, L.D. (1999). The culture of mammalian cells on nanostructured silicon. *Adv. Mater.* 11(4), 318–321.

Belyakov, L.V., Goryachev, D., and Sreseli, O. (2007). Role of singlet oxygen in formation of nanoporous silicon. *Semicond.* 41(12), 1453–1456.

Bimbo, L.M., Sarparanta, M., Santos, H.A. et al. (2010). Biocompatibility of thermally hydrocarbonized porous silicon nanoparticles and their biodistribution in rats. *Acs Nano* 4(6), 3023–3032.

Bimbo, L.M., Mäkilä, E., Laaksonen, T. et al. (2011). Drug permeation across intestinal epithelial cells using porous silicon nanoparticles. *Biomaterials* 32(10), 2625–2633.

Bitar, K.N. and Zakhem, E. (2014). Design strategies of biodegradable scaffolds for tissue regeneration. *Biomed. Eng. Comp. Biol.* 6, 13–20.

Bratlie, K.M., Dang, T.T., Lyle, S. et al. (2010). Rapid biocompatibility analysis of materials via in vivo fluorescence imaging of mouse models. *PLoS ONE* 5, e10032.

Buriak, J.M. (2002). Organometallic chemistry on silicon and germanium surfaces. *Chem. Rev.* 102, 1271–1308.

Burova, E., Borodkina, A., Shatrova, A., and Nikolsky, N. (2013). Sublethal oxidative stress induces the premature senescence of human mesenchymal stem cells derived from endometrium. *Oxid. Med. Cellular Longev.* 2013, 474931.

Buzea, C., Pacheco, I.I., and Robbie, K. (2007). Nanomaterials and nanoparticles: Sources and toxicity. *Biointerphases* 2(4), MR17–MR71.

Canham, L.T. (1990). Silicon quantum wire array fabrication by electrochemical and chemical dissolution of wafers. *App. Phys. Lett.* 57(10), 1046–1048.

Canham, L.T. (1995). Bioactive silicon structure fabrication through nanoetching techniques. *Adv. Mater.* 7(12), 1033–1037.

Canham, L.T. (1997). *Properties of Porous Silicon.* INSPEC, London.

Canham, L.T. (2007). Nanoscale semiconducting silicon as a nutritional food additive. *Nanotechnology* 18(18), 185704.

Canham, L.T. and Ferguson, D. (2014). Porous silicon in brachytherapy. In: Canham L.T. (Ed.) *Handbook of Porous Silicon.* Springer, New York.

Canham, L.T., Reeves, C.L., King, D.O., Branfield, P.J., Crabb, J.G., and Ward, M.C. (1996). Bioactive polycrystalline silicon. *Adv. Mater.* 8(10), 850–852.

Canham, L.T., Reeves, C., Newey, J. et al. (1999). Derivatized mesoporous silicon with dramatically improved stability in simulated human blood plasma. *Adv. Mater.* 11(18), 1505–1507.

Canham, L.T., Stewart, M., Buriak, J. et al. (2000). Derivatized porous silicon mirrors: Implantable optical components with slow resorbability. *Phys. Stat. Sol. (a)* 182(1), 521–525.

Chadwick, E.G., Clarkin, O.M., Raghavendra, R., and Tanner, D.A. (2014). A bioactive metallurgical grade porous silicon–polytetrafluoroethylene sheet for guided bone regeneration applications. *Bio-Med. Mater. Eng.* 24(3), 1563–1574.

Cheng, L., Anglin, E.J., Cunin, F. et al. (2008). Intravitreal properties of porous silicon photonic crystals: A potential self-reporting intraocular drug-delivery vehicle. *Br. J. Ophthalmol.* 92, 705–711.

Clements, L.R., Wang, P.Y., Harding, F., Tsai, W.B., Thissen, H., and Voelcker, N.H. (2011). Mesenchymal stem cell attachment to peptide density gradients on porous silicon generated by electrografting. *Phys. Stat. Sol. (a)* 208(6), 1440–1445.

Dancil, K.-P.S., Greiner, D.P., and Sailor, M.J. (1999). A porous silicon optical biosensor: Detection of reversible binding of IgG to a protein A-modified surface. *J. Am. Chem. Soc.* 121(34), 7925–7930.

Decuzzi, P., Godin, B., Tanaka, T., Lee, S.-Y., Chiappini, C., Liu, X., and Ferrari, M. (2010). Size and shape effects in the biodistribution of intravascularly injected particles. *J. Control. Rel.* 141(3), 320–327.

Freshney, R.I. (2005). *Culture of Specific Cell Types.* Wiley Online Library.

Gentile, F., Chiappini, C., Fine, D. et al. (2008). The effect of shape on the margination dynamics of nonneutrally buoyant particles in two-dimensional shear flows. *J. Biomechanics* 41(10), 2312–2318.

Godin, B., Chiappini, C., Srinivasan, S. et al. (2012). Discoidal porous silicon particles: Fabrication and bio-distribution in breast cancer bearing mice. *Adv. Funct. Mater.* 22(20), 4225–4235.

Goh, A.S.-W., Chung, A.Y.-F., Lo, R.H.-G. et al. (2007). A novel approach to brachytherapy in hepatocellular carcinoma using a phosphorous32 (32P) brachytherapy delivery device—A first-in-man study. *Int. J. Radiation Oncology Biol. Phys.* 67, 786–792.

Gregg, S.J. and Sing, K.S.W. (1983). *Adsorption, Surface Area, and Porosity.* Academic Press, London.

Gregoriadis, G. and Ryman, B.E. (1972). Lysosomal localization of enzyme-containing liposomes injected into rats. *Biochemical J.* 128(4), 142P.

Gülden, M., Jess, A., Kammann, J., Maser, E., and Seibert, H. (2010). Cytotoxic potency of H_2O_2 in cell cultures: Impact of cell concentration and exposure time. *Free Radical Bio. Med.* 49, 1298–1305.

Hong, C., Lee, J., Son, M., Hong, S.S., and Lee, C. (2011). In-vivo cancer cell destruction using porous silicon nanoparticles. *Anticancer Drugs* 22, 971–977.

Jaganathan, H. and Godin, B. (2012). Biocompatibility assessment of Si-based nano- and micro-particles. *Adv. Drug Del. Rev.* 64(15), 1800–1819.

Johansson, F., Wallman, L., Danielsen, N., Schouenborg, J., and Kanje, M. (2009). Porous silicon as a potential electrode material in a nerve repair setting: Tissue reactions. *Acta Biomaterialia* 5(6), 2230–2237.

Jones, C.F. and Grainger, D.W. (2009). In vitro assessments of nanomaterial toxicity. *Adv. Drug Del. Rev.* 61(6), 438–456.

Kafshgari, M.H., Delalat, B., Tong, W.Y. et al. (2015). Oligonucleotide delivery by chitosan-functionalized porous silicon nanoparticles. *Nano Res.* doi:10.1007/s12274-015-0715-0, http://www.thenanoresearch.com.

Kohane, D.S. and Langer, R. (2010). Biocompatibility and drug delivery systems. *Chem. Sci.* 1(4), 441–446.

Kovalev, D., Gross, E., Diener, J., Timoshenko, V. Y., and Fujii, M. (2004). Photodegradation of porous silicon induced by photogenerated singlet oxygen molecules. *App. Phy. Lett.* 85(16), 3590–3592.

Kunzmann, A., Andersson, B., Thurnherr, T., Krug, H., Scheynius, A., and Fadeel, B. (2011). Toxicology of engineered nanomaterials: Focus on biocompatibility, biodistribution and biodegradation. *BBA Gen. Subjects* 1810(3), 361–373.

Laaksonen, T., Santos, H., Vihola, H. et al. (2007). Failure of MTT as a toxicity testing agent for mesoporous silicon microparticles. *Chem. Res. Tox.* 20, 1913–1918.

Lewinski, N., Colvin, V., and Drezek, R. (2008). Cytotoxicity of nanoparticles. *Small* 4(1), 26–49.

Liu, D., Shahbazi, M.-A., Bimbo, L.M., Hirvonen, J., and Santos, H.A. (2014). Biocompatibility of porous silicon for biomedical applications. In: Santos H.A. (Ed.), *Porous Silicon for Biomedical Applications.* Woodhead Publishing, Cambridge, UK, pp. 129–181.

Low, S.P. and Voelcker, N.H. (2014). Cell culture on porous silicon. In: Canham L.T. (Ed.), *Handbook of Porous Silicon.* Springer, New York.

Low, S.P., Williams, K.A., Canham, L.T., and Voelcker, N.H. (2006). Evaluation of mammalian cell adhesion on surface-modified porous silicon. *Biomaterials* 27(26), 4538–4546.

Low, S.P., Voelcker, N.H., Canham, L.T., and Williams, K.A. (2009). The biocompatibility of porous silicon in tissues of the eye. *Biomaterials* 30(15), 2873–2880.

Low, S.P., Williams, K.A., Canham, L. T., and Voelcker, N. H. (2010). Generation of reactive oxygen species from porous silicon microparticles in cell culture medium. *J. Biomed. Mater. Res. A* 93(3), 1124–1131.

Lukas, G., Brindle, S.D., and Greengard, P. (1971). The route of absorption of intraperitoneally administered compounds. *J. Pharmacol. Exp. Therapeutics* 178(3), 562–566.

Maioli, E., Torricelli, C., Fortino, V., Carlucci, F., Tommassini, V., and Pacini, A. (2009). Critical appraisal of the MTT assay in the presence of rottlerin and uncouplers. In: Li S. (Ed.), *Biological Procedures Online.* US NCBI, 11, 227–240.

Masters, J.R.W. (2000). *Animal Cell Culture: A Practical Approach.* Oxford University Press, New York.

Mayne, A.H., Bayliss, S.C., Barr, P., Tobin, M., and Buckberry, L.D. (2000). Biologically interfaced porous silicon devices. *Phys. Stat. Sol. (a)* 182(1), 505–513.

McInnes, S.J.P. and Voelcker, N.H. (2009). Silicon-polymer hybrid materials for drug delivery. *Future Med. Chem.* 1(6), 1051–1074.

McInnes, S.J.P. and Voelcker, N.H. (2014). Porous silicon–polymer composites for cell culture and tissue engineering applications. In: Santos H.A. (Ed.) *Porous Silicon for Biomedical Applications.* Woodhead Publishing, pp. 420–469.

McInnes, S.J.P., Szili, E.J., Al-Bataineh, S.A. et al. (2012). Combination of iCVD and porous silicon for the development of a controlled drug delivery system. *ACS Appl. Mater. Interfaces* 4, 3566–3574.

Mitragotri, S. and Lahann, J. (2009). Physical approaches to biomaterial design. *Nature* 8, 15–23.

Moghimi, S.M., Hunter, A.C., and Murray, J.C. (2005). Nanomedicine: Current status and future prospects. *FASEB Journal* 19, 311–330.

Mori, M., Almeida, P.V., Cola, M., Anselmi, G., Mäkilä, E., Correia, A., Salonen, J., Hirvonen, J., Caramella, C., and Santos, H.A. (2014). In vitro assessment of biopolymer-modified porous silicon microparticles for wound healing applications. *Eur. J. Pharm. Biopharm.* 88, 635–642.

Morita, N., Sovari, A.A., Xie, Y. et al. (2009). Increased susceptibility of aged hearts to ventricular fibrillation during oxidative stress. *Am. J. Physiol. Heart Circ. Physiol.* 297, H1594–H1605.

Nel, A., Xia, T., Madler, L., and Li, N. (2006). Toxic potential of materials at the nano-level. *Science* 311, 622–627.

Noval, A.M., Vaquero, V.S., Quijorna, E.P. et al. (2012). Aging of porous silicon in physiological conditions: Cell adhesion modes on scaled 1D micropatterns. *J. Biomed. Mater. Res. A* 100(6), 1615–1622.

Oberdorster, G. (2010). Safety assessment for nanotechnology and nanomedicine: Concepts of nanotoxicology. *J. Internal Med.* 267, 89–105.

Owens, D.E. and Peppas, N.A. (2006). Opsonization, biodistribution, and pharmacokinetics of polymeric nanoparticles. *Int. J. Pharma.* 307, 93–102.

Pal, A.K., Bello, D., Budhlall, B., and Rogers, E. (2012). Screening for oxidative stress elicited by engineered nanomaterials: Evaluation of acellular DCFH assay. *Dose-Response* 10, 308–330.

Park, J.S. (2010). Studies on the toxicity and cisplatin loading of porous Si microparticles. PhD Dissertation. University of California, San Diego.

Park, J.-H., Gu, L., Von Maltzahn, G., Ruoslahti, E., Bhatia, S.N., and Sailor, M.J. (2009). Biodegradable luminescent porous silicon nanoparticles for in vivo applications. *Nat. Mater.* 8(4), 331–336.

Patel, H.M. (1991). Serum opsonins and liposomes: Their interaction and opsonophagocytosis. *Crit. Rev. Ther. Drug Carrier Sys.* 9(1), 39–90.

Popplewell, J.F., King, S., Day, J. et al. (1998). Kinetics of uptake and elimination of silicic acid by a human subject: A novel application of 32Si and accelerator mass spectrometry. *J. Inorg. Biochem.* 69, 177–180.

Ratner, B.D. and Bryant, S.J. (2004). Biomaterials: Where we have been and where we are going. *Annu. Rev. Biomed. Eng.* 6, 41–75.

Riehemann, K., Schneider, S.W., Luger, T.A., Godin, B., Ferrari, M., and Fuchs, H. (2009). Nanomedicine—Challenges and perspectives. *Angew. Chem. Int.* 48, 872–897.

Rosengren, A., Wallman, L., Bengtsson, M., Laurell, T., Danielsen, N., and Bjursten, L. (2000). Tissue reactions to porous silicon: A comparative biomaterial study. *Phys. Stat. Sol. (a)* 182(1), 527–531.

Ross, P.J., Meenan, J., O' Doherty, M., Calara, J., Palmer, D.H., Heatley, S., and Chow, P.H. (2008). Novel delivery via endoscopic ultrasound of a 32P brachytherapy device in addition to gemcitabine in advanced pancreatic cancer. In: *Abstract book of ASCO Gastronintestinal Cancers Symposium.* January 26–28, American Society for Clinical Oncology's. Abstract No 205.

Sailor, M.J. (2012). Fundamentals of porous silicon preparation. In: Sailor M.J. (Ed.), *Porous Silicon in Practice: Preparation, Characterization and Applications.* Wiley-VCH Verlag GmbH & Co. KGaA. pp. 1–42.

Salonen, J. and Lehto, V.-P. (2008). Fabrication and chemical surface modification of mesoporous silicon for biomedical applications. *Chem. Eng. J.* 137, 162–172.

Salonen, J., Laitinen, L., Kaukonen, A.M. et al. (2005). Mesoporous silicon microparticles for oral drug delivery: Loading and release of five model drugs. *J. Cont. Rel.* 108(2–3), 362–374.

Salonen, J., Kaukonen, A.M., Hirvonen, J., and Lehto, V.P. (2008). Mesoporous silicon in drug delivery applications. *J. Pharm. Sci.* 97(2), 632–653.

Santos, H.A. and Hirvonen, J. (2012). Nanostructured porous silicon materials: Potential candidates for improving drug delivery. *Nanomedicine* 7(9), 1281–1284.

Santos, H.A., Riikonen, J., Salonen, J. et al. (2010). In vitro cytotoxicity of porous silicon microparticles: Effect of the particle concentration, surface chemistry and size. *Acta Biomaterialia* 6(7), 2721–2731.

Santos, H.A., Salonen, J., Bimbo, L.M., Lehto, V.-P., Peltonen, L., and Hirvonen, J. (2011). Mesoporous materials as controlled drug delivery formulations. *J. Drug Del. Sci. Tech.* 21(2), 139–155.

Sapelkin, A.V., Bayliss, S C., Unal, B., and Charalambou, A. (2006). Interaction of B50 rat hippocampal cells with stain-etched porous silicon. *Biomaterials* 27(6), 842–846.

Sarparanta, M., Bimbo, L.M., Rytkönen, J. et al. (2012a). Intravenous delivery of hydrophobin-functionalized porous silicon nanoparticles: Stability, plasma protein adsorption and biodistribution. *Mol. Pharm.* 9(3), 654–663.

Sarparanta, M.P., Bimbo, L.M., Mäkilä, E.M. et al. (2012b). The mucoadhesive and gastroretentive properties of hydrophobin-coated porous silicon nanoparticle oral drug delivery systems. *Biomaterials* 33(11), 3353–3362.

Serda, R.E., Ferrati, S., Godin, B., Tasciotti, E., Liu, X., and Ferrari, M. (2009). Mitotic trafficking of silicon microparticles. *Nanoscale* 1, 250–259.

Serda, R.E., Blanco, E., Mack, A. et al. (2011). Proteomic analysis of serum opsonins impacting biodistribution and cellular association of porous silicon microparticles. *Mol. Imaging* 10(1), 43.

Shah, S., Liu, Y., Hu, W., and Gao, J. (2011). Modeling particle shape-dependent dynamics in nanomedicine. *J. Nanosci. Nanotech.* 11(2), 919.

Shahbazi, M.-A., Hamidi, M., Mäkilä, E.M. et al. (2013). The mechanisms of surface chemistry effects of mesoporous silicon nanoparticles on immunotoxicity and biocompatibility. *Biomaterials* 34(31), 7776–7789.

Sood, A., Salih, S., Roh, D. et al. (2011). Signalling of DNA damage and cytokines across cell barriers exposed to nanoparticles depends on barrier thickness. *Nat. Nanotech.* 6, 824–833.

Sweetman, M.J., Shearer, C.J., Shapter, J.G., and Voelcker, N.H. (2011). Dual silane surface functionalization for the selective attachment of human neuronal cells to porous silicon. *Langmuir* 27(15), 9497–9503.

Tanaka, T., Godin, B., Bhavane, R. et al. (2010a). In vivo evaluation of safety of nanoporous silicon carriers following single and multiple dose intravenous administrations in mice. *Int. J. Pharm.* 402(1), 190–197.

Tanaka, T., Mangala, L.S., Vivas-Mejia, P.E. et al. (2010b). Sustained small interfering RNA delivery by mesoporous silicon particles. *Cancer Res.* 70(9), 3687–3696.

Tanaka, T., Tanigawa, T., Nose, T., Imai, S., and Hayashi, Y. (1994). In vitro cytotoxicity of silicic acid in comparison with that of selenious acid. *J. Trace Elements Exp. Med.* 7(3), 101–111.

Tinsley-Bown, A.M., Canham, L., Hollings, M. et al. (2000). Tuning the pore size and surface chemistry of porous silicon for immunoassays. *Phys. Stat. Sol. (a)* 182(1), 547–553.

Tölli, M.A., Ferreira, M.P.A., Kinnunen, S.M. et al. (2014). In vivo biocompatibility of porous silicon biomaterials for drug delivery to the heart. *Biomaterials* 35(29), 8394–8405.

Uhlir, A. (1956). Electrolytic shaping of germanium and silicon. *Bell System Technical J.* 35(2), 333–347.

Wan, Y., Apostolou, S., Dronov, R., Kuss, B., and Voelcker, N.H. (2014). Cancer-targeting siRNA delivery from porous silicon nanoparticles. *Nanomedicine* 9(15), 2309–2321.

Williams, D.F. (2008). On the mechanisms of biocompatibility. *Biomaterials* 29(20), 2941–2953.

Yang, C.-Y., Huang, L.-Y., Shen, T.-L., and Yeh, J.A. (2010). Cell adhesion, morphology and biochemistry on nano-topographic oxidized silicon surfaces. *Eur. Cell Mater.* 20, 415–430.

Yang, J., Mei, Y., Hook, A.L. et al. (2010). Polymer surface functionalities that control human embryoid body cell adhesion revealed by high throughput surface characterization of combinatorial material microarrays. *Biomaterials* 31(34), 8827–8838.

Yokoi, K., Godin, B., Oborn, C.J. et al. (2013). Porous silicon nanocarriers for dual targeting tumor associated endothelial cells and macrophages in stroma of orthotopic human pancreatic cancers. *Cancer Lett.* 334(2), 319–327.

Zhang, K., Loong, S.L.E., Connor, S. et al. (2005). Complete tumor response following intratumoral 32P BioSilicon on human hepatocellular and pancreatic carcinoma xenografts in nude mice. *Clin. Cancer Res.* 11, 7532–7537.

Applications of Porous Silicon Materials in Drug Delivery

18

Haisheng Peng, Guangtian Wang, Naidan Chang, and Qun Wang

CONTENTS

18.1 Introduction 338

18.2 Methods of PSi and Porous Silica Nanoparticle Synthesis 340

18.3 Loading and Controlled Release of Drugs with PSi 341

    18.3.1 Covalent Attachment 341

    18.3.2 Physical Adsorption 342

    18.3.3 Oxidation 342

18.4 Surface Functionalization 342

    18.4.1 Surface Modification 342

        18.4.1.1 Grafting Process 342

        18.4.1.2 Co-Condensation Process 343

        18.4.1.3 Combined Process 343

        18.4.1.4 Esterification 343

    18.4.2 Functionalized Molecules 344

        18.4.2.1 ε-Poly-L-Lysine 344

        18.4.2.2 Folic Acid 344

        18.4.2.3 Cyclodextrin 344

        18.4.2.4 Polyethylenimine 345

        18.4.2.5 Others 345

18.5	Methods of Drug Loading in the Pores	346
	18.5.1 Changes in the Structure	346
	18.5.2 Changes in Surface Properties	346
18.6	Drug Release	347
	18.6.1 The Influence of Pore Diameter	347
	18.6.2 The Influence of Surface Function	347
	18.6.2.1 Nanoparticles as Gatekeepers	348
	18.6.2.2 Organic Molecules as Gatekeepers	348
	18.6.2.3 Supramolecular Assemblies as Gatekeepers	348
18.7	Internalization of Porous Silicon Nanoparticles by Cells	349
	18.7.1 Influence of Size and Shape	349
	18.7.2 Surface Charge	350
	18.7.3 Trafficking in the Cells	350
18.8	Targeted Delivery	351
	18.8.1 Liver	351
	18.8.2 Vascular Barriers	351
	18.8.3 Lung Cancer	352
18.9	Other Porous Silica Drug Delivery Systems	353
	18.9.1 Microneedle	353
	18.9.2 Scaffolds	354
	18.9.3 Sponges	355
18.10	Clinical Trials and Commercial Marketing	355
18.11	Conclusion and Prospective	355
	References	356

18.1 INTRODUCTION

Strategies for nanoscale drug delivery require multidisciplinary cooperation to resolve problems in health care by using various nanomaterials. Nanomaterials possess feature sizes that are less than 100 nm, and their applications include clinical diagnostics, cellular tracking, biosensing, tissue scaffolds, and drug delivery (Thierry and Textor 2012; Jia et al. 2013; Li et al. 2014a; Liu et al. 2014a,b). Of note, targeted drug delivery is a dream of which the ultimate goal is to control drug release in the affected cells without damaging the surrounding healthy tissues (Farokhzad et al. 2004; Asadishad et al. 2010; Vivero-Escoto et al. 2010; Li et al. 2014b). Experiments with nanoparticles have shown that this dream can be realized. For example, Figure 18.1 shows microscopic images of porous silica particles with hexagonally packed light dots and parallel stripes, which are localized in HeLa cells. It should be noted that there are many small molecules that could be used for drug delivery. However, in addition to poor solid-state stability and low solubility or dissolution, many small molecules exhibit low bioavailability (Santos et al. 2011; Du et al. 2014). These challenges must be overcome before these molecules can be utilized in clinical application.

Porous silicon (PSi)-based nanoparticles in general promise to solve some of these problems. (1) Si-based nanoparticles do not contain heavy metal ions. (2) The surface chemistry is well studied (Lie et al. 2002; Li et al. 2004; Tilley et al. 2005). (3) The robust surface chemistry of Si

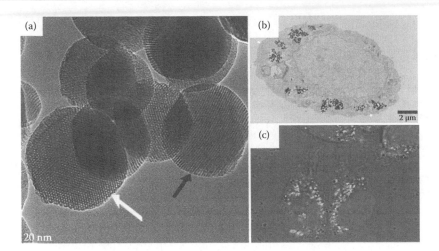

FIGURE 18.1   (a) The TEM image of porous silica nanoparticles, which have the structure of the hexagonally packed light dots (white arrow) and the parallel stripes (black arrow). The internalization (b) and co-localization (c) of silica particles in human cervical cancer (HeLa) by TEM (endosome stained with FM 4-64; nuclear stained with Hoechst 33258). (Vivero-Escoto J.L. et al.: *Small.* 2010, 6, 1952. Copyright Wiley-VCH Verlag GmbH & Co. KGaA. Reproduced with permission.)

particles can greatly simplify the probe used. (4) Si nanoparticles are compatible with the bioconjugation process (Wang et al. 2004). (5) PSi nanoparticles, which exhibit low cytotoxicity, show great promise in terms of compatibility and vitality (Chin et al. 2001). (6) Silicon is a common trace element in humans and a product of the biodegradation of PSi. Orthosilicic acid ($Si(OH)_4$) is the form that is predominantly absorbed by humans and is naturally found in numerous tissues. Furthermore, silicic acid administered to humans is efficiently excreted from the body through urine (Popplewell et al. 1998). (7) PSi-based nanoparticles have a large specific capacity for drug loading. (8) Silicon porosification makes it possible to vary the pore size and to "fine tune" the material for certain drug molecules through chemical modifications of the surface. Owing to these characteristics of PSi-based nanoparticles, in recent years, there has been great interest in their use as carrier systems and therapeutic and diagnostic agents during the treatment of various diseases (Li et al. 2003b; Anglin et al. 2008; Salonen and Lehto 2008; Wu et al. 2011; Jarvis et al. 2012; Savage et al. 2013; Ksenofontova et al. 2014). In the most basic sense, PSi is a network structure of air pores with an internally linked crystalline silicon matrix. Therefore, the free volume inside the pores can entrap drugs that will, in principle, be released into the body with the degradation of the crystalline silicon matrix. By controlling the physical parameters of PSi, the delivery of various types of therapeutics can be regulated. The free volume inside the pores, which may be in excess of 80%, can make room for candidates, including proteins, enzymes (DeLouise and Miller 2004, 2005; DeLouise et al. 2005; Letant et al. 2005; Thomas et al. 2006), drugs (Salonen et al. 2005, 2008; Kaukonen et al. 2007), genes, and even smaller nanoparticles. However, experiments have shown that the pore size, surface chemistry, surface area, and pore structure of PSi materials highly influence their loading efficiency, release rate, and targeting capacity. For example, as it is known, the pore size in PSi may be varied from several nanometers (microporous silicon) up to tens of micrometers (macroporous silicon) (Bisi et al. 2000; Prestidge et al. 2007). This means that all of the abovementioned parameters should be controlled during the preparation of PSi nanoparticles for drug delivery applications.

One should note that regardless of structure and size, the main component of PSi nanoparticles used in drug delivery is usually silica, meaning that silica nanoparticles are representative of PSi materials. PSi nanoparticles have some advantages compared with silica-based materials, but also some disadvantages, such as wider pore size distributions. Similar to PSi, silica nanoparticles do not induce cytotoxicity *in vivo* and may be readily internalized by animal and plant cells. The toxicity of silica-based nanomaterials hinges on several parameters, including particle size, shape, surface chemistry, and porosity (He et al. 2009d; Lin and Haynes 2010; Yu et al. 2012).

18.2 METHODS OF PSi AND POROUS SILICA NANOPARTICLE SYNTHESIS

Experiments have shown that the preparation of PSi nanoparticles and porous silica nanoparticles aimed at drug delivery applications can be achieved using different top-down and bottom-up processes (Bogush et al. 1988; Wilson et al. 1993; Bley and Kauzlarich 1996; Lie et al. 2002; Li et al. 2004; Wang et al. 2004; Tilley et al. 2005). One should note that methods to prepare micron-scale and smaller particles of PSi were developed soon after the discovery of photoluminescence from PSi in the early 1990s (Heinrich et al. 1992; Wilson et al. 1993). The ultrasonication route was the first method used to prepare small PSi particles (Heinrich et al. 1992; Bley et al. 1996). This method is considerably simpler, less expensive, and less time-consuming than lithographic methods (Godin et al. 2012). The general approach involves three steps (Anglin et al. 2008; Qin et al. 2014). First, a PSi layer is etched in a silicon substrate. Single crystalline Si wafers are typically etched with a hydrofluoric acid electrolyte solution. Alternatively, chemical oxidants such as nitric acid can be used to etch the Si surface to produce PSi. A typical TEM image of PSi is shown in Figure 18.2. Next, this PSi layer is removed from the silicon wafer as a freestanding porous film (referred to as "lift-off"). A more detailed discussion of the formation of porous silicon and the removal of the PSi layer from the substrate can be found in *Porous Silicon: Formation and Properties*. Finally, the film is placed in a liquid (usually ethanol or water), and it is fractured through the action of an ultrasonic cleaner. When a freestanding porous layer is subjected to ultrasonication, fracture occurs preferentially along the pore axis in the film. According to Qin et al. (2014), PSi nanoparticles with sizes in the range of 160–350 nm can be prepared by pulsed electrochemical etching of single crystal silicon wafers, followed by ultrasonic fracture of the freestanding porous layer.

It was shown that mechanochemical top-down techniques such as ball-milling could be used as an alternative production method. Russo et al. (2011) reported that using this method, nanometric (<50 nm) PSi powders with high surface area (from 29 up to 100 m^2/g) have been produced. The milling process, which operates at room temperature and atmospheric pressure, could be appealing from a large-scale fabrication point of view, where industrial milling systems are adopted, and are competitive with other synthesis methods. Powders from PSi membranes are reduced in nanometric agglomerates in a very short time without destroying the initial silicon skeleton.

Silica nanoparticles can also be prepared using the abovementioned methods. However, the most common techniques of obtaining such nanoparticles are wet chemical methods. It should be noted that the development of mesoporous silica nanomaterials (MSNs) was a landmark in material science. Monodisperse silica nanoparticles were first chemically synthesized by Stöber and his colleague in 1968 (Stöber and Fink 1968). The synthesis procedure of MSNs follows the well-established Stöber reaction for the preparation of monodisperse nonporous silica spheres. The narrow size distribution is beneficial to the application of MSNs in the field of biomedicine and can be achieved through the precise control of parameters such as nucleation and growth rate of nanoparticles. Strategies for controlling such parameters include the selection of template agents, base- or acid-catalyzed processes, and the use of co-solvents. The structure of MSNs includes MSNs with a hexagonal symmetry, blackberry-like MSNs, chrysanthemum-like MSNs, and amorphous silica.

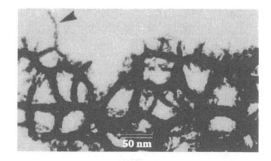

FIGURE 18.2   Non-luminescent samples of high porosity Si layers by TEM. (Reprinted from *Surf. Sci. Rep.* 38, Bisi O. et al., 1, Copyright 2010, with permission from Elsevier.)

Silica nanoparticles can also be prepared by following the process of hydrolysis and condensation of tetraethyl orthosilicate in ethanol, using ammonia as a catalyst. After the reaction is complete, the nanoparticles are collected by high-speed centrifugation, rinsed with ethanol, and dried to prevent continuous reaction. Porous silica nanoparticles can be acquired using the etching process. The silica particles were treated with boron in the presence of silver nitrate and hydrofluoric acid to form porous nanoparticles.

18.3 LOADING AND CONTROLLED RELEASE OF DRUGS WITH PSi

In the design of drug carriers, the drug-loading capacity should be optimized to achieve maximal efficiency. As a host, PSi is similar to dextrin, which provides room for guest molecules or other types of compounds to be delivered to the targeted organ or tissue. Based on the size and dimension of guest molecules, the pore size of PSi can be easily altered to lodge as many drug molecules as possible within it. Drug loading into PSi mesopores could be achieved in several ways (Salonen et al. 2008). The most commonly used method is simple immersion of the PSi particles or layers into the loading solution, in which the desired drug is dissolved in a suitable solvent and the volume of the loading solution is clearly higher than that of the loaded sample. Another method is impregnation, where a controlled amount of drug solution is added to the particles or layers and allowed to infuse into the pores through capillary action (incipient wetness method). Some common advantages of these methods are that the loading can be performed at room temperature and the drug to be loaded is not exposed to harsh chemical environments. In the case of protein and peptide delivery, these features might be essential.

The surface potential and polarity can also be tuned to attract drug molecules to enter the pores of PSi after surface modification (Torney et al. 2007; Chen et al. 2009; Gao et al. 2009; Meng et al. 2010). As shown in Figure 18.3, very large pores in silica particles were synthesized at specific temperatures. In addition to heightening the drug loading capacity, the release profile from the pores can also be regulated by adjusting their structures and modifying their surface properties (He et al. 2009b,c, 2010).

18.3.1 COVALENT ATTACHMENT

Covalent attachment, as a convenient means, not only captures the diagnostic probe but attaches drug molecules through Si-C or Si-O bonds to the pore surface. Long-chain carboxylic acids with terminal alkenes are popular linkers between PSi and drug molecules. Certainly, free carboxyl groups also present opportunities for other modifications with molecules such as poly(ethylene glycol) (PEG) to improve the circulation time of PSi in the vessels. Enzymes or electrolytes may destroy the covalent attachment and release the drug into the tissue.

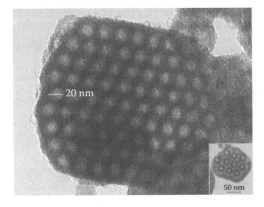

FIGURE 18.3   Very large pores in silica particles were synthesized at 10°C with a double hydrothermal treatment (150 and 140°C); the inset showed the face-cubic centered symmetry structure in a small particle. (Reprinted with permission from Gao F. et al., *J. Phys. Chem. B.* 113, 1796. Copyright 2009 American Chemical Society.)

FIGURE 18.4   The drugs were entrapped in the pore of PSi after oxidation reaction and regulated the release of drug from the pore. (Reprinted from *Drug Deliv. Rev.* 60, Anglin E.J. et al. *Adv.* 1266, Copyright 2008, with permission from Elsevier.)

18.3.2 PHYSICAL ADSORPTION

The polarity of PSi surfaces can be altered when terminates are changed. The hydrophobic hydride on the surface of PSi can increase the adsorption of nonpolar substances, and vice versa. Amphiphilic groups can be regulated by the microenvironment to load various molecules. The low pKa of SiO_2 results in the existence of a layer of negative charges on the PSi, and the electrostatic interactions induced by the surface charges of PSi promote the adsorption of molecules with the opposite charge. The advantage of spontaneous adsorption is that rapid release can be achieved to treat the disease immediately. When the PSi is positively charged, biomolecules with negative charges, such as protein, fibrinogen, and serum albumin, can easily adsorb onto the pore surface.

18.3.3 OXIDATION

Before oxidation, molecules are trapped into the pores of PSi, and the oxidation of PSi shrinks the pores, limiting the release of the molecules from the pores. Aqueous ammonia can be used to induce the oxidized process because of its high pH and nucleophilic nature. Similarly, vapor phase pyridine can initiate the oxidation reaction of PSi. The data supported that the nucleophilic groups on drug molecules such as quinones may be involved in the oxidation reactions. Unfortunately, as shown in Figure 18.4, the shrinkage of the pores caused by oxidation is fairly irreversible, resulting in the endless extension of drug release (Anglin et al. 2008).

18.4 SURFACE FUNCTIONALIZATION

The chemical structure of surfaces is the main factor influencing the interaction of materials with biological systems (Tarn et al. 2013). The surface of porous silica is carpeted with negatively charged silanol groups, which can electrostatically interact with positively charged tetraalkyl-ammonium moieties of the cell membrane, causing cytotoxicity by inhibiting cellular respiration and/or membranolysis (Tao et al. 2008; Tarn et al. 2013). Moreover, rapid aggregation of silica-based nanoparticles can lead to mechanical injury of several important organ capillaries in biological media, resulting in organ failure and even death (Hudson et al. 2008; Yang et al. 2013). Therefore, the functionalization of surface silanol groups with biocompatible molecules is important for improving the biocompatibility of PSi nanoparticles. It was found that surface modification of PSi with different chemical entities changes the physical properties of these nanoparticles, which are crucial for loading efficiency, drug release, and tissue target. Three methods are typically used for the surface modification of nanomaterials: grafting, co-condensation, and combined process.

18.4.1 SURFACE MODIFICATION

18.4.1.1 GRAFTING PROCESS

PSi consists of two different surfaces: the internal surface in the pore and the external surface at the periphery of the PSi particles (Huh et al. 2003). Normally, the internal surface is covered by surfactant molecules before surface modification. The removal of the surfactant makes the

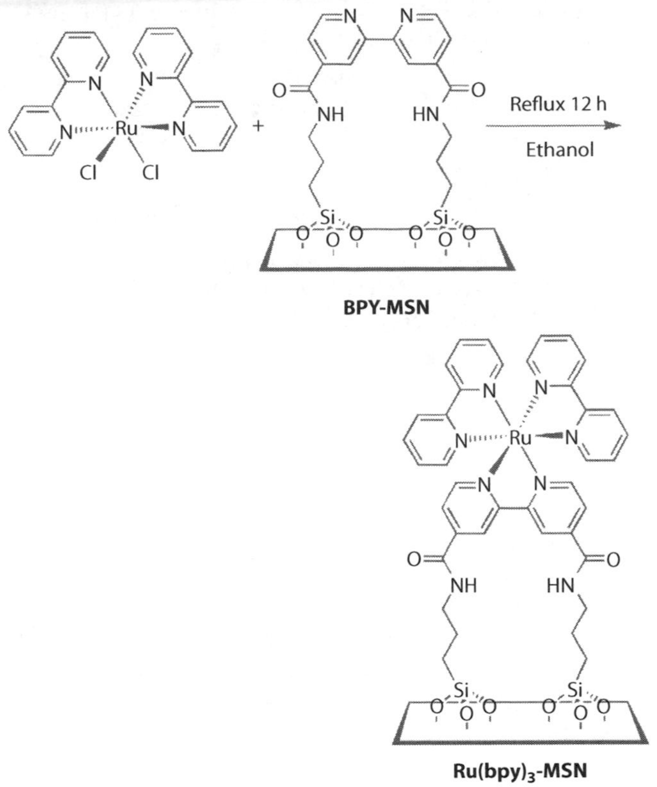

FIGURE 18.5 The surface modification of PSi with excess Ru(bpy)$_2$Cl$_2$. (Reprinted with permission from Kumar R. et al., *Chem. Mater.* 18, 4319. Copyright 2006 American Chemical Society.)

internal surface accessible to further modification via the grafting of molecules. Silanol groups or derivatives such as organoalkoxysilane on the surface can be used for further conjugation. It is more difficult for molecules to be grafted to the internal surface because of the steric hindrance of nanoscale pores (Figure 18.5) (Kumar et al. 2006). Therefore, most conjugated molecules are linked to the external surface or the opening of the pores. The drawback of the grafting process is the deformation of PSi (Lim and Stein 1999).

18.4.1.2 CO-CONDENSATION PROCESS

The co-condensation process of two or more alkoxysilanes can also be used to functionalize the surface of PSi. The existence of silanol precursors (tetramethoxysilane [TMOS]) and cetyl-trimethylammonium bromide (CTAB) can assist the formation of organoalkoxysilanes in the co-condensation process. Various substituents may change the shape of PSi into forms such as spherical or rod-shaped particles based on the entropy of the newly formed structure of PSi (Huh et al. 2003).

18.4.1.3 COMBINED PROCESS

The combination of the two above-mentioned processes produces PSi with a bifunctional surface. Co-condensation can be used to prepare the nanoparticles while the grafting process can be used to modify the free silanols. The surfactant template will be extracted by an acidic alcohol solution to acquire the final PSi (Lin et al. 2000; Liu et al. 2004; Zhang et al. 2004).

18.4.1.4 ESTERIFICATION

Conjugation of undecylenic acid on the PSi surface was performed by hydrosilylation reaction. The PSi surface was conjugated with 3-nitrophenyl undecylenate, resulting in significant oxidation of the surface, while the succinimidyl (NHS) and thioethyl groups of undecylenic acid could not induce the oxidation of the PSi surface. In addition, the hydrophobic end groups can hinder the wetting of the porous surface and stop further reaction. Wojtyk et al. (2002) tried to use ethanol solution similarly to improve the recovery rate of the reaction.

18.4.2 FUNCTIONALIZED MOLECULES

Functional PSi can be achieved via modification of different chemical entities on the PSi surface. The charge and polarity of the surface as well as bioactive molecules such as ligands or enzymatic substrates on the surface could dramatically alter the pharmacokinetics of PSi in the body. In this section, we will elaborate on several typical molecules used for modification.

18.4.2.1 ε-POLY-L-LYSINE

PSi has received much attention, as it is capable of releasing active molecules to certain cells in a controlled fashion. With a homo-polypeptide sequence consisting of approximately 25–30 L-lysine units, ε-poly-L-lysine is a cationic polymer with positively charged hydrophilic amino groups. In addition, ε-poly-L-lysine has been reported to be nontoxic to humans, even at high doses, and is biodegradable by amidases (Shima et al. 1984; Shih et al. 2006). The coating or conjugation of poly-L-lysine on the surface of PSi can increase the attachment of biomolecules on the carriers (Zhou et al. 2011). The polymeric peptide on the surface of MCM-41 is suitable for on-command molecule/drug delivery (Bernardos et al. 2012) and modulates the release kinetics of drugs on PSi (Mondragón et al. 2014).

18.4.2.2 FOLIC ACID

Folic acid, a B vitamin, is a vital component involved in DNA synthesis and repair, also promoting cell division and growth. Many diseases such as cancer, stroke, and heart diseases could be attributed to a deficiency in folic acid. Some cells expressing folic acid receptors can actively uptake ligand-decorated porous silica particles (Hilgenbrink and Low 2005). After uptake, the drug that was loaded into the pores of the PSi particles can be released into the plasma. Several groups have designed mesoporous silica nanoparticles (MSNs) modified with folic acid to deliver chemotherapeutics into tumor cells (Liong et al. 2008; Liu et al. 2011; Li et al. 2012). Some groups have conjugated this vitamin on the surface of MSNs loaded with doxorubicin and evaluated the efficacy of this drug delivery system *in vitro* (Jiang et al. 2010).

18.4.2.3 CYCLODEXTRIN

Cyclodextrin (CD) has a unique structure with a hydrophobic cavity and a hydrophilic rim. Hydrophobic molecules can enter the cavity of CD and form an inclusion complex with the host molecule. The hydrophilic part confers excellent dispersibility in water and can be further functionalized based on the architecture of CD (Callari et al. 2006; Wu et al. 2007; Kollisch et al. 2009; Park et al. 2009; Gonzalez-Campo et al. 2010). Park et al. (2009) plugged β-CD within the pores of PSi, and when β-CD was cleaved from the pores, the drugs were released from within. Dopamine is an important neurotransmitter. The malfunction of dopamine-responsive neurons has been implicated in a number of diseases, including Parkinson disease (Nikolelis et al. 2004). Yu et al. (2011) designed a CD/PSi architecture loaded with dopamine and observed the interference of amino acids on the detection of the drug. The data confirmed that it was a novel and technically simple sensing system. In this system, different amines on the opening of the pores selectively hinder the entrance of amines into the pores of PSi. As shown in Figure 18.6, the system can be used to determine the concentration of dopamine under physiological conditions, with a detection limit of approximately 50 nM (Yu et al. 2011).

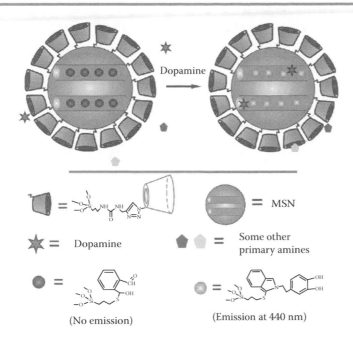

FIGURE 18.6   Mesoporous silica particles for selective detection of dopamine with beta-cyclodextrin as the selective barricade. (From Yu C. et al. *Chem. Commun. (Camb)*. 47, 9086, 2011. Reproduced by permission of The Royal Society of Chemistry.)

18.4.2.4 POLYETHYLENIMINE

Xia et al. (2009) studied the effects of polyethyleneimine (PEI)-coated PSi nanoparticles on the cellular uptake, cytotoxicity, and efficiency of nucleic acid delivery. As cationic polymers, PEIs loaded with siRNA can be easily internalized by cells, making gene therapy possible (Godbey et al. 1999; Urban-Klein et al. 2005). PEIs not only act as a delivery vehicle, but also when coated on nanoparticle surfaces, can attach negative molecules or covalently conjugated active compounds (McBain et al. 2007; Fuller et al. 2008; Park et al. 2008; Elbakry et al. 2009; Liong et al. 2009). In addition, PEIs also regulate endosomal escape for nucleotide delivery (Duan and Nie 2007). PSi nanoparticles have been considered delivery vectors of siRNA. In order to improve the transfection efficiency, the external surfaces of the nanoparticles are normally functionalized with an abundance of amino groups to convert the neutral or negative surface charge to a positive one. Thus, siRNA can easily be complexed with particles through electrostatic forces. However, the coated materials hinder the utility of the pores of the particles. To resolve this problem, strongly dehydrated conditions promoted siRNA to enter the pores of magnetic PSI (M-MSN_siRNA@PEI). The *in vitro* studies have confirmed that this system was good enough to deliver siRNA to specific cells and to release them into the cytoplasm in a highly efficient way.

18.4.2.5 OTHERS

Among numerous polymeric or organosilane surface modification ligands, poly(ethylene glycol) (PEG) is the mostly commonly used because of its biocompatibility and hydrophilic properties (Teow et al. 2011). The amphiphilic block copolymer (F127) containing two hydrophilic PEG blocks and a hydrophobic poly(propylene oxide) (PPO) can facilitate particle formation when it interacts with octyl-modified hydrophobic porous silica nanoparticles in a self-assembly way. The capped block polymer F127 improved the biocompatibility of particles.

18.5 METHODS OF DRUG LOADING IN THE PORES

18.5.1 CHANGES IN THE STRUCTURE

Decades ago, PSi was mainly used as a material with a high specific surface area to load drugs, physically adsorb drugs in the porous surface, and regulate drug release by controlling size or pore form. A series of hollow core/shell or rattle-type structures have been designed by several research groups to encapsulate large numbers of drug molecules into the hollow core. These structures have extraordinarily high drug-loading capacities of greater than 1 g of drugs per gram of nanoparticles (Li et al. 2003a; Zhu et al. 2005a,b,c; He et al. 2009a). Recently, a significant hierarchical porous and hollow structure has been created to simultaneously load two kinds of drugs with different molecular sizes and hydrophilicity or hydrophobicity. A novel core-shell dual-porous structure with smaller pores in the shells and larger ordered pores in the core has been constructed by a simple dual-templating method, which shows a three-step release profile controlled by tuning the shell thickness from 5 nm to 60 nm (Figure 18.7). These core/shell hierarchical structures may provide a useful platform for combined multidrug therapy (Niu et al. 2010).

18.5.2 CHANGES IN SURFACE PROPERTIES

There are two different methods to change the surface properties of the pores to control drug loading: changing the specific surface area of the materials, or adjusting the force between the drug and the surface. Surface functionalization is a method commonly used to modify surface properties, and surfaces with silicon hydroxyl can be coupled with different functional groups. Vallet-Regi et al. (2007) found that the performance of MCM-41, modified with different functional groups, influenced the adsorption and release of ibuprofen. The types of interactions affected the performance, whether it was hydrogen bonding, electrostatic, or hydrophobic interaction. Song et al. (2005) used 3-aminopropyltriethoxysilane (APS) modification and a one-step method of amino functionalization to prepare porous silica and studied drug loading and release, as shown in Figure 18.8. The results showed that amino modification significantly improved the drug-loading rate and induced a long release cycle. The amino functionalization changed the release cycle and extended the solubility of the water-soluble drugs BSA and aspirin.

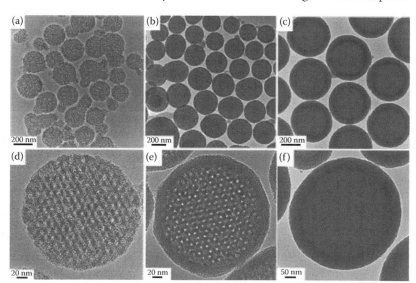

FIGURE 18.7   Shell thicknesses of DMSS by TEM (a, d) 5 nm, (b, e) 25 nm, and (c, f) 60 nm. (Reprinted with permission from Niu D. et al. *J. Am. Chem. Soc.* 132, 15144. Copyright 2010 American Chemical Society.)

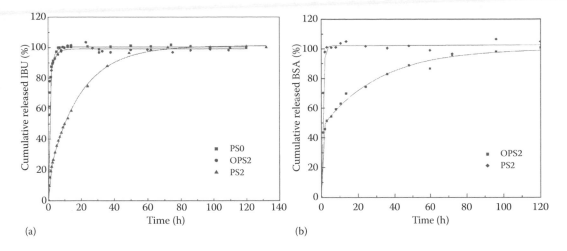

FIGURE 18.8   The release of ibuprofen (a) and bovine serum albumin (b) from various SBA-15 (PS0, OPS2, and PS2). (Reprinted with permission from Song S.W. et al., *Langmuir* 21, 9568. Copyright 2010 American Chemical Society.)

18.6 DRUG RELEASE

To act as a drug carrier, it is highly desired for PSi to exhibit high release after reaching the targeted focal zones. The drug is thus released at specific sites to reduce toxic side effects (Buyuktimkin et al. 2012; Yang et al. 2014b). To reach this goal, great efforts have been made to incorporate various release triggers. Early porous silica particles were mainly based on the physical and textural properties of silica materials such as MCM-41, MCM-48, and SBA-15 with empty pore channels to absorb/encapsulate relatively large numbers of bioactive molecules. The groups of Vallet-Regi and Linden have systematically studied the influence of pore diameter and surface function on drug release. We summarize the results in the following section.

18.6.1 THE INFLUENCE OF PORE DIAMETER

The decrease in pore diameter leads to a decrease in drug loading amount and release rate, which can be linked to steric hindrance effects. These effects were observed when the pore diameters of SBA-15 ranged from 8.2 to 11.4 nm, resulting in the increase in bovine serum albumin loading from 15% up to 27% (Vallet-Regí et al. 2008). In addition, the decrease in pore diameter of MCM-41 or MCM-48 led to a decrease in the release rate of ibuprofen (IBU), erythromycin, and other drugs (Izquierdo-Barba et al. 2005). Qu et al. (2006) pointed out that the pore-size effect could be evaluated only if the morphologies are similar at the microstructure level.

18.6.2 THE INFLUENCE OF SURFACE FUNCTION

As mentioned previously, any drug or gene delivery system must fulfill a list of desirable properties in order to release the content in a suitable concentration, at the desired target, and in a determined amount of time. Silica nanoparticle-based stimuli-responsive systems using the concept of gatekeeping were first developed to achieve these goals (Lai et al. 2003). These systems have the advantage of presenting a variety of chemical entities (such as nanoparticles and organic molecules) as "gatekeepers" to regulate the encapsulation and release of drug molecules, in turn influencing surface function.

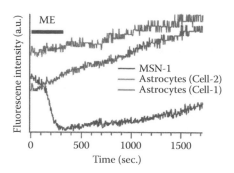

FIGURE 18.9 The graph is a time course of astrocytes and CdS-capped PSi fluorescence incubated with or without mercaptoethanol. (Reprinted with permission from Lai C.-Y. et al. *J. Am. Chem. Soc.* 125, 4451. Copyright 2003 American Chemical Society.)

18.6.2.1 NANOPARTICLES AS GATEKEEPERS

Lai et al. (2003) developed a redox-controlled drug delivery system based on silica particles capped with cadmium sulfide (CdS) nanoparticles. In this system, CdS was chemically attached to particles through a disulfide linker, which was chemically labile and could be cleaved with various disulfide reducing agents such as dithiothreitol (DTT) and mercaptoethanol (ME) (Figure 18.9). As a proof of concept, vancomycin and adenosine triphosphate (ATP) were used as guest molecules. The capping ability of CdS was tested by suspending the CdS-capped silica particles in phosphate buffered saline (PBS). After 12 h of suspension in an aqueous solution, no leaching of guest molecules from the CdS-capped MSNs was observed. The release properties of the CdS-silica particles system were evaluated *in vitro* using different concentrations of reducing agents. After the addition of a disulfide reducing agent (DTT), the CdS-MSN system released 54% of the loaded vancomycin and 28% of the loaded ATP.

The magnetic motor effect is very attractive for the development of site-directing and site-specific drug delivery systems. To accomplish this goal, some groups have developed a redox-controlled drug delivery system that could be realized using a commercial magnet (Giri et al. 2005). The system contained silica nanoparticles functionalized with 3-(propyldisulfanyl)propionic acid. The pores were capped by 3-aminopropyltriethoxysilyl-functionalized superparamagnetic iron oxide nanoparticles (Fe_3O_4). The results indicated that the rate of drug release was dictated by that of the Fe_3O_4 nanoparticles. The combination of controlled release with magnetic motor effects makes this system very attractive for site-specific drug delivery applications (Giri et al. 2005).

18.6.2.2 ORGANIC MOLECULES AS GATEKEEPERS

In addition to solid nanoparticles, flexible organic molecules are able to serve as gatekeepers for porous silica. Tanaka's group first reported a reversible photo-triggered controlled release porous silica (MCM-41) system that could function in organic solvents (Mal et al. 2003). The system took advantage of the photodimerization reaction of coumarin. The pores of MCM-41 were functionalized with 7-[(3-triethoxysilyl)propoxy]coumarin. Given that the size of the coumarinic group was around 1.3 nm, the dimerization of this coumarinic functionality yielded a cyclobutane coumarin dimer, which was large enough to seal the porous opening of MCM-41. The authors showed that both cholestane and phenanthrene could be encapsulated and released in n-hexane solution (Mal et al. 2003).

18.6.2.3 SUPRAMOLECULAR ASSEMBLIES AS GATEKEEPERS

The combination of porous silica materials with supramolecular assemblies has resulted in novel organic/inorganic hybrid materials with improved functionalities. Some groups have developed a dendrimer-capped MSN gene transfection system using a generation 2 poly(amidoamine) dendrimer (G2-PAMAM) (Radu et al. 2004). It was demonstrated that the system could be used for both drug delivery and gene transfection. The group first tethered a G2-PAMAM onto

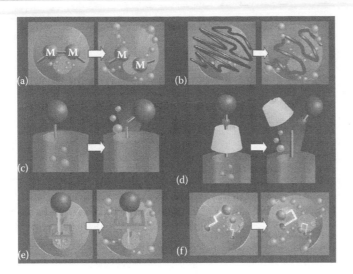

FIGURE 18.10   Pore opening mechanisms of porous silica particle under specific stimuli. (a) Covalent bonds; (b) the swelling and shrinking of polymer chains on the particle surface; (c) existence of metal nanoparticles on the pore openings; (d) with or without a bulky stopper; (e) a supramolecular "nano-valve"; (f) the cis-trans photoisomerization. (From Li Z. et al., *Chem. Soc. Rev.* 41, 2590, 2012. Reproduced by permission of The Royal Society of Chemistry.)

isocyanatopropyl-functionalized silica particles to alter the surface charge of the capped porous silica. This positively charged system was able to bind electrostatically to a plasmid DNA vector encoding the enhanced green fluorescence protein (EGFP). Figure 18.10 shows the pore opening mechanisms with various gatekeepers (Li et al. 2012).

18.7 INTERNALIZATION OF POROUS SILICON NANOPARTICLES BY CELLS

The most important barrier to overcome for the internalization of nanoparticles is the cell membrane. A thorough understanding of the cellular internalization pathways is a central challenge for many applications of nanoparticles in biotechnology and biomedicine. There is a wide variety of pathways in mammalian cells for the internalization of external materials, with more being discovered each year. In general, the internalization mechanisms can be divided into two categories: phagocytosis and pinocytosis (macropinocytosis and endocytosis) (Radu et al. 2004; Mayor and Pagano 2007; Prokop and Davidson 2008). Specialized cells such as macrophages use phagocytosis to engulf microscale particles (>1 μm). The cellular uptake of small particles (<200–300 nm) such as MSNs, polymers, quantum dots, or carbon nanotubes is found to involve endocytosis in the majority of cases (Prokop and Davidson 2008). Previous reports on the intracellular uptake of silica-coated nanoparticles showed that it was possible to internalize these materials through endocytic pathways (Gemeinhart et al. 2005; Xing et al. 2005).

18.7.1 INFLUENCE OF SIZE AND SHAPE

Since the first applications of nanomaterials in intracellular delivery, researchers were aware of the possible effects that the size and shape of nanocarriers could have on the intracellular uptake of nanomaterials. Several studies were carried out to evaluate the effect of size on the uptake of different delivery vehicles (polyplexes, liposomes, and polystyrene) in a wide variety of cell lines (Goula et al. 1998; Zauner et al. 2001; Yamamoto et al. 2002; Nel et al. 2009). The uptake of silica particles by mammalian cells was also influenced by the structural properties of the particles. Particles in the shape of rods (600 nm in length and 100 nm in width) and spheres (80–100 nm in diameter) were studied, and the cellular uptake and kinetics were evaluated in Chinese hamster ovary (CHO) and human fibroblast cells. The data showed that the uptake of particles was dependent on both shape and cell line. CHO cells showed a more efficient uptake capacity for both materials than fibroblast cells did. Recently, the size effect of spherical silica particles on the internalization on HeLa cells was reported. Silica particles were synthesized with similar surface charges but different sizes (30, 50, 110, 170, and 280 nm). In agreement with previous reports, the results showed that a diameter of 50 nm was the optimal size to achieve the most efficient endocytosis (Lu et al. 2009).

18.7.2 SURFACE CHARGE

The cell membrane is a very complex system that consists mainly of lipids, proteins, cholesterol, and receptors. One notable feature of the cell membrane is its net negative surface charge. It is well established that both the surface potential of the nanocarrier and the attachment of cell membrane receptors to the nanocarrier surface can affect endocytic trafficking in mammalian cells (Nel et al. 2009). Considering these properties, Lin and co-workers investigated the effect of charge on the uptake of MSNs by HeLa cells (Slowing et al. 2006; Nel et al. 2009). The surface charge of the silica nanoparticles were tuned based on either amine or guanidine groups. A relationship was discovered between the internalization of the materials and the surface charge. The values for the effective dose at 50% internalization (ED 50) of these materials increased with the increase in the negative value of surface charge. Interestingly, not all of the materials followed the same endocytic pathway. Some of them were internalized through clathrin-mediated pathways while others were internalized through caveolin-mediated pathways.

18.7.3 TRAFFICKING IN THE CELLS

Once the silica particles have overcome the barrier posed by the cell membrane, they are able to release their contents in the cytoplasm. The understanding of the intracellular pathways occurring after endocytosis could support the design of more efficient drug delivery systems. The internalization of the silica particles through endocytosis results in the formation of vesicles, which then undergo a complex series of fusion events and accumulate in specific cytosolic compartments.

The intracellular trafficking pathways of fluorescein-labeled porous silica nanoparticles and the effect of their surface functionalization on their escape from endosomes were investigated by Lin and co-workers (Slowing et al. 2006). The internalization of these materials was monitored by confocal fluorescence microscopy using an endosome marker. The data confirmed that more

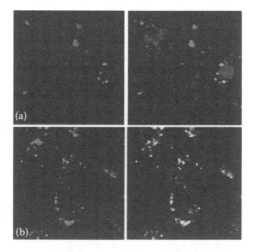

FIGURE 18.11   HeLa cells stained with FM 4-64 and (a) FITC- and (b) FAP-porous silica nanoparticles 6 h after incubation. The images showed the porous silica nanoparticles (white) and FM 4-64-labeled endosomes (gray). (Reprinted with permission from Slowing I. et al., *J. Am. Chem. Soc.* 128, 14792. Copyright 2006 American Chemical Society.)

negatively charged silica nanoparticles were able to escape from the endosomes, while positively charged particles remained trapped within.

These observations were explained by the reversal of surface charge experienced by the MSNs. This phenomenon was due to the transfer of protons from the bulk solution to the surface of the material under acidic conditions, better known as the "proton sponge effect" (Slowing et al. 2006) (see Figure 18.11). The trafficking properties of the silica particles inside the cells were also reported by Chen and co-workers (Huang et al. 2005). The authors investigated the endolysosomal escape of silica particles by using a lysosome-specific label. Moreover, they found that some of the materials localized in the mitochondria according to observations obtained through an organelle-specific probe (Huang et al. 2005).

18.8 TARGETED DELIVERY

In the past decade, the development of silica particle-based nanoscale drug delivery systems for biomedicine applications has been highlighted (Salonen and Lehto 2008; Haidary et al. 2012; Tabasi et al. 2012; Herranz-Blanco et al. 2014). Vital breakthroughs are expected in the future in clinic applications of the diagnosis and treatment of various cancers.

18.8.1 LIVER

Hepatocellular carcinoma (HCC) is a common, malignant disease with a great capability for chemoresistance (Siegel et al. 2012). At least 25,000 newly diagnosed cases of HCC have been reported in the United States in 2012, causing approximately 20,000 deaths. Porous materials with the morphology of particles and controllable sizes have proven to be useful for high loading of therapeutic drugs, protection from environmental interference, control of release behavior, and transport to target sites.

Alginate, a natural polysaccharide, is recognized as the most abundant marine biopolymer in the world and is known for its high biocompatibility and biodegradability (Tonnesen and Karlsen 2002; George and Abraham 2006). Generally, it can chelate with most divalent cations to form a rigid structure (Clark and Ross-Murphy 1987). The structure of alginate could be arranged in many different morphologies such as gels (Novikova et al. 2006), particles (Tan and Takeuchi 2007; Wang et al. 2011), fibers (Knill et al. 2004; Wang et al. 2007), bulk (West and Preece 2009), and films (Srivastava and McShane 2005; Dong et al. 2006). The unique cation-induced gelation properties of alginate enable the encapsulation of any type of nanomaterial with enhanced loading. Researchers have previously combined the advantages of both porous materials and the cation-induced gelation properties of alginate to fabricate porous silica nanoparticle-encapsulated alginate microspheres via the atomization process. The nanocomposite, which has a large surface area for high drug loading, can be applied to the intracellular drug delivery systems against liver cancer cells with sustained release and targeting properties (Liao et al. 2014). Meanwhile, alginate possesses excellent biocompatibility and provides COOH groups for functionalization. The release of rhodamine 6G loaded into the nanocomposite can be prolonged for up to 20 days. For targeted therapy, doxorubicin encapsulated in carriers modified with (lysine)4-tyrosine-arginine-glycine-aspartic acid (K4YRGD) peptides can be used to target hepatocellular carcinoma (HepG2), which are liver cancer cells. In addition, compared with the non-RGD carriers, the RGD modification obviously enhanced the uptake of the nanocomposites. The novel nanocomposite shows great potential as drug vehicles with high biocompatibility, sustained release, and targeting features for future intracellular drug delivery systems.

18.8.2 VASCULAR BARRIERS

Porous silica nanoparticles have recently been found to be one of the most promising carriers for anticancer drug delivery systems (DDSs), and they are expected to achieve enhanced anticancer effects while minimizing harmful side effects to the surrounding normal tissues (Lee et al. 2010;

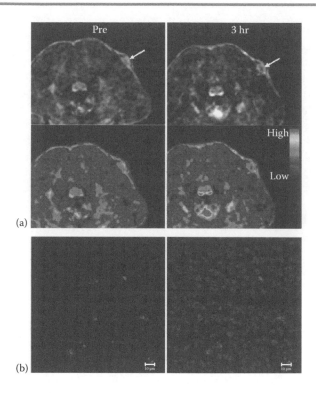

FIGURE 18.12   The image of Fe_3O_4-MSN nanocomposites distribution at tumor site by the *in vivo* imaging system. (a) T2 weighted MR images (first line) and color mapped (second line) images of tumor tissue before and 3 h after administration (arrows indicate tumor site). (b) The images of tumor tissue by confocal laser scanning microscope 24 h after injection. Left: Red fluorescence (white): Fe_3O_4-MSN into the cells. Right: Merged image with DAPI stained nuclei (gray) (scale bar =10 μm). (Reprinted with permission from Lee J.E. et al., *J. Am. Chem. Soc.* 132, 552. Copyright 2010 American Chemical Society.)

Li et al. 2012; Vivero-Escoto et al. 2012; Yang et al. 2012; Chen et al. 2013). As shown via the *in vivo* imaging system in Figure 18.12, Fe_3O_4-MSN nanocomposites were distributed at a tumor site. Unfortunately, effective *in vivo* tumor regression mediated by silica nanoparticle-based drug delivery is still a great challenge because in the process of intravenous DDS-mediated drug administration, there are a series of physiological and biological barriers. To overcome these limitations, a multifaceted targeting strategy is highly desirable to cross successive physiological and biological barriers from blood to organelles for effective intravenous drug delivery. Multifaceted targeting nanomedicine has been functionalized by covalently conjugating RGD and TAT peptides onto the surface of silica nanoparticles. RGD can direct the particles to the tumor vasculature while TAT peptides can promote the particles to cross the barrier of the cell membrane and accumulate in the cell nucleus (Pan et al. 2012, 2013). Therefore, the co-conjugation of RGD and TAT peptides on porous silica nanoparticles could generate a multifaceted vasculature-cell membrane-nucleus targeting system for specific antitumor therapy.

18.8.3 LUNG CANCER

Lung cancer is the main cause of cancerous death in both men and women (Molina et al. 2008). Only 14% of patients suffering from lung cancer can survive for five years, while most of them die within two years (Youlden et al. 2008). One of the leading causes for the poor survival rates among patients is the low efficiency of traditional treatment methods. For lung cancer, pulmonary-targeted drug delivery has been developed to satisfy these requirements (Bi et al. 2008, 2009). Tumor-targeting porous silica nanoparticles were developed for treatment of lung cancer via inhalation. The particles were capable of effectively delivering anticancer drugs (cisplatin and doxorubicin) combined with two types of siRNA to suppress cellular resistance in non–small-cell lung cancer cells. The conjugation of LHRH peptides on the surface of silica particles via PEG linkers can direct the particles to the cancer cells. The combination of delivered anticancer drugs and siRNA induces cell death. Suppression of cancer resistance by siRNA effectively enhanced the concentration of the anticancer drug inside the cancer cells and substantially enhanced the cytotoxicity of the drug. Local delivery of nanocarriers by inhalation led to a stronger accumulation of nanoparticles in mouse lungs, decreasing the escape of particles into systemic circulation and limiting their accumulation in other organs. The data confirmed that the nanocarriers satisfied the major prerequisites for effective treatment of non–small-cell lung carcinoma. Therefore, the lung cancer-targeted silica particles combining drugs and siRNA have high potential in the effective treatment of lung cancer (Taratula et al. 2011).

18.9 OTHER POROUS SILICA DRUG DELIVERY SYSTEMS

18.9.1 MICRONEEDLE

Transdermal delivery is an attractive method to deliver drugs or biological compounds into the human body across the skin because of its distinct advantage of eliminating pain and inconvenient intravenous injections. However, the efficiency of transdermal delivery is greatly limited by the poor permeability of the hard layer of skin at the stratum corneum (SC), which is the outermost layer of skin that forms the primary transport barrier (Ledger 1992). As one of the enhancers, microneedle array devices have been well developed for controlled transdermal drug delivery in a convenient and minimally invasive manner (Lin and Pisano 1999; McAllister et al. 2000; Sivamani et al. 2007). The microneedles are used to penetrate the SC and generate pathways or microchannels for drug delivery into the epidermis layer. No pain is induced as the needles have limited length and do not reach the nerves in the deeper dermis layer.

The microneedles are fabricated by silicon with MEMS technologies. However, the high aspect ratio of the needle structure and the fragility of silicon present several shortcomings. The main problem of the silicon microneedle is that it breaks easily during the insertion process into the skin, which increases the possibility of an infection. A proposed solution is to fabricate the microneedle tip from a biodegradable PSiporous silicon material. This PSiporous silicon is well-known as nanostructured silicon and is adequate for biological applications because of its bioactive and biodegradable properties (Desai et al. 2000; Angelescu et al. 2003). Microneedles have been fabricated to possess macroporous tips using an electrochemical etching process. These porous tips may break off after the drug delivery process and are allowed to remain in the skin, as it can be easily biodegraded within 2–3 weeks.

The researchers fabricated silicon microneedle arrays using the notching effects of reactive ion etching (RIE). As shown in Figure 18.13, the biodegradable tips were successfully realized

FIGURE 18.13   The tip of a microneedle inserted into a pig skin was broken. (With kind permission from Springer Science+Business Media: *Microsyst. Technol.* 14, 2008, 1015. Chen B. et al.)

using the electrochemical anodization process that selectively generated PSiporous silicon only on the top part of the skin. The porous tips can be degraded within a few weeks. The transdermal drug delivery experiments showed that the microneedles could greatly enhance skin permeability for better drug transport. The results indicated the feasibility of microneedles as a more effective enhancer in transdermal drug delivery system with significant clinical potential (Chen et al. 2008).

18.9.2 SCAFFOLDS

Tuberculosis (TB) is a global health problem that continues to present a formidable challenge. According to the World Health Organization, about one-third of the world population is infected with *Mycobacterium tuberculosis*, with millions of new cases being reported annually (Tomioka 2006; Vashishtha 2009). Recently, there has been a spurt in cases of extrapulmonary TB (EPTB) (Huebner and Castro 1995; Raviglione et al. 2001). Therein, osteoarticular TB cases account for approximately 10–11% of EPTB, which amount to approximately 19–38 million cases worldwide (Meier and Beekmann 1995; Puttick et al. 1995; Malaviya and Kotwal 2003). Figure 18.14 shows SEM images of beta-tricalcium phosphate (b-TCP) scaffolds and composite porous silica-b-TCP scaffolds. The scaffolds for treating TB, including osteoarticular TB, mainly promoted multidrug chemotherapy for an appropriate period, which prevented the apparent resistance of TB bacilli to single drugs (Zhu et al. 2011).

Porous silica materials may be a more appropriate choice for controlled and localized antitubercular drug delivery without causing significant changes in pH value. They have been investigated as a drug delivery carrier for over a decade owing to the extensive nanoporous structure of porous silica (Vallet-Regi et al. 2001; Tourne-Peteilh et al. 2003; Doadrio et al. 2004; Shi et al. 2009; Li et al. 2014c; Xin et al. 2014). The pore structure of 2–50 nm in diameter may render good bioactivity and biocompatibility. Compared with solid nanoparticles, the porous silica nanoparticles (MSNs) are apparently a more suitable drug delivery carrier because of their extensive porous structure.

A composite scaffold drug delivery system (CS-DDS) for osteoarticular tuberculosis therapy has been prepared by loading bi-component drugs into a porous beta-tricalcium phosphate scaffold coated with porous silica nanoparticles and an additional bioactive glass coating. Such a CS-DDS showed high performance in the local and sustained delivery of bi-component antitubercular drugs as well as excellent biocompatibility. The composite system is expected to combine the merits of multidrug loading and very sustained and localized drug co-release for effective osteoarticular treatment, in addition to bioactivity for bone repair (Zhu et al. 2011).

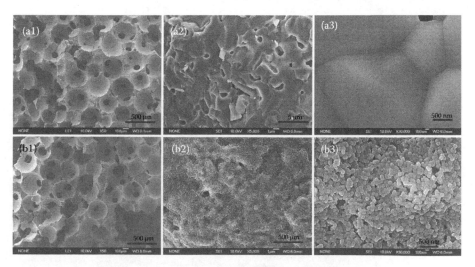

FIGURE 18.14   SEM images of beta-tricalcium phosphate (b-TCP) scaffolds and composite porous silica-b-TCP scaffolds. (Reprinted from *Biomater.* 32, Zhu M. et al., 1986, Copyright 2011, with permission from Elsevier.)

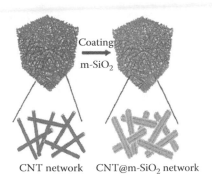

FIGURE 18.15   Schematic illuminaton of CNT@m-SiO$_2$ sponges fabrication. (From Yang Y. et al., *Nanoscale* 6, 3538, 2014. Reproduced by permission of The Royal Society of Chemistry.)

18.9.3 SPONGES

Porous silica materials are highly porous and lightweight, with potential applications in many areas (Corma 1997; Fan et al. 2003; Deng et al. 2008, 2010; Wang et al. 2008, 2013; Wagner et al. 2013). The highly ordered porous structure also gives rise to potential problems in practical applications. Particularly, the structural instability and limited diffusion process through narrow channels hinder the use of porous silica materials in clinical applications of drug delivery.

Carbon nanotubes (CNTs) are nanostructures with excellent mechanical strength. They have been self-assembled into low density, porous materials such as foams, aerogels, and sponges (Gui et al. 2010; Worsley et al. 2010; Zou et al. 2010; Chen et al. 2011; Cong et al. 2012; Kim et al. 2012; Hu et al. 2013; Huang et al. 2013; Nardecchia et al. 2013; Zhao et al. 2013). Data have shown that bulk CNT sponges with three-dimensional CNT networks can be changed into arbitrary shapes and featured large strains with elastic recovery (Gui et al. 2010). The combination of CNT networks with porous silica particles with a high surface area is a promising approach to develop novel porous composites with tailored microstructures and enhanced mechanical properties. Yang et al. (2014a) combined CNT sponges and porous silica nanoparticles through a solution process, significantly enhancing the efficiency of molecular adsorption and drug delivery (Figure 18.15).

18.10 CLINICAL TRIALS AND COMMERCIAL MARKETING

There is still a long way to go before the formulations of PSi-based drug delivery systems can be translated into the clinical market, as sufficient evidence needs to be accumulated to prove their safety and therapeutic efficacy. In addition to scientific efforts, commercial and social drive is also a fundamental requisite. It requires leveraging resources and expertise from a multitude of disciplines, including chemistry, materials science and engineering, biology, pharmaceutical sciences, and clinical/translational research. It is expected that with multidisciplinary efforts for the common objective of clinical benefits, PSi-based nanoformulations can make an exciting breakthrough and direct renovated and individualized therapy in the near future. If encouraging results are acquired from clinical trials, we can envision an era where targeted drug delivery using mechanized nanoparticles will be an important treatment method against various diseases. The advent of personalized medicine will greatly aid the implementation of silica-based drug delivery systems in medicine.

18.11 CONCLUSION AND PROSPECTIVE

This review highlights some recent exciting progress in the synthesis, controlled release, and delivery of drugs based on PSi. Many systems based on porous silica particles have been constructed for controlled drug release and enhanced delivery properties. As summarized, PSi, with its versatile porous structure, has unique advantages that may allow clinically applicable nanoformulations for disease diagnosis and therapy. However, there are still many challenges that need to be overcome in order to make it a reality. One of the biggest challenges in the development of porous silica particles is biosafety, and biodistribution studies need to be carried out in higher mammalian species with the aim of eventually performing clinical trials in humans. It is very important to determine the toxicity and biodistribution of these materials in humans to evaluate the prospect for biomedical applications.

REFERENCES

Angelescu, A., Kleps, I., Mihaela, M. et al. (2003). Porous silicon matrix for applications in biology. *Rev. Adv. Mater. Sci.* 5, 440–449.

Anglin, E.J., Cheng, L., Freeman, W.R., and Sailor, M.J. (2008). Porous silicon in drug delivery devices and materials. *Adv. Drug Deliv. Rev.* 60, 1266–1277.

Asadishad, B., Vossoughi, M., and Alamzadeh, I. (2010). In vitro release behavior and cytotoxicity of doxorubicin-loaded gold nanoparticles in cancerous cells. *Biotechnol. Lett.* 32, 649–654.

Bernardos, A., Mondragon, L., Javakhushvili, I. et al. (2012). Azobenzene polyesters used as gate-like scaffolds in nanoscopic hybrid systems. *Chem. A: Eur. J.* 18, 13068–13078.

Bi, R., Shao, W., Wang, Q., and Zhang, N. (2008). Spray-freeze-dried dry powder inhalation of insulin-loaded liposomes for enhanced pulmonary delivery. *J. Drug Target.* 16, 639–648.

Bi, R., Shao, W., Wang, Q., and Zhang, N. (2009). Solid lipid nanoparticles as insulin inhalation carriers for enhanced pulmonary delivery. *J. Biomed. Nanotechnol.* 5, 84–92.

Bisi, O., Ossicini, S., and Pavesi, L. (2000). Porous silicon: A quantum sponge structure for silicon based optoelectronics. *Surf. Sci. Rep.* 38, 1–126.

Bley, R.A. and Kauzlarich, S.M. (1996). A low-temperature solution phase route for the synthesis of silicon nanoclusters. *J. Am. Chem. Soc.* 118, 12461–12462.

Bley, R.A., Kauzlarich, S.M., Davis, J.E., and Lee, H.W.H. (1996). Characterization of Silicon Nanoparticles Prepared from Porous Silicon. *Chem. Mater.* 8, 1881–1888.

Bogush, H.H., Tracy, M.A., and Zukoski, C.F. (1988). Preparation of monodisperse silica particles: Control of size and mass fraction. *J. Non-Crystall. Solids* 104(1), 95–106.

Buyuktimkin, B., Wang, Q., Kiptoo, P., Stewart, J.M., Berkland, C., and Siahaan, T.J. (2012). Vaccine-like controlled-release delivery of an immunomodulating peptide to treat experimental autoimmune encephalomyelitis. *Mol. Pharm.* 9, 979–985.

Callari, F., Petralia, S., and Sortino, S. (2006). Highly photoresponsive monolayer-protected gold clusters by self-assembly of a cyclodextrin-azobenzene-derived supramolecular complex. *Chem. Commun. (Camb)* 2006, 1009–1011.

Chen, B., Wei, J., Tay, F.E.H., Wong, Y.T., and Iliescu, C. (2008). Silicon microneedle array with biodegradable tips for transdermal drug delivery. *Microsyst. Technol.* 14, 1015–1019.

Chen, A.M., Zhang, M., Wei, D. et al. (2009). Co-delivery of doxorubicin and Bcl-2 siRNA by mesoporous silica nanoparticles enhances the efficacy of chemotherapy in multidrug-resistant cancer cells. *Small.* 5, 2673–2677.

Chen, W., Li, S., Chen, C., and Yan, L. (2011). Self-assembly and embedding of nanoparticles by in situ reduced graphene for preparation of a 3D graphene/nanoparticle aerogel. *Adv. Mater.* 23, 5679–5683.

Chen, Y.P., Chen, C.T., Hung, Y. et al. (2013). A new strategy for intracellular delivery of enzyme using mesoporous silica nanoparticles: Superoxide dismutase. *J. Am. Chem. Soc.* 135, 1516–1523.

Chin, V., Collins, B.E., Sailor, M.J., and Bhatia, S.N. (2001). Compatibility of primary hepatocytes with oxidized nanoporous silicon. *Adv. Mater.* 13, 1877–1880.

Clark, A.H. and Ross-Murphy, S.B. (1987). Structural and mechanical properties of biopolymer gels. *Biopolymers* (*Advances in Polymer Science* Vol. 83). Springer, Berlin, pp. 57–192.

Cong, H.P., Ren, X.C., Wang, P., and Yu, S.H. (2012). Macroscopic multifunctional graphene-based hydrogels and aerogels by a metal ion induced self-assembly process. *Acs. Nan.* 6, 2693–2703.

Corma, A. (1997). From microporous to mesoporous molecular sieve materials and their use in catalysis. *Chem. Rev.* 97, 2373–2420.

DeLouise, L.A. and Miller, B.L. (2004). Quantatitive assessment of enzyme immobilization capacity in porous silicon. *Anal. Chem.* 76, 6915–6920.

DeLouise, L.A. and Miller, B.L. (2005). Enzyme immobilization in porous silicon: Quantitative analysis of the kinetic parameters for glutathione-S-transferases. *Anal. Chem.* 77, 1950–1956.

DeLouise, L.A., Kou, P.M., and Miller, B.L. (2005). Cross-correlation of optical microcavity biosensor response with immobilized enzyme activity. Insights into biosensor sensitivity. *Anal. Chem.* 77, 3222–3230.

Deng, Y., Qi, D., Deng, C., Zhang, X., and Zhao, D. (2008). Superparamagnetic high-magnetization microspheres with an $Fe_3O_4@SiO_2$ core and perpendicularly aligned mesoporous SiO_2 shell for removal of microcystins. *J. Am. Chem. Soc.* 130, 28–29.

Deng, Y., Cai, Y., Sun, Z. et al. (2010). Multifunctional mesoporous composite microspheres with well-designed nanostructure: A highly integrated catalyst system. *J. Am. Chem. Soc.* 132, 8466–8473.

Desai, T.A., Hansford, D.J., Leoni, L., Essenpreis, M., and Ferrari, M. (2000). Nanoporous anti-fouling silicon membranes for biosensor applications. *Biosens. Bioelectron.* 15, 453–462.

Doadrio, A.L., Sousa, E.M., Doadrio, J.C. et al. (2004). Mesoporous SBA-15 HPLC evaluation for controlled gentamicin drug delivery. *J. Control Release* 97, 125–132.

Dong, Z., Wang, Q., and Du, Y. (2006). Alginate/gelatin blend films and their properties for drug controlled release. *J. Membrane Sci.* 280, 37–44.

Du, D., Chang, N., Sun, S. et al. (2014). The role of glucose transporters in the distribution of p-aminophenyl-α-d-mannopyranoside modified liposomes within mice brain. *J. Control Release* 182, 99–110.

Duan, H. and Nie, S. (2007). Cell-penetrating quantum dots based on multivalent and endosome-disrupting surface coatings. *J. Am. Chem. Soc.* 129, 3333–3338.

Elbakry, A., Zaky, A., Liebl, R. et al. (2009). Layer-by-layer assembled gold nanoparticles for siRNA delivery. *Nano. Lett.* 9, 2059–2064.

Fan, J., Yu, C., Gao, F. et al. (2003). Cubic mesoporous silica with large controllable entrance sizes and advanced adsorption properties. *Angew. Chem. Int. Ed. Engl.* 42, 3146–3150.

Farokhzad, O.C., Jon, S., Khademhosseini, A. et al. (2004). Nanoparticle-aptamer bioconjugates: A new approach for targeting prostate cancer cells. *Cancer Res.* 64, 7668–7672.

Fuller, J.E., Zugates, G.T., Ferreira, L.S. et al. (2008). Intracellular delivery of core-shell fluorescent silica nanoparticles. *Biomater.* 29, 1526–1532.

Gao, F., Botella, P., Corma, A., Blesa, J., and Dong, L. (2009). Monodispersed mesoporous silica nanoparticles with very large pores for enhanced adsorption and release of DNA. *J. Phys. Chem. B.* 113, 1796–804.

Gemeinhart, R.A., Luo, D., and Saltzman, W.M. (2005). Cellular fate of a modular DNA delivery system mediated by silica nanoparticles. *Biotechnol. Prog.* 21, 532–537.

George, M. and Abraham, T.E. (2006). Polyionic hydrocolloids for the intestinal delivery of protein drugs: Alginate and chitosan—A review. *J. Control Release* 114, 1–14.

Giri, S., Trewyn, B.G., Stellmaker, M.P., and Lin, V.S. (2005). Stimuli-responsive controlled-release delivery system based on mesoporous silica nanorods capped with magnetic nanoparticles. *Angew. Chem. Int. Ed. Engl.* 44, 5038–5044.

Godbey, W.T., Wu, K.K., and Mikos, A.G. (1999). Tracking the intracellular path of poly(ethylenimine)/DNA complexes for gene delivery. *Proc. Natl. Acad. Sci. U.S.A.* 96, 5177–5181.

Godin, B., Chiappini, C., Srinivasan, S. et al. (2012). Discoidal porous silicon particles: Fabrication and biodistribution in breast cancer bearing mice. *Adv. Funct. Mater.* 22, 4225–4235.

Gonzalez-Campo, A., Hsu, S.H., Puig, L. et al. (2010). Orthogonal covalent and noncovalent functionalization of cyclodextrin-alkyne patterned surfaces. *J. Am. Chem. Soc.* 132, 11434–11436.

Goula, D., Remy, J.S., Erbacher, P. et al. (1998). Size, diffusibility and transfection performance of linear PEI/DNA complexes in the mouse central nervous system. *Gene Ther.* 5, 712–717.

Gui, X., Wei, J., Wang, K. et al. (2010). Carbon nanotube sponges. *Adv. Mater.* 22, 617–621.

Haidary, S.M., Córcoles, E.P., and Ali, N.K. (2012). Nanoporous silicon as drug delivery systems for cancer therapies. *J. Nanomater.* 2012, 18.

He, Q., Guo, L., Cui, F., Chen, Y., Jiang, P., and Shi, J. (2009a). Facile one-pot synthesis and drug storage/release properties of hollow micro/mesoporous organosilica nanospheres. *Mater. Lett.* 63, 1943–1945.

He, Q., Shi, J., Cui, X., Zhao, J., Chen, Y., and Zhou, J. (2009b). Rhodamine B-co-condensed spherical SBA-15 nanoparticles: Facile co-condensation synthesis and excellent fluorescence features. *J. Mater. Chem.* 19, 3395–3403.

He, Q., Shi, J., Zhao, J., Chen, Y., and Chen, F. (2009c). Bottom-up tailoring of nonionic surfactant-templated mesoporous silica nanomaterials by a novel composite liquid crystal templating mechanism. *J. Mater. Chem.* 19, 6498–6503.

He, Q., Zhang, Z., Gao, Y., Shi, J., and Li, Y. (2009d). Intracellular localization and cytotoxicity of spherical mesoporous silica nano- and microparticles. *Small* 5, 2722–2729.

He, Q., Zhang, J., Chen, F., Guo, L., Zhu, Z., and Shi, J. (2010). An anti-ROS/hepatic fibrosis drug delivery system based on salvianolic acid B loaded mesoporous silica nanoparticles. *Biomater.* 31, 7785–7796.

Heinrich, J.L., Curtis, C.L., Credo, G.M., Kavanagh, K.L., and Sailor, M.J. (1992). Luminescent colloidal silicon suspensions from porous silicon. *Science* 255, 66–68.

Herranz-Blanco, B., Arriaga, L.R., Mäkilä, E. et al. (2014). Microfluidic assembly of multistage porous silicon−Lipid vesicles for controlled drug release. *Lab. Chip* 14, 1083–1086.

Hilgenbrink, A.R. and Low, P.S. (2005). Folate receptor-mediated drug targeting: From therapeutics to diagnostics. *J. Pharm. Sci.* 94, 2135–2146.

Huang, D.M., Hung, Y., Ko, B.S. et al. (2005). Highly efficient cellular labeling of mesoporous nanoparticles in human mesenchymal stem cells: Implication for stem cell tracking. *Faseb. J.* 19, 2014–2016.

Huang, H., Chen, P., Zhang, X., Lu, Y., and Zhan, W. (2013). Edge-to-edge assembled graphene oxide aerogels with outstanding mechanical performance and superhigh chemical activity. *Small* 9, 1397–1404.

Hudson, S.P., Padera, R.F., Langer, R., and Kohane, D.S. (2008). The biocompatibility of mesoporous silicates. *Biomater.* 29, 4045–4055.

Huebner, R.E. and Castro, K.G. (1995). The changing face of tuberculosis. *Annu. Rev. Med.* 46, 47–55.

Huh, S., Wiench, J.W., Trewyn, B.G., Song, S., Pruski, M., and Lin, V.S. (2003). Tuning of particle morphology and pore properties in mesoporous silicas with multiple organic functional groups. *Chem. Commun. (Camb)* 2003, 2364–2365.

Hu, H., Zhao, Z., Wan, W., Gogotsi, Y., and Qiu, J. (2013). Ultralight and highly compressible graphene aerogels. *Adv. Mater.* 25, 2219–2223.

Izquierdo-Barba, I., Martinez, A., Doadrio, A.L., Perez-Pariente, J., and Vallet-Regi, M. (2005). Release evaluation of drugs from ordered three-dimensional silica structures. *Eur. J. Pharm. Sci.* 26, 365–373.

Jarvis, K.L., Barnes, T.J., and Prestidge, C.A. (2012). Surface chemistry of porous silicon and implications for drug encapsulation and delivery applications. *Adv. Colloid Interface Sci.* 175, 25–38.

Jia, F., Liu, X., Li, L. et al. (2013). Multifunctional nanoparticles for targeted delivery of immune activating and cancer therapeutic agents. *J. Controll. Release* 172, 1020–1034.

Jiang, X., Ward, T.L., Cheng, Y.S., Liu, J., and Brinker, C.J. (2010). Aerosol fabrication of hollow mesoporous silica nanoparticles and encapsulation of L-methionine as a candidate drug cargo. *Chem. Commun. (Camb)* 46, 3019–3021.

Kaukonen, A.M., Laitinen, L., Salonen, J. et al. (2007). Enhanced in vitro permeation of furosemide loaded into thermally carbonized mesoporous silicon (TCPSi) microparticles. *Eur. J. Pharm. Biopharm.* 66, 348–356.

Kim, K.H., Oh, Y., and Islam, M.F. (2012). Graphene coating makes carbon nanotube aerogels superelastic and resistant to fatigue. *Nat. Nanotechnol.* 7, 562–566.

Knill, C., Kennedy, J., Mistry, J. et al. (2004). Alginate fibres modified with unhydrolysed and hydrolysed chitosans for wound dressings. *Carbohydr. Polym.* 55, 65–76.

Kollisch, H.S., Barner-Kowollik, C., and Ritter, H. (2009). Amphiphilic block copolymers based on cyclodextrin host-guest complexes via RAFT-polymerization in aqueous solution. *Chem. Commun (Camb)* 2009, 1097–1099.

Ksenofontova, O., Vasin, A., Egorov, V. et al. (2014). Porous silicon and its applications in biology and medicine. *Techn. Phys.* 59, 66–77.

Kumar, R., Chen, H.T., Escoto, J.L., Lin, V.S.Y., and Pruski, M. (2006). Template removal and thermal stability of organically functionalized mesoporous silica nanoparticles. *Chem. Mater.* 18, 4319–4327.

Lai, C.Y., Trewyn, B.G., Jeftinija, D.M. et al. (2003). A mesoporous silica nanosphere-based carrier system with chemically removable CdS nanoparticle caps for stimuli-responsive controlled release of neurotransmitters and drug molecules. *J. Am. Chem. Soc.* 125, 4451–4459.

Ledger, P.W. (1992). Skin biological issues in electrically enhanced transdermal delivery. *Adv. Drug Deliv. Rev.* 9, 289–307.

Lee, J.E., Lee, N., Kim, H. et al. (2010). Uniform mesoporous dye-doped silica nanoparticles decorated with multiple magnetite nanocrystals for simultaneous enhanced magnetic resonance imaging, fluorescence imaging, and drug delivery. *J. Am. Chem. Soc.* 132, 552–557.

Letant, S.E., Kane, S.R., Hart, B.R. et al. (2005). Hydrolysis of acetylcholinesterase inhibitors–Organophosphorus acid anhydrolase enzyme immobilization on photoluminescent porous silicon platforms. *Chem. Commun (Camb)* 2005, 851–853.

Li, Y., Shi, J., Hua, Z., Chen, H., Ruan, M., and Yan, D. (2003a). Hollow spheres of mesoporous aluminosilicate with a three-dimensional pore network and extraordinarily high hydrothermal stability. *Nano. Lett.* 3, 609–612.

Li, Y.Y., Cunin, F., Link, J.R. et al. (2003b). Polymer replicas of photonic porous silicon for sensing and drug delivery applications. *Science.* 299, 2045–2047.

Li, X., He, Y., and Swihart, M.T. (2004). Surface functionalization of silicon nanoparticles produced by laser-driven pyrolysis of silane followed by HF–HNO$_3$ etching. *Langmuir* 20, 4720–4727.

Li, Z., Barnes, J.C., Bosoy, A., Stoddart, J.F., and Zink, J.I. (2012). Mesoporous silica nanoparticles in biomedical applications. *Chem. Soc. Rev.* 41, 2590–2605.

Li, M., Deng, H., Peng, H., and Wang, Q. (2014a). Functional Nanoparticles in Targeting Glioma Diagnosis and Therapies. *J. Nanosci. Nanotechnol.* 14, 415–432.

Li, M., Yu, H., Wang, T. et al. (2014b). Tamoxifen embedded in lipid bilayer improves the oncotarget of liposomal daunorubicin in vivo. *J. Mater. Chem. B* 2, 1619–1625.

Li, W., Li, X., Wang, Q. et al. (2014c). Antibacterial activity of nanofibrous mats coated with lysozyme-layered silicate composites via electrospraying. *Carbohydr. Polym.* 99, 218–225.

Liao, Y.T., Wu, K.C., and Yu, J. (2014). Synthesis of mesoporous silica nanoparticle-encapsulated alginate microparticles for sustained release and targeting therapy. *J. Biomed. Mater. Res. B: Appl. Biomater.* 102, 293–302.

Lie, L. H., Duerdin, M., Tuite, E.M., Houlton, A., and Horrocks, B.R. (2002). Preparation and characterisation of luminescent alkylated-silicon quantum dots. *J Electrochem Chem.* 538, pp. 183–190.

Lim, M.H. and Stein, A. (1999). Comparative studies of grafting and direct syntheses of inorganic-organic hybrid mesoporous materials. *Chem. Mater.* 11, 3285–3295.

Lin, L. and Pisano, A.P. (1999). Silicon-processed microneedles. *Microelectromechanical Systems, J.* 8, 78–84.

Lin, Y.S. and Haynes, C.L. (2010). Impacts of mesoporous silica nanoparticle size, pore ordering, and pore integrity on hemolytic activity. *J. Am. Chem. Soc.* 132, 4834–4842.

Lin, H.P., Yang, L.Y., Mou, C.Y., Liu, S.B., and Lee, H.K. (2000). A direct surface silyl modification of acid-synthesized mesoporous silica. *New J. Chem.* 24, 253–255.

Liong, M., Lu, J., Kovochich, M. et al. (2008). Multifunctional inorganic nanoparticles for imaging, targeting, and drug delivery. *Acs. Nano* 2, 889–896.

Liong, M., France, B., Bradley, K.A., and Zink, J.I. (2009). Antimicrobial activity of silver nanocrystals encapsulated in mesoporous silica nanoparticles. *Adv. Mater.* 21, 1684–1689.

Liu, Y.H., Lin, H.P., and Mou, C.Y. (2004). Direct method for surface silyl functionalization of mesoporous silica. *Langmuir.* 20, 3231–3239.

Liu, Y., Mi, Y., Zhao, J., and Feng, S.S. (2011). Multifunctional silica nanoparticles for targeted delivery of hydrophobic imaging and therapeutic agents. *Int. J. Pharm.* 421, 370–378.

Liu, M., Li, M., Sun, S. et al. (2014a). The use of antibody modified liposomes loaded with AMO-1 to deliver oligonucleotides to ischemic myocardium for arrhythmia therapy. *Biomater.* 35, 3697–3707.

Liu, M., Li, M., Wang, G. et al. (2014b). Heart-targeted nanoscale drug delivery systems. *J. Biomed. Nanotechnol.* 10, 2038–2062.

Lu, F., Wu, S.H., Hung, Y., and Mou, C.Y. (2009). Size effect on cell uptake in well-suspended, uniform mesoporous silica nanoparticles. *Small* 5, 1408–1413.

Mal, N.K., Fujiwara, M., and Tanaka, Y. (2003). Photocontrolled reversible release of guest molecules from coumarin-modified mesoporous silica. *Nature* 421, 350–353.

Malaviya, A.N. and Kotwal, P.P. (2003). Arthritis associated with tuberculosis. *Best Pract. Res. Clin. Rheumatol.* 17, 319–343.

Mayor, S. and Pagano, R.E. (2007). Pathways of clathrin-independent endocytosis. *Nat. Rev. Mol. Cell Biol.* 8, 603–612.

McAllister, D.V., Allen, M.G., and Prausnitz, M.R. (2000). Microfabricated microneedles for gene and drug delivery. *Annu. Rev. Biomed. Eng.* 2, 289–313.

McBain, S., Yi, H., El Haj, A., and Dobson, J. (2007). Polyethyleneimine functionalized iron oxide nanoparticles as agents for DNA delivery and transfection. *J. Mater. Chem.* 17, 2561–2565.

Meier, J.L. and Beekmann, S.E. (1995). Mycobacterial and fungal infections of bone and joints. *Curr. Opin. Rheumatol.* 7, 329–336.

Meng, H., Liong, M., Xia, T. et al. (2010). Engineered design of mesoporous silica nanoparticles to deliver doxorubicin and P-glycoprotein siRNA to overcome drug resistance in a cancer cell line. *ACS. Nano.* 4, 4539–4550.

Molina, J.R., Yang, P., Cassivi, S.D., Schild, S.E., and Adjei, A.A. (2008). Non-small cell lung cancer: Epidemiology, risk factors, treatment, and survivorship. *Mayo. Clin. Proc.* 83, 584–594.

Mondragón, L., Mas, N., Ferragud, V. et al. (2014). Enzyme-responsive intracellular-controlled release using silica mesoporous nanoparticles capped with ε-poly-L-lysine. *Chemistry* 20, 5271–5281.

Nardecchia, S., Carriazo, D., Ferrer, M.L., Gutierrez, M.C., and Del Monte, F. (2013). Three dimensional macroporous architectures and aerogels built of carbon nanotubes and/or graphene: Synthesis and applications. *Chem. Soc. Rev.* 42, 794–830.

Nel, A.E., Madler, L., Velegol, D. et al. (2009). Understanding biophysicochemical interactions at the nano-bio interface. *Nat. Mater.* 8, 543–557.

Nikolelis, D.P., Drivelos, D.A., Simantiraki, M.G., and Koinis, S. (2004). An optical spot test for the detection of dopamine in human urine using stabilized in air lipid films. *Anal. Chem.* 76, 2174–2180.

Niu, D., Ma, Z., Li, Y., and Shi, J. (2010). Synthesis of core-shell structured dual-mesoporous silica spheres with tunable pore size and controllable shell thickness. *J. Am. Chem. Soc.* 132, 15144–15147.

Novikova, L.N., Mosahebi, A., Wiberg, M., Terenghi, G., Kellerth, J.O., and Novikov, L.N. (2006). Alginate hydrogel and matrigel as potential cell carriers for neurotransplantation. *J. Biomed. Mater. Res. A.* 77, 242–252.

Pan, L., He, Q., Liu, J. et al. (2012). Nuclear-targeted drug delivery of TAT peptide-conjugated monodisperse mesoporous silica nanoparticles. *J. Am. Chem. Soc.* 134, 5722–5725.

Pan, L., Liu, J., He, Q., Wang, L., and Shi, J. (2013). Overcoming multidrug resistance of cancer cells by direct intranuclear drug delivery using TAT-conjugated mesoporous silica nanoparticles. *Biomaterials.* 34, 2719–2730.

Park, I.Y., Kim, I.Y., Yoo, M.K., Choi, Y.J., Cho, M.H., and Cho, C.S. (2008). Mannosylated polyethylenimine coupled mesoporous silica nanoparticles for receptor-mediated gene delivery. *Int. J. Pharm.* 359, 280–287.

Park, C., Kim, H., Kim, S., and Kim, C. (2009). Enzyme responsive nanocontainers with cyclodextrin gatekeepers and synergistic effects in release of guests. *J. Am. Chem. Soc.* 131, 16614–16615.

Popplewell, J.F., King, S.J., Day, J.P. et al. (1998). Kinetics of uptake and elimination of silicic acid by a human subject: A novel application of ^{32}Si and accelerator mass spectrometry. *J. Inorg. Biochem.* 69, 177–180.

Prestidge, C.A., Barnes, T.J., Lau, C.H., Barnett, C., Loni, A., and Canham, L. (2007). Mesoporous silicon: A platform for the delivery of therapeutics. *Expert Opin. Drug Deliv.* 4, 101–110.

Prokop, A. and Davidson, J.M. (2008). Nanovehicular intracellular delivery systems. *J. Pharm. Sci.* 97, 3518–3590.

Puttick, M.P., Stein, H.B., Chan, R.M. et al. (1995). Soft tissue tuberculosis: A series of 11 cases. *J. Rheumatol.* 22, 1321–1325.

Qin, Z., Joo, J., Gu, L., and Sailor, M.J. (2014). Size control of porous silicon nanoparticles by electrochemical perforation etching. *Part. Part. Syst. Charact.* 31, 252–256.

Qu, F., Zhu, G., Huang, S. et al. (2006). Controlled release of Captopril by regulating the pore size and morphology of ordered mesoporous silica. *Microporous Mesoporous Mater.* 92, 1–9.

Radu, D.R., Lai, C.Y., Jeftinija, K., Rowe, E.W., Jeftinija, S., and Lin, V.S. (2004). A polyamidoamine dendrimer-capped mesoporous silica nanosphere-based gene transfection reagent. *J. Am. Chem. Soc.* 126, 13216–13217.

Raviglione, M.C., Gupta, R., Dye, C.M., and Espinal, M.A. (2001). The burden of drug-resistant tuberculosis and mechanisms for its control. *Ann. N. Y. Acad. Sci.* 953, 88–97.

Russo, L., Colangelo, F., Cioffi, R., Rea, I., and De Stefano, L. (2011). A mechanochemical approach to porous silicon nanoparticles fabrication. *Mater.* 4, 1023–1033.

Salonen, J. and Lehto, V.P. (2008). Fabrication and chemical surface modification of mesoporous silicon for biomedical applications. *Chem. Eng. J.* 137, 162–172.

Salonen, J., Laitinen, L., Kaukonen, A.M. et al. (2005). Mesoporous silicon microparticles for oral drug delivery: Loading and release of five model drugs. *J. Control. Release* 108, 362–374.

Salonen, J., Kaukonen, A.M., Hirvonen, J., and Lehto, V.P. (2008). Mesoporous silicon in drug delivery applications. *J. Pharm. Sci.* 97, 632–653.

Santos, H.A., Bimbo, L.M., Lehto, V.P., Airaksinen, A.J., Salonen, J., and Hirvonen, J. (2011). Multifunctional porous silicon for therapeutic drug delivery and imaging. *Curr. Drug Discov. Technol.* 8, 228–249.

Savage, D.J., Liu, X., Curley, S.A., Ferrari, M., and Serda, R.E. (2013). Porous silicon advances in drug delivery and immunotherapy. *Curr. Opin. Pharmacol.* 13, 834–841.

Shi, X., Wang, Y., Ren, L., Zhao, N., Gong, Y., and Wang, D.A. (2009). Novel mesoporous silica-based antibiotic releasing scaffold for bone repair. *Acta. Biomater.* 5, 1697–1707.

Shih, I.L., Shen, M.H., and Van, Y.T. (2006). Microbial synthesis of poly(epsilon-lysine) and its various applications. *Bioresour. Technol.* 97, 1148–1159.

Shima, S., Matsuoka, H., Iwamoto, T., and Sakai, H. (1984). Antimicrobial action of epsilon-poly-L-lysine. *J. Antibiot. (Tokyo)* 37, 1449–1455.

Siegel, R., Desantis, C., Virgo, K. et al. (2012). Cancer treatment and survivorship statistics, 2012. *CA. Cancer J. Clin.* 62, 220–241.

Sivamani, R.K., Liepmann, D., and Maibach, H.I. (2007). Microneedles and transdermal applications. *Expert Opin. Drug Deliv.* 4, 19–25.

Slowing, I., Trewyn, B.G., and Lin, V.S. (2006). Effect of surface functionalization of MCM-41-type mesoporous silica nanoparticles on the endocytosis by human cancer cells. *J. Am. Chem. Soc.* 128, 14792–14793.

Song, S.W., Hidajat, K., and Kawi, S. (2005). Functionalized SBA-15 materials as carriers for controlled drug delivery: Influence of surface properties on matrix-drug interactions. *Langmuir.* 21, 9568–9575.

Srivastava, R. and McShane, M.J. (2005). Application of self-assembled ultra-thin film coatings to stabilize macromolecule encapsulation in alginate microspheres. *J. Microencapsul.* 22, 397–411.

Stöber, W. and Fink, A. (1968). Controlled growth of monodisperse silica spheres in the micron size range. *J. Colloid Interface Sci.* 26, 62–69.

Tabasi, O., Falamaki, C., and Khalaj, Z. (2012). Functionalized mesoporous silicon for targeted-drug-delivery. *Colloids. Surf. B: Biointerfaces* 98, 18–25.

Tan, W.H. and Takeuchi, S. (2007). Monodisperse alginate hydrogel microbeads for cell encapsulation. *Adv. Mater.* 19, 2696–2701.

Tao, Z., Morrow, M.P., Asefa, T. et al. (2008). Mesoporous silica nanoparticles inhibit cellular respiration. *Nano. Lett.* 8, 1517–1526.

Taratula, O., Garbuzenko, O.B., Chen, A.M., and Minko, T. (2011). Innovative strategy for treatment of lung cancer: Targeted nanotechnology-based inhalation co-delivery of anticancer drugs and siRNA. *J. Drug Target.* 19, 900–914.

Tarn, D., Ashley, C.E., Xue, M., Carnes, E.C., Zink, J.I., and Brinker, C.J. (2013). Mesoporous silica nanoparticle nanocarriers: Biofunctionality and biocompatibility. *Acc. Chem. Res.* 46, 792–801.

Teow, Y., Asharani, P.V., Hande, M.P., and Valiyaveettil, S. (2011). Health impact and safety of engineered nanomaterials. *Chem. Commun. (Camb)* 47, 7025–7038.

Thierry, B. and Textor, M. (2012). Nanomedicine in focus: Opportunities and challenges ahead. *Biointerphases* 7, 19.

Thomas, J.C., Pacholski, C., and Sailor, M.J. (2006). Delivery of nanogram payloads using magnetic porous silicon microcarriers. *Lab. Chip.* 6, 782–787.

Tilley, R.D., Warner, J.H., Yamamoto, K., Matsui, I., and Fujimori, H. (2005). Micro-emulsion synthesis of monodisperse surface stabilized silicon nanocrystals. *Chem. Commun.* 2005, 1833–1835.

Tomioka, H. (2006). Current status of some antituberculosis drugs and the development of new antituberculous agents with special reference to their in vitro and in vivo antimicrobial activities. *Curr. Pharm. Des.* 12, 4047–4070.

Tonnesen, H.H. and Karlsen, J. (2002). Alginate in drug delivery systems. *Drug Dev. Ind. Pharm.* 28, 621–30.

Torney, F., Trewyn, B.G., Lin, V.S., and Wang, K. (2007). Mesoporous silica nanoparticles deliver DNA and chemicals into plants. *Nat. Nanotechnol.* 2, 295–300.

Tourne-Peteilh, C., Lerner, D.A., Charnay, C., Nicole, L., Begu, S., and Devoisselle, J.M. (2003). The potential of ordered mesoporous silica for the storage of drugs: The example of a pentapeptide encapsulated in a MSU-tween 80. *ChemPhysChem.* 4, 281–286.

Urban-Klein, B., Werth, S., Abuharbeid, S., Czubayko, F., and Aigner, A. (2005). RNAi-mediated gene-targeting through systemic application of polyethylenimine (PEI)-complexed siRNA in vivo. *Gene Ther.* 12, 461–466.

Vallet-Regi, M., Ramila, A., Del Real, R., and Perez-pariente, J. (2001). A new property of MCM-41: Drug delivery system. *Chem. Mater.* 13, 308–311.

Vallet-Regi, M., Balas, F., and Arcos, D. (2007). Mesoporous materials for drug delivery. *Angew. Chem. Int. Ed. Engl.* 46, 7548–7558.

Vallet-Regí, M., Balas, F., Colilla, M., and Manzano, M. (2008). Bone-regenerative bioceramic implants with drug and protein controlled delivery capability. *Prog. Solid State Chem.* 36, 163–191.

Vashishtha, V.M. (2009). WHO Global Tuberculosis Control Report 2009: Tuberculosis elimination is a distant dream. *Indian Pediatr.* 46, 401–402.

Vivero-Escoto, J.L., Slowing, I.I, Trewyn, B.G., and Lin, V.S. (2010). Mesoporous silica nanoparticles for intracellular controlled drug delivery. *Small* 6, 1952–1967.

Vivero-Escoto, J.L., Huxford-Phillips, R.C., and Lin, W. (2012). Silica-based nanoprobes for biomedical imaging and theranostic applications. *Chem. Soc. Rev.* 41, 2673–2685.

Wagner, T., Haffer, S., Weinberger, C., Klaus, D., and Tiemann, M. (2013). Mesoporous materials as gas sensors. *Chem. Soc. Rev.* 42, 4036–4053.

Wang, L., Reipa, V., and Blasic, J. (2004). Silicon nanoparticles as a luminescent label to DNA. *Bioconjugate Chem.* 15, 409–412.

Wang, Q., Zhang, N., Hu, X., Yang, J., and Du, Y. (2007). Alginate/polyethylene glycol blend fibers and their properties for drug controlled release. *J. Biomed. Mater. Res. A.* 82, 122–128.

Wang, Q., Wang, L., Detamore, M.S., and Berkland, C. (2008). Biodegradable colloidal gels as moldable tissue engineering scaffolds. *Adv. Mater.* 20, 236–239.

Wang, Q., Jamal, S., Detamore, M.S., and Berkland, C. (2011). PLGA-chitosan/PLGA-alginate nanoparticle blends as biodegradable colloidal gels for seeding human umbilical cord mesenchymal stem cells. *J. Biomed. Mater. Res. A.* 96, 520–527.

Wang, Q., Gu, Z., Jamal, S., Detamore, M.S., and Berkland, C. (2013). Hybrid hydroxyapatite nanoparticle colloidal gels are injectable fillers for bone tissue engineering. *Tissue Eng. A* 19, 2586–2593.

West, T.P. and Preece, J.E. (2009). Bulk alginate encapsulation of Hibiscus moscheutos nodal segments. *Plant Cell Tissue Organ Culture (PCTOC)* 97, 345–351.

Wilson, W.L., Szajowski, P.F., and Brus, L.E. (1993). Quantum confinement in size-selected, surface-oxidized silicon nanocrystals. *Science* 262, 1242–1244.

Wojtyk, J.T., Morin, K.A., Boukherroub, R., and Wayner, D.D. (2002). Modification of porous silicon surfaces with activated ester monolayers. *Langmuir.* 18, 6081–6087.

Worsley, M.A., Pauzauskie, P.J., Olson, T.Y. et al. (2010). Synthesis of graphene aerogel with high electrical conductivity. *J. Am. Chem. Soc.* 132, 14067–14069.

Wu, S., Luo, Y., Zeng, F., Chen, J., Chen, Y., and Tong, Z. (2007). Photoreversible fluorescence modulation of a rhodamine dye by supramolecular complexation with photosensitive cyclodextrin. *Angew. Chem. Int. Ed. Engl.* 46, 7015–7018.

Wu, E.C., Andrew, J.S., Buyanin, A., Kinsella, J.M., and Sailor, M.J. (2011). Suitability of porous silicon microparticles for the long-term delivery of redox-active therapeutics. *Chem. Commun.* 47, 5699–5701.

Xia, T., Kovochich, M., Liong, M. et al. (2009). Polyethyleneimine coating enhances the cellular uptake of mesoporous silica nanoparticles and allows safe delivery of siRNA and DNA constructs. *Acs. Nano* 3, 3273–3286.

Xin, S., Li, X., Wang, Q. et al. (2014). Novel layer-by-layer structured nanofibrous mats coated by protein films for dermal regeneration. *J. Biomed. Nanotechnol.* 10, 803–810.

Xing, X., He, X., Peng, J., Wang, K., and Tan, W. (2005). Uptake of silica-coated nanoparticles by HeLa cells. *J. Nanosci. Nanotechnol.* 5, 1688–1693.

Yamamoto, N., Fukai, F., Ohshima, H., Terada, H., and Makino, K. (2002). Dependence of the phagocytic uptake of polystyrene microspheres by differentiated HL60 upon the size and surface properties of the microspheres. *Colloids Surf. B* 25, 157–162.

Yang, P., Gai, S., and Lin, J. (2012). Functionalized mesoporous silica materials for controlled drug delivery. *Chem. Soc. Rev.* 41, 3679–3698.

Yang, S.T., Liu, Y., Wang, Y.W., and Cao, A. (2013). Biosafety and bioapplication of nanomaterials by designing protein-nanoparticle interactions. *Small* 9, 1635–1653.

Yang, Y., Shi, E., Li, P. et al. (2014a). A compressible mesoporous SiO_2 sponge supported by a carbon nanotube network. *Nanoscale* 6, 3585–3592.

Yang, Y., Wang, S., Wang, Y., Wang, X., Wang, Q., and Chen, M. (2014b). Advances in self-assembled chitosan nanomaterials for drug delivery. *Biotechnol. Adv.* 32, 1301–1316.

Youlden, D.R., Cramb, S.M., and Baade, P.D. (2008). The International epidemiology of lung cancer: Geographical distribution and secular trends. *J. Thorac. Oncol.* 3, 819–831.

Yu, C., Luo, M., Zeng, F., Zheng, F., and Wu, S. (2011). Mesoporous silica particles for selective detection of dopamine with beta-cyclodextrin as the selective barricade. *Chem. Commun. (Camb)* 47, 9086–9088.

Yu, T., Greish, K., Mcgill, L.D., Ray, A., and Ghandehari H. (2012). Influence of geometry, porosity, and surface characteristics of silica nanoparticles on acute toxicity: Their vasculature effect and tolerance threshold. *Acs. Nano* 6, 2289–2301.

Zauner, W., Farrow, N.A., and Haines, A.M. (2001). In vitro uptake of polystyrene microspheres: Effect of particle size, cell line and cell density. *J. Control Release* 71, 39–51.

Zhang, W.H., Lu, X.B., Xiu, J.H. et al. (2004). Synthesis and characterization of bifunctionalized ordered mesoporous materials. *Adv. Funct. Mater.* 14, 544–552.

Zhao, Y., Liu, J., Hu, Y. et al. (2013). Highly compression-tolerant supercapacitor based on polypyrrole-mediated graphene foam electrodes. *Adv. Mater.* 25, 591–595.

Zhou, C., Li, P., Qi, X. et al. (2011). A photopolymerized antimicrobial hydrogel coating derived from epsilon-poly-L-lysine. *Biomater.* 32, 2704–2712.

Zhu, Y.F., Shi, J.L., Li, Y.S., Chen, H.R., Shen, W.H., and Dong, X.P. (2005a). Storage and release of ibuprofen drug molecules in hollow mesoporous silica spheres with modified pore surface. *Microporous Mesoporous Mater.* 85, 75–81.

Zhu, Y., Shi, J., Chen, H., Shen, W., and Dong, X. (2005b). A facile method to synthesize novel hollow mesoporous silica spheres and advanced storage property. *Microporous Mesoporous Mater.* 84, 218–222.

Zhu, Y., Shi, J., Shen, W., Chen, H., Dong, X. and Ruan, M. (2005c). Preparation of novel hollow mesoporous silica spheres and their sustained-release property. *Nanotechnology* 16, 2633.

Zhu, M., Wang, H., Liu, J. et al. (2011). A mesoporous silica nanoparticulate/beta-TCP/BG composite drug delivery system for osteoarticular tuberculosis therapy. *Biomater.* 32, 1986–1995.

Zou, J., Liu, J., Karakoti, A.S. et al. (2010). Ultralight multiwalled carbon nanotube aerogel. *Acs. Nano* 4, 7293–7302.

Tissue Engineering

19

**Pierre-Yves Collart Dutilleul, Csilla Gergely,
Frédérique Cunin, and Frédéric Cuisinier**

CONTENTS

19.1	Introduction and Definition	364
19.2	Porous Materials for Tissue Engineering Applications	365
19.3	Stem Cells	365
	19.3.1 Adult Mesenchymal Stem Cells	367
	19.3.2 Embryonic Stem Cells	367
	19.3.3 Neonatal Stem Cells from the Umbilical Cord	369
	19.3.4 Induced Pluripotent Stem Cells	369
19.4	Interactions of PSi–Stem Cells	369
	19.4.1 Cell Adhesion and Proliferation	369
	19.4.2 Cell Differentiation	371
	19.4.3 PSi Resorption	371
19.5	PSi for Tissue Engineering	372
	19.5.1 Bone Tissue Engineering	372
	19.5.2 Ophthalmic Implants	373
	19.5.3 Nerve Tissue Engineering	374
19.6	Advances in PSi-Based Composites	375
19.7	Legal and Practical Issues	375
References		376

19.1 INTRODUCTION AND DEFINITION

Tissue engineering involves biomaterials used as scaffolds for stem or progenitor cells, combined with bioactive molecules. It is an interdisciplinary domain, applying engineering sciences principles to develop biological substitutes, in order to maintain, repair, or improve tissue or organ functions (Langer and Vacanti 1993). According to the definition given by the European Medicine Agency, tissue engineering products (TEP) are products that contain cells or tissue with properties for human tissue regeneration, replacement, or repair. TEP can contain cells of human or animal origin. These cells can be viable or nonviable. TEP can also contain various substances such as cell products, bioactive molecules, biomaterials, or chemical products (Legislation 2007).

Tissue engineering aims to create functional structures to restore injured tissues. Skin and artificial cartilage are examples of engineered tissues approved by the Food and Drug Administration, even though their clinical applications for human subjects are still restricted (Bioengineering NIoBIa 2014). Tissue engineering is a part of regenerative medicine, where regeneration can be mediated through biological products (i.e., proteins, DNA, growth factors) and/or biomaterials to renew efficient cells for tissue repair. Applications are based on the development of biological substitutes that combine cells and bioactive factors in a defined microenvironment created by a biomaterial scaffold. This biomaterial scaffold should support cell attachment, and be biocompatible and biodegradable at a controlled rate (Wang et al. 2006; Collart Dutilleul et al. 2014a). Porous silicon (PSi) appears to be a promising biomaterial for stem cell-based therapy and tissue engineering as it is both nontoxic and bioresorbable under physiological conditions (Sailor 2012; Collart Dutilleul et al. 2014b). Its dissolution rate into nontoxic silicic acid is dependent on the pore geometry and surface chemical properties (Low et al. 2006; Khung et al. 2008). PSi has a structure that promotes hydroxyapatite growth, suggesting the possible bone implantability of the material (Canham 1995). Its biocompatibility and immunogenicity has already been demonstrated under different conditions (Ainslie et al. 2008). Furthermore, PSi-based scaffolds have been investigated for orthopedic (Whitehead et al. 2008) and ophthalmic implants (Cheng et al. 2008), for controlling the adhesion and proliferation of different cell types (Low et al. 2006; Alvarez et al. 2008; Torres-Costa et al. 2012; Collart Dutilleul et al. 2014a).

To further recreate a living system (to repair injured structures), biological and engineering principles have to be considered: the central component of any living system is the single cell (Vacanti 2010). Each individual cell, to reproduce and function, has simple metabolic requirements, such as food and gas supply, and waste removal. The scaffold design is central to facilitate cell adhesion and growth, until a complex multicellular system. These complex structures are required for engineered tissues to function *in vivo*. In the past decade, regenerative medicine and tissue engineering have focused on the role of the extracellular matrix (ECM) in cell proliferation, migration, and differentiation.

To have an efficient influence on cellular behavior, the engineering of ECM substitutes requires cooperation between biologists, biochemists, and physicists. However, the ECM is not only a static scaffold: there are signaling pathways between cells and ECM that occur through direct cell receptor–ECM ligand interactions, growth factor delivery, and spatial cues and mechanical force transduction (Prestwich 2007). Challenges to engineering such complex-use nanofabricating techniques that create porous, nanometer-organized scaffolds with surface qualities that influence cell fate determination allow regulation of specific protein-expression patterns and encourage cell-specific scaffold remodeling (Heydarkhan-Hagvall et al. 2007). These techniques can regulate surface topography down to the nanometer range and include nanoscale surface pattern fabrication, electrospinning, and self-assembly fabrication (Ayres et al. 2010). Incorporating biological signals in the form of bioactive molecules, mechanical stresses, cell surface receptors, and spatial cues can also influence cellular proliferation, differentiation, and migration (Kelleher and Vacanti 2010).

This chapter focuses on PSi as biomaterial scaffold for tissue engineering, enlightening its properties of being nontoxic, bioresorbable, and inductive for cell differentiation. Indeed, PSi has the ability to degrade in aqueous solutions into nontoxic silicic acid and surface modifications offer control over the degradation rate of PSi (Low et al. 2006).

19.2 POROUS MATERIALS FOR TISSUE ENGINEERING APPLICATIONS

Tissue engineering aims to restore diseased or damaged tissue to its original state and function, reducing the need for transplants and grafts. To achieve this aim, tissue engineering uses biomaterials (scaffolds) to guide and stimulate growth and differentiation of cells to form tissues (Langer and Vacanti 1993).

Multicellular organisms, especially vertebrates, are formed by hierarchical three-dimensional structures with dimensions ranging from nanometers to meters. To repair or replace degenerating tissues, one has to replicate these complex living hierarchical structures. Porous materials have the potential, in tissue engineering applications, to shift treatments from tissue replacement to tissue regeneration. To be considered ideal for tissue engineering applications, porous material must fulfill several conditions: it should act as a template for *in vitro* and eventually *in vivo* tissue growth in three dimensions, and resorb at the same rate as the tissue is repaired, with degradation products that are nontoxic and that can be easily excreted by the body (Holzwarth and Ma 2001; Khan et al. 2012). This biomaterial should be nontoxic (biocompatible) and promote cell adhesion and activity at the genetic level. It should also bond to the host tissue without the formation of scar tissue, and exhibit mechanical properties matching that of the host tissue. Finally, it should be made from a processing technique that can produce irregular shapes to match the defect and have the potential to be commercially producible and sterilizable to the required international standards for clinical use (Jones et al. 2006).

PSi scaffolds have the potential to fulfill all the criteria, as they have a hierarchical porous structure, they can bond to bone and soft tissue, and they release silicon ions that have been found to promote cell functions without any significant inflammatory or toxic effect (Canham 1995). Its structure can be tailored for the required rate of tissue bonding, resorption, and delivery of dissolution products. And the structure and properties of the PSi scaffolds can be optimized with respect to cell response to enable growth of tissue and cells *in vitro* and *in vivo* (Fan et al. 2011; Collart Dutilleul et al. 2014a). Moreover, porous structures have been shown to significantly decrease cell damage and soft tissue reaction, compared to plane surfaces: when implanted *in vivo*, plane surfaces develop a significant increase in capsule thickness compared to porous implants (Rosengren et al. 2002).

Silicon incorporation in biomaterials has been proved to be of interest, particularly due to the dissolution products' direct effect on host cells (Xynos et al. 2001). However, despite numerous advanced techniques in various studies, the reactions of silicon with cell and matrix components are not clearly understood yet. Numerous theories have been tested experimentally and it could be that a combination of any or all of these explain the enhancing effect silicon appears to have when incorporated into biomaterials (Xynos et al. 2001).

As an example, when considering bone reconstructive surgery, current procedures use graft techniques, but there are still many limitations to using these techniques: autografts have low availability and can cause morbidity at the donor site, homografts carry the risk of disease transmission, bone resorption, and rejection, requiring indefinite administration of immunosuppressant drugs to the patient, and xenografts have even larger risks of immune rejection, *in situ* degeneration, and disease transmission. Orthopedic implants used as an alternative have a limited lifespan as they lack the ability to self-repair, to maintain a blood supply, and to modify in response to stimuli such as mechanical load. There is a need for innovative biomaterials allowing the shift from replacement to regeneration, presenting porous and tunable structure, being nontoxic and resorbable (Hench and Polak 2002).

19.3 STEM CELLS

Regenerative medicine involves *in vitro* sorting and expansion of stem cells, before cell grafting. Cells can be of autologous or allogenic origin. Autologous grafts offer the optimal compatibility, but require available cells from the patients in sufficient amount. Therefore, allogenic grafts with cells from alternative allogenic sources are of great interest to provide biomaterials—stem cell complexes.

Indeed, the field of biomaterials has been extensively studied for tissue engineering development, but the role of stem cells is crucial, as *in vitro* studies hold many limitations compared to the complex *in vivo* environment. Grafted cells are able to deal with the numerous *in vivo* chemical and mechanical signals.

Stem cells are found in most tissues of the organism. They are characterized by the ability to go through numerous cycles of cell division, maintaining an undifferentiated state, and having the capacity to differentiate into specialized cell types (Tesche and Gerber 2010). Stem cells are further classified into three categories: totipotent, pluripotent, or multipotent. A totipotent cell has the ability to form an entire organism. Pluripotent stem cells lack the ability to form extraembryonic tissue but can give rise to any cell type from the three germ cell layers (endoderm, mesoderm, or ectoderm). Embryonic stem cells (ESCs) and induced pluripotent stem cells (iPSs) are the main source of pluripotent stem cells (Murry and Keller 2008). Multipotent somatic stem cells are capable of differentiating into a variety of closely related cells within a tissue, but lack the ability to differentiate into tissues originating from the three germ layers (Thomson et al. 1998; Weissman 2000). Adult stem cells or mesenchymal stem cells (MSCs) are examples of multipotent stem cells that are isolated from mature tissues, even though some of them can differentiate into various tissue types from different germ layers (Donovan and Gearhart 2001). Another example of multipotent stem cell is the hematopoietic stem cell (HSC) that can differentiate into all subpopulations of hematopoietic cells (erythrocyte, thrombocyte, and lymphocyte) (Ema and Nakauchi 2003). It is also to be noted that the differentiation potential of stem cells is not limited to cell lineages present in the organ from which they are derived: some have a broader differentiation potential. A schematic representation of these various types of stem cells and their ways of differentiation is shown in Figure 19.1. In this chapter, we explore four main sources of stem cells: adult mesenchymal stem cells, embryonic stem cells, umbilical cord stem cells, and induced pluripotent stem cells.

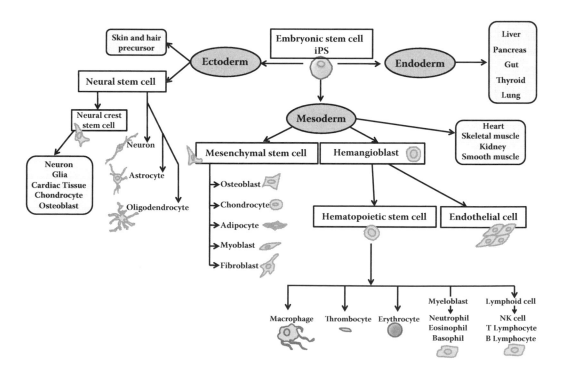

FIGURE 19.1 Schematic representation of differentiation ways, from embryonic stem cells to clinically relevant cells. The three germ layers are represented (endoderm, mesoderm, ectoderm), leading to the various specialized cells.

19.3.1 ADULT MESENCHYMAL STEM CELLS

Plastic adherent cells derived from bone marrow are able to differentiate into a number of mesenchymal cell types including osteoblasts, chondrocytes, and adipocytes. These cells have been named mesenchymal stem cells (MSCs) in reference to their high self-renewing properties, clonogenicity, and ability to form cartilage and bone. They were suggested to be responsible for the normal turnover and maintenance of adult mesenchymal tissues. These MSCs have been defined to be plastic-adherent when maintained in standard culture conditions. They have been shown to express several surface receptors, such as CD105, CD73, and CD90, and lack expression of CD45 (hematopoietic marker). In addition, they have the ability to differentiate into osteoblasts, adipocytes, and chondroblasts *in vitro* (Dominici et al. 2006). Figure 19.2 represents the clonogenic proliferation of MSC, with cell surface markers characterization. MSC's therapeutic potential has generated significant excitement in the field of regenerative medicine, as they can be found in various niches within the human body. The ability of these cells to self-renew and differentiate into multiple tissues makes them an attractive cell source for cellbased regenerative therapies. MSCs have considerable potential for the treatment of musculoskeletal disorders owing to their expansion capacity, immunosuppressive properties, and ability to differentiate into bone and cartilage (Ma 2010).

The most studied source of adult MSCs has been the bone marrow, as it was recognized early that its stroma contained stem cells capable of forming bone and cartilage. These bone marrow stromal cells (BMSCs) must be distinguished from the hematopoietic stem cells (HSCs) also found within bone marrow, as these two types hold different functions and differentiation capacities (see Figure 19.1). BMSC are commonly recovered from pure bone marrow aspirates and isolated according to their adherence and culture on tissue culture plastics. They can be expanded in culture, with an amount varying between patients. Age and marrow aspiration volume influence also the number of isolated stem cells. BMSC can, however, be expanded *in vitro* to large numbers and keep their differentiation potential, especially into osteoblast, even in older patients (Bianco et al. 2001).

Later, other sources of adult stem cells have been described, such as adipose tissue, dental pulp, skeletal muscle, umbilical cord blood, and others (Marolt et al. 2010).

Adipose derived stem cells (ASCs) have been described and characterized from human lipoaspirates. These ASCs represent a multilineage stem cell population, isolated from the stromal fraction of adipose tissue, and able to differentiate into osteoblast, adipocyte, and chondrocyte. They can be recovered from lipoaspirates consecutive to liposuction under anesthesia, and represent an important niche of adult stem cells (Zuk et al. 2002).

Stem cells have also been described within the dental pulp (Gronthos et al. 2000). They have then been studied and characterized as an adult stem-cell population that possesses the properties of high proliferative potential, selfrenewal, and multilineage differentiation. Dental pulp can be easily collected from adult (DPSC) or deciduous teeth (stem cells from human exfoliated deciduous teeth) after dental extraction, when teeth have to be removed. The most frequent case is the collection of normal human third molar extracted for orthodontic reasons (Gronthos et al. 2002). These mesenchymal DPSC hold promises also for neural regeneration as they are originating from the neural crest, during embryological development. They are classified as MSC but are closely related to neural crest stem cells (see Figure 19.1).

19.3.2 EMBRYONIC STEM CELLS

ESCs are pluripotent stem cells obtained from pre-implanted embryos, from the inner-cell mass before the first 2 weeks of development. Rat or other animal ESCs are widely used in research. However, for biomedical applications, ESCs have to be of human origin, which leads to technical and ethical issues. Human ESCs are obtained from extra embryos developed by *in vitro* fertilization techniques. They proliferate *in vitro* while maintaining an undifferentiated state and are capable of differentiating into many somatic cell types. In fact, they have the potential to give rise to any of the hundreds of cell types in the human body, raising exciting new prospects for

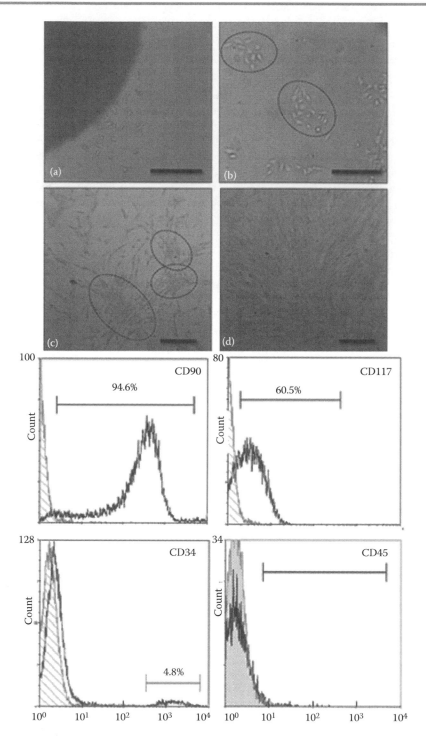

FIGURE 19.2 Human mesenchymal stem cells. (a–d) Optical microscopy evaluation. (a) Stem cells migrating from pulp tissue. (b) Isolated stem cells growing as clones. (c) Stem cell colonies formed after 5 days in culture dishes. (d) Cells reaching confluence after 10 days. Scale bar: 300 μm. (e) Flow cytometry analysis of subconfluent cells (stem cells collected after 5 to 7 days in culture). Mesenchymal stem cells are positive for CD90, partially positive for CD117, and negative for CD45. A subpopulation of CD34+ can be distinguished from the total population.

biomedical research and for regenerative medicine (Murry and Keller 2008). However, their great capacities for cell therapy and tissue engineering create a delicate ethical debate, as they induce destruction of human embryos.

19.3.3 NEONATAL STEM CELLS FROM THE UMBILICAL CORD

Neonatal tissues constitute a stem cell source that can be accessed in a noninvasive and rapid manner during and just after birth. The main source is the umbilical cord, covered by an amniotic epithelium that protects a gelatinous matrix called Wharton's jelly. Stem cells can be recovered during birth from cord blood and from Wharton's jelly (Forraz and McGuckin 2011). On one hand, cord blood can be collected using a collecting needle connected to an anticoagulant-containing bag, and allows the recovery of hematopoietic stem cells. On the other hand, Wharton's jelly is recovered from the whole umbilical cord, and contains an important amount of MSC.

19.3.4 INDUCED PLURIPOTENT STEM CELLS

Pluripotent stem cells can be induced from fibroblasts by retroviral introduction of transcription factor genes: Oct3/4, Sox2, c-Myc, and Klf4 (Takahashi and Yamanaka 2006). These induced pluripotent stem (iPS) cells are similar to ESCs in morphology, proliferation, and differentiation capacities (Okita et al. 2007). They proliferate extensively and differentiate into virtually any desired cell type, providing an unlimited source of replacement cells for human therapy (Ramirez et al. 2010).

These iPSC are obtained by the reprogramming of a wide variety of cell types isolated from the human body. They are ESC-like pluripotent cells, and offer a novel strategy for clinically applicable lineage-specific cells (Yoo et al. 2013).

19.4 INTERACTIONS OF PSi–STEM CELLS

The conversion of bulk Si into its high surface area, biocompatible porous counterpart is commonly achieved by etching in hydrofluoric acid (HF). By simply altering wafer resistivity, HF concentrations, and current densities, different porous structures can be generated (McInnes and Voelcker 2014). These pore dimensions can be precisely controlled and are highly tunable during PSi electrochemical anodization or etching. A variety of pore sizes can be produced: from micropores (<5 nm), mesopores (5–50 nm), to macropores (>50 nm) depending on the preparation conditions (Sailor 2012).

19.4.1 CELL ADHESION AND PROLIFERATION

Microscale topography modulates cellular behavior *in vitro*. However, it is important to consider that cells *in vivo* are in contact with nanoscale as well as microscale topographical features. In addition, even if cells are typically tens of microns in size, the dimensions of subcellular structures tend to the nanometer scale. Furthermore, extracellular supporting tissues present an intricate network of cues at the nanoscale composed of a complex mixture of nanometer-sized pits, pores, protrusions, and fibers (Biggs et al. 2010). Thus, pore size and porosity have an effect on cell growth that is of particular relevance, as it influences the concept of porous biomaterial fabrication.

Substrate topography is known to affect cell functions, such as adhesion, proliferation, migration, and differentiation, especially through cell cytoskeleton modifications (Dalby et al. 2003). Furthermore, cellular organization requires cell microenvironment assessment (extracellular matrix *in vivo* and supporting scaffold *in vitro*) (Buxboim et al. 2010). Cells respond to topographic surfaces in a wide variety of ways, which depend on cell type and pore size, as well as biomaterial physicochemical properties. Diverse topographical features have been assessed for different cell types, at the micrometer and submicrometer scale, and some recent studies on PSi have focused specifically on pore geometry influence on cell adhesion and proliferation (Sapelkin et al. 2006; Gentile et al. 2010; Clements et al. 2012; Wang et al. 2012; Collart Dutilleul et al. 2014a). Pore geometry was clearly shown to affect the cellular response, but each cell type responded differently. Rat hippocampal neurons were observed to preferentially adhere on macroporous surfaces, with a pore size ranging between 50 and 100 nm, rather than on fat silicon surfaces (Sapelkin et al. 2006). Neuroblastoma cells cultured over continuous porous gradient substrates were more likely to develop on surface topography with feature sizes of <20 nm, and substrates with an average pore size of a few hundreds of nanometers restricted cell adhesion and proliferation (Khung et al. 2008). Primary human endothelial cells, mouse fibroblasts, mouse neuroblastoma cells, and human cortical neuron cell lines adhered and proliferated more on mesoporous silicon than on flat silicon, with a tendency to proliferate more on PSi with an average pore size of ≈5 nm, rather than ≈20 nm (Gentile et al. 2012). Rat mesenchymal stem cells' adhesion was enhanced as pore size decreased, with a maximum proliferation for an average pore size of ≈20 nm but responded more strongly to surface chemical changes during short-term culture (Clements et al. 2012), and had a high proliferation rate also on flat silicon (Wang et al. 2012). Another study with ovary cells showed that cells cultured on various PSi nanotopographic structures displayed distinct morphogenesis, adherent responses, and biochemical properties in comparison with nonporous silicon structures. Results demonstrated that cell behaviors could be influenced by the physical characteristic derived from PSi nanotopography, with potential applications for controlling cell development in tissue engineering (Torres-Costa et al. 2012). More recently, primary cultures of human MSC were shown to preferentially adhere and proliferate on PSi with pore diameter ranging from 30 to 40 nm, rather than on PSi with smaller pore diameter (≈10 nm) or larger pore diameter (≈1 μm). These findings showed that nanoporous Si is a good candidate for cell culture and possible development of

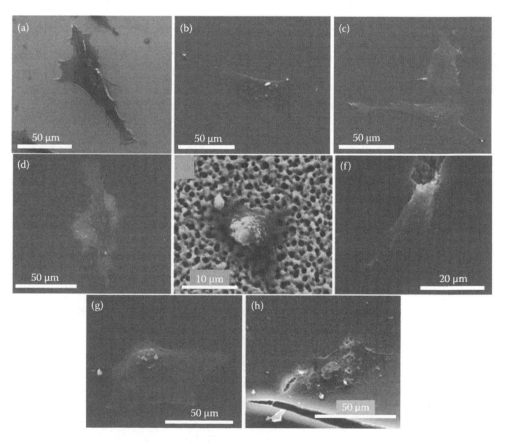

FIGURE 19.3 Scanning electron microscopy of DPSC after 24-h incubation. (a) Glass coverslip, (b) flat Si, (c) PSi 10 nm, (d) PSi 36 nm, (e) PSi 1 μm, (f) undecenoic acid-treated PSi, (g) APTES-treated PSi, and (h) semicarbazide-treated PSi. (Reprinted with permission from Collart-Dutilleul et al., *Appl. Mater. Interfaces* 6, 1719. Copyright 2014 American Chemical Society.)

bio-interfaced devices. It is interesting to note that pore size in the range of 20 to 40 nm is similar to the pore size of biological systems, such as pores in basement membranes, and corresponds to the size of biological pits, pores, protrusions, and fibers (Collart Dutilleul et al. 2014a).

Beyond topography, chemical surface treatment with chemically stable surfaces has been shown to be a key step to promote cell adhesion and growth (Low et al. 2006; Noval et al. 2012). The most common and simple surface treatment was oxidation, which could be performed by ozone, aging, thermal, or chemical treatments. Amine-terminated modifications such as silanization with aminopropyl trimethoxysilane (APTMS) or triethoxysilane (APTES) improved PSi chemical stability and enhanced cell adhesion in comparison to oxidized PSi (Low et al. 2006). Thermal hydrosilylation was also used to graft chemical species to generate a substrate for cell adhesion and proliferation, such as dodecene, undecenoic acid, or oligoethylene glycol (Alvarez et al. 2008). Hydrosilylation with semicarbazide has also been tested and shown to lead to chemical modification that favors cell adhesion and proliferation, especially mitosis after cell adhesion. But such semicarbazide-modified PSi surfaces were stable for only 24 to 48 h, and appeared to be potentially usable for stem cell adhesion and immediate *in vivo* transplantation, whereas stable APTES-treated PSi was more suitable for long term *in vitro* culture, for stem cell proliferation and further differentiation (Collart Dutilleul et al. 2014a). Figure 19.3 shows morphology of human mesenchymal stem cells cultured on PSi with various pore diameters and various chemical surface treatments.

19.4.2 CELL DIFFERENTIATION

Stem cell differentiation has been conducted mainly considering the use of various growth factors or bioactive molecules, but differentiation has recently become increasingly linked to mechanobiological concepts (Reilly and Engler 2010). Indeed, stem cells can begin to differentiate into mature tissue cells when exposed to intrinsic properties of the extracellular matrix (ECM) or supporting biomaterial. Mechano-sensitive pathways can convert biophysical cues into biochemical signals that commit cells to a specific lineage. ECM parameters are extremely dynamic and are spatially controlled during development, suggesting that biomaterials used for tissue engineering have to play a morphogenetic role in guiding differentiation and arrangement of cells. The design of biomimetic biomaterials that may be used to direct the fate of stem cells could have a significant impact on the development of stem cell based therapies (Dolatshahi-Pirouz et al. 2011). As an example, the use of nanoscale disorder has been demonstrated to stimulate human MSC to produce bone mineral *in vitro*, in the absence of any osteogenic supplements (Dalby et al. 2007). Another study showed that MSC cultured on stiff substrates showed more mature focal adhesions, a more contractile cytoskeleton, and an enhanced osteodifferentiation than cells cultured on soft substrates (Engler et al. 2006).

Applying these principles to PSi scaffolds, the variations in pore diameter could affect the number of points at which the cell could contact the substrate, further affecting cytoskeletal tension and subsequently affecting differentiation potential. Some experiments have been conducted to assess cell response to variation in pore size and surface chemical treatment (Collins et al. 2002; Clements et al. 2011, 2012; Wang et al. 2012). These experiments are of particular relevance as cell adhesion, proliferation, and differentiation can be influenced by surface properties, including topography and chemistry.

Thus, osteogenesis was shown to be enhanced by porous topography with a ridge roughness lower than 10 nm, while adipogenesis was enhanced more widely on PSi compared to flat Si substrates (Wang et al. 2012). More studies are needed to define more precisely optimal PSi porosity and chemical modifications enabling differentiation in specific lineage, as it has been done for specific cell attachment.

19.4.3 PSi RESORPTION

PSi surface treatment is necessary not only to improve cell adhesion, but also to permit accessible and stable pores: the inner walls of the porous matrix must be protected from excessive degradation in the aqueous cell environment (*in vitro* or *in vivo*) without eliciting any undesirable effects on the cells (Collart Dutilleul et al. 2014b). PSi surface can be tailored with well-established fabrication methods. Nevertheless, there are some unresolved issues such as deciding whether PSi stabilization is necessary for biological applications and evaluating the effects of PSi stabilization on biological effects. The stabilization process must be considered before functionalization, according to one goal to achieve: surface stability over time has to be defined according to cell attachment, bioactive molecules grafting, immediate *in vivo* use, or long-term *in vitro* culture. Several stabilization processes have been described, such as thermal or ozone oxidation and immersion in H_2O_2, silanization (Naveas et al. 2012). All these stabilizations are highly desirable for biological applications and can be used for the design and optimization of PSi scaffolds, as they can offer control over PSi degradation rates. For tissue engineering, this degradation rate has to be adapted to the regeneration rate of the tissue to be repaired, until the optimal clinical situation where PSi degrades as new tissue forms.

19.5 PSi FOR TISSUE ENGINEERING

Biologic tissues consist of cells, extracellular matrix, and signaling systems. These elements are brought together and are responsible for tissue building, differentiation, and rearrangement (Lanza et al. 2000). Tissue reconstitution, or tissue engineering, aims to imitate natural tissues after injury or disease. This nature imitation could be a biologic replica that exhibits some of the basic properties of the original tissue at the time it is implanted. It could also provide stem cells/progenitors, with the expectation that they will differentiate into the appropriate tissue. Alternatively, it could be nonbiologic replacement. Either of the first two approaches might remedy a deficit and restore functionality in the form of a living tissue or organ.

In this section, we present some applications for which there is no efficient therapeutics thus far, but that PSi-based tissue engineering could potentially treat.

19.5.1 BONE TISSUE ENGINEERING

Nanoscale surface topography has been explored for its influence on cell osteodifferentiation, showing an increased osteogenesis when pore size and roughness decreased (Wang et al. 2012). This impact of pore size on osteodifferentiation is of high importance for the development of scaffold materials that can stimulate stem cell differentiation into osteoblasts in the absence of chemical treatment without compromising material properties (Collart Dutilleul et al. 2014a). Dalby et al. (2007) had already demonstrated the use of nanoscale disorder to stimulate human stem cells to produce bone mineral *in vitro* in the absence of osteogenic supplements. Moreover, controlled dissolution of bioactive silicon-based materials has an impact on stimulation of bone cell function and tissue regeneration (El-Ghannam and Ning 2006). PSi has the ability to degrade, in aqueous solution, into nontoxic silicic acid [Si(OH)4], which is the biologically significant form of Si that is in blood plasma. Silicic acid is vital for normal bone and connective tissue homeostasis. *In vivo*, PSi releases Si(OH)4 with a negligible inflammatory response before being excreted in the urine (Popplewell et al. 1998; Park et al. 2009). The release of Si(OH)4 due to the corrosion of PSi stimulates calcification and collagen growth, which could ideally accelerate new bone regeneration. In addition, it has been demonstrated that Si played an important role in the expression of alkaline phosphatase, a specific biomarker for stem cells' osteodifferentiation (Fan et al. 2011). Figure 19.4 presents results of stem cells' osteodifferentiation on PSi, flat Si, or polystyrene culture plate, highlighting that osteodifferentiation is enhanced by Si release.

In addition, PSi has been reported to stimulate hydroxyapatite growth in stimulated body fluid, inferring its highly efficient potential for bone tissue engineering (Canham 1995). Another study has shown that PSi could be used to promote osteoblast growth, protein–matrix synthesis, and mineralization (Sun et al. 2007). This study was carried out on PSi that had nanoscale (<15 nm) pores, mesoscale (≈50 nm) pores, and macroscale (≈1 μm) pores, and it was found that the mesoporous PSi degraded faster than the macroporous PSi, enhancing PSi potential as a biodegradable scaffold for bone-tissue engineering.

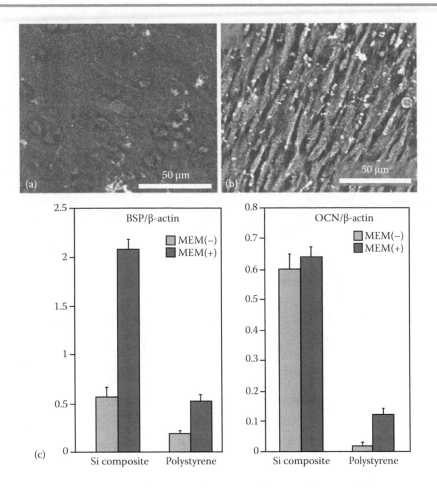

FIGURE 19.4   (a and b) Scanning electron microscopy of stem cells osteodifferentiation and mineralized matrix deposits. (a) On PSi membrane; (b) on cell culture plate (polystyrene). On PSi, osteoblast-like cells are embedded in a mineralized matrix, while cells are more clearly visible on polystyrene, with a less abundant extracellular matrix. (c) mRNA Quantification of bone sialoprotein (BSP) and osteocalcin (OCN) expressed by osteoblast-like cells on pSi-based composite and on cell culture plate (polystyrene), after 2 days of culture in basal medium minimum essential medium MEM(–) or in osteogenic medium MEM(+). Osteogenic markers are expressed at a higher level when cells are cultured in the presence of [Si]. (From Collart-Dutilleul et al., In: *Porous Silicon for Biomedical Applications*, 2014, pp. 486–506. Copyright 2014: Woodhead Publishing. With permission.)

19.5.2 OPHTHALMIC IMPLANTS

Limbal stem cells are self-renewing, highly proliferative cells that have the capacity to reconstruct the entire corneal epithelium in cases of ocular surface injury. Limbal stem cell deficiency can be caused by a variety of conditions, such as primary genetic disorders, chemical and thermal injury, irradiation, inflammatory diseases, and traumatic injuries. It often results in persistent corneal epithelial defect or abnormal reepithelialization by conjunctival epithelial cells, which predisposes to corneal neovascularization and opacification (Gomes et al. 2010). Experiments have been conducted to find alternative source of cells for corneal reconstruction, especially stem cells sharing similar embryologic origin, such as dental pulp stem cells (Monteiro et al. 2009; Gomes et al. 2010). These experiments showed that transplantation of tissue-engineered products carrying efficient stem cells could successfully reconstruct corneal epithelium in animal models. In this context, PSi has been investigated as a suitable biodegradable ophthalmic implant, for both human lens epithelial cells attachment and *in vivo* biocompatibility (Kashanian et al. 2010). For ophthalmology, PSi has been first investigated as an intraocular drug delivery system, demonstrating its good intravitreal biocompatibility (Cheng et al. 2008). Biocompatibility of PSi membranes has been assessed for their potential to support human ocular cells *in vitro* and *in vivo* in rat eyes (Low et al. 2009). Following this development, PSi has been evaluated as implantable microparticles for tissue engineering and drug delivery device (Low et al. 2010). As soft tissues compose the eye, composite polymer-PSi scaffolds have been investigated for ophthalmic application. PSi microparticles have been encapsulated into polycaprolactone fibers produced by electrospinning. This composite biomaterial has been tested for lens epithelial cells attachment and *in vivo* grafting into rats' eyes (Kashanian et al. 2010). This composite biomaterial was found to be compatible to rats' eyes and made the scaffold more flexible than PSi alone. Authors proposed that this composite material could possibly be exploited as a two-stage drug delivery vehicle and for the transfer of primary cells for regenerative medicine, either individually or simultaneously.

Taken altogether, these results suggest a high potential of PSi for eye tissue engineering, as it has been demonstrated to be biocompatible for intravitreal implantation, and capable to support stem cells or specialized lens epithelial cell growth.

19.5.3 NERVE TISSUE ENGINEERING

Nerve injuries induce severe disability and suffering for patients through severe alterations in neurons and the central nervous system. Nerve injuries, particularly where there is a defect, remain a challenge for the surgeon. Reconstruction of nerve injuries with a defect requires utilization of graft material, which can be of various designs. Application of autologous nerve grafts are, thus far, the most common clinical solutions to try to overcome problems with nerve defects (Dahlin et al. 2009). The biocompatibility of implanted devices in the nervous system is a pivotal issue for nerve tissue engineering. Such devices must be nontoxic, should be physically stable, and should ideally provoke no inflammatory reactions. Many mechanisms can trigger inflammatory reactions, such as micromovements when the implant moves into the surrounding tissues, giving rise to shear forces. Micromovements can be minimized by implant design and by manipulating the surface of the implant (chemically or physically) so that the tissue can integrate smoothly with the artificial material (Biran et al. 2007). PSi has been studied for the outgrowth and survival of axons and appeared to be efficient to guide axonal outgrowth, especially when creating a pattern on the PSi surface (Johansson et al. 2005). Moreover, PSi induced minor inflammatory response, enhancing its potential as an implantable device. These two qualities placed PSi as an excellent choice of material for neuroelectronic interfaces, and further studies have been conducted, showing that axons preferred to grow and elongate on PSi surfaces with pores ranging from 150 to 500 nm (Johansson et al. 2008). Such results were confirmed and completed by another study following neuronal cells (neuroblastoma cells) adhesion and spreading on PSi, where neuronal cells appeared to adhere and spread preferentially on PSi with pores ranging from 300 to 1000 nm, and on PSi with pores <50 nm (Khung et al. 2008). Several neural cells have been tested for adhesion and growth onto PSi. Hippocampal cells showed tight adherence on PSi wafers with nanoscale surface topography (Ma et al. 2005). Rat neurons incubated on PSi patterns have revealed the influence of surface topology on neuron network proliferation, suggesting that neuron growth patterns could be achieved using PSi (Sapelkin et al. 2006). PSi has also been shown to act as an extra cellular matrix scaffold, explaining its important role in the neuronal adhesion and neurite outgrowth (Ma et al. 2007). Considering PSi impact on neuronal guidance, chemically patterned PSi surface has been successfully used to direct human neuronal cells attachment (Sweetman et al. 2011). Further *in vivo* studies compared PSi with flat Si as chip-implant surfaces in a nerve regeneration setting, to bridge defects in rat sciatic nerve. Cellular protrusions were observed on PSi with regenerated nerve tissue firmly attached to the PSi surface, while the tissue was hardly attached to the flat Si surface. Moreover, PSi diminished inflammatory response compared to flat Si (Johansson et al. 2009). As previously described, PSi resorption releases silicic acid, which is a component naturally present in some food and excreted in urine (Gonzalez-Muñoz et al. 2008). Silicic acid has been demonstrated to reduce cerebral oxidation, by normalizing the expression of tumor necrotic factor alpha (TNFα) and antioxidant enzymes (Gonzalez-Muñoz et al. 2008). Such a preventive aspect is a positive side effect confirming the promising potential of PSi for nerve tissue engineering, which is a field requiring new and innovative biomaterials.

19.6 ADVANCES IN PSi-BASED COMPOSITES

The combination of a pliable and soft polymeric material with a hard inorganic porous material of high drug-loading capacity such as PSi improves control over degradation and drug release profiles and is beneficial for the preparation of biodegradable scaffolds. Polymers can be easily combined with PSi. This can be done directly, in a bulk fashion without any covalent bonding of the two materials, or via utilizing the surface functionalization techniques discussed previously to create anchoring points. Since the polymer component is often formed at similar or even higher mass fraction to the PSi, we refer to the resulting materials as composites. Due to the limitations of using polymers alone in tissue engineering, the combination of typically brittle inorganic porous scaffolds and flexible polymeric scaffolds has been attempted to yield more robust hybrid scaffolds (McInnes and Voelcker 2014).

Tissue engineering ideally requires the creation of a degradable three-dimensional scaffold structure that has the appropriate physical, chemical, and mechanical properties to enable cell penetration and subsequent tissue formation at the desired site. In the case of bone engineering, porous scaffolds are designed to support cell migration, proliferation, and differentiation and favor cell organization in the proper dimensions. While these scaffolds may be made from a wide variety of both natural and synthetic materials, many of the existing porous biodegradable polymeric systems have been found to have limitations, including inadequate mechanical properties. The alternative of porous ceramic systems is limited by lower fracture toughness relative to cortical bone. PSi possesses several unique features highlighting its utility as a biomaterial when coupled with a biopolymer. Thus, it can be processed into a broad range of self-assembling structures useful to tissue engineering. In addition, it can be produced using electrospinning, which represents a very attractive route to fibrous biomaterial formation, mimicking the size regime of fibers constituting the extracellular matrix of native tissues and organs (Coffer et al. 2005; Whitehead et al. 2007).

Composite sponges of PSi-polycaprolactone have been created by salt leaching and microemulsion/freeze-drying methods. In addition, it has been demonstrated that PSi-containing composites may calcify in an acellular environment and support bone precursor cell proliferation and differentiation (Coffer et al. 2005; Whitehead et al. 2007).

Electrospinning methods have also been used to produce three-dimensional fibrous structures of PSi-polycaprolactone. Electrospinning is an attractive way to fabricate biomaterials for tissue regeneration, as this method offers polymer fibers with high porous surface areas and the possibility of surface modification to improve their cytocompatibility. Moreover, it has been demonstrated that the highly porous and slowly degrading PSi/PCL microfibrous scaffolds can support human mesenchymal stem cell proliferation and differentiation (Fan et al. 2011).

19.7 LEGAL AND PRACTICAL ISSUES

Translation of stem cells and biomaterials research into clinical applications relies on abundant *in vitro* and *in vivo* preclinical data. When it comes to potential therapeutic applications, some restrictions could appear, related to both the stem/progenitor cells and the biomaterials (i.e., PSi).

Human cells for tissue engineering could originate from the patient (autologous) or from a donor (allogenic). Cells of autologous origin bring the less ethical, legal, and practical issues but are often not available in a sufficient amount for cell therapy. Allogenic cells are then required, following strict procedures and respecting informed consent of the donor, after explanations about collection of biological materials, questions about ownership of the collected stem cells, and confidentiality of the information associated with the collected cells (Kato et al. 2012). The constitution of allogenic stem cell banks contains procedures to ensure anonymity, although authorized parties can access some clinically relevant information. The rights of donors and the interests of researchers are protected by incorporating relevant government legislation (ethical committee review) and procedures (e.g., anonymity and consent). It is crucial to explain the use and transfer of cells and data at the time of informed consent, especially highlighting features that distinguish collection for research from collection for clinical tissue engineering applications (Cyranoski

2012). Finally, the physical and intellectual property of biological samples (stem/progenitor cells) collected must be clearly established.

The immunogenicity of PSi has been already studied for its ability to provoke any immune response in human blood derived monocytes, revealing PSi immunogenicity and biocompatibility approximately equivalent to tissue culture polystyrene (Ainslie et al. 2008). The use of PSi for tissue engineering represents a challenge when it is about translating preclinical research into clinical trials. Its manufacturing has to be formalized, following guidance, as proposed by the International Pharmaceutical Excipients Council-Pharmaceutical Quality Group Good Manufacturing Practice (GMP) guide for pharmaceutical excipients (Canham 2014). General factors include the quality management system and the respect of GMP. Some PSi devices are already accepted for clinical trials: the first human clinical safety trial was carried out in 2006, using a PSi-based Brachysil™ device, injected to treat liver tumors. Device toxicity was assessed by the nature, incidence, and severity of adverse events, and no serious events could be observed or attributed to the studied device (Goh et al. 2007).

Taken altogether, *in vitro, in vivo*, and clinical studies showed that PSi could be produced in GMP conditions and rendered biocompatible and biodegradable. This holds promises for future pharmaceutical and medical applications of PSi-based tissue engineering.

REFERENCES

Ainslie, K.M., Tao, S.L., Popat, K.C., and Desai, T.A. (2008). In vitro immunogenicity of silicon-based micro- and nanostructured surfaces. *ACS Nano* **2**, 1076–1084.

Alvarez, S.D., Derfus, A.M., Schwartz, M.P., Bhatia, S.N., and Sailor, M.J. (2008). The compatibility of hepatocytes with chemically modified porous silicon with reference to in vitro biosensors. *Biomater.* **30**, 26–34.

Ayres, C.E., Jha, B.S., Sell, S.A., Bowlin, G.L., and Simpson, D.G. (2010). Nanotechnology in the design of soft tissue scaffolds: Innovations in structure and function. *Wiley Interdiscip. Rev. Nanomed. Nanobiotechnol.* **2**, 20–34.

Bianco, P., Riminucci, M., Gronthos, S., and Robey, P.G. (2001). Bone marrow stromal stem cells: Nature, biology, and potential applications. *Stem Cells* **19**, 180–192.

Biggs, M.J.P., Richards, R.G., and Dalby, M.J. (2010). Nanotopographical modification: A regulator of cellular function through focal adhesions. *Nanomedicine* **6**, 619–633.

Bioengineering NIoBIa (2014). Tissue engineering and regenerative medicine. http://www.nibib.nih.gov/science -education/science-topics/tissue-engineering-and-regenerative-medicine.

Biran, R., Martin, D.C., and Tresco, P.A. (2007). The brain tissue response to implanted silicon microelectrode arrays is increased when the device is tethered to the skull. *J. Biomed. Mater. Res. A* **82**, 169–178.

Buxboim, A., Ivanovska, I.L., and Discher, D.E. (2010). Matrix elasticity, cytoskeletal forces and physics of the nucleus: How deeply do cells "feel" outside and in? *J. Cell. Sci.* **123**, 297–308.

Canham, L.T. (1995). Bioactive silicon structure fabrication through nanoetching techniques. *Adv. Mater.* **7**, 1033–1037.

Canham, L.T. (2014). Porous silicon for medical use: From conception to clinical use. In: Santos H. (Ed.) *Porous Silicon for Biomedical Applications.* Woodhead Publishing, Cambridge, UK, pp. 3–20.

Cheng, L., Anglin, E., Cunin, F. et al. (2008). Intravitreal properties of porous silicon photonic crystals: A potential self-reporting intraocular drug-delivery vehicle. *Br. J. Ophthalmol.* **92**, 705–711.

Clements, L., Wang, P., and Harding, F. (2011). Mesenchymal stem cell attachment to peptide density gradients on porous silicon generated by electrografting. *Phys. Stat. Sol. (a)* **208**, 1440–1445.

Clements, L.R., Wang, P.-Y., Tsai, W.-B., Thissen, H., and Voelcker, N.H. (2012). Electrochemistry-enabled fabrication of orthogonal nanotopography and surface chemistry gradients for high-throughput screening. *Lab. Chip* **12**, 1480–1486.

Coffer, J., Whitehead, M., and Nagesha, D. (2005). Porous silicon-based scaffolds for tissue engineering and other biomedical applications. *Phys. Stat. Sol. (a)* **202**, 1451–1455.

Collart Dutilleul, P.-Y., Secret, E., Panayotov, I. et al. (2014a). Adhesion and proliferation of human mesenchymal stem cells from dental pulp on porous silicon scaffolds. *ACS Appl. Mater. Interfaces* **6**, 1719–1728.

Collart Dutilleul, P.-Y., Deville De Périère, D., Cuisinier, F.J., Cunin, F., and Gergely, C. (2014b). Porous silicon scaffolds for stem cells growth and osteodifferentiation. In: Santos H. (Ed.) *Porous Silicon for Biomedical Applications.* Woodhead Publishing, Cambridge, UK, pp. 486–506.

Collins, B.E., Dancil, K., Abbi, G., and Sailor, M.J. (2002). Determining protein size using an electrochemically machined pore gradient in silicon. *Adv. Func. Mater.* **12**, 187–191.

Cyranoski, D. (2012). Stem-cell pioneer banks on future therapies. *Nature* **488**, 139–142.

Dahlin, L., Johansson, F., Lindwall, C., and Kanje, M. (2009). Future perspective in peripheral nerve reconstruction. *Intern. Rev. Neurobiology* **87**, 507–530.

Dalby, M.J., Riehle, M.O., Yarwood, S.J., Wilkinson, C.D.W., and Curtis, A.S.G. (2003). Nucleus alignment and cell signaling in fibroblasts: Response to a micro-grooved topography. *Exp. Cell Res.* **284**, 274–282.

Dalby, M.J., Gadegaard, N., Tare, R. et al. (2007). The control of human mesenchymal cell differentiation using nanoscale symmetry and disorder. *Nat Mater* **6**, 997–1003.

Dolatshahi-Pirouz, A., Nikkhah, M., and Kolind, K. (2011). Micro- and nanoengineering approaches to control stem cell-biomaterial interactions. *J. Funct. Biomater.* **2**, 88–106.

Dominici, M., Le Blanc, K., Mueller, I. et al. (2006). Minimal criteria for defining multipotent mesenchymal stromal cells. *Cytotherapy* **8**, 315–317.

Donovan, P.J. and Gearhart, J. (2001). The end of the beginning for pluripotent stem cells. *Nature* **414**, 92–97.

El-Ghannam, A. and Ning, C.Q. (2006). Effect of bioactive ceramic dissolution on the mechanism of bone mineralization and guided tissue growth in vitro. *J. Biomed. Mater. Res. A* **76**, 386–397.

Ema, H. and Nakauchi, H. (2003). Self-renewal and lineage restriction of hematopoietic stem cells. *Curr. Opin. Genet. Dev.* **13**, 508–512.

Engler, A.J., Sen, S., Sweeney, H.L., and Discher, D.E. (2006). Matrix elasticity directs stem cell lineage specification. *Cell* **126**, 677–689.

Fan, D., Akkaraju, G.R., Couch, E.F., Canham, L.T., and Coffer, J.L. (2011). The role of nanostructured mesoporous silicon in discriminating in vitro calcification for electrospun composite tissue engineering scaffolds. *Nanoscale* **3**, 354–361.

Forraz, N. and McGuckin, C.P. (2011). The umbilical cord: A rich and ethical stem cell source to advance regenerative medicine. *Cell Prolif* **44**, 60–69.

Gentile, F., La Rocca, R., Marinaro, G. et al. (2010). Differential cell adhesion on mesoporous silicon substrates. *ACS Appl. Mater. Interfaces* **4**, 2903–2911.

Goh, A.S.-W., Chung, A.Y.-F., Lo, R.H.-G. et al. (2007). A novel approach to brachytherapy in hepatocellular carcinoma using a phosphorous32 (32P) brachytherapy delivery device—A first-in-man study. *Int. J. Radiat Oncol. Biol. Phys.* **67**, 786–792.

Gomes, J.A.P., Geraldes Monteiro, B., Melo, G.B. et al. (2010). Corneal reconstruction with tissue-engineered cell sheets composed of human immature dental pulp stem cells. *Invest. Ophthalmol. Vis. Sci.* **51**, 1408–1414.

Gonzalez-Muñoz, M.J., Meseguer, I., Sanchez-Reus, M.I. et al. (2008). Beer consumption reduces cerebral oxidation caused by aluminum toxicity by normalizing gene expression of tumor necrotic factor alpha and several antioxidant enzymes. *Food Chem. Toxicol.* **46**, 1111–1118.

Gronthos, S., Mankani, M., Brahim, J., Robey, P.G., and Shi, S. (2000). Postnatal human dental pulp stem cells (DPSCs) in vitro and in vivo. *Proc. Nat. Acad. Sci. USA* **97**, 13625–13630.

Gronthos, S., Brahim, J., Li, W. et al. (2002). Stem cell properties of human dental pulp stem cells. *J. Dent. Res.* **81**, 531–535.

Hench, L.L. and Polak, J.M. (2002). Third generation of biomedical materials. *Science* **295**, 1014–1018.

Heydarkhan-Hagvall, S., Choi, C.-H., Dunn, J. et al. (2007). Influence of systematically varied nano-scale topography on cell morphology and adhesion. *Cell Commun. Adhes.* **14**, 181–194.

Holzwarth, J.M. and Ma, P.X. (2011). Biomimetic nanofibrous scaffolds for bone tissue engineering. *Biomater.* **32**, 9622–9629.

Jin, G., Prabhakaran, M.P., and Ramakrishna, S. (2011). Stem cell differentiation to epidermal lineages on electrospun nanofibrous substrates for skin tissue engineering. *Acta Biomater.* **7**, 3113–3122.

Johansson, F., Kanje, M., Eriksson, C., and Wallman, L. (2005). Guidance of neurons on porous patterned silicon: Is pore size important? *Phys. Stat. Sol. (c)* **2**, 3258–3262.

Johansson, F., Kanje, M., Linsmeier, C.E., and Wallman, L. (2008). The influence of porous silicon on axonal outgrowth in vitro. *IEEE Trans. Biomed. Eng.* **55**, 1447–1449.

Johansson, F., Wallman, L., Danielsen, N., Schouenborg, J., and Kanje, M. (2009). Porous silicon as a potential electrode material in a nerve repair setting: Tissue reactions. *Acta Biomater.* **5**, 2230–2237.

Jones, J.R., Lee, P.D., and Hench, L.L. (2006). Hierarchical porous materials for tissue engineering. *Philos. Trans. A Math. Phys. Eng. Sci.* **364**, 263–281.

Kashanian, S., Harding, F., Irani, Y. et al. (2010). Evaluation of mesoporous silicon/polycaprolactone composites as ophthalmic implants. *Acta Biomater.* **6**, 3566–3572.

Kato, K., Kimmelman, J., Robert, J., Sipp, D., and Sugarman, J. (2012). Ethical and policy issues in the clinical translation of stem cells: Report of a focus session at the ISSCR Tenth Annual Meeting. *Cell Stem Cell* **11**(6), 765–767.

Kelleher, C.M. and Vacanti, J.P. (2010). Engineering extracellular matrix through nanotechnology. *J. R. Soc. Interface* **7**, 717–729.

Khan, W.S., Longo, U.G., Adesida, A., and Denaro, V. (2012). Stem cell and tissue engineering applications in orthopaedics and musculoskeletal medicine. *Stem Cells Int.* **2012**, 2–4.

Khung, Y.L., Barritt, G., and Voelcker, N.H. (2008). Using continuous porous silicon gradients to study the influence of surface topography on the behaviour of neuroblastoma cells. *Exp. Cell Res.* **314**, 789–800.

Langer, R. and Vacanti, J.P. (1993). Tissue engineering. *Science* **260**, 920–926.

Lanza, R.P., Langer, R., and Vacanti, J. (Eds.) (2000). *Principles of Tissue Engineering*, 2nd ed. Academic Press, New York.

Legislation Regulation (EC) No 1394/2007 of the European Parliament and of the Council. (2007). *Official J. Eur. Union L* 324, 121–131.

Low, S.P., Williams, K.A., Canham, L.T., and Voelcker, N.H. (2006). Evaluation of mammalian cell adhesion on surface-modified porous silicon. *Biomater.* **27**, 4538–4546.

Low, S.P., Voelcker, N.H., Canham, L.T., and Williams, K.A. (2009). The biocompatibility of porous silicon in tissues of the eye. *Biomater.* **30**, 2873–2880.

Low, S.P., Williams, K.A., Canham, L.T., and Voelcker, N.H. (2010). Generation of reactive oxygen species from porous silicon microparticles in cell culture medium. *J. Biomed. Mater. Res. A* **93**, 1124–1131.

Ma, T. (2010). Mesenchymal stem cells: From bench to bedside. *World J. Stem Cells* **2**, 13–17.

Ma, J., Liu, B.F., Xu, Q.Y., and Cui, F.Z. (2005). AFM study of hippocampal cells cultured on silicon wafers with nano-scale surface topograph. *Colloids Surf. B Biointerfaces* **44**, 152–157.

Ma, J., Cui, F.Z., Liu, B.F., and Xu, Q.Y. (2007). Atomic force and confocal microscopy for the study of cortical cells cultured on silicon wafers. *J. Mater. Sci. Mater. Med.* **18**, 851–856.

Marolt, D., Knezevic, M., and Novakovic, G.V. (2010). Bone tissue engineering with human stem cells. *Stem Cell Res. Ther.* **1**, 10–14.

McInnes, S. and Voelcker, N.H. (2014). Porous silicon–polymer composites for cell culture and tissue engineering applications. In: Santos H. (Ed.) *Porous Silicon for Biomedical Applications*. Woodhead Publishing, Cambridge, UK, pp. 420–469.

Monteiro, B.G., Serafim, R.C., Melo, G.B. et al. (2009). Human immature dental pulp stem cells share key characteristic features with limbal stem cells. *Cell Prolif.* **42**, 587–594.

Murry, C.E. and Keller, G. (2008). Differentiation of embryonic stem cells to clinically relevant populations: Lessons from embryonic development. *Cell* **132**, 661–680.

Naveas, N., Costa, V.T., Gallach, D. et al. (2012). Chemical stabilization of porous silicon for enhanced biofunctionalization with immunoglobulin. *Sci. Technol. Adv. Mater.* **13**, 045009.

Noval, A.M., Vaquero, V.S., Quijorna, E.P. et al. (2012). Aging of porous silicon in physiological conditions: Cell adhesion modes on scaled 1D micropatterns. *J. Biomed. Mater. Res. A* **100**, 1615–1622.

Okita, K., Ichisaka, T., and Yamanaka, S. (2007). Generation of germline-competent induced pluripotent stem cells. *Nature* **448**, 313–317.

Park, J.-H., Gu, L., Maltzahn von, G., Ruoslahti, E., Bhatia, S.N., and Sailor, M.J. (2009). Biodegradable luminescent porous silicon nanoparticles for in vivo applications. *Nat. Mater.* **8**, 331–336.

Popplewell, J.F., King, S.J., Day, J.P. et al. (1998). Kinetics of uptake and elimination of silicic acid by a human subject: A novel application of 32Si and accelerator mass spectrometry. *J. Inorg. Biochem.* **69**, 177–180.

Prestwich, G.D. (2007). Simplifying the extracellular matrix for 3-D cell culture and tissue engineering: A pragmatic approach. *J. Cell. Biochem.* **101**, 1370–1383.

Ramirez, J.-M., Bai, Q., Dijon-Grinand, M. et al. (2010). Human pluripotent stem cells: From biology to cell therapy. *World J. Stem Cells* **2**, 24–33.

Reilly, G.C. and Engler, A.J. (2010). Intrinsic extracellular matrix properties regulate stem cell differentiation. *J. Biomech.* **43**, 55–62.

Rosengren, A., Wallman, L., Danielsen, N., Laurell, T., and Bjursten, L.M. (2002). Tissue reactions evoked by porous and plane surfaces made out of silicon and titanium. *IEEE Trans. Biomed. Eng.* **49**, 392–399.

Sailor, M.J. (2012). *Porous Silicon in Practice: Preparation, Characterization and Applications*. Wiley-VCH. John Wiley & Sons, Weinheim, Germany.

Sapelkin, A.V., Bayliss, S.C., Unal, B., and Charalambou, A. (2006). Interaction of B50 rat hippocampal cells with stain-etched porous silicon. *Biomater* **27**, 842–846.

Sun, W., Puzas, J.E., Sheu, T.J., Liu, X., and Fauchet, P.M. (2007). Nano- to microscale porous silicon as a cell interface for bone tissue engineering. *Adv. Mater.* **19**, 921–924.

Sweetman, M.J., Shearer, C.J., Shapter, J.G., and Voelcker, N.H. (2011). Dual silane surface functionalization for the selective attachment of human neuronal cells to porous silicon. *Langmuir* **27**, 9497–9503.

Takahashi, K. and Yamanaka, S. (2006). Induction of pluripotent stem cells from mouse embryonic and adult fibroblast cultures by defined factors. *Cell* **126**, 663–676.

Tesche, L.J. and Gerber, D.A. (2010). Tissue-derived stem and progenitor cells. *Stem Cells Int.* **2010**, 824–826.

Thomson, J.A., Itskovitz-Eldor, J., Shapiro, S.S. et al. (1998). Embryonic stem cell lines derived from human blastocysts. *Science* **282**, 1145–1147.

Torres-Costa, V.V., Martínez-Muñoz, G.G., Sánchez-Vaquero, V.V. et al. (2012). Engineering of silicon surfaces at the micro- and nanoscales for cell adhesion and migration control. *Int. J. Nanomed.* **7**, 623–630.

Vacanti, J. (2010). Tissue engineering and regenerative medicine: from first principles to state of the art. *J. Pediatr. Surg.* **45**, 291–294.

Wang, Y., Kim, H.J., Vunjak-Novakovic, G., and Kaplan, D.L. (2006). Stem cell-based tissue engineering with silk biomaterials. *Biomaterials* **27**, 6064–6082.

Wang, P., Clements, L., Thissen, H., Jane, A., Tsai, W.-B., and Voelcker, N.H. (2012). Screening mesenchymal stem cell attachment and differentiation on porous silicon gradients. *Adv. Func. Mater.* **22**, 3414–3423.

Weissman, I.L. (2000). Translating stem and progenitor cell biology to the clinic: Barriers and opportunities. *Science* **287**, 1442–1446.

Whitehead, M., Fan, D., Akkaraju, G., Canham, L.T., and Coffer, J. (2007). Accelerated calcification in electrically conductive polymer composites comprised of poly ε-caprolactone, polyaniline, and bioactive mesoporous silicon. *J. Biomed. Mater. Res. A* **83**, 225–234.

Whitehead, M.A., Fan, D., Mukherjee, P., Akkaraju, G.R., Canham, L.T., and Coffer, J.L. (2008). High-porosity poly(epsilon-caprolactone)/mesoporous silicon scaffolds: Calcium phosphate deposition and biological response to bone precursor cells. *Tissue Eng. A* **14**, 195–206.

Xynos, I.D., Edgar, A.J., Buttery, L.D.K., Hench, L.L., and Polak, J.M. (2001). Gene expression profiling of human osteoblasts following treatment with the ionic products of Bioglass® 45S5 dissolution. *J. Biomed. Mater. Res.* **55**, 151–157.

Yoo, C.H., Na, H.-J., Lee, D.-S. et al. (2013). Endothelial progenitor cells from human dental pulp-derived iPS cells as a therapeutic target for ischemic vascular diseases. *Biomater.* **34**(33), 8149–8160.

Zuk, P.A., Zhu, M., Ashjian, P. et al. (2002). Human adipose tissue is a source of multipotent stem cells. *Mol. Biol. Cell* **13**, 4279–4295.

Use of Porous Silicon for *In Vivo* Imaging Techniques

Igor Komarov and Sergei Alekseev

20

CONTENTS

20.1	Introduction	382
20.2	General Requirements to the Molecular Probes Used for *In Vivo* Imaging in the Context of PSi Nanostructures	384
	20.2.1 Preparation of PSi Micro- and Nanoparticles and Their Surface Functionalization for *In Vivo* Imaging Applications	384
	20.2.2 Biocompatibility and Biosafety of PSi	387
	20.2.3 *In Vivo* Stability, Biodistribution, and Clearance of PSi	387
20.3	*In Vivo* Imaging Using PSi as the Molecular Probes	389
	20.3.1 PSi in Optical Imaging	389
	20.3.2 Nuclear Imaging with the Aid of PSi-Based Materials	393
	20.3.3 Magnetic Resonance Imaging Using PSi-Containing Nanocomposites	395
20.4	Conclusions and Outlook	397
	References	397

20.1 INTRODUCTION

The term "*in vivo* imaging" implies a number of techniques contemporarily used in medicine and research to visualize living organisms at anatomical, morphological/functional, cellular, and molecular levels, allowing a noninvasive detailed insight into the organisms and helping to understand metabolic processes and disease-related changes in the body. The last three decades witnessed rapid development of *in vivo* imaging techniques, which has had profound impact on drug discovery and development, biomedical and biochemical research, and most importantly on clinical medicine, especially for treatment of cancer and cardiovascular diseases (Ntziachristos et al. 2007; Fass 2008).

Different *in vivo* imaging modalities most widely used today have different underlying physical principles (Figure 20.1).

Ultrasound imaging (sonography) uses high-frequency sound waves to view soft tissues such as muscles and internal organs. Positron emission tomography (PET) and single-photon emission computed tomography (SPECT) are called nuclear imaging methods, as they are based on positron-induced or direct high-energy photon emission of the unstable isotopes introduced in the studied organisms. Optical imaging methods, magnetic resonance imaging (MRI) and X-ray computed tomography (CT) use the responses of the studied objects to electromagnetic irradiation. Despite the differences in underlying principles, almost all the methods use molecular probes (called reporters, tracers, or contrast enhancers, depending on the method involved), which either themselves are the sources of the detected signals (radioactive probes in PET, SPECT) or they take part in conversion of externally applied energy conversion to form the response (contrast enhancers in MRI, ultrasound, CT, reporters in optical imaging). Essential characteristics, including spatial and temporal resolution of some of the *in vivo* imaging modalities, are listed in Table 20.1, and advantages and disadvantages of these methods are listed in Table 20.2.

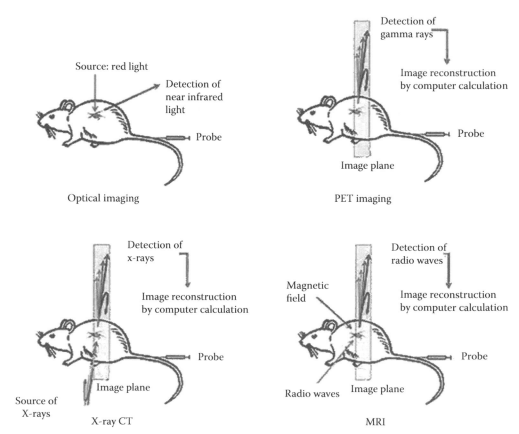

FIGURE 20.1 Schematic representation of the physical principles underlying imaging techniques discussed in this chapter. (With kind permission from Springer Science+Business Media: *Histochem. Cell Biol.* 130, 2008, 845, Debbage P. and Jaschke W.)

TABLE 20.1   Essential Characteristics of Some Modern *In Vivo* Imaging Techniques

Imaging Technique	Radiation Used	Most Used Tracers or Contrast Agents	Spatial Resolution[a]	Depth	Temporal Resolution[b]	Sensitivity[c]
Optical fluorescence imaging	Visible or near-IR	Fluorescent dyes and proteins, II-VI NPs	2–3 mm	<1 cm	Seconds to minutes	~10^{-9}–10^{-12} mole/l
Ultrasound	High-frequency sound	Gas (SF_6, C_3F_8) microbubbles (optimal)	50–500 μm	mm to cm	10 s to minutes	Not well characterized
Positron emission tomography (PET)	High-energy γ rays	^{11}C, ^{18}F, $^{13}NH_3$, $^{82}RbCl$, ^{68}Ga compounds, $H_2^{15}O$	1–2 mm	No limit	10 s to minutes	10^{-11}–10^{-12} mole/l
Single-photon emission computed tomography (SPECT)	Lower-energy γ rays	^{99m}Tc, ^{111}In, ^{123}I, ^{131}I	1–2 mm	No limit	Minutes	10^{-10}–10^{-11} mole/l
Magnetic resonance imaging (MRI)	Radiowaves	Gd^{3+} complexes, Fe_3O_4 and Fe/Pt NPs, Mn^{2+} complexes and NPs	25–100 μm	No limit	Minutes to hours	10^{-3}–10^{-5} mole/l
Computed tomography (CT)	X-rays	$BaSO_4$, iodine compounds (optional)	50–200 μm	No limit	Minutes	Not well characterized

Source: Data extracted from Massoud and Gambhir (2003) and Santos et al. (2013).

[a] Spatial resolution is a measure of the accuracy or detail of graphic display in the images. It is the minimum distance between two independently measured objects that can be distinguished separately.

[b] Temporal resolution relates to the time required to collect enough events to form an image, and to the responsiveness of the imaging system to rates of any change induced by the operator or in the biological system at hand.

[c] Sensitivity, minimum detectable concentration of a molecular probe measured in moles per liter.

TABLE 20.2   Advantages and Disadvantages of Some Modern *In Vivo* Imaging Techniques

Imaging Technique	Advantages	Disadvantages
Optical fluorescence imaging	High sensitivity, detects fluorochrome in live and dead cells	Relatively low spatial resolution, surface-weighted
Ultrasound	Real-time, low cost	Limited spatial resolution, mostly morphological
Positron emission tomography (PET)	High sensitivity, isotopes can substitute naturally occurring atoms, quantitative translational research	PET cyclotron or generator needed, relatively low spatial resolution, radiation to subject
Single-photon emission computed tomography (SPECT)	Many molecular probes available, can image multiple probes simultaneously, may be adapted to clinical imaging systems	Relatively low spatial resolution because of sensitivity, collimation, radiation
Magnetic resonance imaging (MRI)	Highest spatial resolution, combines morphological and functional imaging	Relatively low sensitivity, long scan and postprocessing time, mass quantity of probe may be needed
Computed tomography (CT)	Bone and tumor imaging, anatomical imaging	Limited "molecular" applications, limited soft tissue resolution, radiation

Source: Data extracted from Massoud and Gambhir (2003) and Santos et al. (2013).

While the ultrasound, CT, PET, SPECT, and MRI imaging instruments are now common in clinics as they provide the most reliable information about human organisms at anatomical, functional, and molecular levels, other techniques including optical imaging are more common in research settings, and are used mostly in small animal imaging studies. However, incredible advances in this area will hopefully push relatively novel techniques toward a wider use, thanks to intensive research into their development.

The research works in the *in vivo* imaging area are directed first onto development of novel molecular probes. At the heart of these efforts are nanomaterials, which have been shown to serve as excellent probes for the imaging techniques. However, had this been the only capacity of nanomaterials related to their use in imaging, we would not see such an explosive growth of publications devoted to their study in the last decade. It has now been demonstrated on numerous examples that nanomaterials have a unique combination of properties allowing their multimodal use as nanoplatforms for imaging *and* active influence on living organisms. In medicine, nanomaterials have been used as the platform to construct *theranostic* nanostructures, which provide *thera*peutic action along with diag*nostic*, imaging possibilities. Analysis of recent literature shows that nanomedicine is a very dynamic area of research, especially in cancer therapy and diagnostics. The first efficient therapeutic anticancer agents based on the use of nanoparticles have already reached developmental stages (Peer et al. 2007); theranostic nanostructures are not far behind (Fass 2008).

In this chapter, the use of a particular type of nanomaterial, porous silicon (PSi), as the molecular reporter in the *in vivo* imaging, will be discussed.

The first publication on the synthesis of PSi appeared long ago (Uhlir 1956), but it remained overlooked until the discovery of remarkable photoluminescence properties of this material (Canham 1990; Cullis and Canham 1991). Since then, many research groups exploited other useful properties of PSi, such as high porosity and large surface area, tunable pore size, easy surface modification, biocompatibility, biodegradability, and long fluorescence lifetime. Those PSi properties that are relevant to its use as the reporter in different *in vivo* imaging methods will be discussed here. There are excellent reviews published recently on this subject (He et al. 2010; Santos et al. 2011, 2013, 2014; Gupta et al. 2013). It should be noticed, however, that many important discoveries in this area were done in recent years that will be stressed next.

20.2 GENERAL REQUIREMENTS TO THE MOLECULAR PROBES USED FOR *IN VIVO* IMAGING IN THE CONTEXT OF PSi NANOSTRUCTURES

The pioneering works on the photoluminescence properties of PSi cited above (Canham 1990; Cullis and Canham 1991) sparked numerous studies aimed at the use of this nanomaterial for bioimaging. However, the progress in the *in vivo* imaging based on PSi-based materials as the probes (contrast agents, tracers, reporters) has been slow since then. Promising systems appeared only recently (Santos et al. 2013). The main reason for this is that living organisms impose substantial challenges for a compound or material to be efficiently used *in vivo*, and the imaging probes are not an exception. We first outline the most general requirements to the imaging probes, and analyze how these requirements were met for the PSi-based imaging probes. Specific requirements that depend on the particular imaging method will be discussed later in the corresponding sections.

The general requirements concern such properties of the material as *in vivo* biocompatibility, biosafety, stability in the organism, and biodegradability. All of these properties were found to depend critically on the preparation conditions and surface modification of PSi, which will briefly be reviewed next.

20.2.1 PREPARATION OF PSi MICRO- AND NANOPARTICLES AND THEIR SURFACE FUNCTIONALIZATION FOR *IN VIVO* IMAGING APPLICATIONS

Commonly, the PSi nano- and microparticles applied for *in vivo* bioimaging are prepared by a "top-down" approach consisting of electrochemical or chemical etching of the bulk crystalline Si wafers in the HF-containing solutions. The peculiarities of these processes, the photoluminescent properties of the PSi as well as the PSi-based "photonic crystals" such as the Bragg mirrors or Rugate optical filters are reviewed in *Porous Silicon: Formation and Properties* and *Porous Silicon: Optoelectronics, Microelectronics and Energy Technology Applications*. To get the PSi nanoparticles as well as single Si nanocrystals, the PSi layers can be dispersed by any applicable method (mechanical grinding, ultrasound, etc.) followed by size-separation of the resulted particles using centrifugation or microfiltration. The irregularly shaped particles with more or less wide size distribution, which are formed as a result, were successfully used for bioimaging purposes (Park et al. 2009).

Methods allowing preparation of regularly shaped PSi nanoparticles, which were more frequently used for theranostic applications (combining imaging diagnostic and therapy), usually include a combination of silicon micromachining techniques (photolithography, reactive ion etching, etc.) and electrochemical etching. Thus, highly monodisperse discoidal and quasi-hemispherical PSi NPs were prepared by this route (Chiappini et al. 2010a; Godin et al. 2012) (Figure 20.2a,b). To get PSi nanowires (NWs), the process of metal-assisted stain etching (MASE) was applied (Chiappini et al. 2010b). The nanoparticles of Ag, acting as the internal cathodes in usual electrochemical corrosion, were predeposited on the surface of the Si wafer. Afterward the

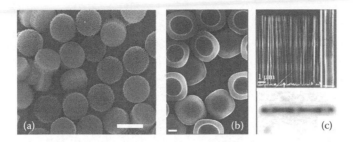

FIGURE 20.2   SEM images of regularly shaped PSi particles. All scale bars are 1 μm. (a) Monodispersed 1000 nm × 400 nm discoidal PSi particles. (Godin B. et al.: *Adv. Funct. Mater.*, 2012. 22. 4225. Copyright Wiley-VCH Verlag GmbH & Co. KGaA. Reproduced with permission.) (b) Bowl-shaped PSi particles. (Chiappini C. et al.: *ChemPhysChem*. 2010. 11, 1029. Copyright Wiley-VCH Verlag GmbH & Co. KGaA. Reproduced with permission.) (c) The porous silicon barcode nanowires obtained by Ag-assisted etch (2.9 M HF/0.05 M H_2O_2 for 5 min followed by 2.9 M HF/0.2 M H_2O_2 for 1 min, repeated three times) of Si wafer (p-type, <0.005 Ω·cm).

wafer was dipped repeatedly into the HF/H_2O_2 etchant with variable H_2O_2 concentration; the PSi barcode nanowires with multiple segments of different porosity and multicolor reflectance were formed as a result (Figure 20.2c).

The dispersibility of the PSi NPs in water media, their stability in biological environment, bio-degradability rate, toxicity, and specific targeting into the body or the cell, all very critical for the bioimaging applications, is determined in the first turn by appropriate surface chemistry of the PSi. Different surface functionalizations of the PSi designed for *in vivo* applications are presented in Figure 20.3.

The surface of as-prepared PSi demonstrates significant hydrophobicity and chemical insta-bility in water media due to the surface coverage mainly by chemically active silane groups ($Si_{4-x}SiH_x$). The silane groups easily react with unsaturated C–C bonds of organic molecules (the hydrosilylation reaction on Figure 20.3) giving hydrolytically stable Si–C bonds. This reaction easily proceeds under thermal or white-light photochemical activation, and, differently from the

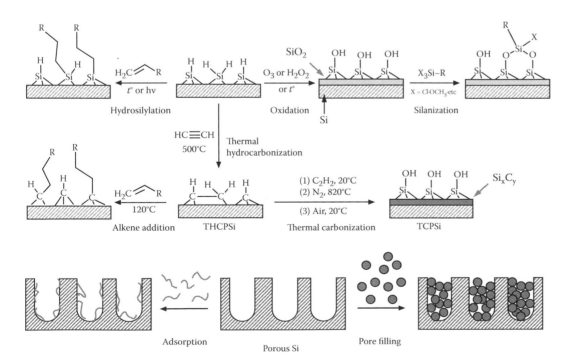

FIGURE 20.3   Schematic representation of different surface functionalizations of the PSi particles applied for *in vivo* bioimaging.

low-molecular silanes, no catalyst is needed. The peculiarities of hydrosilylation as well as other reactions of SiH_x groups and Si–Si bonds on the PSi surface are comprehensively reviewed by Buriak (2002). The oxidation of the PSi by chemical (O_3, H_2O_2, or even water in basic solutions) or thermal (commonly 300°C in air) methods resulted in significant hydrophilization of the PSi due to the formation of a hydroxylated SiO_2 layer. The silanization approach consisted of the reaction of surface hydroxyls with an active silane molecule bearing desired functional group (–R) and an anchor group $-SiX_3$ is commonly used for the SiO_2 functionalization (Figure 20.3). In fact, both the hydrosilylation and silanization routes are applied for the PSi functionalization by relatively simple organic groups such as alkyl ($-C_nH_{2n+1}$), aryl ($-C_6H_5$), aminoalkyl ($-(CH_2)_3NH_2$), carboxylic acid ($-(CH_2)_nCO_2H$) and so on. More complicated biologically related species are commonly grafted on the PSi *via* further conjugation with the above-mentioned simple linkers (Almeida et al. 2014; Wang et al. 2014; Zhang et al. 2014).

A group of authors from Finland developed two procedures of the PSi chemical functionalization by gaseous acetylene resulting in the PSi high stability in biological media (Santos et al. 2014). The procedure (named thermal hydrocarbonization) consisted of the PSi interaction with a continuous flow of C_2H_2 in a temperature range 400–650°C. Highly hydrophobic surface of the resulted thermally hydrocarbonized PSi (THCPSi) is covered mainly by a monolayer of alkyl CH_x groups. The alkenes can be efficiently grafted on the surface of the TCHPSi under the conditions that are identical to that of the thermal hydrosilylation. The reaction proceeds probably due to the interaction of alkenes with strained and unsaturated C–C bonds on the TCHPSi surface (Kovalainen et al. 2012). If the temperature of carbonization increases above the dissociation limit of acetylene and PSi surface restructurization temperature (>650°C), a thin nonstoichiometric SiC-layer is formed on the PSi surface. The surface of the resulted thermally carbonized PSi (TCPSi) is hydrophilic (obviously due to the hydroxylation after ambient air exposition) and it can be further functionalized by the silanization method similarly to the oxidized PSi surface. Table 20.3 gives a survey of different surface modification peculiarities in respect to *in vivo* application and the appropriate references.

PSi surface can also be efficiently functionalized by the surface adsorption or by loading of a substance of interest inside the PSi pores (Figure 20.3). As will be demonstrated later in this chapter, such a functionalization was used most frequently for theranostic PSi applications. Large size of the adsorbed molecules and their affinity to the surface resulted in the increase of adsorbed layer stability. For example, the adsorption of the hydrophobin proteins resulted in hydrophilization of the TCHPSi particles (Bimbo et al. 2011); adsorption of agarose on the aminated PSi nanoparticles enhanced their ability to load and release proteins (De Rosa et al. 2011). The pore loading approach was successfully used to get the composites of the PSi microparticles with magnetic Fe_3O_4 nanoparticles (Kinsella et al. 2011), with gadolinium-based MRI contrast agents (Ananta et al. 2010), and in fact in all extensively studied PSi NPs applications for the medical drugs delivery, which allowed realization of the concept of "magic bullet," that is, the drug that acts "in the proper place in the proper time."

TABLE 20.3 Overview of the Surface Modification Routes Currently Used in the Stabilization of Porous Silicon

Modification	Surface Termination	Stability	Ref.
Native PSi	SiH_x	Low	Canham 1995
Oxidation	Si-O-Si, OSi-H, -OH	Moderate/high	Park et al. 2009; Godin et al. 2012
Hydrocarbonization	CH_x	High	Bimbo et al. 2010; Sarparanta et al. 2012a; Kinnari et al. 2013; Liu et al. 2013
Carbonization	Si-O-Si, -CSi-O, -OH	High	Salonen et al. 2005; Wang et al. 2014
Hydrosilylation	e.g., $-CH_x$ or $-CO_2H$	Moderate/high	Sciacca et al. 2010
Silanization	e.g., Si-O-Si, $-NH_2$	Moderate	Park et al. 2009
Multifunctionalized	$-CH_x + -CO_2H$	High	Sciacca et al. 2010; Kovalainen et al. 2012

Source: Adapted from Santos H.A. et al., *Nanomedicine* 9(4), 535, 2014. Copyright 2014: Future Medicine Ltd. With permission.

20.2.2 BIOCOMPATIBILITY AND BIOSAFETY OF PSi

Biocompatibility, *in vivo* stability and degradation of PSi, dependence of these parameters on the surface chemistry were the focus of many studies, which started back in 1990s. Biocompatibility refers to the ability of a material to perform its desired function within a living body without eliciting any undesirable local or systemic effects, but generating the most appropriate beneficial cellular or tissue response with regard to the desired function (Williams 2008). Biocompatibility is a remarkable property of the PSi, which attracted the interest to this material since the early studies of Canham (1995). Therefore, Chapter 17 of this book is devoted to this subject. Here, we only stress that surface PSi modification is of utmost importance not only for engineering the properties of the particles aimed at their targeting, therapeutic and imaging utility, but at the same time, for ensuring their highest possible biocompatibility and biosafety, critical to their use *in vivo* as imaging probes.

Early *in vitro* studies (Canham 1995; Bayliss et al. 1999; Chin et al. 2001; Angelescu et al. 2003; Low et al. 2006; Alvarez et al. 2009; Fucikova et al. 2009) demonstrated the feasibility of PSi integrating into live organisms for biomedical technologies, including *in vivo* imaging, as they are acceptably compatible with live eukaryotic cells. However, more sophisticated research into biosafety of PSi-based materials, including *in vivo* testing was obviously needed to proceed further and develop safe technologies for imaging in animals and, eventually, in humans. *In vivo* animal toxicity studies were performed first by Park et al. (2009), who demonstrated relatively low overall toxicity of the as-prepared PSi. More recent comprehensive studies were performed by Shahbazi et al. (2013) and Tölli et al. (2014). These recent studies, described in more detail in Chapter 17 in this book, laid the ground for advancement of the PSi to the *in vivo* theranostic applications; however, further studies are needed before the PSi materials reach the clinics.

20.2.3 *IN VIVO* STABILITY, BIODISTRIBUTION, AND CLEARANCE OF PSi

Other highly important issues to be addressed before the PSi materials might enjoy wide *in vivo* clinical applications are their biodistribution in living organisms and the subtle balance between their stability *in vivo* and biodegradation, with the following clearance from the body. It is also very important for *in vivo* imaging molecular probes (tracers, contrast agents) to be delivered to the target efficiently to allow imaging diagnostics (e.g., solid tumor) and then be cleared from the body in a reasonable amount of time. Nonoptimal clearance and low biodegrability is a potential stumbling block for most known nanomaterials on the way to clinical imaging applications (Choi et al. 2007). On one hand, particles with hydrodynamic size less than 5 nm can be too quickly cleared from the organism before they assist the imaging as probes. On the other hand, larger nanoparticles might accumulate in the body for a very long time, thus increasing the toxicity burden on the organism and interfering with therapy or further imaging and diagnostic protocols (e.g., nanoparticles can hamper the use of X-ray-based CT). It was recognized very early that PSi particles have a potential to avoid these problems: with their usual hydrodynamic size of 20–200 nm and higher, they escape renal clearance, but degrade in living organisms to physiologically benign silicic acid.

Pioneering work (Canham 1995) has demonstrated the biodegradability of hydrated porous silicon in simulated body fluid; studies that followed on living organisms confirmed this (Park et al. 2009). As noticed above, very rapid biodegradation of the PSi nanoparticles is a problem for *in vivo* imaging. The PSi nanoparticles might be too quickly cleared through the mononuclear phagocyte system (MPS) (reticuloendothelial system or macrophage system), become entrapped inside the MPS organs like liver and spleen, and degrade to ortho-silicic acid (Varkouhi et al. 2011; Godin et al. 2012). This problem was the focus of many studies, and it was shown that too rapid biodegradation of PSi can be attenuated by proper surface modification. In the work cited previously (Park et al. 2009), the authors used dextran-coated PSi nanoparticles and showed that the coating improved the biodegradability parameters. Great potential in this respect is being uncovered by many scientific groups, which already reported numerous modification of the PSi particle surface by other biopolymers and polymers, for example, by aragose coating (De Rosa et al. 2011), conjugation with peptides hydrophobin II (Bimbo et al. 2011) and E-selectin (Mann et al. 2011), glycosaminoglycan hyaluronic acid (Almeida et al. 2014), by formation of solid lipid nanocomposites (Kallinen et al. 2014), or by covalent conjugation of polyethyleneimine and poly(methylvinyl ether–maleic acid) copolymer (Zhang et al. 2014), as well as by aminoalkyl conjugation (Ahire et al. 2012). Remarkable results were obtained with the PSi nanoparticles covered by bovine serum albumin (Xia et al. 2013). Termination of the Si-H surface groups with long alkyl chains using the microwave-induced hydrosilylation with alkenes followed by encapsulation of bovine serum albumin by hydrophobic interactions yielded "stealth" PSi nanoparticles with good water-dispersibility and long-term stability under physiological conditions. More important, the PSi coated with bovine serum albumin displayed remarkably reduced nonspecific cellular uptake *in vitro* and prolonged blood circulation *in vivo*. A review (Salonen and Lehto 2008) discusses other published PSi surface modifications. All these modifications aimed not only at the optimization of the PSi-based nanomaterial biodegradation rate, but also ensured targeted delivery of the nanoparticles and controlled distribution in targeted objects, which is important for bioimaging applications. For example, PSi modified by tumor-homing peptide which targets

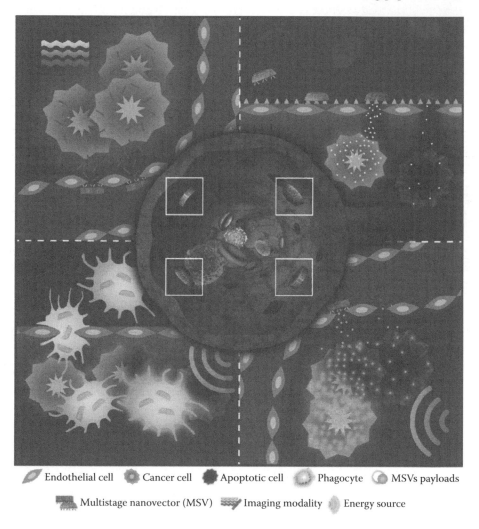

Endothelial cell   Cancer cell   Apoptotic cell   Phagocyte   MSVs payloads

Multistage nanovector (MSV)   Imaging modality   Energy source

FIGURE 20.4 Schematic representation of MSV and their possible mechanisms of action. Central compartment: hemispherical or disk-shaped nanoporous silicon S1MPs are engineered to exhibit an enhanced ability to marginate within blood vessels and adhere to disease-associated endothelium. Once positioned at the disease site, the S1MP can (top right) release the drug/siRNA loaded S2NP to achieve the desired therapeutic effect, prior to complete biodegradation of the carrier particle; release an imaging agent (top left) or external energy activated S2NP (e.g., gold nanoparticles, nanoshells, bottom right). Another possible mechanism of action is cell-based delivery of the MSVs into the disease loci followed by triggered release of the S1MP/S2NP from the cells (bottom left). (Reprinted with permission from Godin B. et al., *Acc. Chem. Res.* 44 (10), 979. Copyright 2011 American Chemical Society.)

the mammary-derived growth inhibitor-expressing cancer cells was shown to specifically accumulate in the cancer cells, both *in vitro* and *in vivo* (Kinnari et al. 2013).

Finally, overall biodistribution profile is very critical for the PSi-based *in vivo* imaging nanoprobes. In a hostile environment inside a living organism, the PSi particles encounter numerous barriers before appropriately distributing throughout the body and reaching the location intended by the researchers as the target, for example, a solid tumor. The study of the nanomaterial distribution in living organisms is an extremely complex task which is nowadays in its infancy, especially when a human body is concerned. Notably, the biodistribution was studied in many works devoted to the *in vivo* imaging with the aid of the PSi nanomaterials. We will discuss these works later in the sections describing specific examples of the PSi-based *in vivo* imaging (Section 20.3). Here, we would only like to highlight the series of works that have introduced a concept of multifunctional multistage delivery systems (MDS) or multistage delivery vectors (MDVs), designed first to achieve their optimal biodistribution, to overcome multiple barriers on the way of the particles and their cargos to the target (Figure 20.4). MDVs are comprised of mesoporous microparticles (first-stage microparticles, S1MPS) loaded with one or more types of stage two nanoparticles (S2NPs), which can in turn carry either active agents or higher-stage particles (Tasciotti et al. 2008; Serda et al. 2010). S1MPS were designed to protect the S2NPs from the biodegradation and to help them in overcoming the biological barriers, and PSi mesoporous microparticles turned out to be ideally suited for this purpose. A successful demonstration of the utility of this concept was done on the example of imaging contrast agents and therapeutics (Chiappini et al. 2011; Godin et al. 2011).

20.3 *IN VIVO* IMAGING USING PSi AS THE MOLECULAR PROBES

Now we turn to discuss the works where PSi particles were used for *in vivo* imaging by different imaging modalities. It should be noted, however, that the vast majority of published works in this area are either proof-of-the-principle demonstrations or the studies carried out on small animals, suitable for preclinical imaging. There are rare exceptions, for example, ^{32}P-loaded PSi microparticles (OncoSil™) currently ongoing clinical testing for human cancer radiotherapy (Goh et al. 2007; Canham and Ferguson 2014). Wider extension of the plethora of the developed imaging technologies to human research and clinical disease diagnosis is still to be awaited. However, the results obtained and the intensity of the ongoing research allow one to hope that translation to the use of PSi in human clinical therapy and diagnosis will be accelerated in the nearest future.

Another aspect should also be stressed: all of the works published by now were not only aimed at imaging. To the contrary, the properties of the PSi described previously prompted scientists to develop nanomaterials for efficient treatment of diseases, or study molecular processes in combination with the imaging, thus discovering truly multimodal, "smart" materials.

20.3.1 PSi IN OPTICAL IMAGING

Optical imaging is the area where PSi was exploited most, first due to its unique optical properties. Along with absorbance spectra favorable for imaging, it demonstrated distinctive photoluminescence properties. The photoluminescence of PSi originates from the quantum confinement effect in the nanocrystals of this semiconductor material and from the defects on its surface, characteristic to its structure (Bisi et al. 2000). In comparison to many other known fluorophores, PSi and other nanostructured materials have higher quantum yields and molecular extinction coefficients of the fluorescence, narrower emission spectra, size-dependent emission, and higher chemical and photostability (Yoffe 2001). These properties have made them ideal fluorescent reporters for bioimaging, delivering information on the cellular and the molecular level.

Optical imaging in living organisms, however, posed a considerable challenge for scientists. In addition to general requirements to an *in vivo* imaging reporter discussed previously (biocompatibility, biosafety, optimal distribution, and clearance), optical *in vivo* imaging needs to be performed with the probes operating in the so-called near-infrared (NIR) window. NIR window is the wavelength range of the light capable to penetrate living tissues and fluids to maximal extent, so the probes that can absorb and emit the NIR light would deliver information from deeper parts of the living organisms as probes operating outside the NIR window. NIR wavelength range is vaguely defined as 650–900 nm because the most abounded light-absorbing components in the living organisms (oxygenated hemoglobin HbO_2, and deoxygenated hemoglobin Hb, water) absorb least in this region (Figure 20.5).

Even in the NIR window, the depth of the light penetration in live tissues is in the range of several centimeters. For example, optical penetration depth for the light at 1100 nm (optimal for skin) is not more than 3.5 cm (Bashkatov et al. 2005) (Figure 20.6). This is a serious limitation of the optical imaging of living organisms. Nevertheless, with the probes that absorb/emit in the NIR window or for the transparent tissue studies, the imaging might offer advantages over the other imaging modalities, first of all, high sensitivity and acceptable spatial resolution, delivering the information on the molecular level (Hilderbrand and Weissleder 2010; see also Table 20.1), ease of use, and cost-efficiency. Promising optical tomography experiments in humans with organic fluorescent dyes as the probes have been published (Corlu et al. 2007).

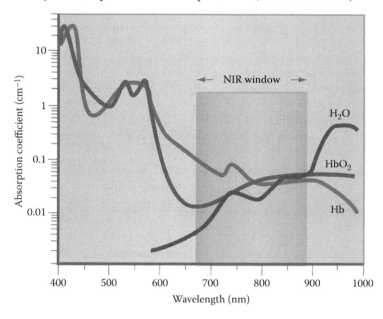

FIGURE 20.5   The NIR window most suitable for optical *in vivo* imaging. (Reprinted by permission from Macmillan Publishers Ltd. *Nat. Biotechnol.* Weissleder R., *Nat. Biotechnol.* 19, 316, copyright 2001.)

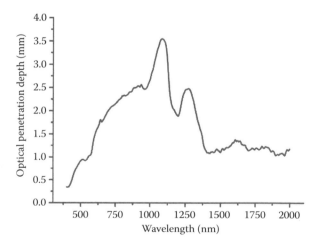

FIGURE 20.6   Optical penetration depth of light into skin over the wavelength range from 400 to 2000 nm. (From Bashkatov A. N. et al., *J. Phys. D: Appl. Phys.* 38, 2543, 2005. Copyright 2005: IOP Publishing Ltd. With permission.)

One of the molecular optical imaging probes, which meets the requirements outlined above almost ideally, is PSi. Absorbance spectrum of a PSi material was advantageously used by Cheng et al. (2008) for studies in the tissue transparent for visible light. The authors investigated the fate of PSi photonic crystals (unmodified, surface alkylated using the hydrosilylation reaction, and oxidized) injected in transparent rabbit vitreous (Figure 20.7). Good biocompatibility and low toxicity of the nanoparticles was observed. Surface alkylation was shown to substantially prolong the particle's half-life in the tissue. It was concluded that properly engineered PSi loaded with a therapeutic agent might be used to treat ocular diseases by the release of the therapeutic agent over a long time period.

Park et al. (2009) were the first who revealed the potential of PSi in the *in vivo* fluorescent imaging, and at the same time to serve as a therapeutic agent delivering system for anticancer treatment. Here, the enhanced permeability and retention effect (EPR-effect) was used to achieve selectivity of the PSi material action against tumor tissue. This effect is a consequence of abnormalities in tumor vasculature, which is characterized by hypervascularization, aberrant vascular architecture, and extensive production of vascular permeability factors that stimulate extravasation of the nanoparticles within tumor tissues (Greish 2010). The concentration of the nanoparticles of the appropriate size and their cytotoxic cargos in tumor buildup reach several-fold higher than that of the plasma concentration for a prolonged time period due to a lack of efficient lymphatic drainage in solid tumor. This is the basis for EPR-based selective anticancer nanotherapy. Combined with the real-time imaging of the tumor, such a therapy represents a powerful instrument of cancer treatment. Medium-sized (126 nm) PSi nanoparticles were prepared in the above-cited work (Park et al. 2009) under optimized conditions to ensure well-defined micro- and mesoporous nanostructure suitable to load an anticancer therapeutic agent doxorubicin and activate the photoluminescence in the NIR region. The particles were shown to be biocompatible and degradable *in vivo*. In order to optimize the *in vivo* stability, the particles were coated with the biopolymer dextran by physisorption. Intravenous administration of the dextran-modified nanoparticles (D-LPSiNPs) into mice allowed the authors to study the biodistribution of the probe in the live animals by observing the fluorescence from the probe (445–490 nm excitation filter and 810–875 nm emission filter). Moreover, the potential of the D-LPSiNP to image solid tumor was evaluated *in vivo*. Intravenous injection of the dextran-coated D-LPSiNPs into a nude mouse bearing an MDA-MB-435 tumor resulted in passive accumulation of the nanomaterial in the tumor, as revealed in the NIR fluorescence image (Figure 20.8).

Similar very promising potential for NIR fluorescent *in vivo* imaging was recently demonstrated by Xia et al. (2014). The authors used BSA-modified "stealth" PSi nanoparticles (BSA/S-PSiNPs) for which they optimized the biodistribution profile, as was referenced above (Xia et al. 2013). Dispersions of the nanoparticles (200 μl, 0.1 mg/ml) were administered into nude mice by subcutaneous injections with a depth of 1 mm, and the mouse was imaged in a fluorescence mode with 430 nm excitation and a 700-nm emission filter. The obtained NIR images were clearly

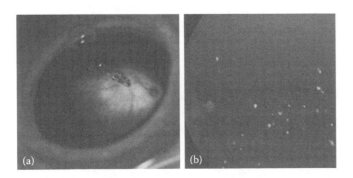

FIGURE 20.7   (a) Photograph of a rabbit eye after intravitreal injection of hydrosilylated PSi particles seen as suspended in the center of the vitreous. (b) Fundus photograph of the particles three months after the injection seen as dispersed in the vitreous, showing partial degradation and dissolution. (From Cheng L. et al., *Br. J. Ophthalmol.* 92, 705, 2008. Copyright 2008: BMJ Publishing Group Ltd. With permission.)

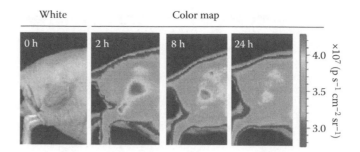

FIGURE 20.8   Representative fluorescence images of a mouse bearing an MDA-MB-435 tumor. The mouse was imaged at the indicated times after intravenous injection of D-LPSiNPs (20 mg kg^{-1}). Strong signal from D-LPSiNPs is observed in the tumor, indicating significant passive accumulation in the tumor by the enhanced permeability and retention (EPR) effect. (Reprinted by permission from Macmillan Publishers Ltd. *Nat Mater* Park J.-H. et al., 8, 331, copyright 2009.)

observed, vanishing with time, which was attributed to the diffusion of the fluorescent probe under the skin (Figure 20.9).

The quality of the NIR fluorescence *in vivo* imaging might be increased by eliminating the inherent fluorescence (autofluorescence) of living tissues and fluids. Although low level of the skin autofluorescence in the NIR region at the visible light excitation was detected in the works discussed above (Park et al. 2009; Xia et al. 2014), in another work on imaging *in vivo* using core-shell semiconductor quantum dots (ZnS-capped CdSe) this was a serious problem when the concentration of the reporter was too low (Gao et al. 2004). Various methods to overcome this problem were proposed, based mainly on spectral differences of the exogenous and endogenous chromophores, but with limited success. An ingenious solution proposed long ago (Andersson-Engels et al. 1990; Cubeddu et al. 1993) consists of the use of the emission lifetime difference between the tissue and exogenous fluorophores. If the emission lifetime for the latter is longer, the autofluorescence effects can be eliminated by the time-delayed detection. However, appropriate biocompatible and biodegradable fluorescence probes have been lacking for a long time. Gu et al. (2013) have found that the photoluminescent PEG-modified PSi nanoparticles (PEG-LPSiNPs) have sufficiently long emission lifetime (5–13 µs) to permit the late time-gated (TG) imaging. The tissue autofluorescence lifetime is in the range of 1–10 ns; therefore, the images acquired 18 ns after excitation contained clear signals from the nanoparticles, but no autofluorescence background. Advantage of this approach over the steady-state, or pseudo continuous wave (CW) fluorescence imaging (without time gating) was demonstrated for the case when the PEG-LPSiNPs in a mouse xenograph tumor was accumulated in low concentrations (Figure 20.10). Previously (Park et al. 2009), it was shown that LPSiNPs are biocompatible, have low systemic toxicity, and are easily biodegradable and cleared from living

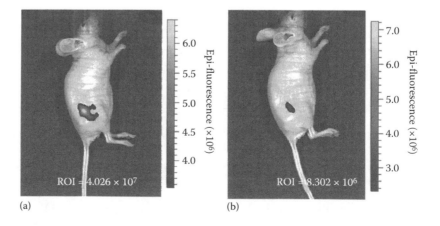

FIGURE 20.9   (a) *In vivo* fluorescence images of a nude mouse subcutaneously injected with BSA/S-PSiNPs (200 µL of 0.1 mg ml^{-1}), and (b) after 120 min. (Adapted from Xia B. et al., *J. Mater. Chem. B.* 2, 8314, 2014. Reproduced by permission of The Royal Society of Chemistry.)

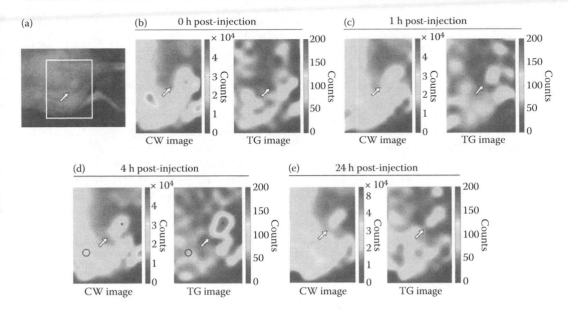

FIGURE 20.10   TG fluorescence images of mouse bearing SKOV3 xenograft tumor after intravenous injection of PEG-LPSiNPs. (a) Bright field image of a nude mouse bearing a tumor at the flank. The arrow indicates the site of the tumor. (b–e) CW and TG fluorescence images of the region indicated with the white box in (a) immediately (b), 1 h (c), 4 h (d), or 24 h (e) postinjection of PEG-LPSiNPs (10 mg kg^{-1} body weight). Note that the tumor was clearly resolved after 4 h postinjection in the TG image, but was obscured in the CW image (d). (Reprinted by permission from Macmillan Publishers Ltd. *Nat Commun*, Gu L. et al., 4, 1, copyright 2013.)

organisms. Therefore, their use as the fluorescent reporters in the TG mode *in vitro* and *in vivo* imaging might enjoy numerous biomedical applications in the nearest future.

In order to be used as a nanocarrier for therapeutic agents, PSi should be prepared with strict control over the porosity, pore size, and shape in order to ensure optimal loading, protection, and release of the cargos. This might compromise the innate fluorescent properties of the PSi, which are used for the optical bioimaging (Canham 1990; Park et al. 2009). An alternative approach was proposed by Tasciotti et al. (2011), who covalently conjugated a near-infrared dye on the surface of the S1MPs. Using the oxidation of the as-prepared particles followed by silanization (see Figure 20.3) with 3-aminopropyltriethoxysilane and reaction with N-hydroxysuccinimide ester activated DyLight 488 (excitation/emission maxima 493 nm/518 nm) or DyLight 750 (excitation/emission maxima 754 nm/776 nm), the organic fluorophores were grafted onto the microparticle surface. With the nanoconstructs obtained, the authors were able to monitor the *in vivo* distribution of the microparticles in healthy mice using an optical imaging system. They determined that the PSi material ends up in liver and spleen in ~24 h after intravenous administration of the modified S1MPs. The imaging data correlated well with the quantification analysis of Si in the organs, verifying the accuracy of the imaging technique for tracing the biodistribution of S1MPs in the body.

20.3.2 NUCLEAR IMAGING WITH THE AID OF PSi-BASED MATERIALS

In the preceding section, we discussed the potential of the PSi use in optical imaging *in vivo*, vividly demonstrated on experimental animals. This potential started to yield valuable results in preclinical studies of PSi-based anticancer drug candidates, with the possible translation to clinical studies in the near future. However, not all the PSi materials and products of their metabolism exhibit fluorescent properties suitable for optical imaging. Therefore, with the aim of more comprehensive *in vivo* pharmacokinetic studies, other imaging modalities involving the use of PSi were considered. Another disadvantage of the optical imaging consists of low anatomical and tissue resolution.

Bimbo et al. were the first to come up with the idea of additional labeling of PSi materials with a radioactive isotope ^{18}F, commonly used in such imaging modality as PET. The decay of the ^{18}F atoms (half-life 1.83 h) proceeds with positron emission, which annihilates with a nearby electron thus generating energetic γ-radiation. Detection of this radiation allows reconstructing the spatial image reflecting the isotope distribution. PET became a very powerful tool in diagnosis of cancer. Many molecular ^{19}F-labeled probes were developed, and imaging instruments are commercially available for their use in clinics (Hamoudeh et al. 2008; Bimbo et al. 2010). In the first (Bimbo et al. 2010) and following publications (Sarparanta et al. 2011), the authors used three type of PSi particles, namely, thermally hydrocarbonized THCPSi, thermally oxidized TOPSi, and thermally carbonized TCPSi for the labeling with [18]F⁻ ion, suggesting that formation of the strong Si-^{18}F bonds will be the driving force facilitating the attachment of the isotope to the Si surface via the mechanisms involving a nucleophilic fluoride ion attack at the silicon (Figure 20.11).

FTIR and XPS studies of the labeled materials confirmed the ^{18}F incorporation, and the radioactivity of the obtained material was sufficient to carry out comprehensive biodistribution studies after enteral and parenteral administration in rats. The studies demonstrated acceptable biosafety of the material, and most importantly, the use of ^{18}F as a tracer for PSi-based drug delivery distribution *in vivo*. Using this tracer, the authors found that PSi materials passed intact through the gastrointestinal tract after oral administration, but its clearance was delayed compared to [18]F⁻ administered to control animals. Therefore, it was concluded that PSi might be used as a carrier for drugs administered orally allowing their steady release in the gastrointestinal tract. Later (Sarparanta et al. 2012b), the ^{18}F-labeled materials were further modified by a protein coating consisting of a class II hydrophobin (HFBII) from fungus *Trichoderma reesei*. [18]F tracer was again used to study the passage of the HFBII-coated THCPSi particles in the rat gastrointestinal tract. It was found that the coated material retained in stomach by gastric mucoadhesion after the oral administration much longer than the uncoated THCPSi. A similar study of the same scientific group (Sarparanta et al. 2012a) aimed at characterization of the biodistribution of ^{18}F-labeled thermally hydrocarbonized porous silicon (THCPSi) nanoparticles after intravenous administration in rats. Again, it was found that covering the nanoparticles with the HFBII altered their biodistribution profile, and resulted in a pronounced change in the degree of plasma protein adsorption to the nanoparticle surface *in vitro*. *In vivo*, the distribution of the nanoparticles between the liver and spleen, the major mononuclear phagocyte system organ in the body, was also altered compared to that of uncoated ^{18}F-THCPSi.

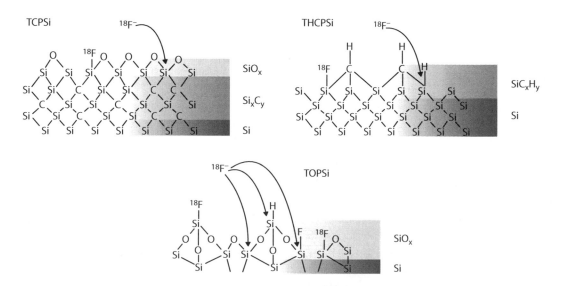

FIGURE 20.11   Plausible sites of ^{18}F incorporation at the PSi surface. (Reprinted with permission from Sarparanta M. et al., *Mol. Pharmaceutics*, 8, 1799. Copyright 2011 American Chemical Society.)

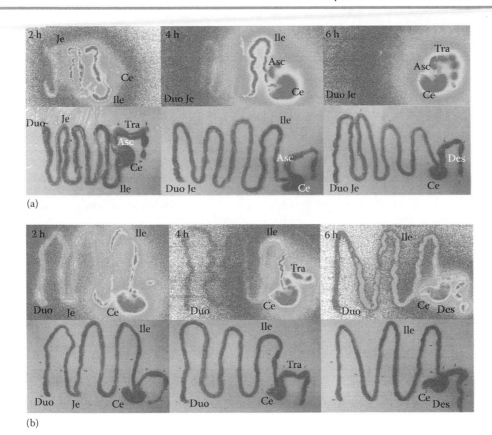

FIGURE 20.12 Macroautoradiographs and respective photographs of the gastrointestinal tracts of rats 2, 4, and 6 h (from left to right) after oral administration of ^{18}F-labeled THCPSi nanoparticles (a) and free ^{18}F-NaF (b). Abbreviations: Duo, duodenum; Je, jejunum; Ile, ileum; Ce, cecum; Asc, ascendens; Tra, transversum; and Des, descendens. (Reprinted with permission from Bimbo L. M. et al., *ACS Nano* 4(6), 3023. Copyright 2010 American Chemical Society.)

Another PSi-based nanocomposite, the THCPSi labeled with ^{18}F as described prevously and encapsulated in solid lipids was prepared and studied *in vivo* using a radioactive tracer (Kallinen et al. 2014). *Ex vivo* autoradiography revealed that the nanocomposites accumulated better in tumor (a murine breast cancer model). At the 7-week point, the tumor-to-liver ratio was more than twice as high for the nanocomposite than for the ^{18}F-THCPSi itself.

Potential use of the ^{18}F-labeled PSi nanoparticles in the PET imaging *in vivo* was mentioned in the works cited previously. In fact, however, the ^{18}F tracer was used *ex vivo*. The studied animals were sacrificed, and the samples from blood, urine, and major organs were collected, weighed, and counted in an automated gamma counter, or macroautoradiographs of the excised rat gastrointestinal tract were recorded (Figure 20.12). The same principle of autoradiography was used in preclinical and clinical studies of ^{32}P-loaded PSi (OncoSil™, ^{32}P is a beta emitter, half-life 14.29 days), mentioned previously (Goh et al. 2007; Canham and Ferguson 2014). The use of the obtained ^{18}F-labeled materials in true noninvasive PET imaging in preclinical and clinical *in vivo* molecular imaging is an obvious step further, and there are good reasons to believe that PSi materials labeled with radioactive isotopes might find a niche in this well-developed field, where other nanoparticles were extensively used (Debbage and Jaschke 2008).

20.3.3 MAGNETIC RESONANCE IMAGING USING PSi-CONTAINING NANOCOMPOSITES

The idea of PSi modification to endow this material with the properties applicable to imaging modalities additional to optical, aiming at multimodal imaging probes, has been realized within the framework of multistage delivery systems (MDS) concept, introduced previously (Tasciotti et al. 2008). Several MDS suitable for use in MRI were obtained.

MRI is an extremely powerful imaging modality currently used worldwide in clinics for diagnosis of numerous diseases (Mansfield 2004; Lentz et al. 2005). It is based on the property of magnetic nuclei (usually protons of water in the living organisms) to exist in different energy states in an external magnetic field, and therefore absorb electromagnetic waves of radiofrequency regions to become excited. Excitation by radiofrequency pulses especially designed for MRI is followed then by longitudinal or spin-lattice relaxation of nuclei to their equilibrium state (characterized by T_1 time constant), and their transverse or spin-spin decay (T_2 time constant). Differences in how the nuclei relax to their equilibrium state are then detected and processed to reconstruct the strength of the signal arising from each imaging unit and to generate an image. In order to enhance the sensitivity of the signals and improve the quality of the images, contrast-enhancing agents are used, which decrease T_1 or T_2. For example, paramagnetic ion complexes of Gd^{3+} are administered to the living organisms, and as a result, the relaxation time T_1 of the protons in tissues is decreased. A brighter region in the image (positive or hyperintense contrast) is then observed. Superparamagnetic iron oxide nanoparticles (SPIONs), to the contrary, are used to decrease predominantly T_2, resulting in localized darker spots (negative or hypointense contrast) in the images.

In the first work published on the use of the MDS concept to construct MRI contrasting agents based on PSi (Ananta et al. 2010), three clinically used Gd-based contrasting agents were loaded into the pores of the first-stage nanoporous PSi microparticles. Magnevist® (Gd^{3+} complex with diethylenetriaminepentaacetate DTPA, of general formula $A_2[Gd(DTPA)(H_2O)]$, where A is aminosugar meglumine), gadofullerenes (Gd^{3+} ion encapsulated by a spherical fullerene cage), and gadonanotubes (single-walled carbon nanotubes loaded with Gd^{3+} clusters) were incorporated into PSi microparticles microfabricated using a combination of photolithography and electrochemical etching. The resulting constructs, quasi-hemispherical particles, with a nominal diameter of 1.6 µm and thickness of ~0.6 µm, and discoidal particles, with a nominal diameter of 1.0 µm and thickness of ~0.4 µm (30–40 nm pore diameter) modified by the Gd-based contrasting agents significantly decreased the proton longitudal relaxation times. Remarkably, the relaxivity values (a measure of the relaxation decrease and thus of imaging quality) for these constructs were 4–50 times larger than for the bare Gd-based contrasting agents. A mechanism of this boost in the relaxivity was proposed, which was based on geometric confinement of the Gd ion in the nanopores of the PSi. Taken into account the biocompatibility and safety of the PSi

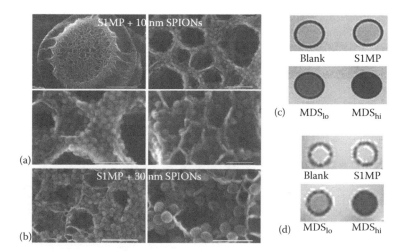

FIGURE 20.13   High-resolution SEM and MRI. S1MPs loaded with SPIONs at increasing magnifications. (a) S1MP loaded with 10 nm SPIONs (25k, 300k, 500k, 600k; bar: 1 mm, 100 nm, 100 nm, 50 nm). (b) S1MP loaded with 30 nm SPIONs (200k, 450k; bar: 200, 100 nm). (c, d) Axial spin (c) and gradient (d) echo MR images of NMR tubes containing PBS (blank), S1MPs, and MDS loaded with low (MDSlo) or high (MDShi) levels of SPIONs. (Serda R. E. et al.: *Small*. 2010. 6(12). 1329. Copyright Wiley-VCH Verlag GmbH & Co. KGaA. Reproduced with permission.)

microparticles, their ability to being loaded with therapeutic agents of other imaging labels, the authors proposed the above-described nanoconstructs as advanced imaging multimodal agents.

It was also shown by Granitzer et al. (2010) that PSi particles can be loaded with SPIONs, and this led to stabilization of the superparamagnetic state of the iron oxide. In the works on developing the MDS concept (Serda et al. 2010; Chiappini et al. 2011), the scientists incorporated SPIONs in the stage one PSi microparticles (S1MPs) and tested them as MRI contrasting agents. The SPION-loaded particles led to shorted T_2 relaxation times and better contrast in model images (Figure 20.13).

Similar results were obtained by Kinsella et al. (2011) who prepared and characterized PSi nanoparticles with a pore morphology that facilitated proximal loading and alignment of superparamagnetic Fe_3O_4 nanoparticles. Pharmacokinetic profile of the material was then studied *in vivo*. The increase in the transverse relaxivity was observed, which was attributed to the greater magnetic strength of the interacting Fe_3O_4 nanoparticles within the PSi matrix. It was concluded that the ability to tailor the magnetic properties of such materials might be useful for *in vivo* imaging, magnetic hyperthermia, or drug-delivery applications.

20.4 CONCLUSIONS AND OUTLOOK

During a relatively short period of intensive research of PSi (since the 1990s), numerous functional materials were developed with very promising practical applications in biology and medicine, including molecular probes for *in vivo* optical, MR, and nuclear imaging. Unique combinations of the PSi properties, and first of all, its biocompatibility and safety, allow one to hope that the PSi-based materials will soon find numerous applications in clinical medicine. The studies described in this section already demonstrated that pharmacokinetic parameters of the PSi-based materials could be fine-tuned to optimize as needed for a specific *in vivo* application. Combining with a possibility of loading PSi with therapeutic agents, targeting molecules, imaging chromophores or tracers, this opens the way to obtain true safe "smart materials" exerting multimodal actions in living organisms. However, the transfer from research to clinics will require further extensive pharmacokinetic studies, and appropriate optimization of the relevant parameters.

REFERENCES

Ahire, J.H., Wang, Q., Coxon, P.R. et al. (2012). Highly luminescent and nontoxic amine-capped nanoparticles from porous silicon: synthesis and their use in biomedical imaging. *ACS Appl. Mater. Interfaces* 4, 3285–3292.

Almeida, P.V., Shahbazi, M.-A., Mäkilä, E. et al. (2014). Amine-modified hyaluronic acid-functionalized porous silicon nanoparticles for targeting breast cancer tumors. *Nanoscale* 6(17), 10377–10387.

Alvarez, S.D., Derfus, A.M., Schwartz, M.P., Bhatia, S.N., and Sailor, M.J. (2009). The compatibility of hepatocytes with chemically modified porous silicon with reference to in vitro biosensors. *Biomaterials* 30, 26–34.

Ananta, J.S., Godin, B., Sethi, R. et al. (2010). Geometrical confinement of gadolinium-based contrast agents in nanoporous particles enhances T1 contrast. *Nat. Nanotechnol.* 5, 815–821.

Andersson-Engels, S., Johansson, J., Stenram, U., Svanberg, K., and Svanberg, S. (1990). Malignant tumor and atherosclerotic plaque diagnosis using laser-induced fluorescence. *IEEE J. Quant. Electron.* 26(12), 2207–2217.

Angelescu, A., Kleps, I., Mihaela, M. et al. (2003). Porous silicon matrix for applications in biology. *Rev. Adv. Mater. Sci. (RAMS)* 5(5), 440–449.

Bashkatov, A.N., Genina, E.A., Kochubey, V.I. and Tuchin, V.V. (2005). Optical properties of human skin, subcutaneous and mucous tissues in the wavelength range from 400 to 2000 nm. *J. Phys. D: Appl. Phys.* 38, 2543–2555.

Bayliss, S.C., Buckberry, L.D., Fletcher, I., and Tobin, M.J. (1999). The culture of neurons on silicon. *Sens. Actuators A* 74, 139–142.

Bimbo, L.M., Sarparanta, M., Santos, H.A. et al. (2010). Biocompatibility of thermally hydrocarbonized porous silicon nanoparticles and their biodistribution in rats. *ACS Nano* 4(6), 3023–3032.

Bimbo, L.M., Mäkilä, E., Raula, J. et al. (2011). Functional hydrophobin-coating of thermally hydrocarbonized porous silicon microparticles. *Biomaterials* 32, 9089–9099.

Bisi, O., Ossicini, S. and Pavesi, L. (2000). Porous silicon: a quantum sponge structure for silicon based optoelectronics. *Surf. Sci. Rep.* 38, 1–126.

Buriak, J.M. (2002). Organometallic chemistry on silicon and germanium surfaces. *Chem. Rev.* 102(5), 1272–1306.

Canham, L.T. (1990). Silicon quantum wire array fabrication by electrochemical and chemical dissolution of wafers. *Appl. Phys. Lett.* 57, 1046–1048

Canham, L.T. (1995). Bioactive silicon structure fabrication through nanoetching techniques. *Adv. Mater.* 7(12), 1033–1037.

Canham, L. and Ferguson, D. (2014). Porous silicon in brachytherapy. In: Canham L. (Ed.) *Handbook of Porous Silicon.* Springer International Publishing, Switzerland.

Cheng, L., Anglin, E., Cunin, F. et al. (2008). Intravitreal properties of porous silicon photonic crystals: A potential self-reporting intraocular drug delivery vehicle. *Br. J. Ophthalmol.* 92, 705–711.

Chiappini, C., Tasciotti, E., Fakhoury, J.R. et al. (2010a). Tailored porous silicon microparticles: Fabrication and properties. *ChemPhysChem.* 11, 1029–1035.

Chiappini, C., Liu, X., Fakhoury, J.R., and Ferrari, M. (2010b). Biodegradable porous silicon barcode nanowires with defined geometry. *Adv. Funct. Mater.* 20, 2231–2239.

Chiappini, C., Tasciotti, E., Serda, R.E., Brousseau, L., Liu, X., and Ferrari, M. (2011). Mesoporous silicon particles as intravascular drug delivery vectors: Fabrication, in-vitro, and in-vivo assessments. *Phys. Status Solidi C* 8(6), 1826–1832.

Chin, V., Collins, B.E., Sailor, M.J., and Bhatia, S.N. (2001). Compatibility of primary hepatocytes with oxidized nanoporous silicon. *Adv. Mater.* 13(24), 1877–1880.

Choi, H.S., Liu, W., Misra, P. et al. (2007). Renal clearance of quantum dots. *Nat. Biotechnol.* 25(10), 1165–1170.

Corlu, A., Choe, R., Durduran, T. et al. (2007). Three-dimensional in vivo fluorescence diffuse optical tomography of breast cancer in humans. *Opt. Express* 15(11), 6696–6716.

Cubeddu, R., Canti, G., Taroni, P., and Valentini, G. (1993). Time-gated fluorescence imaging for the diagnosis of tumours in a murine model. *Photochem. Photobiol.* 57, 480–485.

Cullis, A.G. and Canham, L.T. (1991). Visible light emission due to quantum size effects in highly porous crystalline silicon. *Nature* 353, 335–338.

De Rosa, E., Chiappini, C., Fan, D., Liu, X., Ferrari, M., Tasciotti, E. (2011). Agarose surface coating influences intracellular accumulation and enhances payload stability of a nano-delivery system. *Pharm. Res.* 28, 1520–1530.

Debbage, P. and Jaschke, W. (2008). Molecular imaging with nanoparticles: Giant roles for dwarf actors. *Histochem. Cell Biol.* 130, 845–875.

Fass, L. (2008). Imaging and cancer: A review. *Mol. Oncology,* 2, 115–152.

Fucikova, A., Valenta, J., Pelant, I., and Brezina, V. (2009). Novel use of silicon nanocrystals and nanodiamonds in biology. *Chem. Pap.* 63(6), 704–708.

Gao, X., Cui, Y., Levenson, R.M., Chung, L.W.K., and Nie, S. (2004). *In vivo* cancer targeting and imaging with semiconductor quantum dots. *Nat. Biotech.* 22(8), 969–976.

Godin, B., Tasciotti, E., Liu, X., Serda, R.E., and Ferrari, M. (2011). Multistage nanovectors: From concept to novel imaging contrast agents and therapeutics. *Acc. Chem. Res.* 44(10), 979–989.

Godin, B., Chiappini, C., Srinivasan, S. et al. (2012). Discoidal porous silicon particles: Fabrication and biodistribution in breast cancer bearing mice. *Adv. Funct. Mater.* 22, 4225–4235.

Goh, A.S.-W., Chung, A.Y.F., Lo, R.H.-G. et al. (2007). A novel approach to brachytherapy in hepatocellular carcinoma using a phosphorous-32 (^{32}P) brachytherapy delivery device—A first-in-man study. *Int. J. Radiat. Oncol. Biol. Phys.* 67(3), 786–792.

Granitzer, P., Rumpf, K., Roca, A.G. et al. (2010). Magnetite nanoparticles embedded in biodegradable porous silicon. *J. Magn. Magn. Mat.* 322, 1343–1346.

Greish, K. (2010). Enhanced permeability and retention (EPR) effect for anticancer nanomedicine drug targeting. *Methods Mol. Biol.* 624, 25–37.

Gu, L., Hall, D.J., Qin, Z. et al. (2013). In vivo time-gated fluorescence imaging with biodegradable luminescent porous silicon nanoparticles. *Nat. Commun.* 4, 1–7.

Gupta, B., Zhu, Y., Guan, B., Reece, P.J. and Gooding, J.J. (2013). Functionalised porous silicon as a biosensor: Emphasis on monitoring cells in vivo and in vitro. *Analyst* 138, 3593–3615.

Hamoudeh, M., Kamleh, M. A., Diab, R., and Fessi, H. (2008). Radionuclides delivery systems for nuclear imaging and radiotherapy of cancer. *Adv. Drug Delivery Rev.* 60(12) 1329–1346.

He, X., Gao, J., Gambhir, S.A., and Cheng, Z. (2010). Near-infrared fluorescent nanoprobes for cancer molecular imaging: Status and challenges. *Trends Mol. Med.* 16(12), 574–583.

Hilderbrand, S.A. and Weissleder, R. (2010). Near-infrared fluorescence: Application to in vivo molecular imaging. *Curr. Opin. Chem. Biol.* 14, 71–79.

Kallinen, A.M., Sarparanta, M.P., Liu, D. et al. (2014). In vivo evaluation of porous silicon and porous silicon solid lipid nanocomposites for passive targeting and imaging. *Mol. Pharmaceutics* 11(8), 2876–2886.

Kinnari, P.J., Hyvönen, M.L.K., Mäkilä, E.M. et al. (2013). Tumour homing peptide-functionalized porous silicon nanovectors for cancer therapy. *Biomaterials* 34, 9134–9141.

Kinsella, J.M., Ananda, S., Andrew, J.S. et al. (2011). Enhanced magnetic resonance contrast of Fe_3O_4 nanoparticles trapped in a porous silicon nanoparticle host. *Adv. Mater.* 23, H248–H253.

Kovalainen, M., Mönkäre, J., Mäkilä, E. et al. (2012). Mesoporous silicon (PSi) for sustained peptide delivery: Effect of PSi microparticle surface chemistry on peptide YY3-36 release, *Pharm. Res.* 29, 837–846.

Lentz, M.R., Taylor, J.L., Feldman, D.A., and Cheng, L.L. (2005). Current clinical applications of in vivo magnetic resonance spectroscopy and spectroscopic imaging. *Curr. Med. Im. Rev.*, 1(3), 271–301.

Liu, D., Mäkilä, E., Zhang, H. et al. (2013). Nanostructured porous silicon-solid lipid nanocomposite: Towards enhanced cytocompatibility and stability, reduced cellular association, and prolonged drug release. *Adv. Funct. Mater.* 23, 1893–1902.

Low, S.P., Williams, K.A., Canham, L.T., and Voelcker, N.H. (2006). Evaluation of mammalian cell adhesion on surface-modified porous silicon, *Biomaterials* 27, 4538–4546.

Mann, A.P., Tanaka, T., Somasunderam, A., Liu X., Gorenstein, D.G., and Ferrari, M. (2011). E-selectin-targeted porous silicon particle for nanoparticle delivery to the bone marrow. *Adv. Mater.* 23(36), H278–H282.

Mansfield, P. (2004). Snapshot magnetic resonance imaging (Nobel Lecture). *Angew. Chem. Int. Ed.*, 43, 5456–5464.

Massoud, T.F. and Gambhir, S.S. (2003). Molecular imaging in living subjects: Seeing fundamental biological processes in a new light, *Genes Dev.* 17, 545–580.

Ntziachristos, V., Leroy-Willig, A. and Tavitian, B. (eds) (2007). *Textbook of In Vivo Imaging in Vertebrates*. John Wiley & Sons Ltd., Chichester, UK.

Park, J.-H., Gu, L., von Maltzahn, G., Ruoslahti, E., Bhatia, S.N., and Sailor, M.J. (2009). Biodegradable luminescent porous silicon nanoparticles for in vivo applications. *Nat. Mater.* 8, 331–336.

Peer, D., Karp, J.M., Hong, S., Farokhzad, O.C., Margalit, R., and Langer, R. (2007). Nanocarriers as an emerging platform for cancer therapy. *Nat. Nanotechnol.* 2, 751–760.

Salonen, J. and Lehto, V.-P. (2008). Fabrication and chemical surface modification of mesoporous silicon for biomedical applications. *Chem. Eng. J.* 137, 162–172.

Salonen, J., Laitinen, L., Kaukonen, A.M. et al. (2005). Mesoporous silicon microparticles for oral drug delivery: Loading and release of five model drugs. *J. Controlled Release* 108, 362– 374.

Santos, H.A., Bimbo, L.M., Lehto, V.-P., Airaksinen, A.J., Salonen, J., and Hirvonen, J. (2011). Multifunctional porous silicon for therapeutic drug delivery and imaging. *Curr. Drug Discovery Technol.* 8(3), 228–249.

Santos, H.A., Bimbo, L.M., Herranz, B., Shahbazi, M.-A., Hirvonen, J., and Salonen, J. (2013). Nanostructured porous silicon in preclinical imaging: Moving from bench to bedside. *J. Mater. Res.* 28(2), 152–164.

Santos, H.A., Mäkilä, E., Airaksinen, A.J., Bimbo, L.M., and Hirvonen, J. (2014). Porous silicon nanoparticles for nanomedicine: Preparation and biomedical applications. *Nanomedicine* 9(4), 535–554.

Sarparanta, M., Makila, E., Heikkila, T. et al. (2011). 18F-labeled modified porous silicon particles for investigation of drug delivery carrier distribution in vivo with positron emission tomography. *Mol. Pharmaceutics* 8, 1799–1806.

Sarparanta, M., Bimbo, L.M., Rytkönen, J. et al. (2012a). Intravenous delivery of hydrophobin-functionalized porous silicon nanoparticles: Stability, plasma protein adsorption and biodistribution. *Mol. Pharmaceutics* 9, 654–663.

Sarparanta, M.P., Bimbo, L.M., Mäkilä, E.M. et al. (2012b). The mucoadhesive and gastroretentive properties of hydrophobin-coated porous silicon nanoparticle oral drug delivery systems. *Biomaterials* 33(11), 3353–3362.

Sciacca, B., Alvarez, S.D., Geobaldo, F., and Sailor, M.J. (2010). Bioconjugate functionalization of thermally carbonized porous silicon using a radical coupling reaction. *Dalton Trans.* 39, 10847–10853.

Serda, R.E., Mack, A., Pulikkathara, M. et al. (2010). Cellular association and assembly of a multistage delivery system. *Small* 6(12), 1329–1340.

Shahbazi, M.-A., Hamidi, M., Mäkilä, E.M. et al. (2013). The mechanisms of surface chemistry effects of mesoporous silicon nanoparticles on immunotoxicity and biocompatibility, *Biomaterials* 34(31), 7776–7789.

Tasciotti, E., Liu, X., Bhavane, R. et al. (2008). Mesoporous silicon particles as a multistage delivery system for imaging and therapeutic applications. *Nat. Nanotechnol.* 3, 151–157.

Tasciotti, E., Godin, B., Martinez, J.O. et al. (2011). Near-infrared imaging method for the in vivo assessment of the biodistribution of nanoporous silicon particles. *Mol. Imaging* 10(1), 56–68.

Tölli, M.A., Ferreira, M.P.A., Kinnunen, S.M. et al. (2014). In vivo biocompatibility of porous silicon biomaterials for drug delivery to the heart. *Biomaterials* 35(29), 8394–8405.

Uhlir, A. (1956). Electrolytic shaping of germanium and silicon. *Bell System Tech. J.* 35, 333–347.

Varkouhi, A.K., Scholte, M., Storm, G., and Haisma, H.J. (2011). Endosomal escape pathways for delivery of biologicals. *J. Controlled Release* 151, 220–228.

Wang, C.-F., Mäkilä, E.M., Kaasalainen, M.H. et al. (2014). Copper-free azide–alkyne cycloaddition of targeting peptides to porous silicon nanoparticles for intracellular drug uptake. *Biomaterials* 35, 1257–1266.

Weissleder, R. (2001). A clearer vision for in vivo imaging. *Nat. Biotechnol.* 19, 316–317.

Williams, D.F. (2008). On the mechanisms of biocompatibility. *Biomaterials* 29(20), 2941–2953.

Xia, B., Zhang, W., Shi, J., and Xiao, S.-J. (2013). Engineered stealth porous silicon nanoparticles via surface encapsulation of bovine serum albumin for prolonging blood circulation in vivo. *ACS Appl. Mater. Interfaces* 5, 11718–11724.

Xia, B., Wang, B., Shi, J., Zhang, W., and Xiao, S.-J. (2014). Engineering near-infrared fluorescent styrene terminated porous silicon nanocomposites with bovine serum albumin encapsulation for in vivo imaging. *J. Mater. Chem. B* 2, 8314–8320.

Yoffe, A.D. (2001). Semiconductor quantum dots and related systems: Electronic, optical, luminescence and related properties of low dimensional systems. *Adv. Phys.* 50(1), 1–208.

Zhang, M., Xu, R., Xia, X. et al. (2014). Polycation-functionalized nanoporous silicon particles for gene silencing on breast cancer cells. *Biomaterials* 35(1), 423–431.

Index

Page numbers followed by f and t indicate figures and tables, respectively.

A

Absorption, 14
  of laser energy, 240
Acceleration, 175
Accelerometer
  basic structure, 150, 150f
  classification, 150
  defined, 149
  necessity of, 149
  silicon micromachined, 149–155
Acceptor atom of hard acid, 49
Acetone vapor control, 72
Acidic metal oxides, 55f
Acoustic emission, *see* Ultrasound emitters
Acoustic signal, 198
Activated carbon, 229, 230t
Activated ester termination, 97
Activation energy, 18
  of desorption, 249
Actuation methods, 307f
Adenosine triphosphate (ATP), 348
Adhesion, PSi-stem cells, 369–371, 370f
Adipose derived stem cells (ASCs), 367
Adsorbates, 8
Adsorbents, 226f
  packing, 227
Adsorption
  of ethyl alcohol, 7, 18
  gas preconcentrator and, 224
  monolayer, 288
  of oxygen, 8
  of water molecules, 11
Adsorption-chemisorption process, 11
Advanced porous silicon membrane (APSM)
  process, 142
Advanced silicon etching (ASE) process, 281
Aerogels, 171; *see also* Thermal insulation in
  MEMS
Affinity chips, 251
Aging effect, 29, 30
Air pollutants, 122
Alcohol vapors, 5
Alginate, 351
Allogenic grafts, 365
Alpha-fetoprotein (AFP), 124
Alternating current (AC) electrical
  conductivity, 14
Alternating current (AC)-voltage driving mode,
  204
Ambient gases, 28
Amine doped oxides, 62–63; *see also* Inverse
  hard and soft acid and base (IHSAB)
  principles
Amino acids, 273
Amino functionalization, 346
Aminopropyl triethoxysilane (APTES), 371

3-aminopropyltriethoxysilane (3-APTES), 111,
  127, 346
3-aminopropyltriethoxysilanes modified PSi
  electrode (APTES-PSE), 98, 99f
Aminopropyl trimethoxysilane (APTMS), 371
(3-aminopropyl) trimethoxysilane (APTMS),
  275
Aminosilanization, 309
Amperometric biosensors, 109
Amperometric sensor, 96
Amperometry, 104
Anisotropic chemical wet etch, 210
Anisotropic etchants, properties of, 211t
Anisotropic etching, 102, 213
Anisotropy, 162
Anodic etching process, 102
Anodic stripping voltammetric curves, 110f
Anodization, 88
  current, 284
  current density, 269
  and etching, 283
  membrane fabrication, 280
  process, 220, 233
  of silicon wafer, double-stage, 14
Antibodies (Ab); *see also* Biosensors, PSi-based
  about, 119
  biochemical quantities, control of, 124–125
Antibodies (Ab)-antigens (Ag) interactions, 108
Antigens (Ag); *see also* Biosensors, PSi-based
  about, 119
  biochemical quantities, control of, 124–125
Aspect-ratio (AR) values, 162
Atom force microscopy, 119
Atomic absorption spectroscopy, 273
Atomic force microscopy (AFM), 286
Attomoles, 256
Auger electron spectroscopy (AES), 115
Autografts, 365
Autologous grafts, 365, 374
Automatic braking system (ABS), 149
Auxiliary electrode, 96f

B

Backside anisotropic micromachining, 285
Backside illumination, 280
Bacteria detection, 126–127, 127f
Basin separated by membrane, 311
Benzimidazoledithio, 100
Benzimidazoledithio-modified PSi (BDT-PSi),
  274, 274f
Benzocyclobutene (BCB), 171
Beta-tricalcium phosphate (b-TCP) scaffolds,
  354, 354f
Biocompatibility
  implanted devices, 374
  of PSi, 373– 374, 387

Biocompatibility/bioactivity of PSi
  assessment of biocompatibility
    in vitro assays and assessment methods,
      321
    in vivo assays and assessment methods,
      321–322
  biological responses to PSi, 323–324
  degradation of PSi in aqueous solution,
    322–323
  effect of size/shape/reactivity on
    biocompatibility of PSi, 324–325
  overview, 320
  PSi in clinical trials, 331
  in vitro studies of PSi, 326–327, 326t–327t
  in vivo studies of PSi
    about, 328–329, 329t
    in vivo administration routes, 330–331
Bioconjugation of PSi, 126
Biodistribution, of Psi, 387–389, 388t
Biofunctionalization, PSi surface preparation
  for, 110–113, 112f
Biological responses to PSi, 323–324
Biological substance level in air, control of,
  121–122
Biosafety, of PSi, 387
Biosensors, PSi-based
  biochemical quantities, control of
    Ag/Ab/peptides, control of, 124–125
    biological substance level in air, control
      of, 121–122
    cells analysis, 128–129
    DNA sensors, 127–128
    glucose/appropriate enzyme
      determination, 125–126
    mycotoxin level determination, 122
    registration of Mb concentration,
      119–121, 121t
    retroviral leucosis, diagnostics of,
      122–123, 122t
    viruses/bacteria, detection of, 126–127,
      127f
  biofunctionalization, PSi surface
    preparation for, 110–113, 112f
  immunobiosensors
    basic algorithm of analysis, 114–118
    construction of working devices, 114
    specific signal registration, 114
    structural analysis of PSi samples, 114
  importance/fields of application, 108–110
  lab-on-chip prototype, PSi-based, 130f
  practical application, 113–114
Biotin, 124
Birefringency, 72
Bolometer; *see also* Devices fabricated
  with PSi as thermal insulation
  material; Thermal insulation
  in MEMS

for fabrication, 177
 necessity of, in thermal insulation, 169
 temperature sensitive resistor, 186
Bond dissociation energy (BDE), 57
Bone marrow stromal cells (BMSCs), 367
Bone reconstructive surgery, graft techniques, 365
Bone tissue engineering
 PSi for, 372, 373f
Borofloat glass wafer, 231
Boron diffusion, 188, 218f
Bosch process, 161, 163f, 211
Bovine serum albumin (BSA), 111, 123, 347f
BrachySil, 331
Brachysil™ device, 376
Bragg mirror, 75–76, 76f; see also Gas/vapor sensors, reflectivity-based
Breakdown voltage, 24f, 25
Broadband terahertz optical properties, 273
Bruggeman model, 74
Buffered oxide etchant (BOE), 285
Bulk micromachining, 210, 280
Bulk piezoelectric ceramic accelerometer sensors, 154
Burn-in process, 30

C

Cadmium sulfide (CdS) nanoparticles, 348
Calibration curve, 85f
Calorimeter, 186
Calorimetric biosensor, 109, 114
Cancer therapy
 anticancer therapeutic agent, 391
 nanoparticles in, 383–384
Cantilever process fabrication, 220f
Capacitance curves, 99f
Capacitance sensors, 9–14; see also Gas/vapor sensors, porous silicon-based
Capacitive accelerometers, 152
Capacitive pressure sensor
 benefits, 141
Capacitive sensors, 137
Capillary condensation, 71, 86, 88, 288f
 mechanism, 32
Capillary electrochromatography (CEC) systems, 303
Carbonization of PSi, 13
Carbon molecular sieve, 229, 230t
Carbon nanotubes (CNT), 25, 355
Cell adhesion, 128
 PSi–stem cells, 369–371, 370f
Cell differentiation, PSi–stem cells, 371
Cells analysis, 128–129
Cell types, interactions with PSi materials, 326t
Cetyltrimethylammonium bromide (CTAB), 343
Cetyltrimethyl-ammonium chloride (CTAC), 234, 281
Charge transfer, 19
Chemical etching
 and surface modification, 115
Chemical filters for gas sensors, 227
Chemical functionalization, 113
 of silicon LDI-MS substrates, 251–256; see also Silicon nanostructures for LDI-MS
Chemical hardness/softness, concept, 49
Chemical oxidation of guanine, 127
Chemical reactivity theory (CRT), 49, 51

Chemical sensors, 5, 24
Chemical stain etching, 88
Chemical vapor deposition (CVD), 25, 111
Chemiluminescence, 122
Chinese Hamster Ovary (CHO), 326, 326t
Chromatographic separation columns, 302–307; see also Miniaturized total analysis systems (μTAS)
Chromatographic separation process, 302f
Clearance, of Psi, 387–389, 388t
Cobalt phthalocyanine (CoIIPc), 33
Co-condensation process, 343; see also Surface functionalization
Coefficients of thermal expansion (CTE), 170
Collimator, 83f, 84
Comb-drives, 157
Commercial microbolometers, 177
Complementary metal-oxide-semiconductor (CMOS) devices, 213
Complementary metal oxide semiconductor (CMOS) signal processing, 154
Composites, PSi-based, 375
Composite scaffold drug delivery system (CS-DDS), 354
Computed tomography (CT)
 in vivo imaging technique, 382, 383, 383t
ConCap measurement, 101f
Condensation effect, 7
Conductance of sensor, 104f
Conductivity, measurement of, 96
Conductometric biosensors, 109
Conductometric gas sensors, 46, 51
Conductometric sensors, 14–19, 16t, 96; see also Gas/vapor sensors, porous silicon-based
Configurational diffusion, 288f
Contact angle, 12
Contact potential difference (CPD) measurements, 22, 23f
Controllable vapor generators, PSi-based, 233–235; see also Gas sources for μGC/gas sensor calibration
Conventional emitters, 200t, 201
Copper oxide, 56
Coriolis force, 156, 157
Corona discharges, 25
Correlation coefficient, 103
Covalent attachment, 341
Covalent bonding, 49
Covalent grafting, 102
Cross-flow filtration, 290–291, 291f, 292
Cryogenic etching, 162
Cryogenic process, 211
Current, measurement of, 96
Current-time curve, 281
Current-voltage characteristics, 17, 18, 125
α-cyano-4-hydroxycynnamic acid (CHCA), 248
Cyclic voltammetry, 98f, 100f
Cyclodextrin (CD), 344; see also Functionalized molecules
Cytotoxicity, 321

D

Damping, 152, 158
Dark conductivity, temperature dependence of, 18, 19f
Dark current, 118
Dead-end mode, 290, 291f
Debye length, 36

Deep boron diffusion, 218f
Deep reactive ion etching (DRIE), 138, 141f, 171, 174
 about, 210
 fabrication based on, 300
 fluidic channels and, 226f
 preconcentration and, 231
Degradation of PSi in aqueous solution, 322–323; see also Biocompatibility/bioactivity of PSi
Density function theory (DFT), 49
Depletion capacitance, 19
Desorption, 224f
Desorption-ionization (DI) mechanism, 247–251; see also Silicon nanostructures for LDI-MS
Desorption/ionization on PSi mass spectrometry (DIOS MS), 240, 242
Desorption process, 12
Devices fabricated with PSi as thermal insulation material; see also Thermal insulation in MEMS
 bolometers, 177
 flow sensors, 175–176
 gas sensors, 176–177
 heat-flux sensors/thermoelectric generator, 179–180
 micro-hotplates, 178
 thermoacoustic sound sources, 178–179
Dichloromethane, 113
Dielectric constant, 18
Dielectric permittivity, 11
Differential scanning calorimetry (DSC), 270
Differentiation, cell, 371
Diffusion coefficient, 287
Diffusion membranes
 applications, 293–294
 membrane fabrication, 280–286
 membrane properties, 286–287
 overview, 280
 transport in porous membranes, 287–293
Diffusion systems, 232; see also Gas sources for μGC/gas sensor calibration
Digital counter circuit, 137
Dimethylformamide (DMF), 160
Dimethylsulf-oxide (DMSO), 112, 281
3-(4,5-dimethylthiazol-2-yl)-2,5-diphenyl tetrazolium bromide (MTT) assay, 321
Diode-type detectors, 186
Direct current (DC) electrical conductivity, 14
Direct current (DC)-superimposed driving mode, 204
Direct silanization method, 125
DNA hybridization, 127, 128
DNA sensors, 127–128; see also Biosensors, PSi-based
Donor atom, 49
Donor orbital energy, 56
Doxorubicin, anticancer therapeutic agent, 391
Drift ratchet, 310
Drug delivery, PSi materials in
 clinical trials/commercial marketing, 355
 drug loading in pores
  structure, changes in, 346
  surface properties, changes in, 346–347
 drug release
  pore diameter, influence of, 347
  surface function, influence of, 347

internalization of PSi nanoparticles by cells
   size/shape, influence of, 349–350
   surface charge, 350
   trafficking in cells, 350–351
loading and controlled release of drugs
   with PSi
   covalent attachment, 341
   oxidation, 342
   physical adsorption, 342
overview, 338–339, 339f
porous silica drug delivery systems
   microneedle, 353–354, 353f
   scaffolds, 354
   sponges, 355
PSi and porous silica nanoparticle synthesis
   method, 340–341
surface functionalization
   functionalized molecules, 344–346
   surface modification, 342–344
targeted delivery
   liver, 351
   lung cancer, 352–353
   vascular barriers, 351–352
Drug loading in pores; see also Drug delivery,
   PSi materials in
structure, changes in, 346
surface properties, changes in, 346–347
Drug release; see also Drug delivery, PSi
   materials in
pore diameter, influence of, 347
surface function, influence of
   nanoparticles as gatekeepers, 348
   organic molecules as gatekeepers, 348
   supramolecular assemblies as
      gatekeepers, 348–349
Dry deep reactive ion etching, 210
Dye-labeled fluorescence, 126
Dynamic pumps, 307–308

E

Electric-acoustic effect, 202
Electrical biosensors, 109
Electrical double layer (EDL), 308, 309
Electrical noise, 156
Electric wiring, 190
Electrochemical etching
   of crystalline silicon, 74f
   electrolyte used for, 160
   of n-type silicon wafers, 59
   of silicon, 85
Electrochemical formation process, 292
Electrochemical micromachining (ECM)
   technology, 163
Electrochemical process of silicon dissolution, 313
Electrochemical sensors, PSi-based
   category, 96
   for metal ions, 97–101
   for pH measurements, 101–102
   preparation of, 96–97, 96f
   PSi-based electrochemical gas/vapor
      sensors, 102–104
   types of, 109
Electrodes
   with comb structure, 11
   for position sensing, 153f
   vertical comb, 163f
Electroless process, 34
Electroless techniques, 63
Electromagnetic spectrum, 270, 271f

Electromechanical coupling effect, 158
Electron beam evaporation, 21
Electronegativity, 49, 50–51
Electronic circuit unit (ECU), 149
Electron paramagnetic resonance, 29
Electron transduction, defined, 47
Electron transfer, 7, 51
Electron tunneling, 154
Electro-osmotic flow (EOF), 309
Electro-osmotic pumps (EOP), 308
Electropolishing, 85, 159, 284
Electrospinning methods, for
   PSi-polycaprolactone, 375
Electrospray ionization (ESI), 240, 300, 300f
Electrostatic diaphragms, micromachined, 198
Electrostatistical adhesion, 110
Electrotransduction, 49
ELISA-method, 121, 121t
Ellipsometry, 72
   measurements, 112
Elution of dimethylphthalate, 304f
Elution process, 301
Embryonic stem cells (ESCs), 366, 367, 369
Energy band structure, 188f
Energy dispersive X-ray (EDX), 125
Energy-dispersive X-ray spectroscopy (EDS)
   of palladium nanoparticles, 60f
Energy transfer, 109
Enhanced green fluorescence protein (EGFP),
   349
Enhanced permeability and retention effect
   (EPR-effect), 391
Enzymatic biosensor, 113
Enzymatic sensor, 109
Enzyme determination, 125–126; see also
   Biosensors, PSi-based
E-poly-L-lysine, 344; see also Functionalized
   molecules
Esterification, 343–344; see also Surface
   functionalization
Etchant ethylenediamine pyrocatechol (EDP),
   285
Etching current density, 15f
Etching with gas chopping, 161
Ethanol
   humidity and, causing peak shift, 24
   sensors, 28, 104
   volatilization of, 34f
Ethylenediamine pyrocatechol (EDP), 210
Ethyl magnesium bromide, 113
European Medicine Agency, 364
External surface, 342
Extracellular matrix (ECM)
   engineering of, 364
   parameters, 371
   properties, 371
   role, 364
Extrapulmonary TB (EPTB), 354
Extrinsic semiconductors; see also Inverse
   hard and soft acid and base (IHSAB)
   principles
   alternate, 60
   selective, 46–49

F

Fabrication
   for mechanical micromachining sensors,
      158–164; see also Micromachined
      mechanical sensors

   microchip fabrication with integrated
      porous glass layers, 306f
   with porous nanocrystalline silicon, 285
   process, 128
   of silicon piezoresistive accelerometer, 160f
Fabry-Pérot interferometer, 75, 75f, 76f; see also
   Gas/vapor sensors, reflectivity-based
Fabry-Perot resonator, 4
Fast Fourier transform (FFT), 83
   technique, 65
Fast response time, 187
Femtomoles, 256
Femtosecond laser sources, 270
Fermi-level pinning, 30
Fermi levels, 51
Fickian diffusion, 46
Fick's law, 287, 289
Field-emission array pressure sensor, 144
Field-emission emitter array pressure sensor,
   144f
Field-ionization (FI) gas sensors, 24–27;
   see also Gas/vapor sensors,
   porous silicon-based
Filtering, 312–314; see also Liquid microfluidic
   devices
Filters, preconcentrator and, see
   Preconcentrators and filters
Flow injection analysis (FIA), 86–87, 87f
Flow sensors, 175–176; see also Devices
   fabricated with PSi as thermal
   insulation material
Fluorescence microscopy, 124
Fluorescence plate readers, 258
Fluorescence resonance energy transfer
   (FRET), 108
Fluorescence spectroscopy, 108
Fluorimetric method, 103
Folic acid, 344; see also Functionalized
   molecules
Fourier transform infrared (FTIR)
   spectroscopy, 18
Fowler-Nordheim field emission theory, 144
Fracture strength, 286
Free carboxyl groups, 341
Free molecular diffusion, 288
Freestanding macroporous silicon membranes,
   282
Freestanding porous film, 340
Frustrated Lewis acid/base pairs (FLP), 52, 63;
   see also Inverse hard and soft acid
   and base (IHSAB) principles
Full isolation by porous oxidized silicon
   (FIPOS), 175
Full width half maximum (FWHM) of
   resonance, 77
Functionalized molecules; see also Surface
   functionalization
   cyclodextrin (CD), 344
   ε-poly-L-lysine, 344
   folic acid, 344
   polyethyleneimine (PEI), 345

G

Gas chromatograph (GC), 224, 301
Gas concentration estimation, 6
Gas-phase chromatography preconcentrators,
   168; see also Thermal insulation in
   MEMS
Gas sensitivity, 20

Gas sensors, 168, 176–177; *see also* Devices
    fabricated with PSi as thermal
    insulation material; Thermal
    insulation in MEMS
    arrays, 216
    calibration, *see* Gas sources for μGC/gas
        sensor calibration
    MEMS-based, 225
Gas sources for μGC/gas sensor calibration;
    see also Preconcentrators and filters
    diffusion systems, 232
    permeation systems, 232–233
    PSi-based controllable vapor generators,
        233–235
Gas transport, 291
Gastrointestinal (GI) system, 330
Gas/vapor monitoring by optical reflectivity
    (case study), 79–83, 80f
Gas/vapor sensors
    electrochemical, PSi-based, 102–104
Gas/vapor sensors, porous silicon-based
    about, 4, 5f
    based on Schottky barriers and
        heterostructures, 19–22
    capacitance sensors, 9–14
    combined approach, 23–24
    conductometric sensors, 14–19, 16t
    CPD measurements-based gas sensors, 22,
        23f
    disadvantages of, 27–32
    field-ionization gas sensors, 24–27
    improvement through surface modification
        of porous semiconductors, 32–35,
        32t
    outlook, 35–36
    photoluminescence quenching, employing,
        5–9, 6t
Gas/vapor sensors, reflectivity-based; *see also*
    Optical chemical sensors, PSi-based
    about, 74–75
    Bragg mirror, 75–76, 76f
    Fabry-Pérot interferometer, 75, 75f, 76f
    optical microcavity, 76–78
    photonic quasi-crystal (QC), 78
    rugate filter, 78–79, 79f
Generation 2 poly(amidoamine) dendrimer
    (G2-PAMAM), 348
Geometric capacitance, 19
Geometry and hindrance in sensor
    development, 213
Glass substrates, 170
Glucose determination, 125–126; *see also*
    Biosensors, PSi-based
Glucose oxidase, 125
Glucose oxidase enzyme (GOX), 124
Glutaraldehyde (GL), 111
Glycine amino acid, 273
Glycosaminoglycan hyaluronic acid, 388
Glycyl-Histidyl-Glycyl-Histidine
    (GlyHisGlyHis) peptide, 97, 98f
Gold catalyzed PSi, 54
Grafting process, 342–343; *see also* Surface
    functionalization
Graft techniques, in bone reconstructive
    surgery, 365
Grignard reactions, 89
Guassian fit, 83
Gyroscopes, micromachined, 156–158; *see also*
    Micromachined mechanical sensors
Gyroscopes fabrication, 159

H

Hard acid, 49, 50t, 55f
    hydrogen, 62
Hardness matrix, 50
Heat-flux sensors, 179–180; *see also* Devices
    fabricated with PSi as thermal
    insulation material
Hematopoietic stem cells (HSCs), 366, 367
Hepatocellular carcinoma (HCC), 331, 351
Heterostructure-type sensors
Hexanol, 5, 6f, 7f
Highest occupied molecular orbital (HOMO), 56
Highest occupied molecular orbital-lowest
    unoccupied molecular orbital
    (HOMO-LUMO) gap, 49, 51
Highest occupied molecular orbital-lowest
    unoccupied molecular orbital
    (HOMO-LUMO) mismatch, 51
High-performance liquid chromatography
    instruments (HPLC), 301
Homografts, 365
Hot carrier diodes, 187
Hot carrier thermoemf formation, 188
Human galactosidase, 258
Human immune system, 328
Human lens epithelial cells (HLE), 327t
Human mesenchymal stem cell (hMSC), 327t
Human neuronal cell, 326t
Humidity and ethanol, 24
Humidity sensors, 177, 219
    capacitance of, 11f, 12
    design of, 11
    performance of, 12
    principle of, 11
    PSi structures tested as, 15
    SiC-based, 12
Hybridization of DNA, 127, 128
Hydrofluoric acid (HF), 369
    anodic etching of PSi in, 140
    low concentrated, 280
    silicon wafer in, 159
Hydrogen
    bonding, 12
    desorption, 29
    dissociation, 21
Hydrophilic electrodes, 12
Hydrophobic hydride, 342
Hydrophobicity of surface, 253
Hydrophobin class II (HFBII) nanoparticles,
    330
Hydrosilylation, 89
    thermal, 371
Hydrothermal etching, 13
Hygroscopicity, 12
Hysteresis, 12, 13f

I

Ibuprofen, 347f
IHSAB principles, *see* Inverse hard and soft
    acid and base principles
Immune diffusion test, 123
Immunobiosensors; *see also* Biosensors,
    PSi-based
    basic algorithm of analysis, 114–118
    construction of working devices, 114
    efficiency of, 119
    specific signal registration, 114
    structural analysis of PSi samples, 114

Immunogenicity, of PSi, 376
Implantable PSi-based biomaterials, 329t
Index of refraction, 23
Indium antimonide, 187
Indium arsenide, 272
Induced pluripotent stem cells (iPSs), 366, 369
Inductively coupled plasma (ICP), 162
Infrared absorption vibrational spectroscopy,
    272
In situ conversion of metal oxide sites, 55–59;
    see also Inverse hard and soft acid
    and base (IHSAB) principles
In situ nitridation, 52, 55
In situ sulfidization, 52
Insulating substrate, 170; *see also* Thermal
    insulation in MEMS
Integrated circuit (IC) industry, 137
Integrated PSi microsystems for optical
    sensing, 85–87
Integration strategies, PSi for thermal
    insulation
    geometry, 173
    indirect uses of PSi for thermal insulation,
        174–175
    pretreatments, 172–173
Interactions, of PSi–stem cells, 369–372
    adhesion and proliferation, 369–371, 370f
    differentiation, 371
    resorption, 371–372
Interferometric reflectance spectrum, 33
Interferometry method, 114
Internal energy (IE) transfer, 248
Internalization of PSi nanoparticles by cells;
    see also Drug delivery, PSi materials in
    size/shape, influence of, 349–350
    surface charge, 350
    trafficking in cells, 350–351
Internal surface in pore, 342
Intracellular trafficking, 350
Intraperitoneal injections (I.P.), 330
Intravenous injection, 330; *see also In vivo*
    administration routes
Inverse hard and soft acid and base (IHSAB)
    principles
    alternate extrinsic semiconductors, 60
    basis for nanostructure-directed
        physisorption, 49–54, 50t, 53t
    for catalyst sites and frustrated Lewis acid/
        base pairs
        amine doped oxides, 62–63
        frustrated Lewis acid/base pairs (FLP),
        63
    light-enhanced conductometric sensing,
        59–60
    to nanowire configurations, 60–62
    outlook, 63–65
    selective extrinsic PSi semiconductors,
        46–49
    in situ conversion of metal oxide sites
        via direct nitration/sulfur group
        functionalization, 55–59
In vitro assays and assessment methods, 321
In vitro studies of PSi, 326–327, 326t–327t;
    see also Biocompatibility/bioactivity
    of PSi
In vivo administration routes; *see also*
    Biocompatibility/bioactivity of PSi
    intravenous injection, 330
    oral administration, 330
In vivo assays and assessment methods, 321–322

In vivo imaging techniques, PSi for, 382–397
  advantages and disadvantages, 383t
  essential characteristics, 383t
  general requirements to molecular probes,
      384–389
    biocompatibility and biosafety, 387
    preparation, surface functionalization
      and, 384–386, 385f, 386t
    stability, biodistribution, and clearance,
      387–389, 388t
  as molecular probes, 389–397
    MRI, 395–397, 396f
    nuclear imaging, 393–395, 394f, 395f
    optical imaging, 384–393, 390f, 391f,
      392f, 393f
  overview, 382–384, 382f, 383f
  physical principles, 382f
In vivo studies of PSi; *see also* Biocompatibility/
      bioactivity of PSi
  about, 328–329, 329t
  in vivo administration routes
    intravenous injection, 330
    oral administration, 330
Ionization mechanism, 250
Ion-mobility spectrometer, 224

J

Joule's heat, 199f

K

Kelvin radius, 11, 12
Kelvin-Zissman method, 22
Knudsen diffusion, 11, 288f, 289
Knudsen number, defined, 288
Kohn-Sham orbitals, 49
Koopman's theorem, 51

L

Lab-on-a-Chip (LOC), 280
  device, 85, 85f
Laser ablation, 266
  method, 35
Laser desorption ionization (LDI)
  chemical functionalization, 251
Laser light polarization, 245
Layer liquid fraction (LLF), 71
LDI, *see* Laser desorption ionization (LDI)
Legal and practical issues, tissue engineering,
      375–376
Levitation, defined, 158
Lift-off, 340
Light emitting diodes, 113
Light-enhanced conductometric sensing,
      59–60; *see also* Inverse hard and soft
      acid and base (IHSAB) principles
Light polarization, 249
Limbal stem cells, 373
Limits of detection (LOD), 257
Liquefied petroleum gas (LPG), 22
Liquid chromatography (LC), 301–302; *see also*
      Miniaturized total analysis systems
      (μTAS); Miniaturized total analysis
      systems (μTAS)
Liquid microfluidic devices
  filtering, 312–314
  microfluidic nozzles, 300–301
  miniaturized total analysis systems (μTAS)

chromatographic separation columns,
      302–307
  liquid chromatography, 301–302
  micropumps, 307–310
  separation of particles of different size,
      310–311
Liver, 351; *see also* Targeted delivery
Loading and controlled release of drugs with
      PSi
  covalent attachment, 341
  oxidation, 342
  physical adsorption, 342
Local plasma etching, 139
Lock and key approach, 88–90; *see also*
      Selectivity of sensor
Lowest unoccupied molecular orbital (LUMO),
      56
Low-pressure chemical vapor deposition
      (LPCVD), 170
Low temperature cofired ceramics (LTCC), 170
Luminescence gas sensor, 71
Luminescent PSi nanoparticles (LPSiNP), 324,
      324f
Lung cancer, 352–353; *see also* Targeted
      delivery
(lysine)4-tyrosinearginine-glycine-aspartic acid
      (K4YRGD), 351

M

Macroporous silicon, 31
Macroporous silicon membranes, 174f,
      280–282, 281f
Macroscale gas sensors, 168
Magic bullet, concept of, 386
Magnetic biosensors, 110
Magnetic motor effect, 348
Magnetic resonance imaging (MRI)
      PSi-containing nanocomposites and, 382,
      383, 383t, 395–397, 396f
MARS-RR gyroscope, 157–158, 157f
Masking techniques, 173
Mass spectrometry (MS)
  lateral resolution for, 259
  soft ionization methods in, 240–242, 241t
Material property of thermal insulation, 170;
      see also Thermal insulation in
      MEMS
Matrix-assisted laser desorption/ ionization
      mass spectrometry (MALDI), 240
Maximum tolerated dose (MTD), 329
Mb, *see* Myoglobin
Mechanical biosensors, 109
Mechanical micromachining sensors, *see*
      Micromachined mechanical sensors
Mechanical (reciprocating) micropumps, 307
Mechanical resistance, 172
Mechanochemical top-down techniques, 340
Membrane
  fabrication, 280–286
  properties, 286–287
MEMS, *see* Microelectromechanical system
      (MEMS)
Mercaptoethanol (ME), 348
3-mercaptopropyltriethoxysilane (MPTS), 274
(3-mercaptopropyl) trimethoxysilane
      (MPTMS), 275
Mesenchymal stem cells (MSCs), 366, 367, 368f
Mesoporous silica nanomaterials (MSN), 340,
      344

Mesoporous silicon membrane, 280, 281f,
      282, 285
Metabolomics, 258, 259
Metal-assisted stain etching (MASE), 243, 245
Metal deposition, 31
Metal ion chelating agents, 273
Metal ions, PSi-based electrochemical sensors
      for, 97–101
Metallization process, 31
Metal oxide-PSi (MeOx-PSi) heterostuctures,
      22
Metal oxide semiconductor (MOX) solid-state
      gas sensor, 227
Metal-oxide-semiconductor (MOS) structures,
      191
Metal oxide sensors, 46, 64
Methanol, 5, 6f, 7f, 81f
Microaccelerometers, 151f, 155f
Microbolometers, 177
Microcapillary condensation, 88
Microcavities in silicon wafers, 227
Micro-channels, 231
Microconcentrators, 228
Microelectromechanical system (MEMS);
      see also Thermal insulation in MEMS
  pressure sensors based on, *see* Pressure
      sensors, MEMS-based
  vacuum-encapsulated resonator, 149, 149f
Microfabricated mixer, 304, 305f
Microfabricated preconcentrator devices, 225
Micro-flow injection analysis, 308
Microfluidic nozzles, 300–301; *see also* Liquid
      microfluidic devices
Microfluidics and biological applications,
      168–169; *see also* Thermal insulation
      in MEMS
Microheater, 218f, 219
  measurement setup, 217f
Micro-hotplates, 178; *see also* Devices
      fabricated with PSi as thermal
      insulation material
  applications, 209f
  fabrication, 214t
  with PSi active layer, 217–220;
      see also Porous silicon (PSi) in
      micromachining hotplates
  suspended, 214f
  thermal characterization of, 215t
Micromachined devices, examples of, 208
Micromachined gas chromatography system, 27
Micromachined gyroscopes, 156–158; *see also*
      Micromachined mechanical sensors
Micromachined mechanical sensors
  fabrication technology, 158–164
  micromachined gyroscopes, 156–158
  overview, 148–149, 148f
  silicon micromachined accelerometers,
      149–155
Micromachining technology, 26, 35, 48
  about, 208
Microneedle, 353–354, 353f; *see also* Drug
      delivery, PSi materials in
Micropillar array, 300
Microporous silicon, property of, 4
Micro preconcentrator design, 225–232, 230t;
      see also Preconcentrators and filters
Micropumps, 307–310; *see also* Miniaturized
      total analysis systems (μTAS)
Micro-Raman spectroscopy, 286
Microreactors, 65, 168

Microsensor, 225
Microstructure fabrication process, 159f
Micro total analysis systems (μTAS), 280
Microtraps, 225
Microwave (MW) detection, PSi-based
    overview, 186
    on porous silicon layers, 187–193
    principles, 186–187
Mild air oxidation, 7
Miniaturization of resonant-based sensors, 217
Miniaturized fuel cells, 294
Miniaturized hotplates, 208
Miniaturized total analysis systems (μTAS);
        see also Liquid microfluidic devices
    chromatographic separation columns,
        302–307
    liquid chromatography, 301–302
    micropumps, 307–310
Molecular dynamics (MD), 288
Molecular orbital (MO) theory, 49
Molecular probes, for *in vivo* imaging in
        context of PSi nanostructures
    general requirements, 384–389
        biocompatibility and biosafety, 387
        preparation, surface functionalization
            and, 384–386, 385f, 386t
        stability, biodistribution, and clearance,
            387–389, 388t
    MRI, 395–397, 396f
    nuclear imaging, 393–395, 394f, 395f
    optical imaging, 384–393, 390f, 391f,
        392f, 393f
Monocrystalline silicon (mcSi), 110, 209
Monocrystalline Si-membrane fabrication,
        140f, 141
Monocrystalline Si-membranes fabricated
        using PSi, 139–142; *see also* Pressure
        sensors, MEMS-based
Monofunctional aminosilanes, 111
Mononuclear phagocyte system (MPS), 323, 387
Monte Carlo (MC) simulations, 288
Morphology of silicon nanostructures, used as
        LDI-MS substrates, 242–247
MOSFET gas sensors, 208
MS, *see* Mass spectrometry (MS)
Mulliken electronegativity, 50
Multilayer diffusion, 288f
Multiparameters sensing, 89–90; *see also*
        Selectivity of sensor
Multiparametric PSi sensors, 23, 24
Multipotent somatic stem cells, 366
Multistage delivery systems (MDS), 389,
    395–396
Multistage delivery vectors (MDVs), 389
Multiwalled carbon nanotubes, 124
Multi-walled micro channels (MWμC), 285
Mycobacterium tuberculosis, 354
Mycotoxin level determination, 122; *see also*
        Biosensors, PSi-based
Myoglobin (Mb)
    registration of, 114, 119–121, 121t

N

Nafion-filled porous membranes, 287
NALDI ionization, 243, 244
Nanocomposite, 351
Nanocone LDI substrates, 245
Nanoparticles
    as gatekeepers, 348

Nanoparticles (NPs), Psi
    THCPSi, 394–395
    in vivo imaging, general requirements to
        molecular probes for, 384–389
        biocompatibility and biosafety, 387
        preparation, surface functionalization
            and, 384–386, 385f, 386t
        stability, biodistribution, and clearance,
            387–389, 388t
Nanoporous silicon, 273
Nanosilicon ionization, 240
Nanostructure initiated mass spectrometry
        (NIMS), 255–256, 255f, 259, 259f
Nanowires, 26, 26f
Near-infrared (NIR) window, probes in,
    389–392, 390f
Neonatal stem cells from umbilical cord, 369
Nerve tissue engineering, PSi for, 374
Neuroblastoma cells, 370
Nickel ion detection, 98
Nimzyme assays, 258
Nitrogen oxide, 6–7
Noise equivalent power (NEP), 177
Noise equivalent temperature difference
    (NETD), 177
Non-mechanical (dynamic) micropumps, 307
Nonradiative de-excitation pathways, 8f
Normal phase liquid chromatography (NP-LC),
    302
Nuclear imaging methods
    CT, 382, 383, 383t
    MRI imaging, 382, 383, 383t, 395–397,
        396f
    PET, 382, 383, 383t, 394
    physical principles, 382f
    PSi in, 393–395, 394f, 395f
    SPECT, 382, 383, 383t
    ultrasound, 382, 383, 383t

O

Ohmic contacts, 18, 118
Olyconucleotides, 127
OncoSil™, 389
Ophthalmic implants, PSi for, 373–374
Opsonization, 322, 323
Optical absorption measurements, 112
Optical applications of freestanding porous
        silicon membranes, 294
Optical biosensors, 110
Optical chemical sensors, PSi-based
    approaches in, 70–74
    examples, 83–85, 83f–85f
    gas/vapor monitoring by optical reflectivity
        (case study), 79–83, 80f
    integrated PSi microsystems for optical
        sensing, 85–87
    optimization of
        selectivity, 88–90
        sensitivity, 87–88
        stability, 90–91
    overview, 70
    for reflectivity-based gas and vapor sensors
        about, 74–75
        Bragg mirror, 75–76, 76f
        Fabry-Pérot interferometer, 75, 75f, 76f
        optical microcavity, 76–78
        photonic quasi-crystal (QC), 78
        rugate filter, 78–79, 79f
Optical fiber sensors, 84

Optical imaging, PSi in, 382, 383, 383t,
        384–393, 390f, 391f, 392f, 393f
Optical microcavity, 76–78, 77f; *see also* Gas/
        vapor sensors, reflectivity-based
Optical reflection band, 78
Optical sensors, diaphragm-based, 137
Optical transduction methods, 108
Optimization of PSi-based optical sensors;
        see also Optical chemical sensors,
        PSi-based
    selectivity
        lock and key approach, 88–90
        multiparameters sensing, 89–90
        physical control of surface reactions, 89
        sensors arrays, using, 89
        size exclusion, 89
        specific chemical reactions, using, 89
        surface reactivity control via surface
            functionalizing, 89
    sensitivity, 87
    stability, 90–91
Oral administration of drugs, 330; *see also*
        In vivo administration routes
Organic acids
    decarboxylative/dehydrative coupling of, 63
Organic molecules as gatekeepers, 348
Orthosilicic acid, 322, 339
Osteodifferentiation, cell, 372
Oxidation, 8
    loading/controlled release of drugs, 342
    and pore size reduction, 287
Oxynitride, 53t
Oxysulfides, 52

P

Palladium nanoparticles, 33
Peak broadening, 303
Peak shift
    concomitant, 24
    measured, 23f
    porous silicon microcavity, 24
Pellistor-type gas sensor, 177
Perfluorinated organic functionalities, 252
Perfluorinated siloxanes, 256
Periodic pore arrangement, 280
Permeability
    of mesoporous silicon membranes, 292
    values of hydrogen, 291
Permeation device, 232
Permeation systems, 232–233; *see also* Gas
        sources for μGC/gas sensor calibration
Phase matching, 46
Phase modulation, 80
pH measurements, PSi-based electrochemical
        sensors for, 101–102
Phonon scattering effects, 172
Photoacoustic analysis, 199
Photoactivated chemical modification, 113
Photoactive aryldiazirine cross-linker, 126
Photobleaching, 266
Photochemical passivation, 113
Photocleavage, 254
Photocurrent, 118, 123
Photodetectors (PhD), 117
Photolithography, 220, 235
Photoluminescence (PhL), 108, 110
    intensity, 4, 311f
    quenching, 5–9, 6f, 6t, 7f; *see also* Gas/vapor
        sensors, porous silicon-based

Photon conduction, 287
Photonic band gap (PBG), 76, 77
Photonic quasi-crystal (QC), 78; *see also* Gas/
    vapor sensors, reflectivity-based
Photonic transducers, 72
Photoresistor, 118
Physical adsorption, 342
Physical control of surface reactions, 89;
    see also Selectivity of sensor
Physisorption
    defined, 224
    nanostructure-directed, 49–54, 50t, 53t;
      see also Inverse hard and soft acid
      and base (IHSAB) principles
Piezocrystals, 110
Piezoelectrical-based transducers, 150
Piezoelectric measuring cell (PZE cell), 154
Piezoelectric microaccelerometers, 155f
Piezoelectric pressure sensors, 138
Piezoelectric sensors, advantages of, 154
Piezoelectric thin films, 154
Piezoresistive accelerometer, 151, 151f, 159
Piezoresistive force sensors, 208
Piezoresistive pressure sensor, 137, 138, 141
Piezoresistive transducers, 150
Piezoresistors
    about, 142–144
    cantilever beam integrated with, 218f, 219
    fabrication of, 151
    microcantilevers, 208
Planar microconcentrators, 227
Planar preconcentrator, 227f
Plasma-enhanced chemical vapor deposition
    (PECVD), 170
Point-contact, 189
Polarizability, 10
Polarization interferometry, 72
Poly(ethylene glycol) (PEG), 110, 345
Poly(ethylene glycol) monomethacrylate
    (PEGMA), 111
Polyethyleneimine (PEI), 345; *see also*
    Functionalized molecules
Polyethylene (PE) powder, 271
Polymerase chain reaction (PCR), 128, 168
Polymers, 171; *see also* Thermal insulation in
    MEMS
Poly(propylene oxide) (PPO), 345
Polysilicon
    heater, 210
    passivated, 210
    piezoresistiors, 138
    piezoresistive pressure sensor, 136f
Polystyrene, molecular weight distribution of,
    243
Pore diameter, influence of, 347; *see also* Drug
    release
Pore morphology, 31
Pore opening mechanisms, 349f
Pore size of PSi sensor, 88
Pore size/pore size distribution, 286
Porosification, 32, 36
    of membranes, 313
Porosity, 192
Porous glass (PG), 306f, 307
Porous membrane
    self-supporting, 284
    transport in, 287–293; *see also* Diffusion
      membranes
Porous membrane/porous piezoresistors,
    142–144

Porous nanocrystalline silicon membranes,
    285
Porous polysilicon (PPSi), 143
Porous silica drug delivery systems; *see also*
    Drug delivery, PSi materials in
    microneedle, 353–354, 353f
    scaffolds, 354
    sponges, 355
Porous silicon (PSi)
    based composites, advances in, 375
    carbonization of, 13
    in clinical trials, 331
    degradation of, in aqueous solution,
      322–323
    gold catalyzed, 54
    immunogenicity of, 376
    layers, microwave detection on, 187–193
    legal and practical issues, 376
    microcavity, 24t
    morphology, 172
    photoluminescence properties of, 384
    photoluminescent properties, 384
    and porous silica nanoparticle synthesis
      method, 340–341
    properties of, 384
    for stem cell-based therapy, 364, 365
    stem cells, interactions of, 369–372
      adhesion and proliferation, 369–371,
        370f
      differentiation, 371
      resorption, 371–372
    synthesis, 384
    thermal carbonization (TC) of, 13, 33
    for thermal insulation, 174–175
    for tissue engineering, 372–374
      applications, 365
      bone, 372, 373f
      nerve, 374
      ophthalmic implants, 373–374
    in vitro studies of, 326–327, 326t–327t
    for *in vivo* imaging techniques, 382–397
      advantages and disadvantages, 383t
      biocompatibility and biosafety, 387
      essential characteristics, 383t
      molecular probes, 384–389, 389–397;
        see also Molecular probes
      MRI, 395–397, 396f
      nuclear imaging, 393–395, 394f, 395f
      optical imaging, 384–393, 390f, 391f,
        392f, 393f
      overview, 382–384, 382f, 383f
      physical principles, 382f
      PSi micro- and nanoparticles,
        preparation, 384–386, 385f, 386t
      stability, biodistribution, and clearance,
        387–389, 388t
      surface functionalization, 384–386,
        385f, 386t
      TCPSi, 386
      THCPSi, 386
    in vivo studies of, *see In vivo* studies of PSi
Porous silicon (PSi) as material for thermal
    insulation
    integration strategies
      geometry, 173
      indirect uses of PSi for thermal
        insulation, 174–175
      pretreatments, 172–173
      porous silicon morphology, 172
Porous silicon carbide (PSiC), 32

Porous silicon (PSi) in micromachining
    hotplates
    advantages of, 210–217
    micro hotplates with PSi active layer,
      217–220
    micromachining and applications, 208–210,
      209f
Porous silicon (PSi) in pressure sensor
    fabrication; *see also* Pressure
      sensors, MEMS-based
    field-emission array pressure sensor, 144
    with monocrystalline Si-membranes
      fabricated using PSi, 139–142
    with porous membrane/porous
      piezoresistors, 142–144
    SOI structure pressure sensors, 138–139,
      139f
Porous silicon membranes (PSM), 15
Porous silicon microcavities (PSiMcs), 125
Porous silicon optical transducers for
    reflectivity-based gas/vapor
      sensors, *see* Gas/vapor sensors,
      reflectivity-based
Porous silicon PSi rugate filter (PSRF)
    simulated resonant wavelength of oxidized,
      90f
    vs. refractive index, 87f
Positron emission tomography (PET)
    in vivo imaging technique, 382, 383,
      383t, 394
Potassium bromide (KBr), 271
Potentiometric biosensors, 109, 114
Potentiometric capacitance-voltage
    measurements, 124
Potentiometric sensor, 96
Power consumptions, 49f
Preconcentrators and filters; *see also* Gas
    sources for μGC/gas sensor
      calibration
    micro preconcentrator design, 225–232, 230t
    preconditioning systems (preconcentrators),
      224
Pressure sensor diaphragm, 136f
Pressure sensors, MEMS-based
    porous silicon in pressure sensor fabrication
      field-emission array pressure sensor, 144
      with monocrystalline Si-membranes
        fabricated using PSi, 139–142
      with porous membrane/porous
        piezoresistors, 142–144
      SOI structure pressure sensors,
        138–139, 139f
    silicon-based pressure sensors, 136–138, 136f
Pressure swing process, 229
Proliferation, cell, 369–371, 370f
Proton affinity (PA), 249, 249f
Protonation process, 250–251, 250f
Proton sponge effect, 351
Proximity sensors, ultrasonic transmitters for,
    202
PSi, *see* Porous silicon (PSi)
Pulsed or time-multiplexed etching, 161
Purcell effect, 34
Pyridyldithio-terminated PSi (PDT-PSi), 274

Q

Quantum mechanical process, 109
Quantum tunneling, 24
Quasi-crystal (QC), 78

R

Radical coupling, 89
Raman spectroscopy, 111
    PSi in, 266–270; *see also* Terahertz (THz)
        spectroscopy, PSi in; Trace heavy
        metal ions, pre-enrichment of
Rare earth elements (REE), 275
Rat neural cell, 326t
Rat neural tumor, 326t
Rat pheocheromocytoma cell, 326t
Reactive-ion etching (RIE), 84, 162, 245, 353
Reactive oxygen species (ROS), 321
Real-time capacitance, 104f
Reciprocating pumps, 307–308
Reference electrode, 96f
Refractive index, 23, 24
    contrast, 80
    of PSi, 71, 71f
    in sensing experiment, 82t
    vs. PSi rugate filter (PSRF), 87f
Refractometer, 72, 87
Registration of Mb concentration, 119–121,
        121t; *see also* Biosensors, PSi-based
Reinforced mesoporous silicon membranes, 283
Relative humidity (RH), 12, 24, 29, 29f
Renishaw Ramanscope 2000, 268f
Resistance-capacitance (RC), 18
Resistance-inductance (RL) network, 19
Resistor-type gas sensors, 208
Resistor-type structures, 5f
Resonant pressure sensors, 137
Resonator gyrometer, silicon coupled, 157
Resorption, PSi, 371–372
Response signal, 188, 189f
Retroviral leucosis, diagnostics of, 122–123,
        122t; *see also* Biosensors, PSi-based
Reversed-phase capillary
        electrochromatography, 303
Reverse phase chromatography, 302
Room temperature (RT) sensors, 14
Rugate filter, 78–79, 79f; *see also* Gas/vapor
        sensors, reflectivity-based
Ruthenium bipyridine, 127

S

Sacrificial etching/etchant, 211f, 212t
Scaffolds, 354; *see also* Drug delivery, PSi
        materials in
Scanning electron micrograph, 26f, 27f
    of accelerometers, 152f, 153f
    of piezoresistive tactile sensor, 148
Scanning electron microscope (SEM)
    of macroporous silicon, 314f
    for optical fiber, 84f
    optimized for DIOS, 242f
    of silicon nanostructures, 244
    structural characterization of porous
        membranes, 286
Scanning electron microscopy (SEM)
    of thiol-treated titanium dioxide, 48f
Scanning tunneling microscopy (STM), 115
Schottky barrier-type gas/vapor sensors, 19–22;
        see also Gas/vapor sensors, porous
        silicon-based
Secondary ion mass spectroscopy
        measurements (SIMS), 115
Secondary ions mass-spectrometry (SIMS), 246

Selectivity of sensor; *see also* Optical chemical
        sensors, PSi-based
    lock and key approach, 88–90
    multiparameters sensing, 89–90
    physical control of surface reactions, 89
    sensors arrays, using, 89
    size exclusion, 89
    specific chemical reactions, using, 89
    surface reactivity control via surface
        fuctionalizing, 89
Self-supporting PSi membranes, 287
Semicarbazide, 371
Semiconductor detectors, 186
Sensitivity
    of accelerometers, 159
    of sensor, 87; *see also* Optical chemical
        sensors, PSi-based
Sensor array, 89; *see also* Selectivity of sensor
Sensor sensitivity, 5
Separation of particles of different size,
        310–311; *see also* Liquid microfluidic
        devices
Shear-driven chromatography (SDC) channels,
        303
Short pulse durations, 270
Signal-to-noise (S/N) ratio, 240
Silane gas, 243, 326
Silanes (heptadecafluoro-1,1,2,
        2-tetrahydrodecyl)
        dimethylchlorosilane (PFDS), 287
Silanization technique, 97, 103
Silica gel phase, 275
Silica optical fibers, 84
Silicic acid, for bone and connective tissue
        homeostasis, 372
Silicon
    anodization, 313f
    in biomaterials, 364, 365
Silicon-based pressure sensors, 136–138,
        136f; *see also* Pressure sensors,
        MEMS-based
Silicon carbide membranes, 210
Silicon coupled resonator gyrometer, 157
Silicon micromachined accelerometers,
        149–155; *see also* Micromachined
        mechanical sensors
Silicon micromachining technology, 48
Silicon nanopost arrays, 245
Silicon nanostructures for LDI-MS
    chemical functionalization of silicon
        LDI-MS substrates, 251–256
    desorption-ionization, 247–251
    morphology of, used as LDI-MS substrates,
        242–247
    nSi-LDI-MS, applications of, 256–260
    soft ionization methods in mass
        spectrometry, 240–242, 241t
Silicon nanowires (NW), 240
Silicon nitride, 209, 210, 283
Silicon-on-insulator (SOI)
    for ultra-low-power-consumption devices,
        209
Silicon-on-insulator epitaxial (SOI-Epi) devices,
        213
Silicon-on-insulator (SOI) structure pressure
        sensors, 138–139, 139f; *see also*
        Pressure sensors, MEMS-based
Silicon on insulator (SOI) wafer, 84
Silicon-on-sapphire (SOS) wafers, 139

Silicon oxide, 170–171; *see also* Thermal
        insulation in MEMS
Silicon porosification, 220, 310
Silver, 267, 268f
Single-chip monolithic integration concept, 216
Single crystal membranes, 139
Single-photon emission computed tomography
        (SPECT)
    in vivo imaging technique, 382, 383, 383t
Single strand DNA (ss-DNA), 113
Single transduction methodology, 89
Size exclusion, 89; *see also* Selectivity of sensor
Smart Petri dish, 129
Sodium-enhanced nanostructure initiated
        mass spectrometry, 259
Soft acid, 49, 50t, 55f
Soft ionization methods in mass spectrometry,
        240–242, 241t; *see also* Silicon
        nanostructures for LDI-MS
SOI, *see* Silicon-on-insulator (SOI)
Solid acid catalysts, 62
Sorbents for atmospheric air control, 230t
Specific signal registration, 114
Specific surface area (SSA), 286
Spectrophotometric method, 103
Spectroscopic gas sensors, 73, 74f
Spermicides, 256
Sponges, 355; *see also* Drug delivery, PSi
        materials in
Stability, of Psi, 387–389, 388t
Stability of sensor, 90–91; *see also* Optical
        chemical sensors, PSi-based
Stability tests, 112
Standard organic chemistry, 111
Standard surface analytical techniques, 111
Stem cells, 365–369
    categories, 366
    ESCs, 366, 367, 369
    HSCs, 366, 367
    iPSs, 366, 369
    limbal, 373
    MSCs, 366, 367, 368f
    neonatal stem cells from umbilical cord,
        369
    overview, 365–366, 366f
    PSi, interactions of, 369–372
        adhesion and proliferation, 369–371,
            370f
        differentiation, 371
        resorption, 371–372
        totipotent, 366
Stern–Volmer behavior, 8
Stöber reaction, 340
Straight etches, 75
Stratum corneum (SC), 353
Streptavidin, 124
Sulfosuccinimidyl 4–[N-maleimidomethyl]
        cyclohexane 1–carboxylate
        (Sulfo-SMCC), 127
Suppression of cancer resistance, 353
Supramolecular assemblies as gatekeepers,
        348–349
Surface-assisted laser desorption/ionization
        (SALDI), 240
Surface charge, 3503; *see also* Internalization of
        PSi nanoparticles by cells
Surface chemistry, 99
Surface diffusion, 288f
    coefficient, 289

Surface-enhanced Raman spectroscopy (SERS), 109, 266
Surface flux, 289
Surface function, influence of; *see also* Drug release
  nanoparticles as gatekeepers, 348
  organic molecules as gatekeepers, 348
  supramolecular assemblies as gatekeepers, 348–349
Surface functionalization; *see also* Drug delivery, PSi materials in
  functionalized molecules
    cyclodextrin (CD), 344
    ε-poly-L-lysine, 344
    folic acid, 344
    polyethyleneimine (PEI), 345
    PSi, 384–386, 385f, 386t
  surface modification
    co-condensation process, 343
    combined process, 343
    esterification, 343–344
    grafting process, 342–343
Surface hydrophylicity, 252
Surface micromachining, 211, 211f, 280
Surface modification
  of porous semiconductors, 32–35, 32t; *see also* Gas/vapor sensors, porous silicon-based
  of PSi, 323
Surface postprocessing, 286
Surface reactivity control via surface fuctionalizing, 89; *see also* Selectivity of sensor
Surface roughness, 116
Survival yields (SY), 248
Suspended membranes, 171–172, 174f; *see also* Thermal insulation in MEMS

T

Tape test, 171
Targeted delivery; *see also* Drug delivery, PSi materials in
  liver, 351
  lung cancer, 352–353
  vascular barriers, 351–352
Temperature coefficient of resistivity (TCR), 210
Temperature dependence of dark conductivity, 18, 19f
Terahertz absorption spectroscopy, 272
Terahertz (THz) spectroscopy, PSi in, 270–273; *see also* Raman spectroscopy, PSi in
Terahertz time-domain spectroscopy (THz-TDS), 271, 272
Tetramethoxysilane (TMOS), 343
Tetramethyl ammonium hydroxide (TMAH), 210
Texturization of silicon, 32
TFRE components, 103–104
THCPSi nanoparticles, 323, 325f
Theranostic nanostructures, 383–384
Thermal accelerometer, 175
Thermal carbonization (TC), 13, 33
  with acetylene, 89
  of PSi, 323
Thermal conductivity, 114
  of bulk silicon, 209
  of nano-Si, 247
  of PSi, 179

Thermal cycling, 89
Thermal hydrocarbonization (THC), 16
Thermal insulation in MEMS
  common solutions
    insulating substrate, 170
    material property of thermal insulation, 170
    suspended membranes, 171–172
    thick layers, 170–171
  devices fabricated with PSi
    bolometers, 177
    flow sensors, 175–176
    gas sensors, 176–177
    heat-flux sensors/thermoelectric generator, 179–180
    micro-hotplates, 178
    thermoacoustic sound sources, 178–179
  discussion, 180
  necessity of
    bolometers, 169
    gas-phase chromatography preconcentrators, 168
    gas sensors, 168
    microfluidics and biological applications, 168–169
  PSi as material for
    integration strategies, 172–175
    porous silicon morphology, 172
Thermally carbonized porous silicon (TC-PSi), 33, 386
Thermally hydrocarbonized PSi (THCPSi), 386, 394–395
Thermally induced ultrasound (TIU), 205
Thermally oxidized porous silicon (TOPSi), 327
Thermal oxidation, 13
  of Si, 88
Thermal resistance, 170
Thermo-acoustic (TA) emission, 201
Thermoacoustic sound sources, 178–179; *see also* Devices fabricated with PSi as thermal insulation material
Thermoelectric elements, 186
Thermoelectric generator, 179–180, 179f; *see also* Devices fabricated with PSi as thermal insulation material
Thermoelectromotive force, 187
Thermometer ions, 248
Thermometric sensors, 208
Thermophone, 179
Thermopiles, 175, 176
Thermoresistors, 14
Thermo-vacuum treatments, 28
Thick layers of insulating material; *see also* Thermal insulation in MEMS
  aerogels, 171
  polymers, 171
  silicon oxide, 170–171
Thick macroporous membranes, 281
Thin layer chromatography (TLC), 255
Thiosemicarbazide derivative, 100
Three-dimensional (3D) image sensing, 202
Three-electrode electrochemical sensor, 96f, 97
Thue-Morse Sequence (TMS), 78, 79, 80f, 81, 82f
Time-multiplexed etching, 161
Time-resolved measurements, 85
Tissue engineering, 363–376
  applications, porous materials for, 365
  interactions of PSi–stem cells, 369–372

    adhesion and proliferation, 369–371, 370f
    differentiation, 371
    resorption, 371–372
  legal and practical issues, 375–376
  overview, 364
  products (TEP), defined, 364
  PSi-based composites, advances in, 375
  PSi for, 372–374
    bone, 372, 373f
    nerve, 374
    ophthalmic implants, 373–374
  stem cells, 365–369
    categories, 366
    ESCs, 366, 367, 369
    HSCs, 366, 367
    iPSs, 366, 369
    MSCs, 366, 367, 368f
    neonatal stem cells from umbilical cord, 369
    overview, 365–366, 366f
    totipotent, 366
Titanium dioxide, thiol-treated, 48f
Toluene, 73f
Tortuosity, 289
Totipotent stem cells, 366
Trace heavy metal ions, pre-enrichment of, 273–275
Trafficking in cells, 350–351; *see also* Internalization of PSi nanoparticles by cells
Transdermal delivery, 353
Transduction mechanisms, 137
Transfer matrix method, 77
Transfer of electrons, 47
Transistor-type structure, 5f
Transmembrane pressure, defined, 291
Transmission electron microscopy (TEM), 286, 339
  low-magnification, 61f
  of Pd nanoparticles, 60f
Transport in porous membranes, 287–293; *see also* Diffusion membranes
Trichoderma reesei, 394
Triethylamine, 62
Trifunctional aminosilane, 111
Tunnel accelerometer, 154, 155
Tunneling devices, 150

U

Ultrasonication route, 340
Ultrasonic devices, PSi-based, 201–205
Ultrasonic pressure intensity, 202
Ultrasonic transmitters, 202, 203f
Ultrasound emitters
  and conventional emitter (comparative survey), 200t
  overview, 198
  ultrasonic devices, PSi-based, 201–205
  ultrasound generation by PSi-based emitter, 198–200
Ultrasound generation by PSi-based emitter, 198–200, 200t
Ultrasound imaging
  in vivo imaging technique, 382, 383, 383t
Ultraviolet (UV) light source, 117
Ultraviolet (UV)–vis absorption spectroscopy, 274
Umbilical cord, neonatal stem cells from, 369
Uncooled microfabricated bolometers, 169
Undecylenic acid, 343

V

Vacuum-encapsulated MEMS resonator, 149, 149f
Vanadium oxide (VOx), 169
Vaporization of initiator, 256
Vapor-liquid-solid (VLS) growth, 243, 244
Vapor permeation, 11
Vapor phase stain-etching fabrication, 219f
Variable angle spectroscopic ellipsometry, 74f
Vascular barriers, 351–352; *see also* Targeted delivery
Vertical comb electrodes, 163f
Vibratory gyroscopes, 156, 157f
Virus detection, 126–127, 127f

Viscous flow, 288
Volatile organic compounds (VOC), 84
Voltage, measurement of, 96
Voltage-current characteristics, 191f
Voltage signal formation, 192
Voltammetric biosensors, 109
Voltammetric sensors, 96

W

Water permeability, 292
Water transport, 292
Waveguide-based spectroscopy, 271
Wet bulk micromachining, 210

Wet chemical methods, 111
Wet desorption, 247, 249
Wharton's jelly, 369

X

Xenografts, 365
X-ray diffraction (XRD), 270
    spectrum, 61f
X-ray photoelectron spectroscopy (XPS), 110

Y

Young's modulus, 286